Haiyun Zhou and Xiaolong Qin
Fixed Points of Nonlinear Operators

Also of Interest

Elementary Operator Theory
Marat V.Markin, 2020
ISBN 978-3-11-060096-4, e-ISBN (PDF) 978-3-11-060098-8,
e-ISBN (EPUB) 978-3-11-059888-9

*Variational Methods in Nonlinear Analysis, with Applications in
Optimization and Partial Differential Equations*
Dimitrios C. Kravvaritis, Athanasios N. Yannacopoulos, 2020
ISBN 978-3-11-064736-5, e-ISBN (PDF) 978-3-11-064738-9,
e-ISBN (EPUB) 978-3-11-064745-7

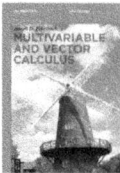

Multivariable and Vector Calculus
Joseph D.Fehribach, 2020
ISBN 978-3-11-066020-3, e-ISBN (PDF) 978-3-11-066060-9,
e-ISBN (EPUB) 978-3-11-066057-9

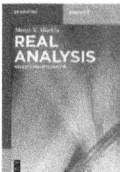

Real Analysis, Measure and Integration
Marat V. Markin, 2019
ISBN 978-3-11-060097-1, e-ISBN (PDF) 978-3-11-060099-5,
e-ISBN (EPUB) 978-3-11-059882-7

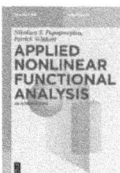

Applied Nonlinear Functional Analysis, An Introduction
Nikolaos S.Papageorgiou, Patrick Winkert, 2020
ISBN 978-3-11-051622-7, e-ISBN (PDF) 978-3-11-053298-2,
e-ISBN (EPUB) 978-3-11-053183-1

Haiyun Zhou and Xiaolong Qin

Fixed Points of Nonlinear Operators

—

DE GRUYTER

国防工业出版社
National Defense Industry Press

Mathematics Subject Classification 2010
47H05, 47H06, 47H09, 47H10, 65K15

Authors
Prof. Haiyun Zhou
Shijiazhuang Mechanical Engineering University
Shijia Zhuang
China

Prof. Xiaolong Qin
Hangzhou Normal University
Hangzhou
China
qxlxajh@163.com

ISBN 978-3-11-066397-6
e-ISBN (PDF) 978-3-11-066709-7
e-ISBN (EPUB) 978-3-11-066401-0

Library of Congress Control Number: 2019957139

Bibliographic information published by the Deutsche Nationalbibliothek
The Deutsche Nationalbibliothek lists this publication in the Deutsche Nationalbibliografie;
detailed bibliographic data are available on the Internet at http://dnb.dnb.de.

Cover image: stevanovicigor / iStock / Getty Images Plus
Typesetting: VTeX UAB, Lithuania
Printing and binding: CPI books GmbH, Leck

www.degruyter.com

Preface

Fixed point theory has a long history. In 1895–1900, a French mathematician H. Poincaré attributed the existence problem of periodic solutions of the restricted three-body problem to the existence problem of fixed points of a continuous transformation under certain conditions. This can be viewed as the origin of the notion of fixed points. In 1910, L. E. Brouwer proved that a continuous mapping on a polyhedron in a finite-dimensional space has at least one fixed point, which initiated the study of fixed point theory. In 1922, a Polish mathematician S. Banach, based on the Picard iteration method, introduced the well-known contraction mapping principle. In 1930, a Polish mathematician J. Schauder extended the Brouwer fixed point theorem from continuous mappings in finite-dimensional spaces to completely continuous mappings in infinite-dimensional spaces.

Then, in 1935, A. Tychonoff extended Schauder fixed point theorem to locally convex topological linear spaces. In 1965, F. E. Browder, W. A. Kirk, and D. Göhde independently established a fixed point theorem for nonexpansive mappings in uniformly convex Banach spaces or reflexive Banach spaces with the normal structure. In 1973, R. H. Martin established a fixed point theorem for continuous pseudocontractive mappings. One year later, K. Deimling improved R.H. Martin's results. Then, in 1999, C. H. Morales and C. E. Chidume further improved the results of Deimling. In 1976, J. Caristi generalized the well-known contraction mapping principle in complete metric spaces, and he proved an amazing fixed point theorem. It is well known that Caristi fixed point theorem is equivalent to the famous Ekeland variational principle. F. E. Browder, a former president of the American Mathematics Association, always highly appraised the Caristi fixed point theorem as one of the most important results in nonlinear analysis.

Under the premise of the existence of a fixed point, the method of constructing the fixed point is an important theme in the research of fixed point theory. Historically, there have been many iteration methods for different mappings, among which the most classical methods are the Picard iterative method, steepest descent methods, the normal Mann iterative method, Ishikawa iterative method, Halpern-type iterative method, Bruck regularization iterative method, the gradient projection method, the extragradient method, hybrid projection method, hybrid steepest descent methods, asymptotic iterative method, Moudafi viscosity iterative method, and Rockafellar approximate proximal method, to name only a few. Mathematicians all over the world apply these iteration methods to successfully construct fixed or zero points of various mappings, and a large number of excellent results with scientific significance and application value have emerged. These achievements are still scattered in mathematics journals at home and abroad, and have not been sorted out yet. This is our original intention of compiling this book. Fixed point iteration methods can be applied to solve practical problems with different physical engineering background, such as res-

https://doi.org/10.1515/9783110667097-201

olution recognition and signal synthesis in signal processing; image restoration and reconstruction in image processing; power control, broadband allocation, and beam imaging in CDMA data networks; image denoising based on wavelet transform; video coding technology; radar antenna mode synthesis; control design of a rocket launch tower; pattern recognition of nuclear submarine, and machine learning and artificial intelligence, etc. These problems can be modeled by convex optimization frameworks, feasible frameworks, or split feasibility frameworks, so they can be solved by fixed point iteration methods or variational inequality iteration methods. We are fully convinced that the fixed point iteration method will be more widely used in the fields of science and technology.

Most of this book material is collected, screened, processed, and sorted out from the vast literature, the proofs of many theorems in the book have been refined and simplified by the authors of the book, and some of the results have not yet been published. Systematic collation of the increasingly expanding literature is indeed one of the purposes of this book, but we absolutely made no attempt to exhaustively survey all the research results. The research of iteration methods is gradually developing in depth, and new ideas, methods, and achievements emerge endlessly, changing with each passing day; it is impossible to include all of them in one book. Therefore, we only select the most fundamental and important results so that readers could appreciate excellent works.

The book is divided into six chapters. In Chapter 1, we give basic knowledge required for this book. In Chapter 2, we discuss iterative methods for fixed points of nonexpansive mappings in Hilbert spaces. In Chapter 3, we consider the iterative methods of fixed points of pseudocontractive mappings and zeros of monotone mappings in Hilbert spaces. In Chapter 4, we study the iteration methods for fixed points of nonexpansive mappings in Banach spaces. In Chapter 5, we investigate the iterative methods for zeros of accretive operators and fixed points of pseudocontractive mappings in Banach spaces. Finally, in Chapter 6, we introduce three kinds of iteration methods for zeros of maximal monotone operators and their corresponding applications in Banach spaces.

To read the book, readers need some prerequisite knowledge of nonlinear functional analysis and convex analysis, especially some knowledge on nonlinear mapping theory and Banach space geometry theory would be helpful. These preliminaries are given directly (without proof) in Chapter 1, and the details where to find proofs are provided in the references.

Due to the limitations of our knowledge, and since the proofs of some theorems are modified by the authors of the book, there may be some inevitably mistakes and shortcomings in the book. We sincerely hope that the readers can notify us.

Shijiazhuang and Hangzhou,   *Haiyun Zhou*
October 2019   *Xiaolong Qin*

Contents

1 Introduction and preliminaries

The purpose of this chapter is to give some essential definitions, symbols, and terminology. We focus on the geometric theory of Hilbert and Banach spaces. Some fundamental concepts in convex analysis are stated. Some important properties of several classes of nonlinear operators are discussed.

1.1 Partial order relations, partial order sets, and Zorn lemma

The purpose of this section is to state the famous Zorn lemma, which is an important tool in many branches of modern mathematics for existence results, such as the Hahn–Banach extension theorem, the maximal extension of monotone or accretive operators, the Kirk fixed point theorem, and some other important results in algebra or topology. In order to state the Zorn lemma, we first review some basic concepts.

Definition 1.1.1. Let X be a nonempty set and let R be a relation on X. If R is
(i) reflexive, i. e., $\langle x, x \rangle \in R, \forall x \in X$;
(ii) antisymmetric, i. e., if $\langle x, y \rangle \in R$ and $\langle y, x \rangle \in R$, then $x = y, \forall x, y \in X$;
(iii) transitive, i,e., if $\langle x, y \rangle \in R$ and $\langle y, z \rangle \in R$, then $\langle x, z \rangle \in R, \forall x, y, z \in X$,

then R is said to be a partial order relation on X, denoted by $\preceq$. If $\langle x, y \rangle \in \preceq$, we denote it as $x \preceq y$. A set X with a partial order relation $\preceq$ is called a partially ordered set, denoted by $\langle X, \preceq \rangle$. Also $x \prec y$ if and only if $x \preceq y$, and $x \neq y$. We say that x is comparable to y if either $x \preceq y$ or $y \preceq x$ holds. For a partially ordered set $\langle X, \preceq \rangle$ and $A \subset X$, if $\forall x, y \in A$, x is comparable to y, then A is a totally ordered set, which is also called a chain.

Definition 1.1.2. Let $\langle X, \preceq \rangle$ be a partially ordered set, $A \subset X$, and $y \in X$.
(i) If $x \in A \Rightarrow x \preceq y, \forall x \in X$, then y is an upper bound of A.
(ii) If $x \in A \Rightarrow y \preceq x, \forall x \in X$, then y is a lower bound of A.

Definition 1.1.3. Let $\langle X, \preceq \rangle$ be a partially ordered set, $A \subset X$, and $y \in A$.
(i)   If $x \in A \Rightarrow y \preceq x, \forall x \in X$, then y is the least element of A.
(ii) If $x \in A \Rightarrow x \preceq y, \forall x \in X$, then y is the greatest element of A.
(iii) If $x \in A \wedge x \preceq y \Rightarrow x = y, \forall x \in X$, then y is the minimal element of A.
(iv) If $x \in A \wedge y \preceq x \Rightarrow x = y, \forall x \in X$, then y is the maximal element of A.

Remark 1.1.1.
(i) A maximal element (or a minimal element) is not necessary the greatest element (or the least element), and a maximal element (or a minimal element) is not necessarily comparable to any element in X.
(ii) For a finite partial order $\langle X, \preceq \rangle$, a maximal element (or a minimal element) must exist, but the greatest element (or the least element) need not exist.

https://doi.org/10.1515/9783110667097-001

(iii) If the greatest element (or the least element) exists, it must be unique. Generally speaking, a maximal element (or a minimal element) is not unique, however, if there is a maximal element (or a minimal element) in X, then it must be the greatest element (or the least element) in X.

Lemma 1.1.1 (Zorn Lemma). *Let $\langle X, \preceq \rangle$ be a partially ordered set. If X has upper bounds for every chain, then it has one maximal element.*

1.2 Metric spaces

In the 1870s, Cantor studied some concepts, which have a close relation with metrics, such as accumulation, interior, exterior, and boundary points, open and closed sets, and so on, in Euclidean spaces. With the development of the classical analysis, Ascoli, Volterra, Alzera, Hadamard, Borel, and many famous mathematicians studied various convergence notions in various function classes. Their convergence analysis is based on the distance between functions. In 1906, Fréchet introduced metric spaces and investigated some basic concepts related to the distance based on Cantor's ideas.

From now on, we use $\mathbb{R}$, $\mathbb{R}^+$, and $\overline{\mathbb{R}}$ to denote $(-\infty, +\infty)$, $[0, +\infty)$, and $[-\infty, +\infty]$, respectively; $\mathbb{N}$ stands for the set of natural numbers.

Definition 1.2.1. Let X be a nonempty set. Let $d : X \times X \rightarrow \mathbb{R}^+$ be a bivariate function satisfying the following three conditions, $\forall x, y, z \in X$,
(d1) $d(x, y) \geq 0$ and $d(x, y) = 0 \Leftrightarrow x = y$;
(d2) $d(x, y) = d(y, x)$;
(d3) $d(x, y) \leq d(x, z) + d(z, y)$.

Then d is called a distance on X, and (X, d) is called a metric space.

The open ball $B_r(x_0)$ with center x_0 and radius r is defined as follows:

$$B_r(x_0) = \{x \in X : d(x_0, x) < r\}.$$

The closed ball $B_r[x_0]$ with center x_0 and radius r is defined as follows

$$B_r[x_0] = \{x \in X : d(x_0, x) \leq r\}.$$

The sphere $S_r(x_0)$ with center x_0 and radius r is defined as follows:

$$S_r(x_0) = \{x \in X : d(x_0, x) = r\}.$$

Let A be a subset of X. The diameter of set A is defined by $\text{Diam}\, A = \sup\{d(x, y) : x, y \in A\}$. Recall that A is open if there exists an open ball $B_{r_x} \subset A$, where $r_x > 0$, $\forall x \in A$. If the complementary set $\wp(A) = X \backslash A$ of A is an open set, then A is said to be a closed set; A is said to be bounded if there exists a closed ball $B_r[0]$ such that $A \subset B_r[0]$.

Let (X, d_1) and (Y, d_2) be metric spaces. Let $f : X \to Y$ be a mapping. Let $\text{Dom}(f)$ denote the domain of f and let $x_0 \in \text{Dom}(f)$ be a fixed element. Recall that f is said to be continuous at x_0 if $\forall \varepsilon > 0$ there exists $\delta = \delta(\varepsilon, x_0) > 0$ such that

$$d_2(f(x), f(x_0)) < \varepsilon$$

whenever $d_1(x, x_0) < \delta$. Recall that $f : \text{Dom}(f) \subset X \to Y$ is said to be continuous if f is continuous at every point in its domain.

Recall that $f : \text{Dom}(f) \subset X \to Y$ is said to be L-Lipschitz if there exists $L > 0$ such that

$$d_2(f(x), f(y)) \le L d_1(x, y), \quad \forall x, y \in \text{Dom}(f).$$

Recall that $f : \text{Dom}(f) \subset X \to Y$ is said to be nonexpansive if the above inequality holds with $L = 1$; also $f : \text{Dom}(f) \subset X \to Y$ is said to be contractive if the above inequality holds with $L \in (0, 1)$. Furthermore, $f : \text{Dom}(f) \subset X \to Y$ is said to be Meir–Keeler contractive if, for any $\varepsilon > 0$, there exists $\delta > 0$ such that

$$d_1(x, x_0) < \delta + \varepsilon \text{ implies } d_2(f(x), f(x_0)) < \varepsilon, \quad \forall x, y \in \text{Dom}(f).$$

Let $\{x_n\}$ be a sequence in a metric space (X, d). Recall that $\{x_n\}$ is said to be convergent in X if, for any $\varepsilon > 0$, there exists a natural number $n_0 \ge 1$ such that $d(x_n, x) < \varepsilon$, where x a point in X, for all $n \ge n_0$. We denote it by $x_n \to x$ or $\lim_{n \to \infty} x_n = x$. Recall that $\{x_n\}$ is said to be a Cauchy sequence if, for any $\varepsilon > 0$, there exists a natural number n_0 such that $d(x_m, x_n) < \varepsilon$ whenever $m, n \ge n_0$. A convergent sequence is a Cauchy sequence and a Cauchy sequence is always bounded. If every Cauchy sequence in X is convergent, then a metric space (X, d) is said to be complete.

Theorem 1.2.1 (Cantor Intersection Theorem). *Let (X, d) be a complete metric space and let $\{F_n\}$ be a countable family of decreasing nonempty closed subsets of X such that $\delta(F_n) = \sup\{d(x, y) : x, y \in F_n\} \to 0$ as $n \to \infty$. Then $\bigcap_{n=1}^{\infty} F_n$ contains exactly one point.*

Using Theorem 1.2.1, one can obtain the following celebrated Banach contraction mapping principle.

Theorem 1.2.2 (Banach Contraction Mapping Principle, 1922). *Let (X, d) be a complete metric space and let $f : X \to X$ be a contraction mapping. Then f has a unique fixed point in X, that is, there exists a unique element $x^* \in X$ such that $x^* = f(x^*)$.*

Remark 1.2.1. The Banach contraction mapping principle is one of the most significant and fundamental fixed point theorems and it has been widely applied to many scientific fields. There are a number of generalization and extensions of the Banach contraction mapping principle, one of which is the following Meir–Keeler fixed point theorem.

Theorem 1.2.3 (Meir–Keeler Fixed Point Theorem, 1969). *Let (X, d) be a complete metric space, and let $f : X \to X$ be a Meir–Keeler contraction mapping. Then f processes a unique fixed point in X.*

Recall that a function $\varphi : X \to \mathbb{R}$ is said to be lower semicontinuous if $\forall x \in X$, $\{x_n\} \subset X$ and $x_n \to x$ imply that

$$\varphi(x) \leq \liminf_{n \to \infty} \varphi(x_n).$$

Using Theorem 1.2.1, one can obtain the following celebrated Caristi fixed point theorem.

Theorem 1.2.4 (Caristi Fixed Point Theorem, 1976). *Let (X, d) be a complete metric space. Let $f : X \to X$ be a mapping (not necessarily continuous) and let $\varphi : X \to \mathbb{R}^+$ be a lower semicontinuous and bounded below function such that*

$$d(x, f(x)) \leq \varphi(x) - \varphi(f(x)), \quad \forall x \in X.$$

Then f has at least one fixed point.

1.3 Topological spaces

The basic concepts in modern analysis are the "limit" and "continuity" of mappings in abstract spaces. It is well known that the concept of "limit" is induced by a "distance" in metric spaces. Based on the "distance", we have open sets, neighborhoods, interior points, accumulation points, subspaces, etc. Convergence with respect to a distance is essential in modern analysis, however, there exist convergence notions, which cannot be characterized by a distance. Can one directly define open sets, neighborhoods, etc., to obtain the "limit" and "continuity"? The answer is positive. There are a number of methods to induce a topology on sets. In 1914, Hausdorff introduced a topology based on neighborhood systems. Since then, many methods, including open set systems, closed set systems, closure operators, etc., were developed to induce a topology. These topologies are uniform. Based on some basic properties of open sets in metric spaces, the theory of topological spaces can be established.

1.3.1 Point sets in topological spaces

Definition 1.3.1. Let X be a nonempty set. Let ζ be a family of subsets of X such that
(O_1) $\emptyset, X \in \zeta$;
(O_2) If $G_1, G_2 \in \zeta$, then $G_1 \cap G_2 \in \zeta$;
(O_3) If $G_i \in \zeta$ $(i \in \Lambda)$, where Λ is an arbitrary index set, then $\bigcup_{i \in \Lambda} G_i \in \zeta$.

Then ζ is said to be a topology on X, (X, ζ) is said to be a topological space, and the elements of ζ are said to be open sets of X.

Example 1.3.1. Let (X, d) be a metric space. Then the family of all open sets of X satisfies Axioms (O_1)–(O_3) in Definition 1.3.1. Hence, every metric space is a topological space.

Example 1.3.2. Let X be an arbitrary set and let $\zeta_\infty = \{A : A \subset X\}$. Then (X, ζ_∞) is a topological space and ζ_∞ is a "discrete topology". Let $\zeta_0 = \{X, \emptyset\}$. Then (X, ζ_0) is also a topological space and ζ_0 is called a "dense topology". This shows that one can induce different topologies on the same set to obtain different topological spaces.

Definition 1.3.2. Let ζ_1 and ζ_2 be two topologies on X. If $\zeta_2 \subset \zeta_1$, then the topology ζ_2 is said to be a coarser (weaker or smaller) topology than ζ_1, and ζ_1 is said to be a finer (stronger or larger) topology than ζ_2.

Definition 1.3.3. Let (X, ζ) be a topological space and $A \subset X$. Let

$$\zeta_A = \{A \cap G : G \in \zeta\}.$$

Then ζ_A is a topology on A. And we call ζ_A a relative topology, and (A, ζ_A) a subspace of (X, ζ).

Definition 1.3.4. Let (X, ζ) be a topological space. Let $A \subset X$ and $U \subset X$. If there exists $G \in \zeta$ such that

$$A \subset G \subset U,$$

then U is said to be a "neighborhood" of A. We use $\pounds(A)$ to denote all the neighborhoods of A. If $A = \{x\}$, U is the neighborhood system of x. In space (X, ζ), the set of all the neighborhood systems of points is said to be the system of neighborhoods of X, which is denoted by $\pounds$, that is,

$$\pounds = \{\pounds(x); x \in X\}.$$

Remark 1.3.1. The neighborhoods of x need not be open subsets containing x. Let $X = \{a, b, c\}$ and $\zeta = \{\emptyset, \{a\}, X\}$. Then (X, ζ) is a topological space. It is obvious that $U = \{a, b\}$ is the neighborhood of a, but it is not an open set. Indeed, U is also not a neighborhood of b. However, an open set is a neighborhood of its any point. The reverse is also true.

Theorem 1.3.1. *Let (X, ζ) be a topological space and let $\pounds$ be a family of neighborhoods of X, reduced by ζ. Then $\pounds$ satisfies the following conditions:*
(i) $\pounds(x) \neq \emptyset, \forall x \in X$;
(ii) $x \in U, \forall U \in \pounds(x)$;
(iii) $U \cap V \in \pounds(x), \forall U, V \in \pounds(x)$;

(iv) $V \in £(x), \forall V \supset U \in £(x)$;

(v) *For any $U \in £(x)$, there exists $V \in £(x)$ such that $U \supset V$ and $V \in £(x)$ for any $y \in V$.*

Definition 1.3.5. Let (X, ζ) be a topological space and let $F \subset X$. If $\wp F = X \backslash F \in \zeta$, where $\wp F$ is the complementary set of F, then F is closed.

Theorem 1.3.2. *Let (X, ζ) be a topological space and let ϑ be the family of all closed sets of X. Then ϑ satisfies the following conditions:*

(i) $\emptyset, X \in \vartheta$;

(ii) *if $F_1, F_2 \in \vartheta$, then $F_1 \cup F_2 \in \vartheta$;*

(iii) *if $F_i \in \vartheta$ $(i \in \Lambda)$, then $\bigcap_{i \in \Lambda} F_i \in \vartheta$.*

Definition 1.3.6. Let (X, ζ) be a topological space and let $A \subset X$. If $x \in A$ has a neighborhood V of x such that $V \subset A$, then x is an interior point of A. The set of all interior points is called the interior of A, denoted by A^0 or Int A; $x \in X$ is said to be an exterior point of A if x is an interior point of $\wp A$. The set of all exterior points is called the external of A, denoted by A^e.

Definition 1.3.7. Let (X, ζ) be a topological space and let $A \subset X$.

(i) If $x \in X$, for all $U \in £(x)$, one has

$$U \cap A \neq \emptyset.$$

Then x is called a closure point of A and the set of all closure points is called the closure of A, denoted by $\overline{A}$ or cl A.

(ii) $x \in X$ is said to be an accumulation point of A if $\forall U \in £(x)$, $U \cap (A \backslash \{x\}) \neq \emptyset$. The set of all accumulation points is called a derived set, denoted by A^d. If $A = A^d$, then A is a completed set.

(iii) $x \in X$ is said to be a boundary point of A if $\forall U \in £(x)$, $U \cap A \neq \emptyset$ and $U \cap \wp A \neq \emptyset$. The set of all boundary points is called the boundary of A, denoted by ∂A.

(iv) x is said to be an isolated point of A if there exists $U \in £(x)$ such that $U \cap A = \{x\}$.

(v) If $B \subset \overline{A}$, then A is dense in B. If $\overline{A} = X$, then A is dense everywhere.

(vi) If there exists a countable dense subset in X, then X is separable. Indeed, $\mathbb{R}^n$, $C[a,b]$, l^p, and $L^p(p \geq 1)$ are all separable.

(vii) If $\text{Int}(\overline{A}) = \emptyset$, then A is a nowhere dense set.

(viii) A subset of a topological space X is said to be of the first category in X if it is a union of countably many nowhere dense subsets. A subset of a topological space X is said to be of the second category in X if it is not of the first category in X. It is known that every complete metric space is of the second category.

Definition 1.3.8. Let (X, ζ) be a topological space. Every mapping $S : \mathbb{N} \to X$ is called a sequence in X, denoted by $\{x_i\}$, where $x_i = S(i) \in X$, $i \in \mathbb{N}$. In particular, $S(i) = a \in X$, $\forall i \in \mathbb{N}$, where $a \in X$ is a fixed point. Then S is a constant sequence, denoted by $\{a\}$.

Definition 1.3.9. Let $\{x_i\}$ be a sequence in the topological space (X, ζ). If for every neighborhood $U \in \mathfrak{L}(x)$ of a point $x \in X$, there exists $i_0 \in \mathbb{N}$ such that $x_i \in U$, when $i \geq i_0$, then x_i is said to be the limit of $\{x_i\}$, denoted by $\lim_i x_i = x$. If $\{x_i\}$ has at least one limit, then $\{x_i\}$ is a convergent sequence.

Definition 1.3.10. Let (X, ζ) be a topological space and let $S, S_1 : \mathbb{N} \to X$ be sequences in X. If there exists a strictly increasing mapping $\varphi : \mathbb{N} \to \mathbb{N}$, such that $S_1 = S \circ \varphi$, then S_1 is said to be a subsequence.

Remark 1.3.2. There is a huge difference regarding the convergence of sequences between topological and metric spaces. For example, any sequence is convergent in a dense topological space. This is not true in metric spaces. However, they also have some similarities.

Theorem 1.3.3. *Let $\{x_i\}$ be a sequence in a topological space (X, ζ). Then*
(i) *If $\{x_i\}$ is a constant sequence that takes value $x \in X$, then $\lim_i x_i = x$;*
(ii) *If $\{x_i\}$ converges to $x \in X$, then every subsequence of $\{x_i\}$ converges to x.*

Theorem 1.3.4. *Let (X, ζ) be a topological space and let $A \subset X$. If there exists a sequence $\{x_i\}$ in $A \backslash \{x\}$ such that $\{x_i\}$ converges to $x \in X$, then x is an accumulation point of A. The reverse may not be true.*

Definition 1.3.11.
(i) If the collection $\{G_\lambda : \lambda \in \Lambda\}$ of X satisfies $X = \bigcup_{\lambda \in \Lambda} G_\lambda$, then $\{G_\lambda : \lambda \in \Lambda\}$ is a cover of X. In particular, if every G_λ is an open set, then $\{G_\lambda : \lambda \in \Lambda\}$ is said to be an open cover of X.
(ii) Let A be a subset of a topological space (X, ζ). We say A is compact if every open cover of A has a finite subcover.
(iii) A is said to be countably compact if every countable open cover has a finite subcover.
(iv) A is said to be sequentially compact if every infinite sequence has a convergent subsequence.

In particular, if $A = X$, it is said to be a compact space, a countably compact space, and a sequentially compact space, respectively. Sequentially compact and compact spaces are mutually independent notions. Both sequentially compact and compact spaces are countably compact, however, the reverses are not true.

Definition 1.3.12. A collection ß of subsets of X is said to have the finite intersection property if the intersection of any finite subcollection of A is nonempty.

Theorem 1.3.5. *The topological space (X, ζ) is compact if and only if any collection of closed subsets of X with the finite intersection property has a nonempty intersection.*

A compact subset in a topological space is not necessarily closed, for example, any nonempty subset in the dense topological space is compact, but not necessarily closed.

Theorem 1.3.6. *Let (X, ζ) be a topological space and let $A, B \subset X$. Then*

(i) $\quad \overline{\emptyset} = \emptyset;$

(ii) $\quad A \subset \overline{A};$

(iii) $\quad \overline{\overline{A}} = \overline{A};$

(iv) $\quad \overline{A \cup B} = \overline{A} \cup \overline{B};$

(v) $\quad \overline{A} = \wp(\wp A)^\circ;$

(vi) $\quad \overline{A} = A \cup A^d;$

(vii) *A is closed if and only if $\overline{A} = A;$*

(viii) *$\overline{A}$ is the minimum closed set containing A.*

Theorem 1.3.7. *Let (X, ζ) be a topological space and let $A, B \subset X$. Then*

(i) $\quad X^\circ = X;$

(ii) $\quad A^\circ \subset A;$

(iii) $\quad A^{\circ\circ} = A;$

(iv) $\quad (A \cap B)^\circ = A^\circ \cup B^\circ;$

(v) $\quad A^\circ = \wp(\overline{\wp A});$

(vi) *A is open if and only if $A^\circ = A;$*

(vii) *A° is the maximum open set included in A.*

Theorem 1.3.8. *Let (X, ζ) be a topological space and let $A \subset X$. Then*

(i) $\quad A^\circ \cap \partial A = A^e \cap \partial A = \emptyset;$

(ii) $\quad X = A^\circ \cup A^e \cup \partial A;$

(iii) $\quad \overline{A} = A^\circ \cup \partial A = A \cup \partial A = A \cup A^d;$

(iv) *A is closed if and only if $\partial A \subset A;$*

(v) *A is open if and only if $A \cap \partial A = \emptyset;$*

(vi) $\quad A^\circ = A \backslash (\wp A)^d = A \backslash \partial A;$

(vii) $\quad \overline{\wp A} = \wp A^\circ;$

(viii) $\quad \partial A = A \cup (\overline{\wp A}) = \overline{A} \backslash A^\circ;$

(ix) $\quad \wp(\partial A) = A^\circ \cup (\wp A)^\circ.$

1.3.2 Topological bases, subbases and neighborhood bases

Based on open sets in usual topological spaces, the set of real numbers has the celebrated "open set construction theorem", that is, any open set in $\mathbb{R}^1$ is of the form $G = \bigcup_{i=1}^m (\alpha_i, \beta_i)$, where m is finite or ∞ and (α_i, β_i) are open intervals with empty intersections. This shows that the open interval as a special open set in $\mathbb{R}^1$ plays a basic role in the construction of collections of open sets. Usually, it is called a collection of "basic open sets" or "open set bases". We can also define "topological bases" in a general topological space.

Definition 1.3.13. Let (X, ζ) be a topological space and let $\mathfrak{B}$ be a family of subsets of ζ. If $G = \bigcup_{B \in \mathfrak{B}} B$, $\forall G \in \zeta$, then $\mathfrak{B}$ is said to be a "topological base" of (X, ζ). Let Φ be a nonempty subset of ζ. If the collection of the intersections of any finite elements in Φ is a topological base of ζ, then Φ is said to be a subbase of ζ.

Theorem 1.3.9. *Let X be a nonempty set and let $\mathfrak{B}$ be a family of subsets of X. Then $\mathfrak{B}$ is some topological base on X if and only if $\mathfrak{B}$ has the following properties:*
(i) *$\forall x \in X$, there exists $G \in \mathfrak{B}$ such that $x \in G$;*
(ii) *If $x \in G_1 \cap G_2$, where $G_i \in \mathfrak{B}(i = 1, 2)$, then there exists $G_3 \in \mathfrak{B}$ such that $x \in G_3 \subset G_1 \cap G_2$.*

Theorem 1.3.10. *A family of sets $\mathfrak{B} \subset \zeta$ is a topological base for the given topology ζ if and only if $\forall G \in \zeta$, $x \in G$, there exists $G_x \in \mathfrak{B}$ such that $x \in G_x \subset G$.*

Definition 1.3.14. Let (X, ζ) be a topological space and let $\mathfrak{B}(x)$ be a family of subsets of $\pounds(x)$. Family $\mathfrak{B}(x)$ is said to be a "neighborhood base" (or "local base" or "fundamental system of neighborhoods") if, for all $U \in \pounds(x)$, there exists $V \in \mathfrak{B}(x)$ such that $V \subset U$.

Definition 1.3.15. Let (X, ζ) be a topological space. Let $\pounds$ be a neighborhood system of X and let $\mathfrak{B}$ be a subfamily of $\pounds$. This $\mathfrak{B}$ is said to be a neighborhood base of (X, ζ) if, $\forall U \in \zeta$, $x \in U$, there exists $B \in \mathfrak{B}$ such that $x \in B \subset U$.

Theorem 1.3.11. *Let $\mathfrak{B}$ be a topological base of a topological space (X, ζ). Then $V \subset X$ is a neighborhood of x if and only if there exists $B \in \mathfrak{B}$ such that $x \in B \subset V$.*

Theorem 1.3.12. *Let $\mathfrak{B}$ be a family of open sets of a topological space (X, ζ). Then $\mathfrak{B}$ is a topological base of X if and only if, for all $x \in X$, $\mathfrak{B}(x) = \{B \in \mathfrak{B} : x \in B\}$ is a neighborhood base of x.*

Theorem 1.3.13. *A family of sets $\mathfrak{B}$ is some topological base on the set $X = \cup\{B : B \in \mathfrak{B}\}$ if and only if $\forall U, V \in \mathfrak{B}$, $\forall x \in U \cap V$, there exists $W \in \mathfrak{B}$ such that $x \in W$ and $W \subset U \cap V$.*

Theorem 1.3.14. *Let φ be a family of nonempty sets. Then the family of all finite intersections of subsets of φ is a topological base of $X = \cup\{S : S \in \varphi\}$.*

Theorem 1.3.15. *Let ζ_1, ζ_2 be two topologies on the set X. Also $\forall x \in X$, let $\mathfrak{B}_1(x)$ and $\mathfrak{B}_2(x)$ be neighborhood bases of x in (X, ζ_1) and (X, ζ_2), respectively. Then $\zeta_2 \subset \zeta_1$ if and only if, $\forall x \in X$, $\forall B_2 \in \mathfrak{B}_2(x)$, there exists $B_1 \in \mathfrak{B}_1(x)$ such that $B_1 \subset B_2$.*

Theorem 1.3.16. *Let $x \in X$ and $\mathfrak{B}(x)$ be a family of subsets of X. Then there exists a topology ζ on X such that $\mathfrak{B}(x)$ is a neighborhood base of x if and only if the following axioms of neighborhood bases hold:*

(i) *For $B_1, B_2, \ldots, B_m \in B(x)$, there exists $B \in \mathcal{B}(x)$ such that $B \subset \cap_{i=1}^{m} B_i$;*
(ii) *If $B \in \mathcal{B}(x)$, then $x \in B$;*
(iii) *If $V \in \mathcal{B}(x)$, then there exists $W \in \mathcal{B}(x)$ such that $W \subset V$, and $\forall y \in W$ there exists $B \in \mathcal{B}(y)$ such that $B \subset V$.*

1.3.3 Directed sets, nets, and convergence of nets

Definition 1.3.16. Let $(S, \preceq)$ be a partially ordered set. If for all $x, y \in S$, there exists $z \in S$ such that $z \succeq x$, $z \succeq y$, then $(S, \preceq)$ is said to be a directed set.

Definition 1.3.17. Let (X, ζ) be a topological space and let $(S, \preceq)$ be a directed set. Then a mapping from S to X is called a net (or directed point sequence) on X, denoted by $\{x_z\}_{z \in S}$, where $x_z \in X$.

Definition 1.3.18. Let (X, ζ) be a topological space and let $(S, \preceq)$ be a directed set. Let $\{x_z\}_{z \in S}$ be a net on X. If $\exists x_0 \in X$, $\forall U \in \pounds(x_0)$, there exists $z_0 \in S$ such that $x_z \in U$ whenever $z \succeq z_0$, then $\{x_z\}_{z \in S}$ is said to be convergent to x_0 (net convergence) and x_0 is said to be a limit of net $\{x_z\}_{z \in S}$, denoted by $x_z \to x_0$ $(z \in S)$ or $\lim_{z \in S} x_z = x_0$.

Definition 1.3.19. Let $(S, \preceq)$ be a directed set and let $C \subset S$. If C is also a directed set on "$\preceq$", then C is said to be a directed subset of S. If $\forall \delta \in S$, there exists $\alpha \in C$, such that $\alpha \succeq \delta$, then C is said to be a cofinal directed subset of S.

Definition 1.3.20. Let $\{x_z\}_{z \in S}$ and $\{y_\alpha\}_{\alpha \in \triangle}$ be two nets. If there exists $J : \triangle \to S$ such that $\forall \alpha \in \triangle$, $y_\alpha = x_{J(\alpha)}$ and it satisfies the following conditions:
(i) $\forall \alpha_1, \alpha_2 \in \triangle$, $\alpha_1 \preceq \alpha_2 \Rightarrow J(\alpha_1) \preceq J(\alpha_2)$;
(ii) $\forall \alpha \in S$, there exists $\delta \in \triangle$, such that $J(\delta) \succeq \alpha$,

then $\{y_\alpha\}_{\beta \in \triangle}$ is called a subnet of $\{x_z\}_{z \in S}$.

Remark 1.3.3. A net on a directed set is not necessarily a subnet. A sequence is always a net, but a net is not necessarily a sequence. A subsequence is always a subnet, but a subnet is not necessarily a subsequence. A sequence may have a subnet which is not its subsequence.

Utilizing the concept of net convergence, one can give characterizations of accumulation and closure points, open, closed, and compact sets, as well as many other important concepts in a topological space (X, ζ).

Theorem 1.3.17. *Let A be a subset of topological space (X, ζ) and $x_0 \in X$. Then the following statements are true:*
(i) *$x_0 \in A^d$ if and only if there exists a net $\{x_\delta\}$ in $A \backslash \{x_0\}$ such that $x_\delta \to x_0$;*
(ii) *$x_0 \in \overline{A}$ if and only if there exists a net $\{x_\delta\}$ in A such that $x_\delta \to x_0$;*
(iii) *A is open if and only if there does not exist any net converging to a point in A;*

(iv) *A is closed if and only if for any net $\{x_\delta\} \subset A$, if $x_\delta \to x$, then $x \in A$;*
(v) *A is compact if and only if each net in A has a subnet converging to a point in A.*

1.3.4 Continuous mappings and homeomorphism mappings

Definition 1.3.21. Let (X, ζ_1) and (Y, ζ_2) be two topological spaces. Let $f : X \to Y$ be a mapping and let $x_0 \in X$. If, for any neighborhood $U_{y_0} \in \mathcal{L}(y_0)$ of $y_0 = f(x_0)$, there exists a neighborhood $V_{x_0} \in \mathcal{L}(x_0)$ such that $f(V_{x_0}) \subset \mathcal{L}(y_0)$, then f is continuous at x_0. If f is continuous at each point $x \in X$, then f is a continuous function. If f is bijection, and both f and f^{-1} are all continuous, then f is a homeomorphism mapping. In such a case, we say that X and Y are homeomorphic. For some property P in a topological space, if it remains unchanged under homeomorphisms, then P is called a topologically invariant property.

Theorem 1.3.18. *Let (X, ζ) and (Y, τ) be two topological spaces. Let $f : X \to Y$ be a mapping and let $\mathcal{B}(x)$ be a neighborhood base of $x \in X$. Let $\mathcal{B}_1(f(x))$ be a neighborhood base of $f(x)$. Then f is continuous at x if and only if $\forall C \in \mathcal{B}_1(f(x))$, there exists $B \in \mathcal{B}(x)$ such that $f(B) \subset C$.*

Theorem 1.3.19. *Let (X, ζ_1) and (Y, ζ_2) be topological spaces and let $f : X \to Y$ be a mapping. Then the following statements are equivalent:*
(i) *f is continuous;*
(ii) *For any open set $G \subset Y$, the preimage set $f^{-1}(G)$ in X is open;*
(iii) *For any closed set $F \subset Y$, the preimage set $f^{-1}(F)$ in X is closed;*
(iv) *For any set $B \subset Y, \overline{f^{-1}(B)} \subset f^{-1}(\overline{B})$;*
(v) *For any $A \subset X, f(\overline{A}) \subset \overline{f(A)}$;*
(vi) *For any net $\{x_\delta\} \subset X$, if $x_\delta \to x$, then $f(x_\delta) \to f(x)$ in Y.*

Theorem 1.3.20. *Let (X, ζ_1) and (Y, ζ_2) be topological spaces and let $f : X \to Y$ be a continuous mapping. If A is a compact subset in X, then $f(A)$ is a compact subset in Y. In particular, if $Y = \mathbb{R}$, there exist $x_0, x_1 \in A$ such that*

$$f(x_0) = \inf\{f(x) : x \in A\} = \min\{f(x) : x \in A\}$$

and

$$f(x_1) = \sup\{f(x) : x \in A\} = \max\{f(x) : x \in A\}.$$

Theorem 1.3.21. *Let $(X, \zeta_1), (Y, \zeta_2), (Z, \zeta_3)$ be topological spaces. If $f : X \to Y$ and $g : Y \to Z$ are continuous mappings, then the composite function $h = gf : X \to Z$ is also a continuous mapping.*

In order to make a topological space more concrete, as close as possible to metric spaces, we need to give some restrictions on the topology. The separation axiom is one of the restrictions.

Definition 1.3.22. Let (X, ζ) be a topological space and let x, y be two distinct points in X. If there exists a neighborhood U of x which does not contain y and there exists a neighborhood V of y which does not contain x, then (X, ζ) is said to be a T_1 topological space.

Theorem 1.3.22. *A topological space (X, ζ) is T_1 if and only if any single point set in X is a closed set.*

Definition 1.3.23. Let (X, ζ) be a topological space. If $\forall x, y \in X$, $x \neq y$, there exist neighborhoods $U \in \pounds(x)$ of x and $V \in \pounds(y)$ of y such that $U \cap V = \emptyset$, then (X, ζ) is said to be a T_2 space or a Hausdorff space.

Theorem 1.3.23. *A topological space (X, ζ) is a T_2 space if and only if every limit of a convergent net in X is unique.*

Definition 1.3.24. Let (X, ζ) be a topological space. If for any given closed set F and any given point $x \notin F$, there exists a neighborhood U of x and a neighborhood V of F such that $U \cap V = \emptyset$, then (X, ζ) is said to be a regular space.

Theorem 1.3.24. *A topological space (X, ζ) is regular if and only if for all $x \in X$, $U \in \pounds(x)$, there exists $V \in \pounds(x)$ such that $x \in V \subset \overline{V} \subset U$.*

Definition 1.3.25. A topological space (X, ζ) is said to be normal if for any disjoint closed sets E and F, there are neighborhoods U of E and V of F such that $U \cap V = \emptyset$.

Theorem 1.3.25. *A topological space (X, ζ) is normal if and only if for any closed set $F \subset X$ and $U \in \pounds(x)$, there exists $V \in \pounds(F)$ such that $F \subset V \subset \overline{V} \subset U$.*

If a regular space satisfies the T_1 axiom, then it is said to be a T_3 space. If a normal space satisfies the T_1 axiom, then it is said to be a T_4 space. It is easy to see that T_4 space $\Rightarrow T_3$ space $\Rightarrow T_2$ space $\Rightarrow T_1$ space, however, the reverse is not true.

Theorem 1.3.26. *Every metric space is a normal space.*

Definition 1.3.26. Let (X, ζ) be a topological space. If ζ is defined by some distance on X, then (X, ζ) is "metrizable".

Another restriction imposed on topological spaces is the "countability axiom".

Axiom A_1 (The First Axiom of Countability). Let (X, ζ) be a topological space. $\forall x \in X$, every neighborhood system of x has an at most countable neighborhood base.

Axiom A_2 (The Second Axiom of Countability). Let (X, ζ) be a topological space. X has an at most countable neighborhood base.

Spaces that satisfy the second axiom of countability also satisfy the first. However, the converse may not be true. Indeed, uncountable spaces with the discrete topology do not satisfy the second axiom of countability. A metric space (X, d) is an A_1 space, and Euclidean space $\mathbb{R}^n$ is an A_2 space.

Theorem 1.3.27. *Let (X, ζ) be an A_2 space. The space X is metrizable if and only if X is an A_3 space.*

Theorem 1.3.28. *Let (X, ζ) be an A_1 space. Then $\forall x \in X$, there exists a countable neighborhood base $\{V_n\}$ $(n = 1, 2, \dots)$ of x such that $V_{n+1} \subset V_n$ $(n = 1, 2, \dots)$.*

Theorem 1.3.29. *Let (X, ζ) be an A_1 space which satisfies the T_1 axiom of separation and let $A \subset X$. Then $x_0 \in A^d$ if and only if there exists a sequence $\{x_n\}$, which is generated by different points in $A \backslash \{x_0\}$, converging to x_0.*

Theorem 1.3.30. *Let (X, ζ) be an A_1 space. Then (X, ζ) is Hausdorff if and only if each limit of a convergent sequence is unique in X.*

Theorem 1.3.31. *Let (X, ζ) be an A_1 space and let (Y, τ) be any topological space. Let $f : X \to Y$ be a mapping from X to Y. Then f is a continuous mapping if and only if for any sequence $x_n \to x_0$ in X, we have $f(x_n) \to f(x_0)$ in Y.*

Definition 1.3.27. Let Γ be an index set, $\{X_\alpha\}_{\alpha \in \Gamma}$ a family of sets, and $x : \Gamma \to \bigcup_{\alpha \in \Gamma} X_\alpha$ a mapping such that $x(\alpha) = x_\alpha \in X_\alpha$. Then the set $\{x \mid x : \Gamma \to \bigcup_{\alpha \in \Gamma} X_\alpha, x_\alpha \in X_\alpha, \alpha \in \Gamma\}$ is called a product set of $X_\alpha (\alpha \in \Gamma)$, denoted by $\prod_{\alpha \in \Gamma} X_\alpha$. Each X_α $(\alpha \in \Gamma)$ is called a factor set of $\prod_{\alpha \in \Gamma} X_\alpha$, the mapping $P_\alpha : \prod_{\alpha \in \Gamma} X_\alpha \to X_\alpha$ defined by $P_\alpha(x) = x_\alpha$ is said to be α-projective, and x_α is called an α-component of x.

Let $\{(x_\alpha, \mathfrak{T}_\alpha)\}_{\alpha \in \Gamma}$ be a family of topological spaces and $X = \prod_{\alpha \in \Gamma} X_\alpha$ a product set. We can introduce a topology on X by means of topology $\mathfrak{T}_\alpha$ on X_α as follows:

For any index $\alpha \in \Gamma$ and any open set G_α in $(X, \mathfrak{T}_\alpha)$, one can construct a family of sets in such manner:

$$\varphi = \{P_\alpha^{-1}(G_\alpha) \mid G_\alpha \in \mathfrak{T}_\alpha, \alpha \in \Gamma\},$$

which induces a topology $\mathfrak{T}$ whose topological base is given by

$$\mathcal{B} = \{P_{\alpha_1}^{-1}(G_{\alpha_1}) \cap \cdots \cap P_{\alpha_n}^{-1}(G_{\alpha_n}) \mid P_{\alpha_i}^{-1}(G_{\alpha_i}) \in \mathfrak{T}, i = 1, 2, \dots, n\}$$

The topology $\mathfrak{T}$ is called the product topology on X and $(X, \mathfrak{T})$ is called the product topological space, in short, product space. It is clear that the product topology is the weakest topology such that each projective mapping $P_\alpha : X \to X_\alpha$ is continuous.

Now we state some basic facts on product spaces. Let $\{X_\alpha\}_{\alpha \in \Gamma}$ be a family of topological spaces and let $X = \prod_{\alpha \in \Gamma} X_\alpha$ be the product space. Then the following facts are known:

(f_1) Each projective mapping $P_\alpha : X \to X_\alpha$ $(\alpha \in \Gamma)$ is a continuous open mapping, that is, P_α is continuous and P_α maps any open set of X into some open set of X_α $(\alpha \in \Gamma)$.

(f_2) Let $\{x_\delta\} \subset X$ be a net. Then $x_\delta \to x \in X$ if and only if for each $\alpha \in \Gamma$, the net $\{(x_\alpha)_\delta\}$ converges to x_α in X_α.

(f_3) Let Y be a topological space and $f : Y \to X$ a mapping. Then f is continuous if and only if for every $\alpha \in \Gamma$, the composite mapping $f_\alpha = P_\alpha f : Y \to X_\alpha$ is continuous.

(f_4) $X = \prod_{\alpha \in \Gamma} X_\alpha$ is a T_2-space if and only if for every $\alpha \in \Gamma$, X_α is a T_2 space.

(f_5) (Tychonoff theorem) $X = \prod_{\alpha \in \Gamma} X_\alpha$ is compact if and only if for every $\alpha \in \Gamma$, X_α is compact.

1.4 Topological linear spaces

When we deal with problems in function spaces, an algebraic operation is also our concern. However, there are only topological structures in topological spaces. In addition to the topological structures, we usually also need other structures, for example, algebraic structures, to deal with many problems in modern analysis.

1.4.1 Linear spaces

Definition 1.4.1. Let X be a nonempty set and let $\mathbb{K}$ be complex field (or real number field). The set X is said to be a linear space if the following conditions hold:

(i) X is an additive commutative group, that is, for all $x, y \in X$, there exists $u \in X$, denoted by $u = x + y$ and called the sum of x, y such that:

   (1) $x + y = y + x$;

   (2) $(x + y) + z = x + (y + z)$;

   (3) There exists a unique element $\theta \in X$ such that for all $x \in X$, $x + \theta = \theta + x$;

   (4) For all $x \in X$, there exists a unique element $x' \in X$ such that $x + x' = \theta$, denoted by $x' = -x$.

(ii) A scalar multiplication with the number field $\mathbb{K}$ is defined, that is, for all $(\alpha, x) \in \mathbb{K} \times X$, there is a product denoted by αx and called α scalar-multiplication with x such that

   (1) $\alpha(\beta x) = (\alpha\beta)x$ (for all $\alpha, \beta \in \mathbb{K}$ and all $x \in X$);

   (2) $1 \cdot x = x$;

   (3) $(\alpha + \beta)x = \alpha x + \beta x$ (for all $\alpha, \beta \in \mathbb{K}$ and all $x \in X$);

   (4) $\alpha(x + y) = \alpha x + \alpha y$ (for all $\alpha \in \mathbb{K}$ and all $x, y \in X$).

An element in a linear space is called a vector. Thus a linear space is also called a vector space.

Linear isomorphism

Let X, Y be linear spaces and let $T : X \to Y$ be a mapping. $T : X \to Y$ is said to be a linear isomorphism if

(i) T is both injective and surjective;

(ii) For all $x, y \in X$ and $\alpha, \beta \in \mathbb{K}$, $T(\alpha x + \beta y) = \alpha Tx + \beta Ty$.

Linear subspace

Let $V \subset X$. Then V is said to be a linear subspace of X if it is a linear space with respect to addition and scalar multiplication on X. It is known that V is a linear subspace of X if and only if V is closed under addition and scalar multiplication, that is, for all $x, y \in V$ and $\alpha, \beta \in \mathbb{K}$, we have $\alpha x + \beta y \in V$. Both X and $\{\theta\}$ are subspaces of X. They are trivial subspaces. Other subspaces are called proper subspaces.

Linear manifold

Let $E \subset X$. If there exist $x_0 \in X$ and a linear subspace $E_0 \subset X$ such that $E = E_0 + x_0 \triangleq \{x + x_0 : x \in E_0\}$, then E is said to be a linear manifold. Indeed, a linear manifold is just a shift of a subspace.

Linear dependence

A set of vectors $x_1, x_2, \ldots, x_n \in X$ is said to be linearly dependent if there exist a finite number of distinct vectors $\lambda_1, \lambda_2, \ldots, \lambda_n \in K$, not all zero, such that $\sum_{i=1}^{n} \lambda_i x_i = \theta$. The set of vectors is said to be linearly independent if no vector in the set can be written as a linear combination of the others. These concepts are central to the definition of dimension.

Dimension

The dimension of a vector space is the cardinality (i. e., the number of vectors of a basis) of the space over its base field, denoted by $\dim X$.

Linear base

If A is a maximal independent set in X, then A is said to be a set of linear bases of X.

Linear closure

Let A be an index set and let $\{x_\lambda \mid \lambda \in \Lambda\}$ be a collection of vectors of X. The set $\{y = \sum_{i=1}^{n} a_i x_{\lambda_i} \mid \lambda_i \in \Lambda, a_i \in K, i = 1, 2, \ldots, n\}$, which is generated by finite linear combinations of $\{x_\lambda \mid \lambda \in \Lambda\}$, is said to be the linear closure of $\{x_\lambda \mid \lambda \in \Lambda\}$. A linear closure is a subspace. Indeed, it is an intersection of all linear subspaces containing $\{x_\lambda \mid \lambda \in \Lambda\}$, and we call the linear closure the linear subspace spanned by $\{x_\lambda \mid \lambda \in \Lambda\}$, denoted by $\text{Span}\{x_\lambda \mid \lambda \in \Lambda\}$.

Linear sum and direct sum

Let X_1, X_2 be subspaces of X. The set $\{x_1 + x_2 \mid x_i \in X, i = 1, 2\}$ is said to be a linear sum of X_1 and X_2, denoted by $X_1 + X_2$. If, in addition, $X_1 \cap X_2 = \{\theta\}$, then the linear sum is said to be a direct sum, denoted by $X_1 \oplus X_2$.

Definition 1.4.2. Let X be a linear space and let $A \subset X$.

(i) If for all $x \in X$, there exists a positive number λ_0 such that for every λ satisfying $|\lambda| \leq \lambda_0$, $\lambda x \in A$, then A is called an absorbing set. In other words, for any direction in X, A must contain a segment with that direction and center θ.

(ii) If λ satisfies $|\lambda| \leq 1$ and is such that $\lambda A \subset A$, then A is a balanced set. For a real linear space X, A is called a balanced set if for all $x \in A$, $[-x, x] \triangleq \{\lambda x | |\lambda| \leq 1\} \subset A$.

(iii) If $A = -A$, then A is called a symmetric set.

Remark 1.4.1. A balanced set is a symmetric set that contains θ.

A set A is said to be convex if, for all x and y in A and all t in the interval $(0, 1)$, the point $(1 - t)x + ty$ also belongs to A. In other words, every point on the line segment connecting x and y lies in A.

The intersection of all convex sets containing A is called the convex hull of A, denoted by $\mathrm{co}(A)$.

Remark 1.4.2.

(i) The convex hull $\mathrm{co}(A)$ of A is the minimum convex set containing A.

(ii) The set $\{y | y = \alpha x, |\alpha| \leq 1\}$ is said to be a balance closure of A, denoted by A_b.

Remark 1.4.3. The balance hull A_b of A is the minimum balance set containing A.

We next list some basic properties about convex sets, convex hulls, and balance hulls.

Property P1. Let A be a convex set. Let $x_i \in A$, $\forall i = 1, 2, \ldots, n$. Then $\sum_{i=1}^{n} \alpha_i x_i$, where $\alpha_i \geq 0$ and $\sum_{i=1}^{n} \alpha_i = 1$, belongs to A.

Property P2. Let A_1, A_2 be convex sets. Then $\lambda_1 A_1 + \lambda_2 A_2$, $\lambda_1, \lambda_2 \in \mathbb{K}$, is also a convex set. This is also true for any finite collection of convex sets.

Property P3. A is a convex set if and only if the shift $x_0 + A$ is a convex set.

Property P4. Let A be a convex set. Let α and β be nonnegative real numbers. Then $\alpha A + \beta A = (\alpha + \beta)A$.

Property P5. Let Λ be an index set and let $\{A_\lambda \mid \lambda \in \Lambda\}$ be a family of convex sets. Then $\bigcap_{\lambda \in \Lambda} A_\lambda$ is also a convex set.

Property P6. Let X, Y are two linear spaces. Let $T : X \rightarrow Y$ be a linear mapping and let A be a convex subset of X. Then $T(A)$ is a convex set of Y. The converse is also true.

Property P7. A is a balance convex set (absolutely convex) if and only if for all $x, y \in A$, $|\alpha| + |\beta| \leq 1$, $\alpha x + \beta y \in A$.

Property P8.

$$\mathrm{co}(A) = \left\{ x | x = \sum_{i=1}^{n} \lambda_i x_i, x_i \in A, \lambda_i \geq 0, \sum_{i=1}^{n} \lambda_i = 1, n \text{ is arbitrary} \right\}.$$

Property P9. The convex hull $\mathrm{co}(A)$ of a balance set A is a balance set.

Property P10. For all $\alpha, \beta \in \mathbb{K}$, $A_1, A_2 \subset X$, we have
  (i) $\mathrm{co}(\alpha A_1 + \beta A_2) = \alpha\mathrm{co}(A_1) + \beta\mathrm{co}(A_2)$;
  (ii) $\mathrm{co}(A_1 \cup A_2) = \cup\{[x, y] \mid x \in \mathrm{co}(A_1), y \in \mathrm{co}(A_2)\}$.

Property P11. The convex hull $\mathrm{co}(A_b)$ of a balance hull A_b of A is called a balance convex set, and

$$\mathrm{co}(A_b) = \left\{ y \mid y = \sum_{i=1}^{n} \alpha_i x_i, \sum_{i=1}^{n} |\alpha_i| \leq 1, x_i \in A, n \text{ is arbitrary} \right\}.$$

1.4.2 Normed linear spaces and Banach spaces

Definition 1.4.3. Let X be a linear space and let $\|\cdot\| : X \to \mathbb{R}^+ = [0, \infty)$ be a mapping. Assume that the following conditions hold:
  (i) $\|x\| \geq 0$, and $\|x\| = 0$ if and only if $x = \theta$ (for all $x \in X$);
  (ii) $\|\alpha x\| = |\alpha|\|x\|$ (for all $\alpha \in K$, $x \in X$);
  (iii) $\|x + y\| \leq \|x\| + \|y\|$ (for all $x, y \in X$).

Then $\|\cdot\|$ is called a norm on X.

A linear space X equipped with a norm is called a normed linear space, denoted by $(X, \|\cdot\|)$. Define $d(x, y) = \|x - y\|$, $\forall x, y \in X$. Then $d : X \times X \to \mathbb{R}^+$ is a distance on X. If X is also complete, then $(X, \|\cdot\|)$ is called a Banach space, that is, a complete normed linear space is called a Banach space. Let $(X, \|\cdot\|_1)$ and $(Y, \|\cdot\|_2)$ be two normed spaces and let $T : X \to Y$ be a linear mapping. Then T is said to be bounded if there exists a constant $M \geq 0$ such that, for all $x \in X$, $\|Tx\| \leq M\|x\|$.

Theorem 1.4.1. *Let $(X, \|\cdot\|_1)$ and $(Y, \|\cdot\|_2)$ be two normed spaces. Let $T : X \to Y$ be a linear mapping. Then the following statements are equivalent:*
  (i)   *T is continuous;*
  (ii) *T is continuous at original point θ;*
  (iii) *T is bounded.*

If $T : X \to Y$ is a bounded linear mapping, then we can define a norm

$$\|T\| = \sup_{\|x\| \leq 1} \|Tx\|.$$

If $X \neq \{\theta\}$, then it is equivalent to

$$\|T\| = \sup_{\|x\| = 1} \|Tx\|.$$

It is not hard to see that

$$\|T\| = \inf\{M : M \geq 0, \|Tx\| \leq M\|x\|, \forall x \in X\}.$$

Then $\|Tx\| \le \|T\|\|x\|, \forall x \in X$.

$$B(X, Y) \triangleq \{T \mid T : X \to Y \text{ is a continuous (or bounded) linear mapping}\}.$$

For $\forall T, S \in B(X, Y)$, we define linear operations:

$$(T + S)x = Tx + Sx, \quad (\alpha T)x = \alpha Tx.$$

Then $B(X, Y)$ is a linear space with the above "sum" and "scalar-multiplication".

Theorem 1.4.2. *If X, Y are normed spaces, then $B(X, Y)$ is also a normed space. If Y is a Banach space, then $B(X, Y)$ is also a Banach space.*

If X is a normed space and $Y = K$, then $B(X, K)$ is a topological dual (or adjoint) space of X, denoted by X^*. From the above theorem, we see that $B(X, K)$ is a Banach space.

For $T \in B(X, Y)$, define the adjoint operator T^* of T by

$$(T^*f)(x) = f(Tx), \quad \forall f \in Y^*, x \in X.$$

It is known that

(i) $T^* : Y^* \to X^*$ is a bounded linear operator and $\|T^*\| = \|T\|$;

(ii) For all $\alpha \in K$, $(\alpha T)^* = \alpha T^*$;

(iii) For all $T_1, T_2 \in B(X, Y)$, $(T_1 + T_2)^* = T_1^* + T_2^*$;

(iv) For all $T_1 \in B(X, Y)$, $T_2 \in B(Y, X)$, $(T_1 T_2)^* = T_2^* T_1^*$;

(v) If T has a bounded inverse operator, then T^* also has a bounded inverse operator, and $(T^*)^{-1} = (T^{-1})^*$.

Theorem 1.4.3 (Hahn–Banach Extension Theorem). *Let V be a subspace of a normed space X and let f be a continuous linear function on V. Then f can be extended to a continuous linear function f_0 on X such that $\|f\| = \|f_0\|$.*

From the Hahn–Banach Extension Theorem, we can obtain the following results.

Corollary 1.4.1. *Let X be a normed space and $x_0 \in X\backslash\{\theta\}$. Then there exists a function $f \in X^*$, which is continuous with $f(x_0) = \|x_0\|$ and $\|f\| = 1$. If $f_0 = \|x_0\|f$, then $\|f_0\| = \|x_0\|$ and $\langle f_0, x_0 \rangle = \|x_0\|^2$.*

Corollary 1.4.2. *Let V be a subspace of a normed space X and $x_0 \notin V$. Then there exists a continuous function $f \in X^*$ such that*

$$f(V) = \{\theta\}, \quad f(x_0) = 1, \quad \text{and} \quad \|f\| = \frac{1}{d},$$

where $d = \inf\{\|x_0 - y\| : y \in V\} > 0$.

Corollary 1.4.3. *For all $x \in X$,*

$$\|x\| = \sup\{|\langle f, x \rangle| : f \in X^*, \|f\| \le 1\} = \max\{|\langle f, x \rangle| : f \in X^*, \|f\| \le 1\}.$$

The normed space X is said to be strictly convex if, for all $x, y \in X$, $\|x\| = \|y\| = 1$ and $x \neq y$,

$$\|tx + (1 - t)y\| < 1, \quad \forall t \in (0, 1).$$

Remark 1.4.4. If X^* is strictly convex, then the functional f_0 given by the Hahn–Banach extension theorem is unique.

Definition 1.4.4. Let X be a real normed space. Let $f; X \to \mathbb{R}^1$ be a linear functional and $f \neq \theta$, $\beta \in \mathbb{R}^1$. Then $H = \{x \in X : f(x) = \alpha\}$ is called a hyperplane in X. Also

$$f^{\leq}(\alpha) = \{x \in X : f(x) \leq \alpha\},$$
$$f^{\geq}(\alpha) = \{x \in X : f(x) \geq \alpha\},$$
$$f^{<}(\alpha) = \{x \in X : f(x) < \alpha\},$$

and

$$f^{>}(\alpha) = \{x \in X : f(x) > \alpha\}$$

are called half-spaces. Indeed, they are convex sets. If $f \in X^*$, then H, $f^{\leq}$, and $f^{\geq}$ are all closed convex sets. In this case, H is said to be a closed hyperplane, and $f^{\leq}$ and $f^{\geq}$ are said to be closed half-spaces. Since $f^{<}$ and $f^{>}$ are open sets, we call $f^{<}$ and $f^{>}$ open half-spaces. It is easy to check that if H is a hyperplane of X, then there exists $x_0 \notin H$ and a nontrivial subspace L of X such that $H = x_0 + L$. Hence, closed hyperplanes do not contain interior points.

Let X be a linear space. Let A and B be nonempty subsets of X, and let $\alpha \in \mathbb{R}^1$. If both $A \subset f^{\leq}(\alpha)$ and $B \subset f^{\geq}(\alpha)$ or both $A \subset f^{\geq}(\alpha)$ and $B \subset f^{\leq}(\alpha)$ (both $A \subset f^{<}(\alpha)$ and $B \subset f^{>}(\alpha)$ or both $A \subset f^{>}(\alpha)$ and $B \subset f^{<}(\alpha)$), then hyperplane H separates (or strictly separates) A and B.

A hyperplane H separates (or strictly separates) sets A and B if and only if for $x \in A$, we have $f(x) \leq \alpha (f(x) < \alpha)$ and for $x \in B$, we have $f(x) \geq \alpha (f(x) > \alpha) \Leftrightarrow \sup_{x \in A} f(x) \leq \inf_{x \in B} f(x) (\sup_{x \in A} f(x) < \inf_{x \in B} f(x))$.

Theorem 1.4.4 (Separating Hyperplane Theorem I). *Let X be a real normed space. Let A and B be two nonempty convex subsets of X. Let $A^0 \neq \emptyset$ and $A^0 \cap B = \emptyset$. Then there exists a closed hyperplane H, which separates A and B. In addition, if both A and B are open, then H strictly separates A and B.*

Theorem 1.4.5 (Separating Hyperplane Theorem II). *Let X be a real normed space. Let A and B be two nonempty convex subsets of X such that A is compact, B is closed and $A \cap B = \emptyset$. Then there exists a closed hyperplane H, which strictly separates A and B.*

Corollary 1.4.4. *Let V be a subspace of a real normed space X. If $\overline{V} \neq X$, then there exists $f \in X^*, f \neq \theta$ such that $\langle f, x \rangle = 0, \forall x \in V$.*

Corollary 1.4.5. *Let X be a real normed space and let C be a nonempty convex subset of X. If $x \notin C$, then there exists $f \in X^*$ such that $f(x) < \inf\{f(y) : y \in C\}$.*

Theorem 1.4.6 (The Resonance Theorem). *Let X be a Banach space and let Y be a normed linear space. Let $\{T_\lambda\}_{\lambda \in \Lambda}$ be a collection of continuous linear operators from X to Y. If there exists a dense subset E of X such that $\sup_{\lambda \in \Lambda} \|T_\lambda(x)\| < +\infty, \forall x \in E$, then there exists a positive real number M such that $\|T_\lambda\| \leq M, \forall \lambda \in \Lambda$.*

Theorem 1.4.7 (The Closed Graph Theorem). *Let X and Y be Banach spaces. Let $T : X \to Y$ be a continuous linear bijection. Then $T^{-1} \in B(Y, X)$ and $T : X \to Y$ is a homeomorphism mapping.*

Theorem 1.4.8 (The Open Mapping Theorem). *Let X and Y be Banach spaces. Let T be a surjective continuous linear operator. Then T is an open mapping, i. e., if U is an open set in X, then $T(U)$ is open in Y.*

Theorem 1.4.9. *Let $\{T_{p,q}\}$ $(p = 1, 2, \dots)$ be a collection of continuous mappings from a Banach space X to a normed linear space Y_q, $q = 1, 2, \dots$ If, for every p, there exists $x_p \in X$ such that $\overline{\lim}_{q \to \infty} \|T_{p,q} x_p\| = \infty$, then*

$$B = \left\{ x \in X : \overline{\lim_{q \to \infty}} \|T_{p,q} x_p\| = \infty, p = 1, 2, \dots \right\}$$

is of the second category set in X.

Theorem 1.4.10 (Bishop–Phelps Theorem). *Let C be a bounded closed convex subset of a Banach space X and let $A = \{f \in X^* : f(x) = \sup f(C), x \in C\}$. Then $\overline{A} = X^*$.*

Definition 1.4.5. Let X be a Banach space and let X^* be a dual space of X. Define a mapping $C : X \to X^{**}$ by $f \to \langle f, x \rangle, \forall x \in X$. Then it is a continuous function from X^* to K, denoted by Cx, i. e., $Cx \in X^{**}$ such that

$$\langle Cx, f \rangle = \langle f, x \rangle, \quad \forall x \in X, f \in X^*.$$

The mapping $C : X \to X^{**}$ is called standard (or "canonical") embedding; $C(X)$ is a closed subspace of X^{**}.

The mapping $C : X \to X^{**}$ may not be surjective. If $C(X) = X^{**}$, then X is said to be reflexive.

Theorem 1.4.11. *Let X be a Banach space. Then X is reflexive if and only if X^* is reflexive.*

Next, we denote the unit spheres of X and X^* by

$$S(X) = \{x \in X : \|x\| = 1\} \text{ and } S(X^*) = \{f \in X^* : \|f\| = 1\}.$$

Theorem 1.4.12 (James Theorem). *Let X be a Banach space. Then X is reflexive if and only if for all $f \in S(X^*)$, there exists $x \in S(X)$ such that $f(x) = 1$.*

Definition 1.4.6. Let X be a normed space. A topology on X is called a strong topology (or norm topology) if it is induced by a distance $d(x,y) = \|x - y\|$, denoted by $\tau_s(X)$.

Definition 1.4.7. Let X be a normed space and let X^* be the dual space of X. For all $f \in X^*$, let $\varphi_f : X \to K$ be a mapping determined by $\varphi_f(x) = f(x)$ and let $\{\varphi_f\}_{f \in X^*}$ be a collection of mappings from X to K for each $f \in X^*$. The weak topology on X is induced by $\{\varphi_f\}_{f \in X^*}$. We denote the weak topology by $\sigma(X, X^*)$, which is the weakest topology such that all mappings $\{\varphi_f\}_{f \in X^*}$ are continuous.

Definition 1.4.8. For all $x \in X$, let $\varphi_f : X^* \to K$ be a mapping determined by $\varphi_f(x) = \langle f, x \rangle$ and let $\{\varphi_x\}_{x \in X}$ be a collection of mappings from X^* to K for each $x \in X$. The weak topology on X^* is induced by $\{\varphi_x\}_{x \in X}$. We denote this weak topology by $\sigma(X^*, X)$, which is the weakest topology such that all mappings $\{\varphi_x\}_{x \in X}$ are continuous.

Remark 1.4.5.
(i) $\sigma(X^*, X) \subseteq \sigma(X^*, X^{**}) \subset \tau_s(X^*)$;
(ii) $\sigma(X, X^*) \subseteq \tau_s(X)$;
(iii) Both $\sigma(X, X^*)$ and $\sigma(X^*, X)$ are Hausdorff;
(iv) Let $x_0 \in X$. Then the neighborhood base of x_0 in the topology $\sigma(X, X^*)$ can be characterized by the collection of sets

$$V = \{x \in X : \langle f_i, x - x_0 \rangle < \varepsilon, \quad \forall i \in I\},$$

where I is a finite set, $f_i \in X^*$ and $\varepsilon > 0$;
(v) Let $f_0 \in X^*$. Then the neighborhood base of f_0 in the topology $\sigma(X^*, X)$ can be characterized by the collection of sets

$$V = \{f \in X^* : \langle f - f_0, x_i \rangle < \varepsilon, \quad \forall i \in I\},$$

where I is a finite set, $x_i \in X$ and $\varepsilon > 0$.

Theorem 1.4.13. *Let X be a normed space and let $\{x_n\}$ be a sequence in X. Then*
$x_n \xrightarrow{w} x$ *in topology $\sigma(X, X^*)$ if and only if $f(x_n) \to f(x)$, $\forall f \in X^*$;*
If $x_n \xrightarrow{s} x$ in norm, then $x_n \xrightarrow{w} x$ in topology $\sigma(X, X^*)$;
If $x_n \xrightarrow{w} x$ in topology $\sigma(X, X^*)$, then $\{\|x_n\|\}$ is bounded and $\|x\| \leq \liminf_{n \to \infty} \|x_n\|$;
If $x_n \xrightarrow{w} x$ in topology $\sigma(X, X^*)$, and $f_n \xrightarrow{s} f$ in norm, then $f_n(x_n) \to f(x)$ $(n \to \infty)$.

Theorem 1.4.14. *Let X be a Banach space and $\{f_n\}$ a sequence in X^*. Then*
$f_n \xrightarrow{w^*} f$ *in topology $\sigma(X^*, X)$ if and only if $f_n(x) \to f(x)$ $(n \to \infty)$, $\forall x \in X$;*
If $f_n \xrightarrow{s} f$ in norm, then $f_n \xrightarrow{w} f$;
If $f_n \xrightarrow{w^*} f$ in topology $\sigma(X^*, X)$, then $\{\|f_n\|\}$ is bounded and $\|f\| \leq \liminf_{n \to \infty} \|f_n\|$;
If $f_n \xrightarrow{w^*} f$ in topology $\sigma(X^*, X)$, and $x_n \xrightarrow{s} x$ in norm, then $f_n(x_n) \to f(x)$ $(n \to \infty)$.

Remark 1.4.6.

(i) With $\{f_n\} \subset X^*, f \in X^*, f_n \xrightarrow{w} f \ (n \to \infty)$, and $\{x_n\} \subset X, x \in X, x_n \xrightarrow{w} x$, we cannot conclude $f_n(x_n) \to f(x)$;

(ii) If $\dim X < +\infty$, then the three topologies on X^* coincide, that is, $\tau_s(X^*) = \sigma(X^*, X) = \sigma(X^*, X^{**})$.

Theorem 1.4.15. *Let X be a normed space and let X^* be the dual space of X. The weak topology $\sigma(X, X^*)$ is equivalent to the strong topology $\tau_s(X)$ if and only if $\dim X < +\infty$.*

Theorem 1.4.16. *Let X be a Banach space and let X^* be the dual space of X. Then the weak topology $\sigma(X^*, X^{**})$ on X^* is equivalent to the weak* topology $\sigma(X^*, X)$ if and only if X is reflexive.*

Theorem 1.4.17. *Let X be a normed space and let X^* be the dual space of X. Then $(X, \sigma(X, X^*))^* = X^*$ and $(X^*, \sigma(X^*, X))^* = X$.*

Next, we use $B_X\{x \in X : \|x\| \leq 1\}$ and $B_{X^*} = \{f \in X^* : \|f\| \leq 1\}$ to denote the closed unit balls of X and X^*, respectively.

Definition 1.4.9. Let X be a normed space and let X^* be the dual space of X. Let A be a subset of X and let B be a subset of X^*. Recall that A is bounded in the norm topology if there exists a closed ball $B_r[\theta]$ such that $A \subset B_r[\theta]$, $r > 0$; A is w-bounded if for any neighborhood V of the origin in $\sigma(X, X^*)$, there exists a constant $\lambda > 0$ such that $\lambda V \supset A$; B is w^*-bounded if for any neighborhood V^* of the origin in $\sigma(X^*, X)$, there exists a constant $s > 0$ such that $sV^* \supset B$.

Theorem 1.4.18. *Let X be a normed space and let X^* be the dual space of X. Let A be a subset of X. Then, A is bounded in the norm topology if and only if A is w-bounded. Let X be a Banach space and B a subset of X^*. Then, B is bounded in norm topology if and only if B is w^*-bounded.*

Theorem 1.4.19. *Let C be a convex subset of a Banach space X. Then, C is strongly closed $\Leftrightarrow$ C is weakly closed $\Leftrightarrow$ C is sequentially weakly closed.*

Theorem 1.4.20. *Let X be a normed space and let B_X be a closed unit ball in X. Then B_X is compact in the norm topology if and only if $\dim X < +\infty$.*

Theorem 1.4.21 (Alaoglu Theorem). *Let X be a Banach space and let X^* be the topological conjugate space of X. Then the closed unit ball B_{X^*} of X^* is weak* compact.*

Theorem 1.4.22 (Eberlein–Smulian Theorem). *Let X be a Banach space and let A be a subset of X. Then A is weakly compact if and only if A is weakly sequentially compact.*

Theorem 1.4.23. *Let X be a Banach space. Then*

(i) *the closed unit ball B_{X^*} in X^* is metrizable in topology $\sigma(X^*, X)$ if and only if X is separable;*

(ii) *the closed unit ball B_X in X is metrizable in topology $\sigma(X, X^*)$ if and only if X^* is separable.*

Theorem 1.4.24. *Let X be a Banach space. Then the following statements are equivalent:*

(i)   *X is reflexive;*

(ii)   *any bounded closed convex subset of X is weakly compact;*

(iii)   *any bounded sequence in X has a weakly convergent sequence;*

(iv)   *the closed unit ball B_X in X is weakly compact;*

(v)   *the closed unit ball B_X in X is weakly sequentially compact;*

(vi)   *every closed subspace in X is reflexive;*

(vii)   *every closed convex subset in X has a minimum-norm element;*

(viii) *every closed convex subset of X is approximable, that is, for every $x \in X$, there exists $y \in A$ such that $\|x - y\| = d(x, A)$;*

(ix)   *every decreasing sequence of bounded closed convex sets of X has a nonempty intersection;*

(x)   *every w-closed set is w^*-closed;*

(xi)   *every pair of disjoint closed convex sets, one of which is bounded, can be strictly separated by a hyperplane;*

(xii)   *every pair of disjoint bounded closed convex sets can be strictly separated by a hyperplane;*

(xiii) *$X^{**} = (X^*, \sigma(X^*, X))^*$;*

(xiv) *for all $f \in X^*, f$ attains its norm on B_X;*

(xv)   *every nontrivial quotient space of X is reflexive;*

(xvi) *M and X/M are reflexive, where M is any closed subspace of X;*

(xvii) *M and X/M are reflexive, where M is some closed subspace X.*

1.4.3 Hilbert spaces

Definition 1.4.10. Let H be a linear space on a number field $\mathbb{K}$ and let $\langle \cdot, \cdot \rangle : H \times H \to \mathbb{K}$ be a mapping. Assume that

(i) $\langle x, x \rangle \geq 0$ and $\langle x, x \rangle = 0 \Leftrightarrow x = \theta, \forall x \in H$;

(ii) $\langle x, y \rangle = \overline{\langle y, x \rangle}, \forall x, y \in H$;

(iii) $\langle \alpha x + \beta y, z \rangle = \alpha \langle x, z \rangle + \beta \langle y, z \rangle, \forall x, y, z \in H$, and $\forall \alpha, \beta \in \mathbb{K}$.

Then $\langle x, y \rangle$ is said to be an inner product of x and y. A linear space equipped with the inner product is said to be an inner product space.

Let H be an inner product space. Define

$$\|x\| = \sqrt{\langle x, x \rangle}, \quad \forall x \in H.$$

Then $\| \cdot \| : H \to \mathbb{R}^+$ is said to be a norm on H and the norm is induced by the inner product. If the inner product space H is complete in the norm topology, then H is called a Hilbert space. If $\langle x, y \rangle = 0$, then x and y are orthogonal, denoted by $x \perp y$. Let $M \subset H$, $x \in H$. If $x \perp y$, $\forall y \in M$ then x and M are orthogonal, denoted by $x \perp M$. Put $M^\perp = \{x \in H : x \perp M\}$. The set $M^\perp$ is said to be the orthogonal complement of M. Let $M, L \subset H$. If $\forall x \in M$, $x \perp L$, then M and L are orthogonal, denoted by $M \perp L$.

Theorem 1.4.25. *Let H be an inner product space. Then*

(i) $|\langle x, y \rangle|^2 \leq \langle x, x \rangle \langle y, y \rangle$, *for any $x, y \in H$. The "=" holds if and only if x and y are linearly dependent. The above inequality is called the Cauchy–Schwarz inequality;*

(ii) $\langle \cdot, \cdot \rangle : H \times H \to \mathbb{K}$ *is continuous, that is, $\forall \{x_n\}, \{y_n\} \subset H$, $x_n \to x$, $y_n \to y \Leftrightarrow \langle x_n, y_n \rangle \to \langle x, y \rangle$. If $x_n \to x$ and $y_n \to y$ (or $x_n \to x$ and $y_n \to y$), then $\langle x_n, y_n \rangle \to \langle x, y \rangle$;*

(iii) $\|x \pm y\|^2 = \|x\|^2 \pm 2\operatorname{Re}\langle x, y \rangle + \|y\|^2$, $\forall x, y \in H$;

(iv) *(Parallelogram Law)* $\|x + y\|^2 + \|x - y\|^2 = 2(\|x\|^2 + \|y\|^2)$, $\forall x, y \in H$;

(v) *(Pythagorean Theorem) If $x \perp y$, then* $\|x + y\|^2 = \|x\|^2 + \|y\|^2$, $\forall x, y \in H$;

(vi) *(Polarization Identity)*

$$\langle x, y \rangle = \frac{1}{4}(\|x + y\|^2 - \|x - y\|^2 + i\|x + iy\|^2 - i\|x - iy\|^2), \quad \forall x, y \in H,$$

$$\operatorname{Re}\langle x, y \rangle = \frac{1}{4}(\|x + y\|^2 - \|x - y\|^2), \quad \forall x, y \in H;$$

(vii) $\|(1 - \lambda)y + \lambda x\|^2 = (1 - \lambda)\|y\|^2 - \lambda(1 - \lambda)\|x - y\|^2 + \lambda\|x\|^2$, $\forall x, y \in H$ and $\forall \lambda \in \mathbb{R}$;

(viii) *(Opial Condition [60]) Let $\{x_n\}$ be a sequence in a Hilbert space with $x_n \xrightarrow{w} x$. Then, for all $y \neq x$, $\liminf_{n\to\infty} \|x_n - x\| < \liminf_{n\to\infty} \|x_n - y\|$, which is also equivalent to $\limsup_{n\to\infty} \|x_n - x\| < \limsup_{n\to\infty} \|x_n - y\|$;*

(ix) *Let H be a Hilbert space. For any $y \in H$, define a functional f by*

$$f_y(x) = \langle x, y \rangle, \quad x \in H.$$

Then $f_y : H \to \mathbb{R}$ is a bounded linear functional and $\|f_y\| = \|y\|$. On the contrary, for any $f \in H^*$, there exists a unique $y \in H$ such that

$$f(x) = \langle x, y \rangle, \forall x \in H;$$

(x) *Let H_1 and H_2 be two Hilbert spaces. Let $\langle \cdot, \cdot \rangle_1$ and $\langle \cdot, \cdot \rangle_2$ be the inner products of H_1 and H_2, respectively. For any $x = (x_1, x_2)$, $y = (y_1, y_2) \in H_1 \times H_2$, define the sum and the scalar product in $H_1 \times H_2$ as follows:*

$$x + y = (x_1, x_2) + (y_1, y_2) = (x_1 + y_1, x_2 + y_2),$$

$$\alpha x = \alpha(x_1, x_2) = (\alpha x_1, \alpha x_2),$$

and

$$\langle x, y \rangle = \langle x_1, y_1 \rangle_1 + \langle x_2, y_2 \rangle_2$$

Then $H_1 \times H_2$ is a Hilbert space.

Let $g : H_1 \times H_2 \to \mathbb{R}$ be a bounded linear functional. From the above, we find that there exists a unique $(z_1, z_2) \in H_1 \times H_2$ such that, for all $(x_1, x_2) \in H_1 \times H_2$,

$$g\langle x_1, x_2 \rangle = \langle (x_1, x_2), (z_1, z_2) \rangle = \langle x_1, z_1 \rangle_1 + \langle x_2, z_2 \rangle_2.$$

Theorem 1.4.26. *Let X be a normed space. If the norm satisfies the parallelogram law, then we can define an inner product such that the new norm induced by the inner product coincides the original norm of X.*

Theorem 1.4.27. *Let C be a nonempty closed convex subset of a Hilbert space H. Then, for any $x \in H$, there exists a unique $x_0 \in C$ such that*

$$\|x - x_0\| = \inf_{y \in C}\{\|x - y\|\} = d(x, C).$$

Theorem 1.4.28. *Let C be a nonempty closed convex subset of a Hilbert space H. Let $x \in H$ and $x_0 \in C$. Then*

$$\|x - x_0\| = d(x, C) \Leftrightarrow \langle x - x_0, y - x_0 \rangle \le 0, \quad \forall y \in C.$$

Let $x_0 = P_C x$. Then $P_C : H \to C$ is said to be the metric projection from H onto C. If C is simple enough, then P_C has an analytic expression.
(i) If $C = \{x \in H : \|x - \bar{x}\| \le r\}$ is a closed ball centered at $\bar{x} \in H$ with a radius $r > 0$, then

$$P_C = \begin{cases} \bar{x} + r\frac{x - \bar{x}}{\|x - \bar{x}\|}, & x \notin C, \\ x, & x \in C. \end{cases}$$

(ii) If $C = [a, b]$, where $a = (a_1, a_2, \ldots, a_n)^T$ and $b = (b_1, b_2, \ldots, b_n)^T$ are closed rectangles in an n-dimensional Euclidean space, then

$$(P_C x)_i = \begin{cases} a_i, & x_i < a_i, \\ x_i, & x_i \in [a_i, b_i], \\ b_i, & x_i > b_i. \end{cases}$$

(iii) If $C = \{y \in H : \langle u, y \rangle = \alpha\}$, where $u \ne \theta$ and $\alpha \in \mathbb{R}$ is a hyperplane, then

$$P_C x = x + \frac{\alpha - \langle u, x \rangle}{\|u\|^2}u.$$

(iv) If $C = \{y \in H : \langle u, y \rangle \le \alpha\}$, where $u \ne \theta$ and $\alpha \in \mathbb{R}$ is a closed half-space, then

$$P_C x = \begin{cases} x + \frac{\alpha - \langle u, x \rangle}{\|u\|^2}u, & \langle u, x \rangle > \alpha, \\ x, & \langle u, x \rangle \le \alpha. \end{cases}$$

(v) If $C = \text{Ran}(A)$, where $A \in \mathbb{R}^{m \times n}$ is an $m \times n$ matrix and $r(A) = n$, then

$$P_C x = AA^+ x,$$

where $A^+ = (A^T A)^{-1} A^T$ is a generalized P–M inverse of A.

Theorem 1.4.29. *Let C be a nonempty closed convex subset of a Hilbert space H. The metric projection P_C has the following properties:*
(i)

$$\|P_C x - P_C y\|^2 \leq \langle x - y, P_C x - P_C y \rangle, \quad \forall x, y \in H,$$

$$\left\|(I - P_C)x - (I - P_C)y\right\|^2 \leq \langle x - y, (I - P_C)x - (I - P_C)y \rangle, \quad \forall x, y \in H.$$

In particular,

$$\left\|(I - P_C)x - (I - P_C)y\right\| \leq \|x - y\|, \quad \forall x, y \in H,$$

and

$$\|P_C x - P_C y\| \leq \|x - y\|, \quad \forall x, y \in H.$$

(ii)

$$\|P_C x - P_C y\|^2 \leq \|x - y\|^2 - \left\|(I - P_C)x - (I - P_C)y\right\|^2, \quad \forall x, y \in H.$$

In particular,

$$\|y - P_C x\|^2 + \|x - P_C x\|^2 \leq \|x - y\|^2, \quad \forall x \in H, y \in C.$$

(iii) *Let $\{x_n\} \subset H$ and $x \in H$. If $x_n \xrightarrow{w} x$ and $P_C x_n \to y$, then $P_C x = y$.*
(iv) $P_C = \frac{1}{2}I + \frac{1}{2}S$, *where $S : H \to H$ is a nonexpansive mapping.*

Theorem 1.4.30. *Let M be a closed subspace of a Hilbert space H. Then the projection $P_M : H \to M$ is a bounded linear operator and $\|P_M\| = 1$.*

Definition 1.4.11. Let G be a closed subspace of a Banach space H and let L be a subspace of X. Subspace L is said to be a complementary subspace of G if
(i) L is a closed set;
(ii) $X = G \oplus L$.

Remark 1.4.7.
(i) Any finite dimensional subspace G has a topologically complementary subspace.
(ii) In a Hilbert space H, any closed subspace has a topologically complementary subspace.

(iii) Any closed subspace G with a finite codimension $\text{codim}(G) = \dim(X/G)$ has a topologically complementary subspace.

Theorem 1.4.31 (Lindenstrauss and Tzafriri [41]). *Let X be a Banach space. If every nonempty closed subspace of X is topologically complementary, then Banach space X is isomorphic to some Hilbert space.*

Theorem 1.4.32 (Kakutani [34]). *Let X be a Banach space with $\dim X \geq 3$. If every closed subspace with the complementary dimension 1 is equal to the range of the projection operator with the norm 1, then X is an inner product space.*

Theorem 1.4.33 (Bruck [14]). *Let X be a Banach space. Every nonempty closed convex subset of X is a nonexpansive retract of X if and only if either $\dim X = 2$ or X is a Hilbert space.*

Corollary 1.4.6. *Let X be a Banach space with $\dim X > 3$. Then every nonempty closed convex subset of X is a nonexpansive retract of X if and only if X is a Hilbert space.*

1.4.4 Topological linear spaces

Definition 1.4.12. Let $\mathbb{K}$ be a number field and let X be a linear space over $\mathbb{K}$. Let ζ be a topology on X. If the following conditions are satisfied:
(i) X is a T_1 space, that is, every single point set is closed in X;
(ii) The linear operation in X is continuous with respect to ζ, that is, the addition operation $(x,y) \to x + y$ as a mapping of $(X,\zeta) \times (X,\zeta) \to (X,\zeta)$ is continuous; the multiplication operation $(\alpha, x) \to \alpha x$ as a mapping of $(\mathbb{K}, \tau_K) \times (X, \zeta) \to (X,\zeta)$, where τ_K is a topology according to the usual distance, is continuous. Then (X,ζ) is said to be a topological linear space. If every neighborhood of the zero point contains a convex neighborhood, then the topological space (X,ζ) is said to be locally convex. Normed linear spaces $(X, \|\cdot\|)$, $(X, \sigma(X,X^*))$, $(X^*, \sigma(X^*, X))$ all are locally convex, regular, and Hausdorff linear spaces.

Definition 1.4.13. Let $(X, \mathfrak{T})$ be a topological linear space and let A be a subset of X. A is said to be $\mathfrak{T}$-bounded if and only if for any neighborhood V of A, there exists a constant $\lambda > 0$ such that $A \subset \lambda V$.

Theorem 1.4.34. *Let $(X, \mathfrak{T})$ be a topological linear space and let A be a subset of X. A is $\mathfrak{T}$-bounded if and only if $\forall \lambda_n \to 0$ and $\forall \{x_n\} \subset A$, $\lambda_n x_n \xrightarrow{\mathfrak{T}} \theta \ (n \to \infty)$.*

Definition 1.4.14. Let $(X, \mathfrak{T}_1)$ and $(Y, \mathfrak{T}_2)$ be two topological linear spaces. Let $T : \text{Dom}(T) \subset X \to Y$ be a mapping. This T is said to be bounded if and only if T maps any bounded subset of $\text{Dom}(T)$ into some bounded subset of Y. In particular, T is said to be a bounded functional on X if and only if $Y = \mathbb{R}$ or $\mathbb{K}$.

Theorem 1.4.35. *Let $(X, \mathfrak{T}_1)$ and $(Y, \mathfrak{T}_2)$ be two topological linear spaces. Let $T :$ $\mathrm{Dom}(T) \subset X \to Y$ be a linear mapping. If T is continuous, then it must be bounded. In particular, if $T : \mathrm{Dom}(T) \subset X \to Y$, where $Y = \mathbb{R}$ or $\mathbb{K}$, is continuous, then it must be a bounded linear functional.*

We also remark here that the inverse of the above theorem may be not true.

Theorem 1.4.36. *If $(X, \mathfrak{T})$ is an A_1 topological linear space, then every bounded linear functional is continuous. Hence, the boundedness and continuity of a bounded linear functional are equivalent in the framework of topological spaces satisfying the first axiom of countability.*

Theorem 1.4.37 (Separation Theorem). *Let X be a topological linear space. Let A and B be two disjoint nonempty convex sets of X.*
(i) *If A is an open set, then there exist $f \in X^*$ and $r \in \mathbb{R}$ such that*

$$\mathrm{Re}\, f(x) < r \leq \mathrm{Re}\, f(y), \quad \forall x \in A, y \in B.$$

(ii) *If A is compact, B is closed, and X is locally convex, then there exist $f \in X^*$ and $r_1, r_2 \in \mathbb{R}$ such that*

$$\mathrm{Re}\, f(x) < r_1 < r_2 < \mathrm{Re}\, f(y), \quad \forall x \in A, y \in B.$$

Theorem 1.4.38 (Krein–Milman Theorem). *Let X be a topological linear space and let X^* be the conjugate space of X. Assume that there exists a point $f \in X^*$ such that $f(x_1) \neq f(x_2)$, $\forall x_1, x_2 \in X$, $x_1 \neq x_2$. Let K be a nonempty compact convex subset of X. Then $K = \overline{\mathrm{co}}(\mathrm{ext}\, K)$, that is, K is a closed convex hull of its extreme point set.*

Theorem 1.4.39 (Tychonoff Fixed Point Theorem). *Let X be a locally convex topological vector space. Let C be a nonempty compact convex subset of X and let $T : C \to C$ be a continuous mapping. Then T has a fixed point in C.*

From the Tychonoff fixed point theorem, we can prove the Schauder and Brouwer fixed point theorems.

Theorem 1.4.40 (Schauder Fixed Point Theorem). *Let X be a normed linear space. Let C be a nonempty compact convex subset of X and let $T : C \to C$ be a continuous mapping. Then T has a fixed point in C.*

Theorem 1.4.41 (Brouwer Fixed Point Theorem). *Let $\mathbb{R}^n$ be an n-dimensional Euclidean space. Let C be a nonempty bounded closed convex subset of $\mathbb{R}^n$ and let $T : C \to C$ be a continuous mapping. Then T has a fixed point in C.*

Theorem 1.4.42 (Banach–Alaoglu Theorem). *Let X be a topological linear space and let V be a neighborhood of θ in X. Let X^* be the topological conjugate space of X. Then the set $K = \{f \in X^* : |f(x)| \leq 1, x \in V\}$ is w^*-compact.*

1.5 Lower semicontinuity, convexity, and conjugate functions

Definition 1.5.1. Let X be a topological space and let $f : X \to \overline{\mathbb{R}} = (-\infty, \infty]$ be a function. Then f is said to be lower semicontinuous if $\forall \alpha \in \mathbb{R}$, the level set $\{x \in X : f(x) \leq \alpha\}$ is a closed set in X. If $f \not\equiv \infty$, then f is said to be proper. It is clear that f is proper if and only if $\mathrm{Dom}(f) \neq \emptyset$.

Theorem 1.5.1. Let X be a compact topological space and let $f : X \to \overline{\mathbb{R}}$ be a lower semicontinuous function. Then there exists $x_0 \in X$ such that

$$f(x_0) = \min\{f(x) : x \in X\}.$$

Let X be a topological space and let $f : X \to \overline{\mathbb{R}}$ be a function. Let $\{x_\alpha\}$ be a net in X. Define $\liminf_\alpha f(x_\alpha) = \sup_\alpha \inf_{\alpha \leq \beta} f(x_\beta)$ and $\limsup_\alpha f(x_\alpha) = \inf_\alpha \sup_{\alpha \leq \beta} f(x_\beta)$.

Theorem 1.5.2. Let X be a topological space and let $f : X \to \overline{\mathbb{R}}$ be a function. Then $f : X \to \overline{\mathbb{R}}$ is lower semicontinuous if and only if

$$x_\alpha \to x_0 \Rightarrow f(x_0) \leq \liminf_\alpha f(x_\alpha), \quad \forall x_0 \in X.$$

Hence, if f is lower semicontinuous and $x_n \to x$, then $f(x) \leq \liminf_{n \to \infty} f(x_n)$.

Definition 1.5.2. The set $\mathrm{epi}(f) = \{[x, \lambda] \in X \times \overline{\mathbb{R}} : f(x) \leq \lambda\}$ is said to be the epigraph of f.

Theorem 1.5.3. Let X be a topological space and let $f : X \to \overline{\mathbb{R}}$ be a function. Then f is lower semicontinuous if and only if $\mathrm{epi}(f)$ is a closed subset of $X \times \mathbb{R}$.

Theorem 1.5.4. Let X be a topological space and let $\{f_\lambda\}_{\lambda \in \Lambda}$ be a family of lower semicontinuous functions from X to $\overline{\mathbb{R}}$. Define a function $g : X \times \overline{\mathbb{R}}$ by

$$g(x) = \sup_{\lambda \in \Lambda} f_\lambda(x), \quad \forall x \in X.$$

Then g is lower semicontinuous and g is called a coenvelope of $\{f_\lambda\}_{\lambda \in \Lambda}$.

Theorem 1.5.5. Let X be a topological space. Let $f, g : X \to \overline{\mathbb{R}}$ be lower semicontinuous function. Then, $\forall \alpha, \beta \in \mathbb{R}^+$, $\alpha f + \beta g$ is lower semicontinuous.

Theorem 1.5.6. Let X, Y be topological spaces. Let $\beta : X \to \mathbb{R}^+ = [0, \infty)$ be a continuous function and let $f : Y \to \overline{\mathbb{R}}$ be a lower semicontinuous function. Define a function $\beta f : X \times Y \to \overline{\mathbb{R}}$ by

$$(\beta f)(x, y) = \beta(x) f(y), \quad (x, y) \in X \times Y.$$

Then βf is lower semicontinuous.

Definition 1.5.3. Let X be a linear space. A function $f : \text{Dom}(f) \subset X \rightarrow \overline{\mathbb{R}}$ is said to be convex if

$$f[tx + (1-t)y] \leq tf(x) + (1-t)f(y), \quad \forall x, y \in \text{Dom}(f), \forall t \in [0,1].$$

If the above inequality strictly holds for $x \neq y$, then f is said to be strictly convex. Recall that f is uniformly convex on $\text{Dom}(f)$ if there exists a function $\mu : \mathbb{R}^+ \rightarrow \mathbb{R}^+$ with $\mu(t) = 0 \Leftrightarrow t = 0$ and

$$f[tx + (1-t)y] \leq tf(x) + (1-t)f(y) - t(1-t)\mu(\|x-y\|),$$

$\forall t \in [0,1], \forall x, y \in \text{Dom}(f)$. If the above inequality holds for $t = \frac{1}{2}$, then f is said to be uniformly convex at the center. It is well known that f is uniformly convex at the center if and only if f is uniformly convex on $\text{Dom}(f)$. Recall also that f is said to be quasi-convex if the following inequality holds for $\forall t \in [0,1]$ and $\forall x, y \in \text{Dom}(f)$:

$$f[tx + (1-t)y] \leq \max\{f(x), f(y)\}.$$

Furthermore, if the above inequality strictly holds for $x \neq y$, then f is said to be strictly quasi-convex.

Clearly, if f is uniformly convex on $\text{Dom}(f)$, then it is strictly convex on $\text{Dom}(f)$; if f is (strictly) convex on $\text{Dom}(f)$, then it is (strictly) quasi-convex on $\text{Dom}(f)$.

Theorem 1.5.7. *Let X be a linear space and $f : X \rightarrow \overline{\mathbb{R}}$ a function. Then the following assertions are true:*

(i) *f is a convex function on X if and only if $\text{epi}(f)$ is a convex subset of $X \times \mathbb{R}^+$;*

(ii) *If f is a convex function on X, then $\forall \lambda \in \mathbb{R}$ the level set $\{x \in X : f(x) \leq \lambda\}$ is convex. But the converse is not true;*

(iii) *If f_1 and f_2 are convex functions on X, then for any $c_1 \geq 0, c_2 \geq 0, c_1 f_1 + c_2 f_2$ is also convex;*

(iv) *If $\{f_\lambda\}_{\lambda \in \Lambda}$ is a family of convex functions on X, then the coenvelope of $\{f_\lambda\}_{\lambda \in \Lambda}$ is also convex;*

(v) *Let $f : X \rightarrow \overline{\mathbb{R}}$ be a convex function and $g : \mathbb{R} \rightarrow \overline{\mathbb{R}}$ an increasing convex function. Then the composite function of f and g, $h(x) = g(f(x))$, is also a convex function, here we require that $g(+\infty) = +\infty$;*

(vi) *If X is a real normed space and $f : X \rightarrow \overline{\mathbb{R}}$ a convex function, then every local minimum is also global one;*

(vii) *$f : X \rightarrow \mathbb{R}$ is convex if and only if for any fixed $x, y \in \text{Dom}(f)$, the univariate function $\phi(t) = f(tx + (1-t)y)$ is convex in $t \in [0,1]$.*

Next, we assume that X is a normed linear space.

Definition 1.5.4. Let $f : X \rightarrow \overline{\mathbb{R}}$ be a proper function. The function defined by $f^*(x^*) = \sup_{x \in X}\{\langle x^*, x \rangle - f(x)\}, x^* \in X^*$ is said to be the conjugate function of f.

By means of Definition 1.5.4, it is easy to check that the following are true.
(1) $f^* : X^* \to \overline{\mathbb{R}}$ is lower semicontinuous and convex.
(2) If $f, g : X \to \overline{\mathbb{R}}$ satisfy the relation $f \le g$, then $f^* \ge g^*$.
(3) (Young inequality) For all $x \in X$ and $x^* \in X^*$,

$$f(x) + f^*(x^*) \ge \langle x^*, x \rangle.$$

Theorem 1.5.8. *Let $f : X \to \overline{\mathbb{R}}$ be a proper and lower semicontinuous convex function. Then $f^* \not\equiv \infty$.*

The function $f^{**} : X \to \overline{\mathbb{R}}$ defined by

$$f^{**}(x) = \sup_{x^* \in X^*} \{ \langle x^*, x \rangle - f^*(x^*) \}$$

is said to be the second conjugate function of f. We denote by $\overline{co}(f)$ the function whose epigraph is $\overline{co}(\text{epi} f)$.

Example 1.5.1. Let X be a normed space and let $p > 1$ be a fixed number. Then the function $f : X \to \mathbb{R}$ defined by

$$f(x) = \frac{1}{p} \|x\|^p, \quad x \in X,$$

is a convex function,

$$f^*(x^*) = \frac{1}{q} \|x^*\|^q, \quad x^* \in X^*,$$

and $f^{**} = f$, where $\frac{1}{p} + \frac{1}{q} = 1$.

Theorem 1.5.9 (Fenchel–Moreau–Rockafellar). *Let $f : X \to \overline{\mathbb{R}}$ be a function. Then*

$$f^{**} = \overline{co} f,$$

in particular, if f is a proper and lower semicontinuous convex function, then $f^{**} = f$.

Theorem 1.5.10 (Fenchel–Rockafellar). *Let $f, g : X \to \overline{\mathbb{R}}$ be two convex functions. Assume that there exists $x_0 \in X$ such that $f(x_0) < +\infty$ and $g(x_0) < +\infty$. If f is continuous at x_0, then*

$$\inf_{x \in X} \{ f(x) + g(x) \} = \sup_{x^* \in X^*} \{ -f^*(-x^*) - g^*(x^*) \}$$
$$= \max_{x^* \in X^*} \{ -f^*(-x^*) - g^*(x^*) \}.$$

Definition 1.5.5. Let C be a nonempty closed convex subset of X and define

$$I_C(x) = \begin{cases} +\infty, & x \notin C, \\ 0, & x \in C. \end{cases}$$

Then I_C is an indicator function of C.

It is easy to see that I_C is a convex, proper, lower semicontinuous function, and its conjugate function I_C^* is called the supporting function of C, denoted by S_C, that is,

$$S_C(x^*) = \sup\{\langle x^*, x\rangle : x \in C, \, x^* \in X^*\}.$$

It is clear that $S_C^*(x) = I_C(x)$.

Theorem 1.5.11. $d(x_0, C) = \max_{x^* \in X^*, \|x^*\| \leq 1}\{\langle x^*, x_0\rangle - I_C^*(x^*)\}, \, \forall x_0 \in X$.

Theorem 1.5.12. *Let C be a nonempty closed convex subset of a Banach space X. Let $f : X \to \overline{\mathbb{R}}$ be a convex function. Then f is lower semicontinuous in the norm topology if and only if f is lower semicontinuous in the weak topology.*

Theorem 1.5.13. *Let C be a nonempty weakly compact convex subset of a Banach space X. Let $f : X \to \overline{\mathbb{R}}$ be a proper, lower semicontinuous, and convex function. Then there exists $x_0 \in X$ such that $f(x_0) < +\infty$ and $f(x_0) = \min\{f(x) : x \in C\}$.*

Theorem 1.5.14. *Let C be a nonempty closed convex subset of a reflexive Banach space X. Let $f : X \to \overline{\mathbb{R}}$ be a proper, lower semicontinuous, and convex function. Assume that $f(x) \to +\infty$ as $\|x\| \to +\infty$. Then there exists $x_0 \in X$ such that $f(x_0) < +\infty$ and $f(x_0) = \min\{f(x) : x \in C\}$.*

1.6 Differential calculus on Banach spaces

Let X and Y be two normed linear spaces. Let $T : \mathrm{Dom}(T) \subset X \to Y$ be a mapping.

Definition 1.6.1. A mapping $T : \mathrm{Dom}(T) \subset X \to Y$ is said to be
(i)   bounded if T maps any bounded subset of $\mathrm{Dom}(T)$ to a bounded subset of Y;
(ii)   locally bounded if $\forall x_0 \in \mathrm{Dom}(T)$ there exists a neighborhood U of x_0 such that T is bounded on U;
(iii) continuous at $x_0 \in \mathrm{Dom}(T)$ if

$$x_n \to x_0 \Rightarrow Tx_n \to Tx_0 \ (n \to \infty), \quad \forall\{x_n\} \subset \mathrm{Dom}(T);$$

(iv)   uniformly continuous on $\mathrm{Dom}(T)$ if

$$x_n - y_n \to \theta \Rightarrow Tx_n - Ty_n \to \theta \ (n \to \infty), \quad \forall\{x_n\}, \{y_n\} \subset \mathrm{Dom}(T);$$

(v)   demicontinuous at $x_0 \in \mathrm{Dom}(T)$ if

$$x_n \to x_0 \Rightarrow Tx_n \xrightarrow{w} Tx_0 \ (n \to \infty), \quad \forall\{x_n\} \subset \mathrm{Dom}(T);$$

(vi)   weakly continuous at $x_0 \in \mathrm{Dom}(T)$ if

$$x_n \xrightarrow{w} x_0 \Rightarrow Tx_n \xrightarrow{w} Tx_0 \ (n \to \infty), \quad \forall\{x_n\} \subset \mathrm{Dom}(T);$$

(vii) strongly continuous at $x_0 \in \text{Dom}(T)$ if

$$x_n \xrightarrow{w} x_0 \Rightarrow Tx_n \to Tx_0 \ (n \to \infty), \quad \forall \{x_n\} \subset \text{Dom}(T).$$

Let $Y = X^*$ and $T : \text{Dom}(T) \subset X \to X^*$. Then T is said to be hemicontinuous at $x_0 \in \text{Dom}(T)$ if, for all $h \in X$, $t_n \to 0$ as $t_n \to 0$, and $x_0 + t_n h \in \text{Dom}(T)$, $T(x_0 + t_n h) \xrightarrow{w^*} Tx_0 \ (n \to \infty)$.

The implication relations for various kinds of continuity are listed as follows:

$$\text{strongly continuous} \underset{\text{weakly continuous}}{\overset{\text{continuous}}{\underset{\searrow}{\overset{\nearrow}{\diagdown}}}} \text{demicontinuous} \to \text{hemicontinuous}$$

The implication relations for uniform continuity, continuity, boundedness, and local boundedness are listed as follows:

$$\text{Continuous} \longrightarrow \text{Locally bounded}$$
$$\uparrow \qquad\qquad\qquad \uparrow\downarrow \dim X < +\infty$$
$$\text{Uniformly continuous} \quad \longrightarrow \quad \text{Bounded}$$
$$\text{Dom}(T) \text{ is convex}$$

Definition 1.6.2. Let X be a real linear space and let Y be a real normed linear space. Let $T : \text{Dom}(T) \subset X \to Y$ be a mapping and $x_0 \in \text{Dom}(T)$. For all $h \in X$, there exists $\alpha_h > 0$ such that $x_0 + h \in \text{Dom}(T)$ when $|t| < \alpha_h$. If $\lim_{t \to 0} \frac{T(x_0+th)-Tx_0}{t}$ exists, then T is said to be Gâteaux differentiable at x_0 and the limit is called the Gâteaux differential of T at x_0 along direction h, denoted by $D[Tx_0, h]$. If there exists a bounded linear operator $B : X \to Y$ such that Gâteaux differential can be represented as $D[Tx_0, h] = Bh$, then T has a bounded linear Gâteaux differential (G differential) at x_0 and B is called a Gâteaux derivative (G derivative) at x_0, denoted by $T'(x_0)$.

It is clear that if T is Gâteaux differentiable at x_0, then T is hemicontinuous at x_0, but not necessarily continuous at x_0.

Definition 1.6.3. Let X be a real normed linear space and let Ω be an open set of X. Let $f : \Omega \to \mathbb{R}^1$ be a function. If f has a bounded linear Gâteaux differential on Ω, denoted by $F(x) = f'(x)$, then the mapping $F : \Omega \to X^*$ is said to be the gradient mapping of f, denoted by $F(x) = \text{grad} f(x)$ or $F = \text{grad} f = \nabla f$. We say that f is the potential of F, and the gradient mapping satisfies

$$\lim_{t \to 0} \frac{f(x + th) - f(x)}{t} = F(x)h, \quad \forall h \in X, x \in \Omega.$$

Definition 1.6.4. Let X, Y be real normed linear spaces and let $T : \text{Dom}(T) \subset X \to Y$ be a mapping. Let x_0 be an interior point of $\text{Dom}(T)$. Then T is said to be Fréchet differentiable at x_0 if there exists a bounded linear operator $B : X \to Y$ such that

$$\lim_{\|h\| \to 0} \frac{\|T(x_0 + h) - Tx_0 - Bh\|}{\|h\|} = 0.$$

Operator B is said to be the Fréchet derivative at x_0, denoted by $T'(x_0)$, and Bh is the F differential of T at x_0. If T is F differentiable at each point of $\mathrm{Dom}(T)$, then T is said to be F differentiable on $\mathrm{Dom}(T)$.

It is clear that if T is Fréchet differentiable at x_0, then T is continuous at x_0.

Theorem 1.6.1. *Let Ω be an open set in X and $T : \mathrm{Dom}(T) \to Y$. Let $x_0 \in \mathrm{Dom}(T)$. Then the following assertions hold:*
(i) *If T is F differentiable at x_0, then T must have a bounded linear G differential at x_0, and the G derivative of T coincides with the F derivative of T at x_0;*
(ii) *If T has a bounded linear G derivative everywhere in some neighborhood of x_0, and the G derivative $T'(x)$ is continuous at $x = x_0$, then T is F differentiable at x_0.*

Theorem 1.6.2 (Chain rule). *Let X, Y, Z be real normed linear spaces and let $U \subset X, V \subset Y$ be open subsets. Let $T : U \to V$ be F differentiable at x_0 and let $S : V \to Z$ be F differentiable at $T(x_0)$. Then $S \circ T : U \to Z$ is F differentiable at x_0 and*

$$(S \circ T)'(x_0) = S' \circ T(x_0)T'(x_0).$$

Example 1.6.1. Let X be a real reflexive Banach space. Let $A \in L(X, X^*)$ be a positive operator, that is, $\langle Ax, x \rangle \geq 0, \forall x \in X$, denoted by $\varphi(x) = \langle Ax, x \rangle$. Then $\mathrm{grad}\, \varphi(x) = (A + A^*)x, \forall x \in X$, where A^* is the conjugate operator of A. In particular, If $X = H$, a Hilbert space, and A is an identity mapping, then

$$\mathrm{grad}\, \|x\|^2 = 2x, \ x \in H, \quad \mathrm{grad}\, \|x\| = \frac{x}{\|x\|} \ (x \neq \theta).$$

Example 1.6.2. Let Ω be a Lebesgue measurable set in $\mathbb{R}^n$ and $X = L^p(\Omega), p > 1$. Then

$$\mathrm{grad}\, \|x\|^p = p|x(s)|^{p-2}x(s), \quad x \neq \theta$$

and

$$\mathrm{grad}\, \|x\| = \|x\|^{1-p}|x(s)|^{p-2}x(s), \quad x \neq \theta.$$

Example 1.6.3. Let $X = l^p$, where $p > 1$. Then $\mathrm{grad}\, \|x\| = \|x\|^{1-p}z, x \neq \theta$, where $x = (x_1, x_2, \dots) \in l^p, z = (|x_1|^{p-2}x_1, |x_2|^{p-2}x_2, \dots) \in l^q, \frac{1}{p} + \frac{1}{q} = 1$.

1.7 Subdifferentials of convex functions

Definition 1.7.1. Let X be a Banach space and let $\varphi : X \to \overline{\mathbb{R}}$ be a proper convex function. Let $x_0 \in X$ be fixed vector. Then $x^* \in X^*$ is said to be a subgradient of φ at x_0 if $\varphi(x_0) + \langle x^*, x - x_0 \rangle \leq \varphi(x)$, for any $x \in X$.

The family of all the subgradients of φ at x_0 is said to be the subdifferential of φ at x_0, denoted by $\partial\varphi(x_0)$. If $\partial\varphi(x_0) \neq \emptyset$, then φ is said to be subdifferentiable at x_0. If $\partial\varphi(x_0) = \emptyset$, then φ is said to be not subdifferentiable at x_0.

From the definition of the subdifferential, one has

$$\partial\varphi(x_0) = \bigcap_{x \in X}\{x^* \in X^* : \varphi(x_0) + \langle x^*, x - x_0 \rangle \leq \varphi(x)\}, \quad \forall x_0 \in X.$$

Generally speaking, $\partial\varphi$ is multi-valued.

For convex functions on $\mathbb{R}$, the subgradient at x_0 contains the tangent slopes through x_0.

Example 1.7.1. Let $a, b \in \mathbb{R}$ with $a < b$. Let $f : (a, b) \to \mathbb{R}$ be convex. Then

$$\partial f(x_0) = [f'_-(x_0), f'_+(x_0)], \quad \forall x_0 \in (a, b).$$

Example 1.7.2. Let $X = \mathbb{R}$ and consider

$$\varphi(x) = \begin{cases} +\infty, & |x| \geq 1, \\ 0, & |x| < 1. \end{cases}$$

If $|x_0| \geq 1$, then $\partial\varphi(x_0) = \emptyset$.

Example 1.7.3. Let $\varphi(x) = |x|$, $x \in \mathbb{R}$. Then

$$\partial\varphi(x) = \begin{cases} \frac{x}{|x|}, & x \neq \theta, \\ [-1, 1], & x = \theta. \end{cases}$$

The following theorems show the relation between the subdifferential and the G differential.

Theorem 1.7.1. *Let $\varphi : X \to \overline{\mathbb{R}}$ be a proper, convex function and continuous at $x \in X$. If φ is G differentiable at $x \in X$, then φ is subdifferentiable at $x \in X$, and $\partial\varphi(x) = \{\varphi'(x)\}$.*

Theorem 1.7.2. *Let $\varphi : X \to \overline{\mathbb{R}}$ be a convex function. Assume that φ is continuous at $x \in \text{Dom}(\varphi)$ and $\partial\varphi(x)$ contains only one subgradient. Then φ is G differentiable at $x \in \text{Dom}(\varphi)$.*

Example 1.7.4. Let $X = H$ be a Hilbert space. Let $\varphi(x) = \frac{1}{2}\|x\|^2$, $\forall x \in H$. Then $\partial\varphi(x) = \{\nabla\varphi(x)\} = \{x\}$.

Example 1.7.5. Let $X = H$ be a Hilbert space. Let $\varphi(x) = \|x\|$, $\forall x \in H$. Then

$$\partial\|x\| = \begin{cases} \frac{x}{\|x\|}, & x \neq \theta, \\ B_1[\theta], & x = \theta, \end{cases}$$

where $B_1[\theta] = \{x \in H : \|x\| \leq 1\}$ is a closed unit ball in H.

Definition 1.7.2. Let C be a closed convex subset of a real Banach space X and $x \in C$. The set

$$N_C(x) = \{x^* \in X : \langle x^*, y - x \rangle \leq 0, \ \forall y \in C\}$$

is said to be a normal cone of C at x.

Example 1.7.6. $\partial I_C(x) = N_C(x), \ \forall x \in C$.

Next, we list some basic properties of the subdifferential of φ in the following theorem.

Theorem 1.7.3. *Let $\varphi : X \to \bar{\mathbb{R}}$ be a convex function. Then*
(i)   *$\partial\varphi(x_0)$ is a w^*-closed convex subset of X^*, $\forall x_0 \in X$.*
(ii) *$\mathrm{Dom}(\partial\varphi) \subset \mathrm{Dom}(\varphi)$ and $\mathrm{Ran}(\partial\varphi) \subset \mathrm{Dom}(\varphi^*)$.*
(iii) *φ attains its minimum at $x \in \mathrm{Dom}(\varphi)$ if and only if $\theta \in \partial\varphi(x)$.*
(iv) *$\partial(\lambda\varphi) = \lambda\partial\varphi, \ \forall\lambda > 0$.*
(v)   *If $\varphi : X \to \bar{\mathbb{R}}$ is proper, lower semicontinuous and convex, and $(\mathrm{Dom}(\varphi))^0 \neq \emptyset$, then φ is continuous and subdifferentiable on $(\mathrm{Dom}(\varphi))^0$, and $\partial\varphi(x_0)$ is w^*-compact, $\forall x_0 \in (\mathrm{Dom}(\varphi))^0$.*
(vi) *If $\varphi_1, \varphi_2 : X \to \bar{\mathbb{R}}$ are convex functions satisfying $\mathrm{Dom}(\varphi_1) \cap \mathrm{Dom}(\varphi_2) \neq \emptyset$ and there exists $x_0 \in \mathrm{Dom}(\varphi_1) \cap \mathrm{Dom}(\varphi_2)$ such that φ_1 or φ_2 is continuous at x_0, then $\partial(\varphi_1 + \varphi_2) = \partial\varphi_1 + \partial\varphi_2$.*
(vii) *$x^* \in \partial\varphi(x_0) \Leftrightarrow \varphi(x_0) + \langle x^*, x - x_0 \rangle = 0$ is a supporting hyperplane of $\mathrm{epi}(\varphi)$ at $(x_0, \varphi(x_0))$.*
(viii) *Let $\varphi : X \to \bar{\mathbb{R}}$ be a proper, lower semicontinuous convex function. Then*

$$\varphi(x) + \varphi^*(x^*) = \langle x^*, x \rangle \Leftrightarrow x^* \in \partial\varphi(x).$$

(ix) *Let $\varphi : X \to \bar{\mathbb{R}}$ be a proper and lower semicontinuous convex function. Then $x^* \in \partial\varphi(x) \Leftrightarrow x \in \partial\varphi^*(x^*)$, which implies $\partial\varphi^* = (\partial\varphi)^{-1}$.*
(x)   *Let X be a real reflexive Banach space, $f : X \to \bar{\mathbb{R}}$ and $g : \mathbb{R}^+ \to \bar{\mathbb{R}^+}$ be two proper, lower semicontinuous convex functions. Then the following are equivalent:*
     (a) *$\forall x^* \in \partial f(x), y \in \mathrm{Dom}(f)$,*

$$f(y) \geq f(x) + \langle y - x, x^* \rangle + g(\|y - x\|);$$

     (b) *$\forall x^* \in \partial f(x), y^* \in \mathrm{Dom}(f^*)$,*

$$f^*(y^*) \leq f^*(x^*) + \langle x, y^* - x^* \rangle + g^*(\|y^* - x^*\|).$$

(xi) *(Mean-Value Formula) Let $f : X \to \bar{\mathbb{R}}$ be proper, lower semicontinuous and convex. Let $x, y \in X$ be two distinct points. Write $x_t := ty + (1 - t)x$ for all $t \in [0, 1]$ and define the univariate function $\varphi(t) := f(x_t)$. Then the following expressions are valid:*
     (1) *$\partial\varphi(t) = \{\langle s, y - x \rangle : s \in \partial f(x_t)\}$.*

(2) *There exist $t \in (0,1)$ and $s \in \partial f(x_t)$ such that*

$$f(y) - f(x) = \langle s, y - x \rangle.$$

(3) *For any $s_t \in \partial f(x_t)$,*

$$f(y) - f(x) = \int_0^1 \langle s_t, y - x \rangle \, dt.$$

(xii) *(Moreau–Rockafellar Theorem) Let X, Y be Banach spaces. Let $f : X \to \overline{\mathbb{R}}$, $g : Y \to \overline{\mathbb{R}}$ be proper, lower semicontinuous, and convex functions. Let $A : X \to Y$ be a bounded linear operator and $F(x) = f(x) + g(Ax)$. Assume that the regularity condition $\theta \in \mathrm{Int}\{A(\mathrm{Dom}(f)) - \mathrm{Dom}(g)\}$ holds. Then, for all $x_0 \in \mathrm{Dom}(F)$,*

$$\partial F(x_0) = \partial f(x_0) + A^*[\partial g(Ax_0)],$$

where A^* is the adjoint operator of A.

Example 1.7.7. Let $X = H$ be a Hilbert space and let $f(x) = \frac{1}{2}\|(I - P_C)x\|^2$, $\forall x \in H$. Then $\partial f(x) = \{\nabla f(x)\}$, $\nabla f(x) = (I - P_C)x$, $\forall x \in H$.

Example 1.7.8. Let $X = H$ be a Hilbert space. Let $A : H \to H$ be a bounded linear operator and $g(x) = \frac{1}{2}\|(I - P_C)Ax\|^2$, $\forall x \in H$. Then $\partial g(x) = \{\nabla g(x)\}$, $\nabla g(x) = A^*(I - P_C)Ax$, $x \in H$.

1.8 Geometry of Banach spaces

1.8.1 Convexity of Banach spaces

Definition 1.8.1. Let X be a Banach space. Then X is said to be strictly convex if for any $x, y \in X$, which are linearly independent, $\|x + y\| < \|x\| + \|y\|$; X is said to be uniformly convex if for any $\{x_n\}, \{y_n\} \subset X$ with $\|x_n\| = \|y_n\| = 1$ and $\|x_n + y_n\| \to 2$, $\|x_n - y_n\| \to 0$ as $n \to \infty$; X is said to be locally uniformly convex if for any $\varepsilon > 0$ and a given point $x \in X$, $x \in S(X) = \{x \in X : \|x\| = 1\}$, there exists $\delta = \delta(x, \varepsilon) > 0$, such that $\|x - y\| \geq \varepsilon \Rightarrow \|\frac{x+y}{2}\| \leq 1 - \delta$, $\forall y \in S(X)$.

Theorem 1.8.1. *Let X be a Banach space. Then the following assertions are equivalent:*
(i) *X is uniformly convex;*
(ii) *$\forall \{x_n\}, \{y_n\} \subset X$, $\|x_n\| \to 1$ and $\|y_n\| \to 1$, $\|x_n + y_n\| \to 2 \Rightarrow x_n - y_n \to \theta$ as $n \to \infty$;*
(iii) *$\forall \varepsilon \in (0, 2]$, there exists $\delta = \delta(\varepsilon) > 0$ such that $\frac{1}{2}\|x + y\| \leq 1 - \delta$.*

The following theorem is not hard to derive.

Theorem 1.8.2. *Let X be a Banach space. Then X is locally uniformly convex if and only if, for any $x \in S(X)$ and $\{x_n\} \subset S(X)$, $\|x_n - x\| \to 0$ as $\|x_n + x\| \to 2$; X is strictly convex if and only if, for any $f \in X^*$, f attains its maximum at most one point on $S(X) \Leftrightarrow \|\frac{x+y}{2}\| < 1$, $\forall x, y \in S(X), x \neq y$.*

We have the following implications:

uniformly convex $\Longrightarrow$ locally uniformly convex $\Longrightarrow$ strictly convex.

Note that any inverse of above implications is not true in general. However, for a finite-dimensional normed linear space X, they are true, that is, X is uniformly convex if and only if X is locally uniformly convex if and only if X is strictly convex.

Example 1.8.1. Fix $\mu > 0$ and consider the norm $\|x\|_\mu = \|x\|_0 + \mu(\int_0^1 x^2(t)dt)^{\frac{1}{2}}$ on $C[0,1]$, where $\|x\|_0$ is the usual maximum norm. Then, for any $x \in C[0,1]$, $\|x\|_0 \leq \|x\|_\mu \leq (1+\mu)\|x\|_0$, and $(C[0,1], \|\cdot\|_\mu)$ is strictly convex, but not uniformly convex.

Example 1.8.2. $l^1, l^\infty, c_0, L^1, L^\infty, C[a,b]$ are not strictly convex. Hence, they are also not uniformly convex.

Example 1.8.3. Both l^p and L^p, where $p > 1$, are uniformly convex. Hence, they are also locally uniformly convex and strictly convex.

Definition 1.8.2. Let X be a normed linear space with $\dim X \geq 2$. Define the modulus of convexity $\delta_X : (0,2] \to [0,1]$ of X by $\delta_X(\varepsilon) = \inf\{1 - \|\frac{x+y}{2}\| : \|x\| = \|y\| = 1; \varepsilon = \|x - y\|\}$, where $\delta_X(0) = 0$. It is obvious that $\delta_X : [0,2] \to [0,1]$ is continuous and monotonically increasing, while $\frac{\delta_X(\varepsilon)}{\varepsilon}$ is nondecreasing on $(0,2]$. In addition, we have the following equivalent expressions:

$$\delta_X(\varepsilon) = \inf\left\{1 - \left\|\frac{x+y}{2}\right\| : \|x\| \leq 1, \|y\| \leq 1, \varepsilon \leq \|x - y\|\right\}$$
$$= \inf\left\{1 - \left\|\frac{x+y}{2}\right\| : \|x\| \leq 1, \|y\| \leq 1, \varepsilon = \|x - y\|\right\}.$$

Definition 1.8.3. Let X be a normed space and let $p > 1$ be a fixed real number. Then X is said to be p-uniformly convex if there exists a constant $c > 0$ such that

$$\delta(\varepsilon) \geq c\varepsilon^p, \quad \forall \varepsilon \in [0,2].$$

Theorem 1.8.3. *Let X be a Banach space and let $p > 1$ be a fixed real number. Then X is p-uniformly convex if and only if there exists a constant $c \in (0,1]$ such that*

$$\frac{1}{2}(\|x + y\|^p + \|x - y\|^p) \geq \|x\|^p + c^p\|y\|^p, \quad \forall x, y \in X.$$

Theorem 1.8.4 (Xu [98], 1991). *Let X be a Banach space and let $p > 1$ be a fixed real number. Then X is p-uniformly convex if and only if there exists a constant $c_p > 0$ such that, $\forall x, y \in X, \lambda \in [0, 1]$,*

$$\|(1 - \lambda)y + \lambda x\|^p \leq (1 - \lambda)\|y\|^p - c_p w_p(\lambda)\|x - y\|^p + \lambda\|x\|^p,$$

where $w_p(\lambda) = \lambda^p(1 - \lambda) + \lambda(1 - \lambda)^p$.

Example 1.8.4. Let $X = H$ be a Hilbert space. Then

$$\delta_H(\varepsilon) = 1 - \sqrt{1 - \frac{\varepsilon^2}{4}} \geq \frac{1}{8}\varepsilon^2.$$

For any Banach space, one has $\delta_X(\varepsilon) \leq \delta_H(\varepsilon)$. In addition, if $\delta_X(\varepsilon) \geq c\varepsilon^p$, $c > 0$, then $p \geq 2$.

Example 1.8.5. Let $X = L^p$ where $p > 1$. Then

$$\delta_{L^p}(\varepsilon) = \begin{cases} 1 - [1 - (\frac{\varepsilon}{2})^p]^{\frac{1}{p}} \geq \frac{1}{p}(\frac{\varepsilon}{2})^p, & p \geq 2, \\ [1 - (\frac{\varepsilon}{2})^q]^{\frac{1}{q}} \geq \frac{p-1}{8}\varepsilon^2, & p \in (1, 2], \end{cases}$$

where p, q are conjugate numbers with $\frac{1}{p} + \frac{1}{q} = 1$.

From the above example, we see that L^p is $\max\{p, 2\}$-uniformly convex and Hilbert space is 2-uniformly convex, in particular, L^2 is 2-uniformly convex.

Theorem 1.8.5. *Let X be a Banach space. Then X is uniformly convex if and only if $\delta_X(\varepsilon) > 0, \forall \varepsilon > 0$.*

Theorem 1.8.6. *Let X be a uniformly convex Banach space. Then, $\forall r \geq \varepsilon > 0$, $\|x\| \leq r$, $\|y\| \leq r$ and $\|x - y\| \geq \varepsilon > 0$ imply $\delta_X(\frac{\varepsilon}{r}) > 0$ and*

$$\frac{1}{2}\|x + y\| \leq r\left\{1 - \delta_X\left(\frac{\varepsilon}{r}\right)\right\}.$$

Theorem 1.8.7 (Bruck [13], 1978). *Let X be a uniformly convex Banach space. Then, $\forall x, y \in B_1[\theta], t \in [0, 1]$,*

$$\|tx + (1 - t)y\| \leq 1 - 2\min\{t, 1 - t\}\delta_X(\|x - y\|).$$

Theorem 1.8.8 (Zeidle [112], 1985). *Let X be a uniformly convex Banach space. Let $\{t_n\}$ be a real number sequence such that $0 < a \leq t_n \leq b < 1$ and let $\{x_n\}$ and $\{y_n\}$ be sequences in X such that $\limsup_{n\to\infty} \|x_n\| \leq c$, $\limsup_{n\to\infty} \|y_n\| \leq c$ and*

$$\lim_n \|t_n x_n + (1 - t_n)y_n\| = c.$$

Then $x_n - y_n \to \theta$ as $n \to \infty$.

Theorem 1.8.9 (Xu [98], 1991). *Let X be a uniformly convex Banach space and let r and p be two real numbers such that $r > 0$ and $p > 1$. Then there exists a continuous strictly increasing convex function $g : \mathbb{R}^+ \to \mathbb{R}^+$ with $g(0) = 0$, such that, $\forall x, y \in B_r[\theta]$, $\forall \lambda, \mu \in [0,1]$, $\lambda + \mu = 1$,*

$$\|\lambda x + \mu y\|^p \leq \lambda\|x\|^p + \mu\|y\|^p - \omega_p(\lambda)g(\|x - y\|),$$

where $\omega_p(\lambda) = \lambda^p\mu + \lambda\mu^p$. In particular, if $p = 2$, then

$$\|\lambda x + \mu y\|^2 \leq \lambda\|x\|^2 + \mu\|y\|^2 - \lambda(1 - \lambda)(\lambda)g(\|x - y\|).$$

Let X be a Banach space and let $\{x_n\}$ be a sequence in X. Then X is said to have the Kadec–Klee property if $x_n \rightharpoonup x$ and $\|x_n\| \to \|x\|$ imply $x_n \to x$ as $n \to \infty$.

Theorem 1.8.10. *Let X be a locally uniformly convex Banach space. Then X has the Kadec–Klee property.*

In particular, uniformly convex Banach spaces have the Kadec–Klee property.

Theorem 1.8.11. *Uniformly convex Banach spaces are reflexive.*

So, the class of uniformly convex Banach spaces is a class of significant spaces between the class of Hilbert spaces and the class of reflexive spaces.

Theorem 1.8.12. *Let X be a uniformly convex Banach space and let C be a nonempty closed subset of X. Then, $\forall x \in X$, there exists unique point $x_0 \in C$ such that*

$$\|x - x_0\| = d(x, C) \triangleq \inf\{\|x - y\| : y \in C\}.$$

1.8.2 Duality mappings

Most results in this subsection are taken from [1, 87, 98].

Definition 1.8.4. Let φ be a continuous and strictly increasing function with $\varphi(0) = 0$ and $\varphi(t) \to +\infty$ $(t \to \infty)$. Then $\varphi : \mathbb{R}^+ \to \mathbb{R}^+$ is said to be a gauge function.

Theorem 1.8.13. *Let $\varphi : \mathbb{R}^+ \to \mathbb{R}^+$ be a gauge function and let $\psi(t) = \int_0^t \varphi(s)ds$. Then ψ is a convex function.*

Definition 1.8.5. Let φ be a gauge function. Define a mapping $J_\varphi : X \to 2^{X^*}$ by

$$J_\varphi x := \{x^* \in X^* : \langle x^*, x \rangle = \|x^*\|\|x\|; \|x^*\| = \varphi(\|x\|)\}.$$

Then J_φ is said to be a generalized duality mapping with gauge function φ. In particular, duality mapping $J = J_\varphi$ is called the normal duality mapping if $\varphi(t) = t$.

From the Hahn–Banach theorem, one sees that $J_\varphi x \neq \emptyset$, $\forall x \in X$. So $\mathrm{Dom}(J_\varphi) = X$. Next, we use $j_\varphi(x)$ to denote any element of $J_\varphi(x)$ and list some basic properties of J_φ as follows:

(i)   For any $x \in X$, $J_\varphi x$ is a ω^*-closed convex subset of X^*.

(ii) For any $x, y \in X$, $x^* \in J_\varphi x$, $y^* \in J_\varphi y$, one has $\langle x - y, x^* - y^* \rangle \geq 0$.

(iii) For any $x \neq \theta \in X$, $\lambda \in \mathbb{R}$, one has

$$J_\varphi(\lambda x) = \mathrm{sign}(\lambda) \frac{\varphi(|\lambda| \|x\|)}{\varphi(\|x\|)} J_\varphi(x),$$

and $J_\varphi(\theta) = \{\theta\}$. In particular, one has $J(\lambda x) = \lambda J x$ and $J(\theta) = \{\theta\}$, $\forall x \in X$, $\lambda \in \mathbb{R}$.

(iv) J is a bounded operator, that is, J maps any bounded subset of X into a bounded subset of X^*.

(v)   For any $x \in X$, $J_\varphi(x) = \partial \psi(\|x\|)$, where $\psi(t) = \int_0^t \varphi(s) ds$. In particular, for the generalized duality mapping $J_\varphi(x) = \|x\|^{p-2} J x$, $x \neq 0$, and the normal duality mapping with gauge functions $\varphi(t) = t^{p-1}$ and $\varphi(t) = t$, respectively, one has

$$\begin{cases} J_\varphi(x) = \partial(\frac{1}{p} \|x\|^p), & \forall p > 1, \\ J(x) = \partial(\frac{1}{2} \|x\|^2). \end{cases}$$

(vi) For any $x, y \in X$, $y^* \in J y$, $\|x\|^p - \|y\|^p \geq p \langle x - y, y^* \rangle$, $\|x\|^2 - \|y\|^2 \geq 2 \langle x - y, y^* \rangle$, we find that

$$\begin{cases} \|x + y\|^p \leq \|x\|^p + p \langle y, j(x + y) \rangle, & \forall j(x + y) \in J(x + y), \\ \|x + y\|^2 \leq \|x\|^2 + 2 \langle y, j(x + y) \rangle, & \forall j(x + y) \in J(x + y), \end{cases}$$

which are called the subdifferential inequalities.

(vii) Let $X = L^p$ (or l^p), where $p > 1$. Then the normal duality mappings on X are

$$\begin{cases} Jx = \|x\|^{2-p} |x(s)|^{p-2} x(s), & x \in L^p, \\ Jx = \|x\|^{2-p} z, & x \in l^p, \end{cases}$$

where $x = (x_1, x_2, x_3, \dots)$, $z = (|x_1|^{p-2} x_1, |x_2|^{p-2} x_2, |x_3|^{p-3} x_3, \dots) \in l^q$, $\frac{1}{p} + \frac{1}{q} = 1$.

(viii) Let X be a Banach space with a weakly continuous duality J_φ, that is, there exists some gauge function $\varphi : \mathbb{R}^+ \to \mathbb{R}^+$ such that J_φ is single-valued and sequentially continuous in the ω-topology of X and the ω^*-topology of X^*. Then, $\forall \{x_n\} \subset X$, $y \in X$, $x_n \xrightarrow{w} x$,

$$\limsup_{n \to \infty} \psi(\|x_n - y\|) = \limsup_{n \to \infty} \psi(\|x_n - x\|) + \psi(\|x - y\|).$$

So X satisfies the Opial condition,

$$\limsup_{n \to \infty} \|x_n - y\| > \limsup_{n \to \infty} \|x_n - x\|, \quad \forall y \neq x.$$

This also shows that Banach spaces with the weakly continuous duality J_φ must satisfy the Opial condition, however, the inverse may not be true.

(ix) Let X be a real smooth Banach space. Then, $\forall x, y \in X$,

$$\psi(\|x + y\|) = \psi(\|x\|) + \int_0^1 \langle y, j_\varphi(x_t)\rangle dt, \quad \forall j_\varphi(x_t) \in J_\varphi(x_t),$$

where $x_t = ty + (1 - t)x$, $t \in [0, 1]$ and $\psi(t) = \int_0^t \varphi(s)dx$

We remark here that l^p has the weakly continuous duality J_φ with $\varphi(t) = t^{p-1}$, where $p > 1$, however, L^p has no weakly continuous duality mapping J_φ unless $p = 2$. So, l^p satisfies the Opial condition, however, L^p does not satisfy the Opial condition if $p \neq 2$.

Theorem 1.8.14 (Takahashi [87], 2000). *Let X be a Banach space. Then the following assertions hold:*
(i)   *If X is strictly convex, then J_φ is injective, that is, $x \neq y \Rightarrow J_\varphi(x) \cap J_\varphi(y) = \emptyset$.*
(ii) *If X is reflexive, then J_φ is surjective, that is, $J_\varphi(X) = X^*$.*
(iii) *If X^* is strictly convex, then J_φ is single-valued.*

Theorem 1.8.15 (Takahashi [87], 2000). *Let X be a Banach space. Then X is strictly convex $\Leftrightarrow x^* \in Jx$, $y^* \in Jy$, and $x \neq y \Rightarrow \langle x - y, x^* - y^*\rangle > 0$.*

Definition 1.8.6. Let $\omega : \mathbb{R}^+ \to \mathbb{R}^+$ be a function. It is said to belong to the Γ class if it satisfies the following conditions:
(i) $\omega(0) = 0$;
(ii) $\omega(r) > 0$, $\forall r > 0$;
(iii) $t \leq s \Rightarrow \omega(t) \leq \omega(s)$.

Theorem 1.8.16 (Takahashi [87], 2000). *Let X be a uniformly convex Banach space. Then $\forall R > 0$ there exists a function $\omega_R \in \Gamma$ such that, $\forall x, y \in B_R[\theta]$ with $x^* \in Jx$ and $y^* \in Jy$,*

$$\langle x - y, x^* - y^*\rangle \geq \omega_R(\|x - y\|)\|x - y\|.$$

By means of the generalized duality mapping, we can give the following characteristic inequalities for p-uniformly convex Banach spaces.

Theorem 1.8.17 (Xu [98], 1991). *Let $p > 1$ be a given real number and let X be a Banach space. Then the following statements are equivalent:*
(i)   *X is p-uniformly convex.*
(ii) *There exists a constant $c > 0$ such that, for any $x, y \in X$, $j_p(x) \in J_p(x)$ and $j_p(y) \in J_p(y)$, the following inequality holds:*

$$\|x + y\|^p \geq \|x\|^p + p\langle y, j_p(x)\rangle + c\|y\|^p.$$

(iii) *There exists a constant $c_1 > 0$ such that, for any $x, y \in X$, $j_p(x) \in J_p(x)$ and $j_p(y) \in J_p(y)$, the following inequality holds:*

$$\langle x - y, j_p(x) - j_p(y) \rangle \geq c_1 \|x - y\|^p.$$

1.8.3 Differentiability of norms and smoothness of spaces

Let X be a Banach space and let $S(X) = \{x \in X : \|x\| = 1\}$.

Definition 1.8.7.
(i) The norm of X is said to be G-differentiable (or X is said to be smooth) if the limit

$$\lim_{t \to 0} \frac{\|x + ty\| - \|x\|}{t} \tag{L}$$

exists for each $x, y \in S(X)$.
(ii) The norm of X is said to be uniformly G-differentiable if, for each $y \in S(X)$, limit (L) is attained uniformly over $x \in S(X)$.
(iii) The norm of X is said to be F-differentiable, if for each $x \in S(E)$, limit (L) is attained uniformly over $y \in S(E)$.
(iv) The norm of X is said to be uniformly F-differentiable (or X is said to be uniformly smooth), limit (L) is attained uniformly over $x, y \in S(E)$.

Theorem 1.8.18. *A Banach space is smooth if and only if J_φ is single-valued.*

Theorem 1.8.19. *Let X^* be a uniformly convex Banach space. Then J_φ is single-valued and is uniformly continuous on bounded subsets of X.*

Theorem 1.8.20.
(i) *Let X be a Banach space. If the norm of X is F-differentiable, then J_φ is continuous.*
(ii) *Let X be a reflexive Banach space and let X^* be strictly convex. Then $J_\varphi : X \to X^*$ is demicontinuous.*
(iii) *If X is reflexive and X^* is locally uniformly convex, then $J_\varphi : X \to X^*$ is continuous.*

Theorem 1.8.21. *Let X be a Banach space. If the norm of X is uniformly G-differentiable, then J_φ is $\| \cdot \|$-ω^* uniformly continuous on bounded subsets of X.*

Theorem 1.8.22. *Space X^* is uniformly convex if and only if the norm of X is uniformly F-differentiable.*

Theorem 1.8.23. *Let X be a Banach space. Then*
(i) *If X^* is strictly convex, then X is smooth;*
(ii) *If X^* is smooth, then X is strictly convex.*

Theorem 1.8.24. *Let X be a reflexive Banach space. Then*
(i) *X is strictly convex if and only if X^* is smooth;*
(ii) *X is smooth if and only if X^* is strictly convex.*

Theorem 1.8.25. *l^1, l^∞, L^1 and L^∞ are not smooth.*

Definition 1.8.8. Let X be a Banach space. Then X is said to be uniformly smooth if $\forall \varepsilon > 0$ there exists $\delta = \delta(\varepsilon) > 0$ such that, $\forall x, y \in X$, $x \in S(X)$ and $\|y\| \le \delta$ imply

$$\frac{\|x - y\| + \|x + y\|}{2} - 1 < \varepsilon\|y\|.$$

Definition 1.8.9. Let X be a Banach space. Define a function $\rho_X(\cdot) : \mathbb{R}^+ \to \mathbb{R}^+$ by

$$\rho_X(\tau) = \sup\left\{\frac{\|x - y\| + \|x + y\|}{2} - 1 : x \in S(X), \|y\| \le \tau\right\}.$$

Function ρ_X is said to be the smoothness module of X and has the following properties:
(i) $\rho_X(0) = 0, \rho_X(\tau) \le \tau$;
(ii) $\rho_X(\cdot)$ is continuous, convex, and monotonically increasing;
(iii) $\frac{\rho_X(\tau)}{\tau}$ is monotonically increasing;
(iv) $\forall 0 < \tau \le \eta \Rightarrow \frac{\rho_\eta}{\eta^2} \le c\frac{\rho_X(\tau)}{\tau^2}$, where $c > 0$ is fixed constant;
(v) $\rho_X(c\tau) \le c^2\rho_X(\tau), \forall \tau > 0$.

Remark 1.8.1.
(i)

$$\rho_X(\tau) = \sup\left\{\frac{\|x + y\| + \|x - y\|}{2} - 1 : \|x\| \le 1, \|y\| \le \tau\right\}$$

$$= \sup\left\{\frac{\|x + y\| + \|x - y\|}{2} - 1 : \|x\| = 1, \|y\| = \tau\right\}.$$

(ii) Finite-dimensional Banach spaces are smooth if and only if they are uniformly smooth.

Theorem 1.8.26. *Let X be a Banach space. Then X is uniformly smooth if and only if $\frac{\rho_X(\tau)}{\tau} \to 0$ as $\tau \to 0$.*

Theorem 1.8.27. *Let X be a Banach space. Then the following assertions are mutually equivalent:*
(i) *X is uniformly smooth;*
(ii) *X^* is uniformly convex;*
(iii) *$J_\varphi : S(X) \to S(X^*)$ is uniformly continuous in the norm topology;*
(iv) *The norm of X is uniformly F-differentiable.*

Hence, if X is uniformly smooth, then X is smooth and reflexive.

Theorem 1.8.28 (Reich [72], 1978). *Let X be a real uniformly smooth Banach space. Then there exists a continuous monotonically increasing function $b : \mathbb{R}^+ \to \mathbb{R}^+$ with $b(0) = 0$ and $b(ct) \leq cb(t)$, $\forall c \geq 1$, such that*

$$\|x + y\|^2 \leq \|x\|^2 + \max\{\|x\|, 1\}\|y\|b(\|y\|) + 2\langle y, Jx\rangle, \quad \forall x, y \in X. \tag{RI}$$

Remark 1.8.2. The function b appearing in (RI) indeed is defined by

$$b(t) = \sup\left\{\frac{\|x + ty\|^2 - 1}{t} - 2\langle y, J(x)\rangle : x, y \in S(X)\right\}.$$

Further, we can require that function b in Reich inequality (RI) be strictly increasing.

Theorem 1.8.29 (Xu and Roach [106], 1991). *Let X be a real uniformly smooth Banach space. Then there exist two constants $k, c > 0$ such that*

$$\|x + y\|^2 \leq \|x\|^2 + 2\langle y, Jx\rangle + k\max\left\{\|x\| + \|y\|, \frac{c}{2}\right\}\rho_X(\|y\|), \quad \forall x, y \in X, \tag{XRI}$$

where $\rho_X(\cdot)$ is the smoothness module of X.

Theorem 1.8.30 (Alber [2], 1996). *Let X be a real uniformly smooth Banach space. Then the following inequality holds:*

$$\|x + y\|^2 \leq \|x\|^2 + c\rho_X(\|y\|) + 2\langle y, Jx\rangle, \quad \forall x, y \in X, \tag{AI}$$

where $c = 48\max\{L, \|x\|, \|y\|\}$ and $L \in (1, 1.7)$ is the Figiel constant.

Theorem 1.8.31 (Xu [98], 1991). *Let X be a real Banach space. Let $q > 1$ and $r > 0$. Then X is uniformly smooth Banach space if and only if there exists a continuous strictly increasing convex function $\beta : \mathbb{R}^+ \to \mathbb{R}^+$ with $\beta(0) = 0$ such that*

$$\|x + y\|^q \leq \|x\|^q + q\langle y, J_q(x)\rangle + \beta(\|y\|), \tag{XI}$$

where

$$J_q(x) = \{x^* \in X^* : \langle x, x^*\rangle = \|x\|^q, \|x^*\| = \|x\|^{q-1}\}$$

is the generalized duality mapping.

Theorem 1.8.32. *Let X be a Banach space. Then*

$$\rho_{X^*}(\tau) = \sup\left\{\frac{\tau\varepsilon}{2} - \delta_X(\varepsilon) : 0 \leq \varepsilon \leq 2\right\}$$

and

$$\rho_X(\tau) = \sup\left\{\frac{\tau\varepsilon}{2} - \delta_{X^*}(\varepsilon) : 0 \leq \varepsilon \leq 2\right\}$$

Example 1.8.6.

$$\rho_H(\tau) = (1+\tau^2)^{\frac{1}{2}} - 1,$$

$$\rho_{l^p}(\tau) = \rho_{L^p}(\tau) = \rho_{W_m^p}(\tau) = \begin{cases} (1+\tau^p)^{\frac{1}{p}} - 1 < \frac{1}{p}\tau^p, & 1 < p < 2, \\ \frac{p-1}{2}\tau^2 + o(\tau^2) < \frac{p-1}{2}\tau^2, & p \geq 2. \end{cases}$$

It is clear that Hilbert spaces are 2-uniformly smooth and l^p, L^p and W_m^p are $\min\{p, 2\}$-uniformly smooth.

Theorem 1.8.33. *For any Banach space X, we have $\rho_X(\tau) \geq \rho_H(\tau) = \sqrt{1+\tau^2} - 1$. If $\rho_X(\tau) \leq c\tau^q$, then $q \leq 2$.*

Definition 1.8.10. Let X be a Banach space and $q > 1$ be a given real number. Then X is said to be q-uniformly smooth if there exists some constant $c > 0$ such that $\rho_X(\tau) \leq c\tau^q$.

Theorem 1.8.34 (Xu [98], 1991). *Let X be a real Banach space. Then the following assertions are mutually equivalent:*

(i) *X is q-uniformly smooth;*

(ii) *There exists a constant $c > 0$ such that*

$$\|x + y\|^q \leq \|x\|^q + q\langle y, J_q(x)\rangle + c\|y\|^q, \quad \forall x, y \in X.$$

(iii) *There exists a constant $c_1 > 0$ such that*

$$\langle x - y, J_q(x) - J_q(y)\rangle \leq c_1\|x - y\|^q, \quad \forall x, y \in X.$$

Theorem 1.8.35. *Let $X = L^p$ (or l^p), $p > 1$. Then we have the following conclusions:*

(i) *If $2 < p < +\infty$, for all $x, y \in X$, $\lambda \in [0, 1]$, we have*

    (1) $\|(1 - \lambda)y + \lambda x\|^p \leq (1 - \lambda)\|y\|^p - w_p(\lambda)c_p\|x - y\|^p + \lambda\|x\|^p$;

    (2) $\|(1 - \lambda)y + \lambda x\|^2 \geq (1 - \lambda)\|y\|^2 - \lambda(1 - \lambda)(p - 1)\|x - y\|^2 + \lambda\|x\|^2$;

    (3) $\|x + y\|^p \geq \|x\|^p + p\langle y, J_p(x)\rangle + c_p\|y\|^p$;

    (4) $\|x + y\|^2 \leq \|x\|^2 + 2\langle y, J_p(x)\rangle + (p - 1)\|y\|^2$;

    (5) $\langle J_p(x) - J_p(y), x - y\rangle \geq 2p^{-1}c_p\|x - y\|^p$;

    (6) $\langle Jx - Jy, x - y\rangle \leq (p - 1)\|x - y\|^2$.

(ii) *If $1 < q \leq 2$, for all $x, y \in L^q$, $\lambda \in [0, 1]$, then*

    (1) $\|(1 - \lambda)y + \lambda x\|^2 \leq (1 - \lambda)\|y\|^2 - \lambda(1 - \lambda)(q - 1)\|x - y\|^2 + \lambda\|x\|^2$;

    (2) $\|(1 - \lambda)y + \lambda x\|^q \geq (1 - \lambda)\|y\|^q - w_q(\lambda)c_q\|x - y\|^q + \lambda\|x\|^q$;

    (3) $\|x + y\|^2 \geq \|x\|^2 + 2\langle y, J_q(x)\rangle + c_q\|y\|^2$;

    (4) $\|x + y\|^q \leq \|x\|^q + q\langle y, J_q(x)\rangle + c_q\|y\|^q$;

    (5) $\langle Jx - Jy, x - y\rangle \geq (q - 1)\|x - y\|^2$;

    (6) $\langle J_q(x) - J_q(y), x - y\rangle \leq 2q^{-1}c_q\|x - y\|^q$,

    where $c_p, c_q > 0$ are fixed constants.

1.8.4 Geometry constants in Banach spaces

Geometric constants of Banach spaces play an important role in fixed point problems of nonlinear Lipschitz mappings. In this subsection, we consider asymptotic centers, normal structures, and uniform normal structure constants.

Let X be a Banach space. Let C be a nonempty subset of X and let $\{x_n\}$ be a bounded sequence in X. Define a functional $r : X \to \mathbb{R}^+$ by

$$r(x) = \limsup_{n\to\infty} \|x_n - x\|, \quad \forall x \in X,$$

where
$$r(C, \{x_n\}) := \inf\{r(y) : y \in C\},$$
$$A(C, \{x_n\}) := \{z \in C : r(z) = \min\{r(x) : x \in C\}\}.$$

Definition 1.8.11. $r(C, \{x_n\})$ is called the asymptotic radius of $\{x_n\}$ on C. If $z \in A(C, \{x_n\})$, then $z \in C$ is called the asymptotic center of $\{x_n\}$ on C.

It is worth mentioning that $A(C, \{x_n\})$ may be an empty set or a singleton, or a set with infinitely many points. In fact, if $x_n \to x$ as $n \to \infty$, then $A(C, \{x_n\}) = \{x\}$ for $x \in C$ and $A(C, \{x_n\}) = \{y \in C : \|x - y\| = d(x, C)\}$ for $x \notin C$.

What we are mainly concerned is the case when $A(C, \{x_n\})$ is a singleton.

Theorem 1.8.36. *Let C be a nonempty closed convex subset of a uniformly convex Banach space X and let $\{x_n\}$ be a bounded sequence in X. Then $A(C, \{x_n\})$ is a singleton.*

Let X be a uniformly convex Banach space. From the above theorem, we see that any bounded sequence has a unique asymptotic center with respect to a nonempty closed convex subset C of X.

Next theorem further characterizes the property of asymptotic centers.

Theorem 1.8.37. *Let C be a nonempty closed convex subset of a uniformly convex Banach space X. Then each bounded sequence $\{x_n\}$ with respect to C has a unique asymptotic center $z \in C$, $\{z\} = A(C, \{x_n\})$, and*

$$\limsup_{n\to\infty} \|x_n - z\| < \limsup_{n\to\infty} \|x_n - x\|, \quad \forall x \ne z.$$

From the above theorem, one can immediately conclude the following result.

Theorem 1.8.38. *Let X be a uniformly convex Banach space satisfying the Opial condition. Let C be a nonempty closed convex subset of X. Let $\{x_n\}$ be a sequence in C such that $x_n \overset{w}{\longrightarrow} z$. Then $\{z\} = A(C, \{x_n\})$.*

Definition 1.8.12. Let C be a nonempty bounded subset of a Banach space X. $\operatorname{diam}(C) = \sup\{\|x - y\| : x, y \in C\}$ is said to be the diameter of C. We say that $x_0 \in C$ is the
(i)   diameter point of C if $\sup\{\|x_0 - x\| : x \in C\} = \operatorname{diam}(C)$;
(ii) non-diameter point of C if $\sup\{\|x_0 - x\| : x \in C\} < \operatorname{diam}(C)$.

Definition 1.8.13. Let C be a nonempty convex subset of a Banach space X. Then C is said to have the normal structure if each bounded subset D of C that contains at least two points has a non-diameter point in C, that is, there exists $x_0 \in D$ such that

$$\sup\{\|x_0 - x\| : x \in D\} < \mathrm{diam}(D).$$

We say that X has the normal structure if each any nonempty bounded convex subset C with a positive diameter $(\mathrm{diam}(C) > 0)$ has the normal structure in X.

Theorem 1.8.39.
(i)   *Each compact convex subset C of a Banach space X has the normal structure.*
(ii) *Each bounded closed convex subset of a uniformly convex Banach space X has the normal structure.*
(iii) *Any uniformly convex Banach space has the normal structure.*
(iv) *If a Banach space X satisfies the Opial condition, then X has the normal structure, however, the inverse is not true.*

Definition 1.8.14. Let X be a Banach space. Let C be a bounded closed convex subset of X and $\mathrm{diam}(C) > 0$. Then

$$N(X) := \inf\left\{ \frac{\mathrm{diam}(C)}{r(C)} : \mathrm{diam}(C) > 0\right\}$$

is said to be the normal structure coefficient of X, where $r(C) := \{r_x(C) : x \in C\}$, and $r_x(C) := \sup\{\|x - y\| : y \in C\}$. It is obvious that $N(X) \geq 1$. Also X is said to have the uniformly normal structure if $N(X) > 1$. For a Hilbert space H, one has $N(H) = \sqrt{2}$.

Theorem 1.8.40. *Let X be a uniformly smooth Banach space. Then X has the uniform normal structure. Hence, it also has the normal structure.*

1.8.5 Banach limits

It is known that the limit operation of real number sequences is linear, but, the limit of bounded sequences does not always exist. This motivates us to consider superior and inferior limits. It is not true that both superior limit and inferior limit are not linear. Can one find a linear operator to replace the superior/inferior limit operation? The answer is positive. The Banach limit is such an operation that keeps the usual linear structure and is more efficient than the superior/inferior limit operation.

Definition 1.8.15. Let μ be a bounded linear function on l^∞. If it has the following properties:
(L1) $\|\mu\| = \mu(e) = 1$, where $e = (1, 1, 1, \dots) \in l^\infty$;
(L2) $\mu(x_{n+1}) = \mu(x_n)$, for all $x_n = (x_1, x_2, x_3, \dots) \in l^\infty$,

then μ is called a Banach limit. Sometimes, we also use μ_n, or LIM, to denote the Banach limit.

Using the Tychonoff fixed point theorem or the Hahn–Banach theorem, we can show the existence of the Banach limit.

Theorem 1.8.41. *Let μ be a Banach limit. Then*

$$\liminf_{n\to\infty} x_n \leq \mu(x) \leq \limsup_{n\to\infty} x_n, \quad \forall x = (x_1, x_2, \dots) \in l^\infty.$$

In particular, if $x_n \to a$, then $\mu(x) = a$. But, the converse may not be true.

Example 1.8.7. Let $x = (1, 0, 1, 0, \dots) \in l^\infty$. Then $\liminf_{n\to\infty} x_n = 0$ and $\limsup_{n\to\infty} x_n = 1$. Hence, $\lim_{n\to\infty} x_n$ does not exist, however, for any Banach limit μ, one has $\mu(x_n) = \frac{1}{2}$.

Theorem 1.8.42. *Let $x = (x_n) \in l^\infty$ and $a \in \mathbb{R}$. If for all Banach limits μ, $\mu(x_n) \leq a$, and $\limsup_{n\to\infty}(x_{n+1} - x_n) \leq 0$, then $\limsup_{n\to\infty} x_n \leq a$.*

Theorem 1.8.43. *Let X be a reflexive Banach space and let $\{x_n\}$ be a bounded sequence in X. Let μ be a Banach limit. Then there exists $x_0 \in X$ such that*

$$\mu_n \langle x_n, x^* \rangle = \langle x_0, x^* \rangle, \quad \forall x^* \in X^*.$$

Theorem 1.8.44. *Let X be a reflexive Banach space and let C be a nonempty closed convex subset of X. Let $\{x_n\}$ be a bounded sequence in C and let μ_n be a Banach limit. Define $\varphi(x) = \mu_n \|x_n - x\|^2$, $\forall x \in C$. Then the set defined by*

$$K = \left\{ z \in C : \mu_n \|x_n - z\| = \inf_{x \in C} \varphi(x) \right\},$$

is a nonempty, bounded, closed, and convex subset of C. If X is a uniformly convex Banach space, then K is a singleton.

Theorem 1.8.45. *Let X be a Banach space with a uniformly G-differentiable norm. Let C be a nonempty, closed, and convex subset of X, and let $\{x_n\}$ be a bounded sequence in C. Let μ_n be a Banach limit, and let $z \in C$ be fixed element. Then*

$$z \in K \Leftrightarrow \mu_n \langle y - z, J(x_n - z) \rangle \leq 0, \quad \forall y \in C.$$

We remark here that most of the results presented in this subsection are taken from Takahashi [87].

1.8.6 Projection mappings

Projection mappings play an important role in various iterative methods. It is necessary to understand their basic properties. This subsection is devoted to four kinds of projection mappings: the metric projection, the generalized projection, the sunny nonexpansive projection, and the sunny generalized nonexpansive projection.

Let X be a real reflexive, smooth, and strictly convex Banach space. Let C be a nonempty, closed, and convex subset of X. Then $\forall x \in X$ there exists unique $x_0 \in C$ such that

$$\|x - x_0\| = \min\{\|x - y\| : y \in C\}.$$

Definition 1.8.16. Let $x_0 = P_C(x)$. Then $P_C : X \to C$ is uniquely defined and it is said to be the metric projection from X onto C.

Using $Ju = \frac{1}{2}\,\mathrm{grad}\,\|u\|^2$, we have the following characterization.

Theorem 1.8.46. *Let $x \in X$ and $x_0 \in C$. Then $x_0 = P_C(x)$ if and only if $\langle x_0 - y, J(x - x_0)\rangle \geq 0$, $\forall y \in C$.*

Consider a bivariate function

$$\varphi(x, y) = \|x\|^2 - 2\langle x, Jy\rangle + \|y\|^2, \quad \forall x, y \in X.$$

Fixing $y \in X$, we see that $\varphi(\cdot, y)$ is a continuous convex function with $\varphi(x, y) \to +\infty$ ($\|x\| \to +\infty$). Hence, for any $x \in X$, there exists $x_0 \in C$ such that

$$\varphi(x_0, y) = \min\{\varphi(x, y) : x \in C\}.$$

The uniqueness of x_0 is guaranteed by the strict convexity of X.

Definition 1.8.17. Let $x_0 = \Pi_C x$. Then $\Pi_C : X \to C$ is uniquely defined and it is said to be generalized projection from X onto C.

Theorem 1.8.47. *Let $x \in X$ and $x_0 \in C$. Then $x_0 = \Pi_C x$ if and only if $\langle x_0 - y, J(x - x_0)\rangle \geq 0$, $\forall y \in C$.*

It is known that the metric projection P_C, mapping Hilbert space H onto a nonempty closed convex subset C, has the following property:

$$P_C[\lambda x + (1 - \lambda)P_C(x)] = P_C x, \quad \forall x \in H, \lambda \in [0, 1].$$

This shows that each point on the segment $\mathrm{seg}\{x, P_C x\}$ can be the best approximation of $P_C x$. This motives us to introduce the following general concepts.

Definition 1.8.18. Let X be a Banach space. Let C and D be subsets of X. Let $P : C \to D$ be a mapping. Then P is said to be sunny if $P[tx + (1 - t)Px] = Px$, for all $x \in C$, $t \geq 0$ and $tx + (1 - t)Px \in C$.

Let $D \subset C$. We say that a continuous mapping $P : C \to D$ is a retraction if $Px = x$, $\forall x \in D$, that is, $P^2 = P$. The set D is called a retract of C. In addition, if P is nonexpansive, then P is said to be a nonexpansive retraction, and the set D is called a nonexpansive retract.

Example 1.8.8. Every closed convex subset of a Euclidean space $\mathbb{R}^n$ (or a Hilbert space H) is a retract of $\mathbb{R}^n$ (or H).

Example 1.8.9. Every closed convex subset of a uniformly convex Banach space X is a retract of X, however, it may not be a nonexpansive retract.

Theorem 1.8.48. *Let C be a nonempty convex subset of a smooth Banach space X. Let D be a nonempty subset of C and let $P : C \to D$ be a retraction mapping such that*

$$\langle x - Px, J(y - Px) \rangle \leq 0, \quad \forall x \in C, y \in D.$$

Then $P : C \to D$ is a sunny nonexpansive retraction.

Theorem 1.8.49. *Let C be a nonempty convex subset of a smooth Banach space X. Let D be a nonempty subset of C and let $P : C \to D$ be a retraction mapping. Then the following assertions are mutually equivalent:*
(i) *P is a sunny nonexpansive retraction mapping;*
(ii) *$\langle x - Px, J(y - Px) \rangle \leq 0, \forall x \in C, y \in D$;*
(iii) *$\langle x - y, J(Px - Py) \rangle \geq \|Px - Py\|^2, \forall x, y \in C$.*

Remark 1.8.3.
(i) There is at most one sunny nonexpansive retraction mapping in a smooth Banach space X.
(ii) The concept of the nonexpansive retract generalizes two results: one is a linear ergodic theorem in the framework of reflexive Banach spaces, and the other is the metric projection in the framework of Hilbert spaces.

Example 1.8.10. Let X be a reflexive Banach space and let $L : X \to X$ be a linear operator with $\|L\| \leq 1$. Then the mean ergodic theorem guarantees that the mean-value operator $L_n = n^{-1} \sum_{j=1}^{n} L^j$ pointwise converges to projection P from X to $\mathrm{Ker}(I - L)$ with $\|P\| \leq 1$ and $\mathrm{Ker}(I - L) = \mathrm{Fix}(L)$.

Example 1.8.11. Let H be a Hilbert space and let C be a nonempty closed and convex subset of H. Let $T : C \to C$ be a nonexpansive mapping, that is, $\|Tx - Ty\| \leq \|x - y\|$, for all $x, y \in C$. Denote the fixed point set of T by $\mathrm{Fix}(T)$. Then $\mathrm{Fix}(T)$ is a closed convex subset of X. If $\mathrm{Fix}(T) \neq \emptyset$, then $P_{\mathrm{Fix}(T)}$ is nonexpansive. Indeed, $P_{\mathrm{Fix}(T)}$ is firmly nonexpansive, that is,

$$\|P_{\mathrm{Fix}(T)}x - P_{\mathrm{Fix}(T)}y\|^2 \leq \langle x - y, P_{\mathrm{Fix}(T)}x - P_{\mathrm{Fix}(T)}y \rangle, \quad \forall x, y \in C.$$

Definition 1.8.19. Let D be a nonempty closed convex subset of a Banach space X. Let $R : X \to D$ be a mapping. Then R is said to be generalized nonexpansive if $\mathrm{Fix}(R) \neq \emptyset$, and $\varphi(Rx, y) \leq \varphi(x, y), \forall x \in X, y \in \mathrm{Fix}(R)$, where

$$\varphi(x, y) = \|x\|^2 - 2\langle y, Jx \rangle + \|y\|^2, \quad \forall x, y \in X.$$

Theorem 1.8.50. *Let X be a reflexive, smooth, and strictly convex Banach space. Let C be a nonempty closed subset of X and let $R_C : X \to C$ be a retraction mapping. Then $R_C : X \to C$ is a sunny generalized nonexpansive mapping if and only if*

$$\langle x - R_C x, J(R_C x) - J(y) \rangle \geq 0, \quad \forall x \in X, y \in X.$$

In a reflexive, smooth, and strictly convex Banach space X, there are four kinds of projection mappings from X to a nonempty closed convex subset C of X. They are metric projection P_C, generalized projection Π_C, sunny nonexpansive projection Q_C, and sunny generalized nonexpansive projection R_c. Also $\forall x \in X$, $x_0 \in C$, we have
(i) $x_0 = P_C x \Leftrightarrow \langle J(x - x_0), x_0 - y \rangle \geq 0, \forall y \in C;$
(ii) $x_0 = \Pi_C x \Leftrightarrow \langle Jx - Jx_0, x_0 - y \rangle \geq 0, \forall y \in C;$
(iii) $x_0 = Q_C x \Leftrightarrow \langle x - x_0, J(x_0 - y) \rangle \geq 0, \forall y \in C;$
(iv) $x_0 = R_C x \Leftrightarrow \langle x - x_0, J(x_0 - Jy) \rangle \geq 0, \forall y \in C.$

If $X = H$ is a Hilbert space, then $P_C = \Pi_C = Q_C = R_C$. In addition, P_C and Π_C uniquely exist, however, Q_C and R_C may not exist.

1.9 Some classes of nonlinear mappings

In this section, we introduce several classes of important nonlinear mappings: the class of nonexpansive mappings, the class of accretive operators, the class of monotone mappings, and the class of pseudocontractive mappings. Several demiclosed principles and fixed point theorems are also investigated in this section.

From Bruck's viewpoint, the class of nonexpansive mappings, which is a natural extension of the class of contractive mappings, is one of most important nonlinear mappings.
(i)   The concept of a nonexpansive mapping has a close relation with the monotone methods that were investigated in 1960s. Browder–Kirk–Göhde fixed point theorem is the first fixed point theorem based on geometric properties of Banach spaces instead of compactness.
(ii) Nonexpansive mappings have an intimate relation with the following initial value problem of the differential inclusion:

$$\theta \in \frac{du}{dt} + T(t)u,$$

where $\{T(t)\}_{t \geq 0}$ is a family of dissipative mappings or multi-valued continuous accretive operators.

On the other hand, nonexpansive mappings have an intimate relation with accretive mappings, monotone mappings, nonlinear semigroups, and nonlinear evolution equations.

(i)   Let $T : \mathrm{Dom}(T) \subset X \to X$ be a nonexpansive mapping. Let $U = I - T$ and $\mathrm{Dom}(U) = \mathrm{Dom}(T)$. Then $U : \mathrm{Dom}(U) \to X$ is accretive.

(ii) Let $\{U(t)\}_{t \geq 0}$ be a nonlinear semigroup and T its generator. Then $U(t)$ is nonexpansive $\Leftrightarrow -T$ is accretive $\Leftrightarrow T$ is dissipative.

The class of pseudocontractive mappings is a generalization of the class of nonexpansive mappings. It also has an intimate relation with the class of accretive operators, that is, T is pseudocontractive if and only if $I - T$ is accretive. Since $\mathrm{Fix}(T) = U^{-1}(\theta)$, we can convert zero point problems of accretive operators to fixed point problems of pseudocontractive mappings.

Now, we give some necessary notations, which will be used in the subsequent discussion and statements.

In Sections 1.7 and 1.8, we discussed two multi-valued mappings $\partial\varphi$ and J_φ. The reasons for studying multi-valued mappings are aspective and sufficient.

Let X and Y be two real normed linear spaces and let X^* be the topological conjugate of X. Let $A : X \to 2^Y$ be a multi-valued mapping. Then we use the following notations:

$\mathrm{Dom}(A) = \{x \in X : Ax \neq \emptyset\}$: the effective domain of A,

$\mathrm{Ran}(A) = \bigcup\{Ax : x \in X\}$: the range of A,

$\mathrm{Graph}(A) = \{[x, y] \in X \times Y : x \in \mathrm{Dom}(A), y \in \mathrm{Ran}(A)\}$: the graph of A.

We may view a multi-valued mapping $A : X \to 2^Y$ as the $\mathrm{Graph}(A)$ of A, and use the notation $A \subset X \times Y$ to denote the multi-valued mapping $A : X \to 2^Y$.

Let $A, B : X \to 2^Y$ be multi-valued mappings and let $\lambda \in \mathbb{R}$. Then

$A + B := \{[x, y + z] \in X \times Y : [x, y] \in A, [x, z] \in B\}$,

$\lambda A := \{[x, \lambda y] \in X \times Y : [x, y] \in A\}$,

$A^{-1} := \{[x, y] \in X \times Y : [y, x] \in A\}$.

Let X, Y and Z be real normed linear spaces. Let $A \subset X \times Y$ and $B \subset Y \times Z$ be multi-valued mappings. Then define a multi-valued mapping $BA \subset X \times Z$ by

$$BA := \{[x, z] \in X \times Z : \text{there exists } y \in Y \text{ such that } [x, y] \in A \text{ and } [y, z] \in B\}.$$

From above notations, we have the following conclusions:

(i)   Let $A, B \subset X \times Y$. Then
   (a) $\mathrm{Dom}(A + B) = \mathrm{Dom}(A) \cap \mathrm{Dom}(B)$,
   (b) $\mathrm{Dom}(\lambda A) = \mathrm{Dom}(A)$,
   (c) $\mathrm{Dom}(A^{-1}) = \mathrm{Ran}(A)$.

(ii) Let $A \subset X \times Y$ and $B \subset Y \times Z$. Then

$$\mathrm{Dom}(BA) = \{x \in X : x \in \mathrm{Dom}(A), Ax \in \mathrm{Dom}(B)\}.$$

(iii) Let $A, B \subset X \times Y$. Then

$$(A + B)x = Ax + Bx = \{y + z \in Y : y \in Ax, z \in Bx\}, \quad \forall x \in \mathrm{Dom}(A + B),$$

(iv) $(\lambda A)x = \{\lambda y \in Y : y \in Ax\}, \forall x \in \mathrm{Dom}(\lambda A),$

(v) $A^{-1}(x) = \{y \in X : x \in Ay\}, \forall x \in \mathrm{Dom}(A^{-1}),$

(vi) Let $A \subset X \times Y$ and $B \in Y \times Z$. Then

$$(BA)x = B(Ax) = \{z \in Z : \exists y \in \mathrm{Dom}(B), y \in Ax, z \in By\} \quad \forall x \in \mathrm{Dom}(BA).$$

Let $A \subset X \times Y$ be multi-valued mapping and let $T : X \to X$ be a single-valued mapping. We also use the following notations:

(i) $A^{-1}(\theta) = N(A) := \{x \in \mathrm{Dom}(A) : \theta \in Ax\}$: the set of zeros of A;

(ii) $\mathrm{Fix}(T) := \{x \in \mathrm{Dom}(A) : x = Tx\}$: the set of fixed points of T;

1.9.1 Nonexpansive mappings

Definition 1.9.1. Let X be a normed space and let C be a nonempty subset of X. Let $T : C \subset X \to X$ be a mapping. Then T is said to be

(i) L-Lipschitz if there exists $L > 0$ such that

$$\|Tx - Ty\| \le L\|x - y\|, \quad \forall x, y \in C;$$

(ii) strictly contractive if $L \in [0, 1)$;

(iii) nonexpansive if $L = 1$;

(iv) averaged nonexpansive (or averaged) if

$$T = (1 - \lambda)I + \lambda S,$$

where S is a nonexpansive mapping on C and λ is a real number in $(0, 1)$;

(v) firmly nonexpansive if

$$\|Tx - Ty\| \le \|r(x - y) + (1 - r)(Tx - Ty)\|, \quad \forall x, y \in C, r > 0;$$

(vi) quasi-nonexpansive if $\mathrm{Fix}(T) \ne \emptyset$ and

$$\|Tx - y\| \le \|x - y\|, \quad \forall x \in C, y \in \mathrm{Fix}(T);$$

(vii) strongly nonexpansive if

$$\lim_{n \to \infty} ((x_n - y_n) - (Tx_n - Ty_n)) = \theta,$$

whenever $\{x_n\}$ and $\{y_n\}$ are sequences in C such that $\{x_n - y_n\}$ is bounded and $\lim_{n \to \infty} (\|x_n - y_n\| - \|Tx_n - Ty_n\|) = 0$.

Proposition 1.9.1.

(i)   *T is firmly nonexpansive if and only if there exists $j(Tx - Ty) \in J(Tx - Ty)$ such that*

$$\langle x - y, j(Tx - Ty) \rangle \geq \|Tx - Ty\|^2, \quad \forall x, y \in C.$$

(ii)   *T is strongly nonexpansive if and only if $\forall M > 0$ there exists a strictly increasing function $y : [0, 2M] \to [0, M]$ such that, $\forall r > 0, y(r) > 0$, and $\forall x, y \in C, \|x - y\| \leq M$,*

$$y(\|x - y - (Tx - Ty)\|) + \|Tx - Ty\| \leq \|x - y\|.$$

(iii) *A linear projection mapping P with $\|P\| = 1$ is firmly nonexpansive. The metric projection $P_C : H \to C$ from a Hilbert space H onto its nonempty closed and convex subset C is firmly nonexpansive and it is also a $\frac{1}{2}$-averaged mapping.*

(iv)   *If X is uniformly convex, then firmly nonexpansive mappings and averaged mappings are strongly nonexpansive.*

(v)   *The composition of a finite family of nonexpansive mappings (if well defined) is still nonexpansive.*

(vi)   *The composition of a finite family of averaged mappings (if well defined) is still averaged. In particular, let T_1 be a λ_1-averaged mapping, where $\lambda_1 \in (0, 1)$ and let T_2 be a λ_2-averaged mapping, where $\lambda_2 \in (0, 1)$. If both $T_1 T_2$ and $T_2 T_1$ are well defined, then both $T_1 T_2$ and $T_2 T_1$ are λ-averaged, where $\lambda = \lambda_1 + \lambda_2 - \lambda_1 \lambda_2$.*

(vii) *Let C be nonempty subset of a strictly convex Banach space X. Let $T_1, T_2, T_3, \ldots, T_r$, where r is some positive integer, be averaged mappings from C to itself with a nonempty common fixed point set $F = \bigcap_{i=1}^r \text{Fix}(T_i) \neq \emptyset$. Then*

$$F = \text{Fix}(T_r T_{r-1} \cdots T_1) = \text{Fix}(T_1 T_r \cdots T_2) = \cdots = \text{Fix}(T_{r-1} T_{r-2} \cdots T_r).$$

(viii) *Let X be a Banach space and let C be a nonempty subset of X. Let $T_1, T_2, T_3, \ldots, T_r$, where r is some positive integer, be strongly nonexpansive mappings from C to itself with a nonempty common fixed point set $F = \bigcap_{i=1}^r \text{Fix}(T_i) \neq \emptyset$. Then*

$$F = \text{Fix}(T_r T_{r-1} \cdots T_1) = \text{Fix}(T_1 T_r \cdots T_2) = \cdots = \text{Fix}(T_{r-1} T_{r-2} \cdots T_r).$$

(ix)   *A convex combination of a nonexpansive mapping and an averaged mapping is an averaged mapping.*

(x)   *Let X be a uniformly convex Banach space. Let T_0 be a strongly nonexpansive mapping and let T_1 be a nonexpansive mapping. Let $S = (1 - c)T_0 + cT_1$, where c is a real number in $(0, 1)$. Then S is strongly nonexpansive.*

(xi)   *Let X be a strictly convex Banach space and let $\{T_i\}_{i=1}^\infty$ be an infinite family of nonexpansive mappings. Let $Tx = \sum_{i=1}^\infty \lambda_i T_i x, \forall x \in X$, where $\lambda_i \in (0, 1)$ for each $i = 1, 2, \ldots$ with $\sum_{i=1}^\infty \lambda_i = 1$. Then T is nonexpansive. If $\bigcap_{i=1}^\infty \text{Fix}(T_i) \neq \emptyset$, then $\bigcap_{i=1}^\infty \text{Fix}(T_i) = \text{Fix}(T)$.*

For nonexpansive mappings, we have the following known result.

Theorem 1.9.1. *Let X be a Banach space and let X^* be the dual space of X. Let C be a nonempty, closed, and convex subset of X, and let $T : C \to C$ be a nonexpansive mapping. Then there exists some $f \in S(X^*)$ such that*

$$\lim_{n \to \infty} f\left(\frac{T^n x}{n}\right) = \lim_{n \to \infty}\left\|\frac{T^n x}{n}\right\| = \inf_{y \in C}\{\|y - Ty\|\} \triangleq d, \quad \forall x \in C.$$

For averaged mappings, we have the following known result.

Theorem 1.9.2. *Let X be a Banach space and let C be a nonempty, closed, and convex subset of X. Let $T : C \to C$ be an averaged mapping. Then*

$$\lim_{n}\|T^{n+1}x - T^n x\| = \frac{\lim_{n}\|T^{n+k}x - T^n x\|}{k} = \lim_{n}\left\|\frac{T^n x}{n}\right\| = d, \quad \forall k \geq 1, x \in C,$$

where d is defined as above.

If $\{T^n x\}$ is bounded, then

$$T^{n+1}x - T^n x \to \theta \quad \text{as } n \to \infty.$$

In this case, we say that T is asymptotically regular at $x \in C$. If T is asymptotically regular at every point of C, then T is asymptotically regular on C. The asymptotical regularity is a necessary, but not sufficient condition for convergence of $\{T^n x\}$.

Example 1.9.1. Let $X = l^1$ and let $C = \{x_n \in l^1 : x_n = (1, \frac{1}{2}, \ldots, \frac{1}{n}, 0, \ldots)\}$. Define $Tx_n = x_{n+1}$. Then $T : C \to C$ is a nonexpansive mapping, which is also asymptotically regular. Since

$$\|x_n\| = \sum_{i=1}^{n} \frac{1}{i} \to \infty,$$

we see that $\{T^n x\}$ is not convergent for all $x \in C$.

Theorem 1.9.3 (Browder [7], 1965). *Let C be a nonempty convex and closed subset of a uniformly convex Banach space X. Let $T : C \to X$ be a nonexpansive mapping. Then, for any $\{x_n\} \subset C$ satisfying $x_n \rightharpoonup x$ and $(I - T)x_n \to y \in X$, we have $x \in C$ and $(I - T)x = y$. In addition, $(I - T)(C)$ is a closed subset of X. We say that $I - T$ is demiclosed at $y \in X$. In particular, if $y = \theta$, then $I - T$ is demiclosed at the origin θ.*

Using the above demiclosedness principle, the following celebrated Brower fixed-point theorem is not hard to derive.

Theorem 1.9.4 (Browder [7], 1965). *Let X be a uniformly convex Banach space. Let C be a nonempty, bounded, closed, and convex subset of X, and let $T : C \to C$ be a nonexpansive mapping. Then $\mathrm{Fix}(T)$ is a nonempty and closed convex subset of C.*

In 1965, Kirk proved the following existence result.

Theorem 1.9.5 (Kirk [38], 1965). *Let X be a reflexive Banach space which has the normal structure. Let C be a nonempty, bounded, closed, and convex subset of X, and let $T : C \to C$ be a nonexpansive mapping. Then T has a fixed point.*

Since every nonempty, closed, convex, and bounded subset of a uniformly convex Banach space has the normal structure, Kirk's existence theorem generalizes the Browder's existence theorem. Let D be a nonempty set of X. Indeed, compact convex sets always have the normal structure, however, Alspach's example [3] shows that weakly compact convex sets need not have the normal structure. Recall that $D \subset C$ is said to have the fixed point property (in short, f. p. p.) if every nonexpansive mapping defined on weakly compact convex subsets of D has fixed points. Uniformly convex Banach spaces have the fixed point property and reflexive Banach spaces with the normal structure also have the fixed point property. A Banach space X is said to have the weak fixed point property if for each weakly compact convex subset $D \subset E$, and a nonexpansive mapping $T : D \to D$, D contains a fixed point for T.

Theorem 1.9.6. *Let X be a Banach space and let C be a nonempty weakly compact convex subset of X, which also has the normal structure. Let $T = \{T_1, T_2, \dots, T_r\}$ be a finite commutative family of nonexpansive mappings from C to itself. Then there exists $x_0 \in C$ such that $x_0 = T_i x_0$, $i = 1, 2, \dots, r$, that is, $\bigcap_{i=1}^r \mathrm{Fix}(T_i) \neq \emptyset$.*

Theorem 1.9.7. *Let X be a strictly convex Banach space and let C be a weakly compact convex subset of X, which also has the normal structure. Let $\mathbb{R} = \{T_i : i \in \Lambda\}$ be a commutative family of nonexpansive mappings from C to itself. Then there exists $x_0 \in C$ such that $x_0 = T_i x_0$, $i \in \Lambda$, that is, $\bigcap_{i \in \Lambda} \mathrm{Fix}(T_i) \neq \emptyset$.*

Let $\{x_n\}$ be a sequence in X. We use $\omega_w(x_n)$ to denote the weak limit set of $\{x_n\}$, that is,

$$\omega_w(x_n) = \{x \in X : \exists \{x_{n_i}\} \subset \{x_n\} \text{ such that } x_{n_i} \rightharpoonup x\}.$$

Let X be a Banach space and let C be a nonempty, closed, and convex subset of X. Let $T : C \to C$ is an averaged nonexpansive mapping with $\mathrm{Fix}(T) \neq \emptyset$. From the above theorems, we see that, for any $x \in C$, $T^{n+1}x \to T^n x \to \theta$. Putting $x_n = T^n x$, we have $Tx_n - x_n \to \theta$. Sequence $\{x_n\}$ is called a sequence of an approximate fixed points of T. Let X be a uniformly convex Banach space. Form the Browder demiclosedness principle, we see that $\omega_w(x_n) \subset \mathrm{Fix}(T)$. If $\omega_w(x_n)$ is a singleton, then $\{x_n\}$ converges weakly to some fixed point of T. Hence, it is necessary to give conditions that guarantee $\omega_w(x_n)$ is a singleton.

Theorem 1.9.8. *Let C be a nonempty closed convex subset of a Banach space X. Let $\{T_n\}_{n \geq 1} : C \to C$ be a family of Lipschitz mappings with the Lipschitz coefficient $L_n \geq 1$, for each $n \geq 1$. Assume that $\bigcap_{n \geq 1} \mathrm{Fix}(T_n)$ is nonempty and $\sum_{n \geq 1}(L_n - 1) < \infty$. Let $\{x_n\}$ be a sequence generated by the iterative process $x_1 \in C$, $x_{n+1} = T_n x_n$. Then*

(i) *if X is a uniformly convex Banach space, then, $\forall u, v \in \bigcap_{n\geq 1} \text{Fix}(T_n)$, $\forall t \in [0,1]$, $\lim_n \|tx_n + (1-t)u - v\|$ exists;*

(ii) *if X is a uniformly convex Banach space with the F-differentiable norm, then $\forall u, v \in \bigcap_{n\geq 1} \text{Fix}(T_n)$, $\lim_n \langle x_n, j(u-v) \rangle$ exists. In particular, $\langle p-q, j(u-v) \rangle = 0, \forall p, q \in \omega_w(x_n)$. If $\omega_w(x_n) \subset \bigcap_{n\geq 1} \text{Fix}(T_n)$, then $p = q$. So $\omega_w(x_n)$ is a singleton.*

Theorem 1.9.9. *Let X be a reflexive Banach space and let X^* be the dual space. Assume that X^* has the Kadec–Klee property and $\{x_n\}$ is a bounded sequence in X. If, for some $u, v \in \omega_w(x_n)$, $\lim_{n\to\infty} \|tx_n + (1-t)u - v\|$, $\forall t \in [0,1]$ exists, then $u = v$.*

Let X be a uniformly convex Banach space and let X^* be its dual space. Assume that X^* has the Kadec–Klee property and $\omega_w(x_n) \subset \bigcap_{n\geq 1} \text{Fix}(T_n)$. Then $\omega_w(x_n)$ must be a singleton.

Definition 1.9.2. Let C be a nonempty closed convex subset of a Banach space X. If $\|x_{n+1} - u\| \leq \|x_n - u\|$, $\forall u \in C$, then the sequence $\{x_n\} \subset X$ is said to be Fejér monotone on C.

Theorem 1.9.10. *Let X be a Banach space satisfying the Opial condition and let C be a nonempty, closed, and convex subset of X. Assume that $\{x_n\} \subset K$ is Fejér monotone on C, and $\omega_w(x_n) \subset C$. Then $\omega_w(x_n)$ is a singleton.*

1.9.2 Accretive operators

The class of accretive operators was introduced independently by Browder [8] and Kato [37] in 1967. Interest in accretive operators stems mainly from their connections with the existence theory of solutions for nonlinear evolution equations in Banach spaces. It is well known that many real problems with physical background can be modeled in terms of an initial value problem of the form:

$$\begin{cases} \frac{du}{dt} + Au = \theta, \\ u(0) = u_0, \end{cases} \tag{IVP}$$

where A is an accretive operator in an appropriate Banach space. Typical examples of such evolution equations may be found in modes involving the heat, wave or Schrödinger equations (see, e. g., [112, 113] and the references therein).

An early significant and fundamental result in the theory of accretive operators, due to Browder [8], states that equation (IVP) is solvable when A is locally Lipschitz continuous and accretive on a Banach space X. Utilizing the existence result for equation (IVP), Browder [8] further proved that if $A : X \to X$ is locally Lipschitz continuous and accretive, then A is m-accretive. Afterwards, Martin [48] extended the result of Browder to the case where A is continuous and accretive. Deimling [27], Kartsos [36],

Morales [51, 52], and others extensively developed the results of Browder and Martin. Within the past 50 years or so, the accretive operator theory has been developed rapidly and extensively. The achievements of the accretive operator theory are rich and colorful. It is very difficult for one to cover every aspect of the theory. Now we pick up some important and basic results which will be needed in the subsequent chapters.

Let X be a Banach space. Let $A : \text{Dom}(A) \subset X \to 2^X$ be a set-valued mapping. Sometimes, we write set-valued mapping by $A \subset X \times X$.

$$\text{Dom}(A) = \{x \in X : Ax \neq \emptyset\}, \quad \text{Ran}(A) = \cup\{Ax : x \in \text{Dom}(A)\},$$
$$\text{Graph}(A) = \{[x, y] \in X \times X : x \in \text{Dom}(A), y \in A(x)\}$$

Definition 1.9.3. Let $A \subset X \times X$ be a set-valued mapping. Then A is said to be
(i) accretive if, for all $x, y \in \text{Dom}(A)$,

$$\|x - y\| \leq \|x - y + s(u - v)\|, \quad \forall s > 0, \forall u \in Ax, \forall v \in Ay.$$

Equivalently, A is accretive if, for all $x, y \in \text{Dom}(A)$ there exists some $j(x - y) \in J(x - y)$ such that

$$\langle u - v, j(x - y) \rangle \geq 0, \quad \forall u \in Ax, \forall v \in Ay.$$

(ii) η-strongly accretive if there exist a positive real number η and $j(x - y) \in J(x - y)$ such that, for all $x, y \in \text{Dom}(A)$,

$$\langle u - v, j(x - y) \rangle \geq \eta\|x - y\|^2, \quad \forall u \in Ax, \forall v \in Ay.$$

(iii) v-inverse-strongly accretive if there exist a positive real number v and $j(x - y) \in J(x - y)$ such that, for all $x, y \in \text{Dom}(A)$,

$$\langle u - v, j(x - y) \rangle \geq v\|u - v\|^2, \quad \forall u \in Ax, \forall v \in Ay.$$

(iv) g-strongly accretive if there exist a function $g : \mathbb{R}^+ \to \mathbb{R}^+$ with $g(0) = 0, \forall r_0 > 0$, $\lim\inf_{r \to r_0} g(r) > 0, \lim\sup_{r \to \infty} g(r) > 0$ and $j(x - y) \in J(x - y)$ such that, for all $x, y \in \text{Dom}(A)$,

$$\langle u - v, j(x - y) \rangle \geq g(\|x - y\|)\|x - y\|, \quad \forall u \in Ax, \forall v \in Ay.$$

(v) m-accretive if A is accretive and there exists some $\lambda > 0$ such that $Ran(I + \lambda A) = X$.
(vi) maximal accretive if A is accretive and for any accretive operator $B \supset A$, we have $A = B$.

Theorem 1.9.11. *A set-valued mapping A is m-accretive if and only if A is accretive and* $\text{Ran}(I + \lambda A) = X, \forall \lambda > 0$.

Example 1.9.2. Let C be a nonempty, closed, and convex subset of a Banach space X, and let $T : C \to C$ be a nonexpansive operator. Let $A := I - T$. Then $A : C \to X$ is an accretive mapping and

$$C = \text{Dom}(A) \subset \bigcap_{r>0} \text{Ran}(I + rA).$$

In such a case, we say that A satisfies the "range condition".

Example 1.9.3. Let H be a Hilbert space and let $f : H \to \overline{\mathbb{R}}$ be a proper, lower semi-continuous convex function. Then $\partial f : H \to 2^H$ is an m-accretive operator.

Let $A \subset X \times X$ be an accretive operator and let $J_\lambda := (I + \lambda A)^{-1}$, where λ is a positive real number. Then $J_\lambda : \text{Ran}(I + \lambda A) \to \text{Dom}(A)$ is said to be the resolvent of A, denoted by $A_\lambda := \lambda^{-1}(I - J_\lambda)$, $\forall \lambda > 0$; $A_\lambda : \text{Ran}(I + \lambda A) \to \text{Dom}(A)$ is said to the Yosida approximation. Both J_λ and A_λ are single-valued. Also $\text{Dom}(J_\lambda) = \text{Ran}(I + \lambda A)$ and $\text{Ran}(J_\lambda) = \text{Dom}(A)$.

We list some basic properties of J_λ and A_λ as follows.

Theorem 1.9.12.

(i) $\|J_\lambda x - J_\lambda y\| \leq \|x - y\|$, $\forall x, y \in \text{Ran}(I + \lambda A)$.

(ii) A_λ is accretive and

$$\|A_\lambda x - A_\lambda y\| \leq \frac{2}{\lambda}\|x - y\|, \quad \forall x, y \in \text{Ran}(I + \lambda A).$$

(iii) $A_\lambda x \in AJ_\lambda x$, $\forall x \in \text{Ran}(I + \lambda A)$.

(iv) $\|A_\lambda x\| \leq |Ax|$, $\forall x \in \text{Dom}(A) \cap \text{Ran}(I + \lambda A)$, where $|Ax| = \inf\{\|z\| : z \in Ax\}$.

(v) $A^{-1}(\theta) = \{x \in \text{Dom}(A) : \theta \in A(x)\} = \text{Fix}(J_\lambda)$, $\forall \lambda > 0$.

(vi) For all $0 < \mu \leq \lambda$, and $x \in \text{Ran}(I - \lambda A)$, we have $\|A_\lambda x\| \leq \|A_\mu x\|$.

(vii) For all $\lambda, \mu > 0$ and $x \in \text{Dom}(J_\lambda)$, we have

$$\frac{\mu}{\lambda}x + \frac{\lambda - \mu}{\lambda}J_\lambda x \in \text{Dom}(J_\mu), \quad J_\lambda x = J_\mu\left(\frac{\mu}{\lambda}x + \frac{\lambda - \mu}{\lambda}J_\lambda x\right).$$

(viii) Let $c_2 \geq c_1 > 0$. Then $\|J_{c_1}x - x\| \leq \|J_{c_2}x - x\|$, $\forall x \in \text{Dom}(J_\lambda)$.

(ix) Let $\overline{\text{Dom}(J_\lambda)} \subset \bigcap_{\lambda>0} \text{Ran}(I - \lambda A)$. Then $\forall x \in \text{Dom}(A)$, $\lambda, \mu > 0$, and $m, n > 0$,

$$\|J_\lambda^n x - J_\mu^m x\| \leq [(n\lambda - m\mu)^2 + n\lambda^2 + m\mu^2]^{\frac{1}{2}}|Ax|.$$

Theorem 1.9.13 (Reich [74], 1980; Takahashi [88], 1984). *Let X be a reflexive Banach space whose norm is uniformly G-differentiable. Let $A \subset X \times X$ be an accretive operator satisfying the range condition $\overline{\text{Dom}(A)} \subset \bigcap_{r>0} \text{Ran}(I + rA)$. Let C be a nonempty, closed, and convex subset of X satisfying $\overline{\text{Dom}(A)} \subset C \subset \bigcap_{r>0} \text{Ran}(I + rA)$. Suppose that, for nonexpansive mappings, each weakly compact convex subset of C has the f. p. p., $\theta \in \text{Ran}(A)$. Then, for any $x \in C$, $\lim_{t\to\infty} J_t x$ exists and belongs to $A^{-1}(\theta)$. In addition, if we define $Qx = \lim_{t\to\infty} J_t x$, then Q is a unique sunny nonexpansive retraction from Q onto $A^{-1}(\theta)$.*

Theorem 1.9.14 (Morales and Chidume [53], 1999). *Let X be a Banach space and let C be a nonempty, closed, and convex subset of X. Let A : C → X be a continuous g-strongly accretive operator satisfying the "flow-invariance" condition:*

$$\lim_{\lambda \to 0^+} \frac{d(x - \lambda Ax, C)}{\lambda} = 0.$$

If there exists some point $x_0 \in C$ such that $\liminf_{r \to r_0} g(r) > \|Ax_0\|$, then there exists a unique $z \in C$ such that $Az = \theta$.

Theorem 1.9.15 (Martin [47], 1973). *Let X be a Banach space and let D be a nonempty closed set of X. Let B : D → X be a continuous η-strongly accretive mapping. If B satisfies the "flow-invariance" condition*

$$\lim_{h \to 0^+} \frac{d(x - hBx, D)}{h} = 0, \quad \forall x \in D,$$

then there exists a unique $x^* \in D$ such that $Bx^* = \theta$.

Theorem 1.9.16 (Deimling [27], 1974). *Let X be a Banach space and let D be a nonempty closed set of X. Let A : D → X be a continuous mapping. If*

$$\langle Ax - Ay, j(x - y) \rangle \geq \alpha(\|x - y\|)\|x - y\|, \quad \forall x, y \in D, j(x - y) \in J(x - y),$$

where $\alpha : \mathbb{R}^+ \to \mathbb{R}^+$ is a continuous function, and
(i) $\alpha(0) = 0$;
(ii) $\alpha(r) > 0, \forall r > 0$;
(iii) $\alpha(r) \to \infty$ *as* $r \to \infty$;
(iv) $\liminf_{r \to 0} \alpha(r)/r > 0$.
 If

$$\lim_{h \to 0^+} \frac{d(x - hBx, D)}{h} = 0, \quad \forall x \in D,$$

then there exists a unique $x^* \in D$ such that $Ax^* = \theta$.

Theorem 1.9.17 (Ray [71], 1980). *Let X be a Banach space and let D be a nonempty, closed, and convex subset of X. Assume that D has the f. p. p. for nonexpansive mappings, and A : D → X is a continuous accretive operator. If A satisfies the "flow-invariance" condition, then $\theta \in A(D)$.*

Generally speaking, the sum of two accretive operators may not be accretive in a Banach space since we cannot find the same $j(x - y) \in J(x - y)$ for the two accretive operators at the same time. Morales [54] proved the following result.

Theorem 1.9.18 (Morales [54], 2007). *Let X be a Banach space and let C be a nonempty convex subset of X. Let A : C → X be a continuous accretive operator. If A satisfies the* *"flow-invariance" condition:*

$$\lim_{h \to 0^+} \frac{d(x - hAx, C)}{h} = 0, \quad \forall x \in C,$$

then $\langle Ax - Ay, j(x - y) \rangle \geq 0, \forall x, y \in C, \forall j(x - y) \in J(x - y)$.

Theorem 1.9.19 (Garcéa-Falset and Morales [29], 2005). *Let X be a Banach space. Let $A \subset X \times X$ be an m-accretive operator and let $B : \mathrm{Dom}(B) = X \to X$ be a continuous g-strongly accretive operator. Then, $\forall \mu, \lambda > 0$, both $A + \mu B$ and $B + \lambda A$ are surjective, that is, $\mathrm{Ran}(A + \mu B) = \mathrm{Ran}(B + \lambda A) = X$. Hence, for any $\lambda > 0$, there exists a unique continuous path $\{x_\lambda\} \subset X$ such that, $\forall z \in X, z \in Bx_\lambda + \lambda Ax_\lambda$.*

Theorem 1.9.20. *Let X be a uniformly smooth Banach space. Let $A \subset X \times X$ be an m-accretive operator and let $B : \mathrm{Dom}(B) = X \to X$ be a demicontinuous accretive operator. Then $A + B$ is m-accretive.*

1.9.3 Monotone mappings

The terminology of mappings of monotone type was introduced by Browder, Minty, and Brézis et al. in the early 1960s. The main motivation for the study of mappings of monotone type was inspired by a lot of problems from pure mathematics and applied sciences. The interest and importance of this class of mappings lies in the fact that both theoretical and practical problems can be modeled in terms of mappings of monotone type, and solving these problems usually reduces to finding a zero of a maximal monotone mapping. Examples of mappings of monotone type are rich and countless, for instance, under appropriate assumptions on a Banach space X, the normalized duality mapping $J : X \to 2^{X^*}$ is monotone, strictly monotone, η-strongly monotone, φ-strongly monotone, and maximal monotone. Another important example is the subdifferential ∂f of a lower semicontinuous proper convex function $f : X \to \overline{\mathbb{R}}$. It is well known that if a Banach space X is reflexive, then ∂f is maximal monotone. More examples of mappings of monotone type can be found in Rockafellar [76, 77], Browder [8], Pascali and Sburlan [61], and Zeidle [112, 113]. The achievements of the theory of mappings of monotone type are flourishing and perfected, especially in reflexive Banach spaces. Indeed, the theory of monotone mappings, which has a close relation with optimization theory, variational inequalities, and equilibrium problems, finds lots of applications in the boundary value problems of elliptic partial differential equations and parabolic partial differential equations, in the solubility of the Hammerstein non-linear integral equations.

In this subsection, we pick up some basic concepts and important results for mappings of monotone type, which will be needed in the subsequent chapters. All of the following concepts and results can be found in the references mentioned above.

In the sequel, we always use X to denote a real Banach space and X^* the topological conjugate of X, unless otherwise indicated. Also, we use $A \subset X \times X^*$ to denote a multi-valued mapping from X into 2^{X^*}, and $A : X \to X^*$ a single-valued mapping from X into X^*.

Definition 1.9.4. Let X be a real Banach space and let X^* be the conjugate space of X. Let $T \subset X \times X^*$ be a mapping. Then T is said to be
(i)   monotone if, for all $x, y \in \mathrm{Dom}(T)$ and $u \in Tx$, $v \in Ty$,

$$\langle u - v, x - y \rangle \geq 0.$$

If $\langle u - v, x - y \rangle = 0$ implies $x = y$, then T is said to be strictly monotone.
(ii) η-strongly monotone if there exists a positive real number $\eta > 0$ such that, for all $x, y \in \mathrm{Dom}(T)$ and $u \in Tx$, $v \in Ty$,

$$\langle u - v, x - y \rangle \geq \eta \|x - y\|^2.$$

(iii) φ-strongly monotone if there exists a strictly increasing function $\varphi : \mathbb{R}^+ \to \mathbb{R}^+$ with $\varphi(0) = 0$ such that $\langle u - v, x - y \rangle \geq \varphi(\|x - y\|)\|x - y\|$ for all $x, y \in \mathrm{Dom}(T)$, $u \in Tx$, $v \in Ty$;
(iv) v-inverse-strongly monotone if there exists a positive real number $v > 0$ such that, for all $x, y \in \mathrm{Dom}(T)$ and $u \in Tx$, $v \in Ty$,

$$\langle u - v, x - y \rangle \geq v \|u - v\|^2.$$

(v)   maximal monotone if $[x, y] \in X \times X^*$, $\langle x - u, y - v \rangle \geq 0$, for all $[u, v] \in \mathrm{Graph}(T) \Rightarrow [x, y] \in \mathrm{Graph}(T)$.
(vi) generalized pseudomonotone if $[x_n, f_n] \in \mathrm{Graph}(T)$, $x_n \to x$, $f_n \rightharpoonup f$ and

$$\limsup_{n \to \infty} \langle f_n, x_n - x \rangle \leq 0 \text{ implies that } [x, f] \in \mathrm{Graph}(T) \quad \text{and} \quad \lim_{n \to \infty} \langle f_n, x_n \rangle = \langle f, x \rangle;$$

(vii) of type (M) if $[x_n, f_n] \in \mathrm{Graph}(T)$, $x_n \rightharpoonup x$, $f_n \to f$, and $\limsup_{n \to \infty} \langle f_n, x_n - x \rangle \leq 0$ imply $[x, f] \in \mathrm{Graph}(T)$;

Clearly, the zero mapping defined on the whole space X is maximal monotone, however, the zero mapping defined on a proper subset of X is monotone but not necessarily maximal monotone.

In order to introduce the concept of pseudomonotone mappings, we give the so-called upper semicontinuity concept for a multi-valued mapping in a general topological space.

Definition 1.9.5. Let X and Y be topological spaces and let $T \subset X \times Y$ be a multi-valued mapping. Then T is said to be upper semicontinuous at $x \in \mathrm{Dom}(T)$ if, for any neighborhood V of $Tx \subset Y$, there exists some neighborhood U of X such that

$$T(U) = \{f \in Y : f \in Ty, y \in \mathrm{Dom}(T) \cap U\} \subset V.$$

Furthermore, T is said to be upper semicontinuous if T is upper semicontinuous at every $x \in \text{Dom}(T)$.

We remark here that a single-valued mapping is upper semicontinuous if and only if it is continuous.

Definition 1.9.6. Let X be a real reflexive Banach space and let X^* be the topological conjugate of X. A multi-valued mapping $T \subset X \times X^*$ is said to be pseudomonotone if the following conditions are satisfied:

(C1) for every $x \in \text{Dom}(T)$, Tx is a closed convex subset of X^*;
(C2) $\{x_n\} \subset \text{Dom}(T)$, $x_n \rightharpoonup x$, $f_n \in Tx_n$ and $\limsup_{n \to \infty} \langle f_n, x_n - x \rangle \le 0$ imply that, $\forall y \in \text{Dom}(T)$, there exists $f(y) \in Tx$ such that

$$\langle f(y), x - y \rangle \le \liminf_{n \to \infty} \langle f_n, x_n - y \rangle;$$

(C3) for every subspace W of X with $\dim W < \infty$, the restriction $T|_W : \text{Dom}(T) \cap W \to 2^{X^*}$ is upper semicontinuous in the w-topology $\sigma(X^*, X)$.

Mappings defined above are said to be of monotone type according to their usage. By means of above definitions, it is not difficult to show the following implications: (here we assume that X is a real reflexive Banach space, and $T \subset X \times X^*$ is a mapping such that $\text{Dom}(T) = X$):

(R$_1$) T being maximal monotone implies that T is pseudomonotone;
(R$_2$) T being pseudomonotone implies that T is generalized pseudomonotone;
(R$_3$) T being generalized pseudomonotone implies that T is of type (M);
(R$_4$) T being maximal monotone implies that T is monotone;
(R$_5$) T being η-strongly monotone implies that T is φ-strongly monotone; furthermore, being both η-strongly monotone and φ-strongly monotone implies being strictly monotone;
(R$_6$) T being v-inverse-strongly monotone implies that T is $1/v$-Lipschitz continuous and monotone;
(R$_7$) T being strictly monotone implies that T is monotone.

We remark in passing that the converse implications above do not hold in general.

Now we present some examples of monotone type mappings.

Example 1.9.4. Let $\varphi : \mathbb{R} \to \mathbb{R}$ be a monotonically increasing function. Then, the mapping $T : \mathbb{R} \to 2^{\mathbb{R}}$ defined by

$$Tx = [\varphi(x - 0), \varphi(x + 0)], \quad \forall x \in \mathbb{R},$$

is maximal monotone, where $\varphi(x - 0)$ and $\varphi(x + 0)$ stand for the left- and right-limit of φ at x, respectively.

Example 1.9.5. Let $X = H = l^2$ and define a mapping $T : l^2 \to l^2$ by

$$Tx_n = \alpha_n x_n, \quad \forall \{x_n\} \in l^2,$$

where $\alpha_n > 0$ and $\lim_{n \to \infty} \alpha_n = 0$. Then T is strictly monotone, but not strongly monotone.

Example 1.9.6. Let $X = \mathbb{R}$ and define a mapping $T : \mathbb{R} \to \mathbb{R}$ by

$$Tx = \begin{cases} t, & t \leq 0, \\ t + 1, & t > 0. \end{cases}$$

Then $T : \mathbb{R} \to \mathbb{R}$ is monotone, but not maximal monotone.

Example 1.9.7. Let A be a semipositive definite matrix of order n, $n \in \mathbb{N}$. Then A is monotone, and A is η-strongly monotone if A is a positive definite matrix of order n, where $\eta = \min_{1 \leq i \leq n} \{\lambda_i\}$, $\{\lambda_i\}$ are all eigenvalues of A.

Example 1.9.8. Let X be a real reflexive Banach space. Let $\varphi : X \to \overline{\mathbb{R}}$ be a lower semicontinuous, proper, and convex function. Then $\partial \varphi \subset X \times X^*$ is maximal monotone.

Example 1.9.9. Let $X = H$ be a real Hilbert space and let $T : H \to H$ be a nonexpansive mapping. Set $A = I - T$. Then $A \subset H \times H$ is a 2-Lipschitz continuous and $\frac{1}{2}$-inverse-strongly monotone mapping.

Example 1.9.10. Let X be a real Banach space and let $J \subset X \times X^*$ be the normalized duality mapping. Then the following assertions are known:
(a_1) J is monotone;
(a_2) if X is reflexive, then J is surjective, i. e., $JX = X^*$;
(a_3) if X is smooth, then $J : X \to X^*$ is single-valued;
(a_4) if X is strictly convex, then J is strictly monotone;
(a_5) if X is reflexive and smooth, then J is maximal monotone;
(a_6) if X is real uniformly convex, then J is ω_R-strongly monotone on any closed ball

$$B_R[\theta] = \{x \in X : \|x\| \leq R\}, \quad R > 0,$$

where $\omega_R : \mathbb{R} \to \mathbb{R}$ is a function given in Theorem 1.8.16.
(a_7) let $X = L^p$, $1 < p \leq 2$. Then J is η-strongly monotone with $\eta = p - 1 \in (0, 1]$.

For the generalized duality mapping J_p, we have similar conclusions.

Example 1.9.11. Let X be a real Banach space. Let C be a nonempty, closed, and convex subset of X, and let $A : C \to X^*$ be a hemicontinuous and monotone mapping. Then the mapping $T \subset X \times X^*$ defined by

$$Tx = \begin{cases} N_C(x) + Ax, & x \in C, \\ \emptyset, & x \notin C \end{cases}$$

is maximal monotone. Furthermore, $N(T) = VI(C, A)$, where

$$N_C(x) = \{x \in X^* : \langle y - x, x^* \rangle \leq 0, y \in C\}$$

denotes the normal cone of C at x,

$$VI(C, A) = \{u \in C : \langle Au, y - u \rangle \geq 0 \ \forall y \in C\}$$

denotes the solution set of the variational inequality problem

$$\langle Au, y - u \rangle \geq 0 \quad \forall y \in C.$$

Now we have the following observations:

(O_1) under translation and positive multiplicative transformations, the monotonicity of a mapping of monotone type stays unchanged; more precisely, if A enjoys some kind monotonicity, then the mapping $\hat{A} \subset X \times X^*$ defined by

$$\hat{A}x = A(x + x_0), \quad x \in \mathrm{Dom}(\hat{A}) = \mathrm{Dom}(A) - x_0$$

and $x_0 \in \mathrm{Dom}(A)$ is a fixed vector, enjoys the same kind monotonicity; for a fixed vector $z \in X^*$, the mapping $\tilde{A} \subset X \times X^*$ defined by

$$\tilde{A}x = Ax - z, \quad x \in \mathrm{Dom}(\tilde{A}) = \mathrm{Dom}(A),$$

enjoys the same kind monotonicity; for $\lambda > 0$, λA enjoys same kind monotonicity.

(O_2) if $A_i \subset X \times X^*$ ($i = 1, 2, \dots, n$) are monotone (pseudomonotone), so is $\sum_{i=1}^{n} A_i$;

(O_3) if X is a real reflexive Banach space, then A is monotone (maximal monotone) if and only if A^{-1} is monotone (maximal monotone).

In order to study and understand various properties of monotone type mappings, it is necessary to give some related concepts.

Definition 1.9.7. A mapping $A \subset X \times X^*$ is said to be locally bounded at $x_0 \in \mathrm{Dom}(A)$ if there exists some neighborhood U of x_0 such that

$$A(U) = \{f \in X^* : [x, f] \in \mathrm{Graph}(A), x \in U\}$$

is a bounded subset of X^*; A is said to be locally bounded on $\mathrm{Dom}(A)$ if A is locally bounded at every $x \in \mathrm{Dom}(A)$; $A^{-1} \subset X^* \times X$ is said to be locally bounded on X^* if for any $f \in X^*$ there exist some $r > 0$ and open ball $B_r(f)$ such that set $\{x \in X : Ax \cap B_r(f) \neq \emptyset\}$ is a bounded subset of X.

It is evident that A is locally bounded at $x_0 \in \mathrm{Dom}(A)$ if and only if for any $\{f_n\} \subset X^*, f_n \in Ax_0$, if $x_n \to x_0 \in \mathrm{Dom}(A)$, then $\{f_n\}$ remains bounded. It is also clear that A^{-1} is locally bounded on X^* if and only if for any $\{x_n\} \subset X, y_n \in Ax_n$, if $y_n \to f \in X^*$, then $\{x_n\}$ remains bounded.

We remark here that a mapping which is locally bounded on $\mathrm{Dom}(A)$ is certainly bounded in a normed linear space X with $\dim X < \infty$.

Definition 1.9.8. Let $A, B \subset X \times X^*$ be multi-valued mappings. Then B is said to be an extension of A, denoted by $A \subset B$, if $\mathrm{Dom}(A) \subset \mathrm{Dom}(B)$ and $Ax \subset Bx$ for all $x \in \mathrm{Dom}(A)$.

Definition 1.9.9. A set $M \subset X \times X^*$ is called monotone if

$$\langle f - g, x - y \rangle \geq 0, \quad \forall [x, f], [y, g] \in M.$$

Furthermore, M is called a maximal monotone set if M is a monotone set and it is not a proper subset of any monotone set in $X \times X^*$.

It is evident that $A \subset X \times X^*$ is monotone (maximal monotone) if and only if $\mathrm{Graph}(A)$ is a monotone (maximal monotone) set of $X \times X^*$.

Definition 1.9.10. Let C be a nonempty subset of X. Then a set $G \subset C \times X^*$ is said to be maximal monotone if it is monotone and it is not a proper subset of any monotone set in $C \times X^*$; $T : \mathrm{Dom}(T) \subset C \to 2^{X^*}$ is said to be maximal monotone with respect to C if $\mathrm{Graph}(T)$ is a maximal monotone set of $C \times X^*$.

Definition 1.9.11. A mapping $T \subset X \times X^*$ is said to be coercive if there exists a function $c : \mathbb{R}^+ \to \mathbb{R}$ with $c(r) \to \infty$ as $r \to \infty$ such that

$$\langle f, x \rangle \geq c(\|x\|)\|x\|, \quad \forall [x, f] \in \mathrm{Graph}(T).$$

Furthermore, T is said to be coercive with respect to $h \in X^*$ if there exists $r > 0$ such that

$$\langle f - h, x \rangle > 0, \quad \text{for all } [x, f] \in \mathrm{Graph}(T) \text{ and } \|x\| > r.$$

Using these concepts, we can state some fundamental and important results from the theory of mappings of monotone type.

Theorem 1.9.21. *Let $A \subset X \times X^*$ be a monotone mapping. Then the following assertions hold true:*
(a_1) *A is locally bounded on $\mathrm{int}(\mathrm{Dom}(A))$;*
(a_2) *if $A : \mathrm{Dom}(A) \subset X \to X^*$ is single-valued and hemicontinuous, then A is demicontinuous on $\mathrm{int}(\mathrm{Dom}(A))$;*
(a_3) *A has a maximal monotone extension $\tilde{A}$; if X is a real reflexive Banach space and C a nonempty, closed, and convex subset of X, then there exists a maximal monotone extension $\tilde{A}$ of $A|_C$ such that $\mathrm{Dom}(\tilde{A}) \subset C$, where $A|_C$ is the restriction of A on C, i. e.,*

$$\mathrm{Graph}(A|_C) = \{[x, f] \in \mathrm{Graph}(A) : x \in C \cap \mathrm{Dom}(A)\};$$

(a_4) *if A is maximal monotone, then*
   (i) *A is $s\text{-}w^*$-upper semicontinuous on $\mathrm{int}(\mathrm{Dom}(A))$;*
   (ii) *$Ax = \bigcap_{[y,g] \in \mathrm{Graph}(A)} \{f \in X^* : \langle f - g, x - y \rangle \geq 0\}$ for all $x \in \mathrm{Dom}(A)$;*

(iii) *Ax is w^*-closed convex for all $x \in \mathrm{Dom}(A)$;*
(iv) *Ax is bounded for all $x \in \mathrm{int}(\mathrm{Dom}(A))$;*
(v) *Ax is w^*-compact for all $x \in \mathrm{int}(\mathrm{Dom}(A))$;*
(vi) *Graph(A) is demiclosed, i. e., either*
$\quad$ ($\mathrm{dc_1}$) *for any $\{x_n\} \subset \mathrm{Dom}(A)$, $x_n \rightharpoonup x, y_n \in Ax_n$ and $y_n \to y$ imply $[x,y] \in$*
$\quad$ *Graph(A) or*
$\quad$ ($\mathrm{dc_2}$) *for any $\{x_n\} \subset \mathrm{Dom}(A)$, $x_n \to x$, $y_n \in Ax_n$ and $y_n \rightharpoonup y$ imply $[x,y] \in$*
$\quad$ *Graph(A).*

In the following theorems, X is supposed to be a real reflexive Banach space.

Theorem 1.9.22. *Let $A \subset X \times X^*$ be a pseudomonotone mapping. Then Ax is bounded for all $x \in \mathrm{int}(\mathrm{Dom}(A))$.*

The next fundamental result is closely related to the variational inequality problems with multi-valued mappings of monotone type.

Theorem 1.9.23. *Let C be a closed convex subset of X. Let $A \subset X \times X^*$ be a monotone mapping with $[\theta, \theta^*] \in \mathrm{Graph}(A|_C)$ and let $P : C \to X^*$ be pseudomonotone, bounded, and coercive with respect to $h \in X^*$. Then there exists $x_0 \in C$ such that*

$$\langle f + Px_0 - h, x - x_0 \rangle \geq 0, \text{ for all } [x,f] \in \mathrm{Graph}(A|_C).$$

Thinking of (O_1), we see that condition "$[\theta, \theta^*] \in \mathrm{Graph}(A|_C)$" can be replaced by $\mathrm{Dom}(A) \cap C \neq \emptyset$.

Theorem 1.9.24. *Let $A \subset X \times X^*$ be a maximal monotone mapping and let C a closed convex subset of X with $\theta \in \mathrm{Dom}(A)$ and $\mathrm{Dom}(A) \subset C$. Let $P : C \to X^*$ be pseudomonotone, bounded, and coercive. Then, for any $h \in X^*$, there exists $x_0 \in C$ such that*

$$h \in (A + P)x_0,$$

which implies

$$\mathrm{Graph}(A + P) = X^*.$$

We remark that condition "$\theta \in \mathrm{Dom}(A)$" can be dropped.
In the following theorems, X is supposed to be real reflexive and smooth.

Theorem 1.9.25. *Let $A \subset X \times X^*$ be a maximal monotone mapping. Then*

$$\mathrm{Ran}(A + \lambda J) = X^*, \quad \forall \lambda > 0,$$

where $J : X \to X^*$ is the normalized duality mapping.

Theorem 1.9.26. *Let $A \subset X \times X^*$ be maximal monotone and coercive. Then $\mathrm{Ran}(A) = X^*$.*

Theorem 1.9.27. *Let $A : \text{Dom}(A) = X \to X^*$ be monotone hemicontinuous. Then A is maximal monotone.*

Theorem 1.9.28. *Let $A : \text{Dom}(A) = X \to X^*$ be hemicontinuous and φ-strongly monotone such that $\varphi(r) \to \infty$ as $r \to \infty$. Then $\text{Ran}(A) = X^*$.*

Theorem 1.9.29. *Let $A, B \subset X \times X^*$ be monotone. Then A is maximal monotone whenever $A + B$ is maximal monotone.*

In the following theorems, X is supposed to be real reflexive, smooth and strict convex.

Theorem 1.9.30. *Let $A \subset X \times X^*$ be monotone. If $\text{Ran}(A + \lambda J) = X^*$ for $\lambda > 0$, then $(A + \lambda J)^{-1} : X^* \to X$ is single-valued, demicontinuous, and maximal monotone.*

The next theorem presents a criterion for a monotone mapping to be maximal monotone.

Theorem 1.9.31. *Let $A \subset X \times X^*$ be a monotone mapping. Then A is maximal monotone if and only if there exists $\lambda > 0$ such that*

$$\text{Ran}(A + \lambda J) = X^*.$$

By means of Theorem 1.9.31, we can prove the following important result.

Theorem 1.9.32. *Let $A \subset X \times X^*$ be a maximal monotone mapping. Then the following conclusions hold true:*
$(c)_1$ *for arbitrarily given $x \in X$ and $\lambda > 0$, there exists a unique solution $R_\lambda x \in \text{Dom}(A)$ of the inclusion $\theta \in J(y - x) + \lambda Ay$, which yields*

$$R_\lambda x = (I + \lambda J^{-1} A)^{-1} x.$$

Write $A_\lambda x := \lambda^{-1} J(x - R_\lambda)x$. Then

$$A_\lambda x = (\lambda J^{-1} + A^{-1})^{-1} x, \quad A_\lambda x \in AR_\lambda x$$

and

$$J(R_\lambda x - x) + \lambda A_\lambda x = \theta.$$

The single-valued mappings $R_\lambda : X \to \text{Dom}(A)$ and $A_\lambda : X \to X^*$ are called resolvent and Yosida approximation of A, respectively.
$(c)_2$ $R_\lambda x \to x$ *on $\overline{\text{co}}(\text{Dom}(A))$ as $\lambda \to 0^+$, $\overline{\text{Dom}(A)}$ and $\overline{\text{Ran}(A)}$ are convex, $\|A_\lambda x\| \to \infty$ as $\lambda \to 0^+$ if $x \notin \overline{\text{Dom}(A)}$ and $A_\lambda x \to A°x$ on $\text{Dom}(A)$ as $\lambda \to 0^+$, where $A°x$ is the unique element of Ax having minimal norm, i. e., $\|A°x\| = d(\theta, Ax)$. Furthermore, A_λ is a bounded demicontinuous maximal monotone mapping with $A_\lambda(\theta) = \theta$ if $\theta \in A\theta$. If X^* is locally uniformly convex, then $A_\lambda x \to A°x$ on $\text{Dom}(A)$ as $\lambda \to 0^+$.*

Theorem 1.9.33. *Let X be a real reflexive, smooth, and strictly convex Banach space whose topological conjugate space X^* is locally uniformly convex. Let $A \subset X \times X^*$ be maximal monotone. Then the following statement holds: for any $x_\lambda, x \in \mathrm{Dom}(A)$, if $x_\lambda \to x$ as $\lambda \to 0^+$, $A^\circ x_\lambda \to A^\circ x$ as $\lambda \to 0^+$, then $A_\lambda x_\lambda \to A^\circ x$ as $\lambda \to 0^+$.*

Next theorem presents a criterion for a maximal monotone to be surjective.

Theorem 1.9.34. *Let X be a real reflexive, smooth, and strictly convex Banach space and let $A \subset X \times X^*$ be a maximal monotone mapping. Then $\mathrm{Ran}(A) = X^*$ if and only if A^{-1} is locally bounded on X^*.*

Theorem 1.9.35. *Let X be a real reflexive, smooth, and strictly convex Banach space. Let $A \subset X \times X^*$ be a maximal monotone mapping. Suppose that there exists $r > 0$ and $\bar{x} \in \mathrm{Dom}(A)$ such that*

$$\langle f, x - \bar{x} \rangle \geq 0, \quad \forall [x,f] \in \mathrm{Graph}(A) \text{ with } \|x - \bar{x}\| \geq r.$$

Then there exists $x_0 \in \mathrm{Dom}(A) \cap B_r[\bar{x}]$ satisfying

$$\theta^* \in Ax_0.$$

Usually, Theorem 1.9.35 is called the acute angle principle for monotone mappings. As a corollary of Theorem 1.9.35, we have the following acute angle principle for single-valued monotone mappings.

Theorem 1.9.36. *Let X be real reflexive, smooth, and strictly convex Banach space. Let $A : \mathrm{Dom}(A) = X \to X^*$ be monotone and hemicontinuous. Suppose that there exists $r > 0$ and $\bar{x} \in X$ such that*

$$\langle Ax, x - \bar{x} \rangle \geq 0, \quad \forall x \in \partial B_r(\bar{x}).$$

Then there exists $x_0 \in B_r[\bar{x}]$ satisfying $Ax_0 = \theta^*$.

Generally speaking, the sum of two maximal monotone mappings is not necessarily maximal monotone, however, we have the following known result.

Theorem 1.9.37 (Rockafellar [77]). *Let X be a real reflexive, smooth, and strictly convex Banach space with the topological conjugate space X^*. Let $A, B \subset X \times X^*$ be maximal monotone mappings. If one of the following conditions holds:*
(1) $\mathrm{Dom}(A) \cap \mathrm{int}(\mathrm{Dom}(B)) \neq \emptyset$;
(2) there exists $x \in \overline{\mathrm{Dom}(A)} \cap \overline{\mathrm{Dom}(B)}$ such that B is locally bounded at x,

then $A + B$ is maximal monotone.

The following three results, which are from Takahashi [87], are closely related to the monotone type single-valued variational inequality problems.

Theorem 1.9.38. *Let C be a nonempty convex subset of a topological linear space X, and let T be a hemicontinuous and monotone mapping of C into X^*. Let $x_0 \in C$ be an element of C. Then the following statements are equivalent each other:*

(1) $\langle Tx, x - x_0 \rangle \geq 0, \forall x \in C$;

(2) $\langle Tx_0, x - x_0 \rangle \geq 0, \forall x \in C$.

Theorem 1.9.39. *Let C be a nonempty compact convex subset of a topological linear space X, and let T be a monotone mapping of C into X^*. Then there exists $x_0 \in C$ such that $\langle Tx, x - x_0 \rangle \geq 0, \forall x \in C$.*

Theorem 1.9.40. *Let C be a nonempty compact convex subset of a topological linear space X, and let T be a monotone and hemicontinuous mapping of C into X^*. Then there exists $x_0 \in C$ such that $\langle Tx_0, x - x_0 \rangle \geq 0, \forall x \in C$.*

As a consequence of Theorem 1.9.40, we have the following known result.

Theorem 1.9.41. *Let C be a nonempty bounded and closed convex subset of a real reflexive Banach space X, and let T be a monotone and hemicontinuous mapping of C into X^*. Then there exists $x_0 \in C$ such that $\langle Tx_0, x - x_0 \rangle \geq 0 \ \forall x \in C$.*

Applying Theorem 1.9.41 to a closed ball $B_r[\theta]$, $r > 0$, we can obtain the following interesting result.

Theorem 1.9.42. *Let X be a real reflexive and smooth Banach space. Let $T : B_r[\theta] \to X^*$ be monotone and hemicontinuous. Then the following conclusions hold:*

(1) *there exists $x_0 \in B_r[\theta]$ such that*

$$\langle Tx_0, x - x_0 \rangle \geq 0 \quad \forall x \in B_r[\theta];$$

(2) *if $x_0 \in B_r(\theta)$, then $Tx_0 = \theta^*$;*

(3) *if $x_0 \in \partial B_r[\theta]$ and $Tx_0 \neq \theta^*$, then there exists $\lambda > 0$ such that $Tx_0 = -\lambda Jx_0$, equivalently, if for any $t \geq 0$ and $x \in \partial B_r[\theta]$, one has $Tx + tJx \neq \theta^*$, then $Tx_0 = \theta^*$.*

When C is not bounded, we have at least two ways to prove the existence of solutions for the above monotone type variational inequality problem. One way is to apply the following theorems and the other is to apply the previous Example 1.9.11.

Theorem 1.9.43 (Baiocchi and Capelo [5], 1984). *Let C be a nonempty, closed, and convex subset of a real reflexive Banach space X, and let T be a monotone and hemicontinuous mapping of C into X^*. Suppose that T is coercive in the following sense: there exists $w \in C$ such that*

$$\frac{\langle Tx, x - w \rangle}{\|x\|} \to \infty \quad as \ \|x\| \to \infty,$$

where $w \in C$ is a fixed element. Then the following conclusions hold:

(1) *for any $h \in X^*$, there exists $x_0 \in C$ such that*

$$\langle Tx_0 - h, x - x_0 \rangle \geq 0, \quad \forall x \in C;$$

(2) *if T is strictly monotone, then above $x_0 \in C$ is a unique.*

Theorem 1.9.44 (Baiocchi and Capelo [5], 1984). *Let C be a nonempty, closed, and convex subset of a real reflexive Banach space X, and let T be a η-strongly monotone and hemicontinuous mapping of C into X^*. Then for any $h \in X^*$, there exists a unique $x_0 \in C$ such that*

$$\langle Tx_0 - h, x - x_0 \rangle \geq 0, \quad \forall x \in C.$$

In 1966, Browder [9] established the following known result. In 1967, Mosco [55] gave a new proof to this fact by combining the previous a result due to Browder with the product-space technique.

Theorem 1.9.45. *Let T be a monotone hemicontinuous mapping from a real reflexive Banach space X to its dual space X^*. Let f be a proper, convex, and lower semicontinuous function from X to $\overline{\mathbb{R}}$ with $f(\theta) = 0$ such that*

$$\frac{\langle Tu, u \rangle + f(u)}{\|u\|} \to \infty \quad \text{as } \|u\| \to \infty.$$

Then for any given $h \in X^*$, there exists a solution $u_0 \in X$ of the variational inequalities

$$\langle Tu_0 - h, x - u_0 \rangle \geq f(u_0) - f(x) \quad \text{for all } x \in X.$$

In particular, If $f \equiv \theta$, then $h = Tu_0$. Hence $\mathrm{Ran}(T) = X^*$. If $T \equiv \theta$ and $h = \theta^*$, then

$$f(u_0) = \min_{x \in X} f(x).$$

Finally, we present more examples of monotone type mappings to support our conclusions.

Example 1.9.12. Let $X = \mathbb{R}^2$. Then the mapping defined by

$$\varphi(x, y) = xy^2(x^2 + y^4)^{-1}, \quad (x, y) \in \mathbb{R}^2,$$

is hemicontinuous, but not demicontinuous.

Example 1.9.13. Let $X = \mathbb{R}^2$, $n \geq 2$, $\mathrm{Dom}(A) = B_1(\theta)$, $\{x^i : i \in \mathbb{N}\} \subset \partial B_1(\theta)$. Then the mapping $A : B_1(\theta) \to \mathbb{R}^n$ by

$$Ax = \begin{cases} x, & x \neq x^i, \\ (i+1)x^i, & x = x^i, \end{cases}$$

is unbounded on $\partial B_1(\theta)$.

Example 1.9.14. Let $X = \mathbb{R}^n$ and let $f : \mathbb{R}^n \to \mathbb{R}$ be a function defined by

$$f(x) = \begin{cases} -(1 - \|x\|^2)^{\frac{1}{2}}, & \|x\| \leq 1, \\ \infty, & \|x\| > 1. \end{cases}$$

Then f is proper, lower semicontinuous, and convex function. It is also clear that $f(x)$ is differentiable as long as $\|x\| < 1$ and

$$\nabla f(x) = \frac{x}{\sqrt{1 - \|x\|^2}}.$$

Furthermore, $\partial f(x) = \emptyset$ as long as $\|x\| \geq 1$. Thus

$$\partial f(x) = \begin{cases} \left\{ \frac{x}{\sqrt{1-\|x\|^2}} \right\}, & \|x\| < 1, \\ \emptyset, & \|x\| \geq 1. \end{cases}$$

Notice that $\mathrm{Dom}(f) = B_1[\theta]$ and $\mathrm{Dom}(\partial f) = B_1(\theta)$. We see that $\partial f : B_1(\theta) \to \mathbb{R}^n$ is maximal monotone, unbounded, and $\mathrm{Ran}(\partial f) = \mathbb{R}^n$.

Example 1.9.15. Let $X = L^2(\mathbb{R})$, $\mathrm{Dom}(A) = \{x \in C(\mathbb{R}) : \lim_{t \to \infty} x(t) = 0 \text{ and } x' \in X\}$, $Ax = x'$, $\mathrm{Dom}(B) = \mathrm{Dom}(A)$, and $Bx = -x'$. Then both A and B are maximal monotone, however, the sum $A + B$ is not maximal monotone.

Example 1.9.16. Let X be a real Banach space and let $f : X \to \mathbb{R}^+$ be a function defined by

$$f(x) = \|x\|, \quad \forall x \in X.$$

Then

$$\partial f(x) = \mathrm{argmax}\{\langle x^*, x \rangle : x^* \in B_{X^*}\} = \{x^* \in B_{X^*} : \langle x^*, x \rangle = \|x\|\}.$$

Clearly, $\partial\|\theta\| = B_{X^*}$ and

$$\partial f(x) = \{x^* \in X^* : \|x^*\| = 1, \langle x^*, x \rangle = \|x\|\}, \quad x \neq \theta.$$

Hence,

$$\partial\|x\| = \begin{cases} B_{X^*}, & x = \theta, \\ \{x^* \in X^* : \|x^*\| = 1, \langle x^*, x \rangle = \|x\|\}, & x \neq \theta. \end{cases}$$

Thus, $\partial f : X \to 2^{X^*}$ is a bounded maximal monotone mapping with $\mathrm{Dom}(\partial f) = X$. If X^* is strictly convex, then

$$\{x^* \in X^* : \|x^*\| = 1, \langle x^*, x \rangle = \|x\|\}$$

is a singleton. In particular, if X is a real Hilbert space, then

$$\partial\|x\| = \begin{cases} B_H, & x = \theta, \\ \frac{x}{\|x\|}, & x \neq \theta. \end{cases}$$

Example 1.9.17. Let X be a real reflexive, smooth, and strictly convex Banach space. If $A : \mathrm{Dom}(A) = X \to X^*$ is a hemicontinuous monotone mapping such that $\|Ax\| \to \infty$ as $\|x\| \to \infty$, then $\mathrm{Ran}(A) = X^*$.

We remark that the condition "$\|Ax\| \to \infty$ as $\|x\| \to \infty$" is weaker than the coercivity condition

$$\frac{\langle Ax, x \rangle}{\|x\|} \to \infty \quad \text{as } \|x\| \to \infty.$$

Conversely, we have the following interesting fact.

Example 1.9.18. Let $A : \mathbb{R}^n \to \mathbb{R}^n$ be monotone and $\mathrm{Ran}(A) = \mathbb{R}^n$. Then $\|Ax\| \to \infty$ as $\|x\| \to \infty$.

Example 1.9.19. Let H be a real Hilbert space and let $A : \mathrm{Dom} \subset H \to H$ be a bounded maximal monotone mapping. Then $\overline{\mathrm{Dom}(A)} = H$.

Example 1.9.20. Let H be a real Hilbert space and let $L : \mathrm{Dom}(L) \subset H \to H$ be linear and monotone. Then L is maximal monotone if and only if $\overline{\mathrm{Dom}(L)} = H$ and L is maximal in the family of all linear monotone $K : \mathrm{Dom}(K) \to H$, where $\mathrm{Dom}(K)$ is a subspace of H.

Example 1.9.21. Let X be a real reflexive Banach space and let $A \subset X \times X^*$ be a linear and monotone mapping, i. e., $\mathrm{Graph}(A)$ is a subspace of $X \times X^*$ and $\langle f, x \rangle \geq 0, \forall [x, f] \in Ax$. Then the following assertions hold:
(1) if $\overline{\mathrm{Dom}(A)} = X$, then A is single-valued;
(2) A is maximal monotone if and only if $\mathrm{Graph}(A)$ is closed and A^* is monotone, where $A^* : \mathrm{Dom}(A^*) \subset X \to 2^{X^*}$ is the adjoint of A, defined by $x^* \in A^*x$ if and only if

$$\langle x^*, y \rangle = \langle f, x \rangle$$

for all $[x, f] \in \mathrm{Graph}(A)$.

Example 1.9.22. Let $H = L^2(0, 2\pi)$. Define a mapping $A : \mathrm{Dom}(A) \subset H \to H$ by

$$(Ax)(t) = x'(t), \quad \mathrm{Dom}(A) = \{x \in H : x'(t) \text{ exists and } x'(t) \in H\}$$

and mapping $B : \mathrm{Dom}(B) \subset H \to H$ by

$$(Bx)(t) = x'(t), \quad \mathrm{Dom}(B) = \{x \in H : x(0) = x(2\pi), x'(t) \text{ exists and } x' \in H\}.$$

Then $\mathrm{Dom}(B) \subset \mathrm{Dom}(A)$, $Ax = Bx$ if $x \in \mathrm{Dom}(B)$, $\mathrm{Graph}(B) \subset \mathrm{Graph}(A)$, A and B are all unbounded maximal monotone mappings.

1.9.4 Pseudocontractive mappings

The terminology and concept of pseudocontractive mappings were introduced and used by Browder [8] in 1967. A characterization of this class of mappings is given by Browder [8]. He observed that T is pseudocontractive if and only if $A := I - T$ is accretive. Apart from being a generalization of nonexpansive mappings, interest in the pseudocontractive mapping theory stems mainly from their firm connection with the class of accretive operators. The achievements of the fixed point theory for pseudocontractive mappings are fruitful, and so it is very difficult to cover every aspect of the theory. Therefore, we only collect and reorganize some important and useful concepts and facts of the fixed point theory for pseudocontractive mappings.

Definition 1.9.12. Let X be a Banach space and let $T : \text{Dom}(T) \subset X \to X$ be a mapping. Then T is said to be
(i)   pseudocontractive if there exists $j(x - y) \in J(x - y)$ such that

$$\langle Tx - Ty, j(x - y) \rangle \le \|x - y\|^2, \quad \forall x, y \in \text{Dom}(T);$$

(ii) β-strongly pseudocontractive if there exist $\beta \in [0, 1)$ and $j(x - y) \in J(x - y)$ such that

$$\langle Tx - Ty, j(x - y) \rangle \le \beta \|x - y\|^2, \quad \forall x, y \in \text{Dom}(T);$$

(iii) k-strictly pseudocontractive if there exist $k \in [0, 1)$ and $j(x - y) \in J(x - y)$ such that

$$\langle Tx - Ty, j(x - y) \rangle \le \|x - y\|^2 - k\|(I - T)x - (I - T)y\|^2, \quad \forall x, y \in \text{Dom}(T);$$

(iv) g-strongly pseudocontractive if there exist $g : \mathbb{R}^+ \to \mathbb{R}^+$ with $g(0) = 0, \forall r_0 > 0$, $\liminf_{r \to r_0} g(r) > 0, \limsup_{r \to \infty} g(r) > 0$, and $j(x - y) \in J(x - y)$ such that

$$\langle Tx - Ty, j(x - y) \rangle \le \|x - y\|^2 - g(\|x - y\|)\|x - y\|, \quad \forall x, y \in \text{Dom}(T).$$

Remark 1.9.1.
(i) Let $A := I - T$. Then T is g-strongly pseudocontractive if and only if A is g-strongly accretive.
(ii) If T is k-strictly pseudocontractive, then T is L-Lipschitz, where $L = \frac{1+k}{k} \ge 1$.

Definition 1.9.13. Let C be a nonempty convex subset of a Banach space X. Then $I_C(x) := \{y \in X : y = x + \lambda(u - x), u \in C, \lambda \ge 0\}, \forall x \in C$, is said to be an inward set. Let $T : C \to X$ be a mapping. If $Tx \in I_C(x), \forall x \in C$, mapping T is said to satisfy the inward condition. If $Tx \in \overline{I_C(x)}, \forall x \in C$, mapping $T : C \to X$ is said to satisfy the weak inward condition.

In 1976, Caristi [17] established the equivalence between the "weak inward" condition (WIC) and the "flow-invariance" condition (FIC).

Theorem 1.9.46 (Caristi [17], 1976). *Let C be a nonempty, closed, and convex subset of a Banach space X, and let $A : C \to X$ be a mapping. Then A satisfies the "flow-invariance" condition if and only if $T = I - A$ satisfies the "weak inward" condition.*

One can give a equivalent characterization of the "weak inward" condition. It is useful to see whether a mapping $T : C \to X$ satisfies the "weak inward" condition.

Theorem 1.9.47 (Caristi [17], 1976). *Let C be a real nonempty closed convex subset of a Banach space X. Let $T : C \to X$ be a mapping. Then T satisfies the "weak inward" condition if and only if*

$$x \in \partial C, \ x^* \in X^*, \quad x^*(x) = \sup\{x^*(y) : y \in C\} \Rightarrow x^*(x - Tx) \geq 0.$$

From Theorems 1.9.46 and 1.9.47, the following fixed point theorems are not hard to derive.

Theorem 1.9.48 (Morales–Chidume [53], 1999). *Let C be a nonempty closed convex subset of a Banach space X. Let $T : C \to X$ be a continuous g-strongly pseudocontractive mapping satisfying the "weak inward" condition. If there exists some point $x_0 \in C$ such that $\liminf_{r \to \infty} g(r) > \|x_0 - Tx_0\|$, then T has unique fixed point in X.*

Theorem 1.9.49 (Caristi [17], 1976). *Let X be a Banach space. Let C be a nonempty closed convex subset such that it has the f. p. p. for nonexpansive mappings. If $T : C \to X$ is a continuous and pseudocontractive mapping satisfying the "weak inward" condition, then T at least has one fixed point in X.*

Theorem 1.9.50 (Zhou [114], 2008). *Let X be a uniformly convex Banach space and let C be a nonempty closed convex subset of X. Let $T : C \to X$ be a continuous and pseudo-contractive mapping satisfying the "weak inward" condition. Then $I - T$ is demiclosed at the origin.*

Theorem 1.9.51 (Zhou [114], 2008). *Let X be a reflexive Banach space satisfying the Optial condition, and let C be a nonempty closed convex subset of X. Let $T : C \to X$ be a continuous and pseudocontractive mapping satisfying the "weak inward" condition. Then $I - T$ is demiclosed at the origin.*

1.10 Some useful lemmas

Lemma 1.10.1 (Tan and Xu [92], 1993). *Let $\{a_n\}$, $\{b_n\}$, and $\{\sigma_n\}$ be nonnegative sequences of real numbers. Assume that*

$$a_{n+1} \leq (1 + \sigma_n)a_n + b_n, \quad \forall n \geq 1.$$

If $\sum_{n=1}^{\infty} b_n < \infty$ and $\sum_{n=1}^{\infty} \sigma_n < \infty$, then $\lim_{n \to \infty} a_n$ exists. If sequence $\{a_n\}$ has a subsequence $\{a_{n_i}\}$ such that $a_{n_i} \to 0$ as $i \to \infty$, then $a_n \to 0$ as $n \to \infty$.

Lemma 1.10.2 (Liu [43], 1995; Xu [99], 2002). *Let $\{a_n\}$ be a nonnegative sequence of real numbers. Assume that*

$$a_{n+1} \le (1 - t_n)a_n + t_n\delta_n + b_n, \quad \forall n \ge 1,$$

where $t_n \in [0,1]$, $\delta_n \in \mathbb{R}$ and $b_n \in \mathbb{R}^+$ satisfy
(i) $t_n \to 0$ and $\sum_{n=1}^{\infty} t_n = \infty$;
(ii) $\limsup_{n\to\infty} \delta_n \le 0$;
(iii) $\sum_{n=1}^{\infty} b_n < \infty$.

Then $a_n \to 0$ as $n \to \infty$.

Lemma 1.10.3 (Suzuki [82], 2005). *Let $\{x_n\}$ and $\{y_n\}$ be bounded sequences of a Banach space X. Let $\{\lambda_n\}$ be a real number sequence in $(0,1)$ such that*

$$0 < \liminf_{n\to\infty} \lambda_n \le \limsup_{n\to\infty} \lambda_n < 1.$$

Let $x_{n+1} = (1 - \lambda_n)x_n + \lambda_n y_n$, $\forall n \ge 1$. If $\liminf_{n\to\infty}(\|y_{n+1} - y_n\| - \|x_{n+1} - x_n\|) \le 0$, then $\lim_{n\to\infty} \|y_n - x_n\| = 0$.

Lemma 1.10.4 (Maingé [45], 2010). *Let $\{\Gamma_n\}$ be a real number sequence. Assume that there exists some subsequence $\{\Gamma_{n_j}\}_{j\ge 0} \subset \{\Gamma_n\}$ satisfying $\Gamma_{n_j} < \Gamma_{n_j+1}$, $\forall j \ge 0$. Define a sequence $\{\tau_n\}_{n\ge n_0}$ by*

$$\tau(n) := \max\{k \le n \mid \Gamma_k < \Gamma_{k+1}\}.$$

Then the following assertions hold:
(i) $\{\tau_n\}_{n\ge n_0}$ *is a nondecreasing sequence with $\tau(n) \to \infty$ as $n \to \infty$;*
(ii) $\Gamma_{\tau(n)} \le \Gamma_{\tau(n)+1}$, $\forall n \ge n_0$;
(iii) $\Gamma_n \le \Gamma_{\tau(n)+1}$, $\forall n \ge n_0$.

Lemma 1.10.5 (Maingé [46], 2014). *Let $\{a_n\}$ be a nonnegative sequence of real numbers such that*

$$a_{n+1} \le (1 - \gamma_n)a_n + \gamma_n r_n, \quad \forall n \ge 1,$$

where $\{r_n\} \subset (-\infty, +\infty)$ is bounded above and $\{\gamma_n\} \subset [0,1]$ satisfies $\sum_{n=1}^{\infty} \gamma_n = \infty$. Then $\liminf_{n\to\infty} a_n \le \limsup_{n\to\infty} r_n$.

1.11 Exercises

1. Let X be a real normed linear space with the topological conjugate space X^*. Let A be a bounded subset of X in the weak topology $\sigma(X, X^*)$. Prove that A is bounded in the norm topology.

2. Let X be a real Banach space with the topological conjugate space X^*. Let B be a bounded subset of X^* in the weak star topology $\sigma(X^*, X)$. Prove that B is bounded in the norm topology. Try to give an example to show that the above conclusion is not necessarily true in a normed linear space X.

3. Let X be a real Banach space with the topological conjugate space X^*. Let A be a subset of X^*. Prove that A is bounded in the norm topology $\Leftrightarrow A$ is bounded in the weak topology $\sigma(X^*, X) \Leftrightarrow A$ is bounded in the weak star topology $\sigma(X^*, X)$.

4. Let X be a real Banach space with the topological conjugate space X^*. Prove that every w^*-bounded and w^*-closed subset of X^* is w^*-compact. Present an example to show that above conclusion is not necessarily true for the weak topology.

5. Let X be a real Banach space and X^* the topological conjugate space of X. Let A be a subset of X^*. Prove that if A is w^*-sequentially compact, then it is w^*-compact. Try to give an example to show that a w^*-compact subset in X^* is not necessarily w^*-sequentially compact.

6. Let X be a real Banach space and $S(X)$ denote the unit sphere. Prove that $\overline{S(X)}^w = B_X = \{x \in X : \|x\| \leq 1\}$, where $\overline{S(X)}^w$ denotes the weak closure in the weak topology $\sigma(X, X^*)$.

7. Use Theorem 1.8.28 to establish the relation between $b(t)$ and $\rho_X(t)$.

8. Let X be a real Banach space and $T : X \to 2^{X^*}$ a maximal monotone mapping. Prove that the function defined by

$$\varphi(x) = \inf\{\|f\| : f \in Tx\}$$

is lower semicontinuous.

9. Let X be a real reflexive and smooth Banach space, $T : B_r[\theta] \to X^*$ a hemicontinuous and monotone mapping. Prove that the following conclusions hold:
   (i) There exists some $x_0 \in B_r[\theta]$ such that

$$\langle Tx_0, x - x_0 \rangle \geq 0, \ \forall x \in B_r[\theta].$$

   (ii) If $x_0 \in B_r(\theta)$, then $Tx_0 = \theta^*$.
   (iii) If $\|x_0\| = r$, and $Tx_0 \neq \theta^*$, then there exists some $s > 0$ such that $Tx_0 = -sJx_0$.

10. Let X be a real Banach space and C a nonempty closed convex subset of X. Define a function $d : X \to \mathbb{R}^+$ by

$$d(x) = \text{dist}\,(x, C) = \inf\{\|x - y\| : y \in C\}.$$

Prove that the following assertions hold:
   (i) $d : X \to \mathbb{R}^+$ is nonexpansive and convex.
   (ii) If X is reflexive, then $\forall x \in X$, there exists $\bar{x} \in C$ such that $d(x) = \|x - \bar{x}\|$. If X is also strictly convex, then above $\bar{x} \in C$ is unique.
   (iii) $\partial d(x_0) = N_C(\bar{x}) \cap \{x^* \in X^* : x^* \in \partial\|x_0 - \bar{x}\|\}$, where $N_C(x) = \{x^* \in X^* : \langle x^*, y - x \rangle \leq 0, \forall y \in C\}$ is the normal cone of C at x.

(iv) If $X = H$, a real Hilbert space, then

$$\partial d(x_0) = \begin{cases} N_C(x_0) \cap B_H, & \text{if } x_0 \in C, \\ \dfrac{x_0 - P_C x_0}{\|x_0 - P_C x_0\|}, & \text{if } x_0 \notin C. \end{cases}$$

11. Let $f : \mathbb{R}^n \to \mathbb{R}$ be a convex function. Prove that $\forall x \in \mathbb{R}^n$, $\partial f(x)$ exists and $\partial f : \mathbb{R} \to 2^{\mathbb{R}^n}$ is a bounded operator, that is, $\partial f(A)$ is bounded subset of $\mathbb{R}^n$ whenever A is a bounded subset of $\mathbb{R}^n$.

12. Consider a function $f : \mathbb{R}^n \to \overline{\mathbb{R}}$ given by

$$f(x) = \begin{cases} -(1 - \|x\|^2)^{\frac{1}{2}}, & \|x\| \leq 1, \\ +\infty, & \|x\| > 1. \end{cases}$$

Compute $\partial f(x)$.

13. Let X be a real Banach space and let $\varphi : X \to \mathbb{R}$ be a continuous convex function. Prove that $\forall x \in X$, $\partial \varphi(x) \neq \emptyset$ and $\partial \varphi(x)$ is w^*-compact.

14. Let X be a real Hilbert space and C a nonempty closed convex subset of H. Define a function $f : H \to \mathbb{R}$ by

$$f(x) = \frac{1}{2}\|(I - P_C)x\|^2, \quad x \in H.$$

Compute $\nabla f(x)$. Let $A : H \to H$ be a bounded linear operator. Define a function $g : H \to \mathbb{R}$ by

$$g(x) = \frac{1}{2}\|(I - P_C)Ax\|^2, \quad x \in H.$$

Compute $\nabla g(x)$.

15. Let X be a real Banach space. For $p > 1$, define a function $\varphi : X \to \mathbb{R}$ by

$$\varphi(x) = \frac{1}{p}\|x\|^p, \quad x \in X.$$

Compute $\varphi^*(x^*)$.

16. Let H be a real Hilbert space and let $f : H \to \mathbb{R}$ be a convex function such that $\nabla f(x)$ exists for all $x \in H$ and $\nabla f(x)$ is L-Lipschitz continuous. Prove that $\nabla f : H \to H$ is $\frac{1}{L}$-inverse-strongly monotone.

17. Let $X = l^p$ ($p > 1$) and J_φ the generalized duality mapping with the gauge function $\varphi(t) = t^{p-1}$. Prove that $J_\varphi : l^p \to l^q$ is weakly sequentially continuous. Present an example to show that above conclusion does not hold for L^p ($p \neq 2$).

18. Let X be a real Banach space and let $f : X \to \overline{\mathbb{R}}$ be a proper lower semicontinuous convex function. Show that there exist $x^* \in X^*$ and $\mu \in \mathbb{R}$ such that $f(x) \geq \langle x, x^* \rangle + \mu$ for all $x \in X$.

19. Let H be a real Hilbert space, let $f : H \to \overline{\mathbb{R}}$ be a proper lower semicontinuous convex function, and let $x_0^* \in H$. If $g : H \to \overline{\mathbb{R}}$ is defined by

$$g(x) = \frac{1}{2}\|x\|^2 + f(x) - \langle x, x_0^* \rangle, \quad \forall x \in H,$$

then show that g is proper, lower semicontinuous and convex. Further, show that $\|z_n\| \to \infty \Rightarrow g(z_n) \to \infty$.

20. Let $\varphi : \mathbb{R}^+ \to \mathbb{R}^+$ satisfy the following conditions:
    (i) $t_1 > t_2 \Rightarrow \varphi(t_1) \geq \varphi(t_2)$;
    (ii) $\varphi(t) = 0 \Leftrightarrow t = 0$.
    Let $\{b_n\} \subset \mathbb{R}^+$ satisfy $\lim_{n\to\infty} b_n = 0$. Then show that any sequence $\{a_n\} \subset \mathbb{R}^+$ defined by

$$a_{n+1} \leq a_n - \varphi(a_n) + b_{n+1}, \quad n = 0, 1, 2, \ldots$$

    converges to 0.

21. Let (X, d) be a metric space and let K be a nonempty compact subset of X. Let $T : K \to K$ be a strictly nonexpansive mapping, that is,

$$d(Tx, Ty) \leq d(x, y),$$

    for all $x, y \in K$, and "=" holds if and only if $x = y$. Prove that T has a unique fixed point in K.

22. Let H be a real Hilbert space and let $n \geq 2$, $x_i \in H$ and $\lambda_i \in [0, 1]$ with $\sum_{i=1}^n \lambda_i = 1$. Prove that

$$\left\| \sum_{i=1}^n \lambda_i x_i \right\|^2 = \sum_{i=1}^n \lambda_i \|x_i\|^2 - \sum_{1 \leq i < j \leq n} \lambda_i \lambda_j \|x_i - x_j\|^2.$$

23. Let X be a real Banach space and $\varphi : X \to \mathbb{R}$ be a continuous convex function. Prove that φ is strictly convex if and only if $\partial\varphi \subset X \times X^*$ is strictly monotone.

24. Let X be a real reflexive Banach space and $f : X \to \mathbb{R}$ a continuous convex function. Prove that if f is uniformly convex on X, then $\partial f \subset X \times X^*$ is ψ-strongly monotone and $\mathrm{Ran}(\partial f) = X^* = \mathrm{Dom}(f^*)$.

2 Iterative methods for fixed points of nonexpansive mappings in Hilbert spaces

The purpose of this chapter is to discuss several popular iterative methods for fixed points of nonexpansive mappings in Hilbert spaces. They cover Banach–Picard iterative methods, Krasnosel'skiĭ–Man iterative methods, Halpern-type iterative methods, and Moudafi–Halpern-type iterative methods. We also focus on the weak–strong convergence analysis of these iterative methods. Finally, we give some of their applications in optimization problems, variational inequality problems, and split feasibility problems.

In this chapter, we use H to denote a real Hilbert space with inner product $\langle x, y \rangle$ and induced norm $\|x\| = \sqrt{\langle x, x \rangle}$, $\forall x, y \in H$. The letter I is used to denote the identity mapping on H. Further, we use "$\to$" and "$\rightharpoonup$" to denote the strong and weak convergence, respectively; "$\implies$" and "$\iff$" stand for "implies" and "if and only if", respectively. We also use $\mathbb{R}$ and $\mathbb{N}$ to denote the real and natural numbers, respectively. Let $T : H \to H$ be a mapping. We use $\mathrm{Dom}(T)$ and $\mathrm{Ran}(T)$ to denote the domain and range of T, respectively, and use $\mathrm{Fix}(T)$ to denote the fixed points set of T, that is, $\mathrm{Fix}(T) := \{x \in \mathrm{Dom}(T) : x = Tx\}$.

2.1 Basic properties for nonexpansive mappings and their subclasses

In this section, we present some basic properties for nonexpansive mappings and their subclasses, in particular, we give various equivalent relations between firmly nonexpansive mappings and averaged nonexpansive mappings.

Proposition 2.1.1. *Let* $T : \mathrm{Dom}(T) \subset H \to H$ *and* $A := I - T$. *Then*
(1) $\mathrm{Dom}(A) = \mathrm{Dom}(T)$.
(2) *T is nonexpansive if and only if*

$$\langle Ax - Ay, x - y \rangle \geq \frac{1}{2}\|Ax - Ay\|^2, \quad \forall x, y \in \mathrm{Dom}(T).$$

(3) *T is firmly nonexpansive if and only if A is firmly nonexpansive.*
(4) *T is firmly nonexpansive if and only if*

$$\|Tx - Ty\|^2 \leq \|x - y\|^2 - \|Ax - Ay\|^2, \quad \forall x, y \in \mathrm{Dom}(T).$$

(5) *T is firmly nonexpansive if and only if T is $\frac{1}{2}$-averaged nonexpansive.*

https://doi.org/10.1515/9783110667097-002

Proof.

(1) It is obvious.

(2) Using the norm and scalar product properties on H, one sees that

$$\|Tx - Ty\|^2 - \|x - y\|^2 = \|Ax - Ay\|^2 - 2\langle Ax - Ay, x - y \rangle, \quad \forall x, y \in \text{Dom}(T).$$

Hence, we obtain assertion (2).

(3) As similar as (2), one sees that

$$\|Ax - Ay\|^2 - \langle Ax - Ay, x - y \rangle = \|Tx - Ty\|^2 - \langle Tx - Ty, x - y \rangle, \quad \forall x, y \in \text{Dom}(T).$$

Combining with the definition of firmly nonexpansive mappings, we have the conclusion.

(4) Again using the properties of the norm and scalar product on H, one sees that

$$\|x - y\|^2 - \|Ax - Ay\|^2 = 2\langle Tx - Ty, x - y \rangle - \|Tx - Ty\|^2, \quad \forall x, y \in \text{Dom}(T).$$

Combining with the definition of a firmly nonexpansive mapping, we have the conclusion.

(5) Denote $S := 2T - I$. Then T is $\frac{1}{2}$-averaged nonexpansive if and only if S is nonexpansive. Note the following fact:

$$\begin{aligned} \|Sx - Sy\|^2 &= \|2Tx - x - 2Ty + y\|^2 \\ &= \|x - y - 2(Tx - Ty)\|^2 \\ &= \|x - y\|^2 - 4\langle Tx - Ty, x - y \rangle + 4\|Tx - Ty\|^2. \end{aligned}$$

Hence, S is nonexpansive if and only if T is firmly nonexpansive. This completes the proof. □

We know that the Banach–Picard iteration $\{T^n x\}$ of a nonexpansive mapping $T : C \to C$ for any $x \in C$ is not necessarily asymptotically regular, that is, it does not necessarily mean that

$$T^{n+1}x - T^n x \to \theta \quad (n \to \infty).$$

Example 2.1.1. Take $H = \mathbb{R}$, $C = [-1, 1]$ and define $T : C \to C$ as follows:

$$Tx = -x, \quad \forall x \in C.$$

Then $T : C \to C$ is a nonexpansive mapping with $\text{Fix}(T) = \{0\}$. Letting $x_0 = 1 \in C$, we have

$$T^{n+1}x_0 - T^n x_0 = (-1)^{n+1} - (-1)^n \nrightarrow \theta \quad (n \to \infty).$$

In addition, if we consider the mapping

$$T_\lambda := (1 - \lambda)I + \lambda T, \quad \forall \lambda \in (0, 1),$$

then it follows, for all $x \in C$, that

$$T_\lambda^{n+1} x - T_\lambda^n x \to \theta \quad (n \to \infty).$$

In fact, we also have

$$T_\lambda x = (1 - 2\lambda)x \implies T_\lambda^n x = (1 - 2\lambda)^n x \to \theta \quad (n \to \infty),$$

that is, the Banach–Picard iterative sequence $\{T_\lambda^n x\}$ convergence to the (unique) fixed point of T. In particular, we have

$$T_\lambda^{n+1} x - T_\lambda^n x \to \theta \quad (n \to \infty).$$

The following two propositions reveal important properties of averaged nonexpansive mappings.

Proposition 2.1.2. *Let $T : \mathrm{Dom}(T) \subset H \to H$ be a λ-averaged nonexpansive mapping. Then, for all $x, y \in \mathrm{Dom}(T)$, we have*

$$\|Tx - Ty\|^2 \le \|x - y\|^2 - \frac{1 - \lambda}{\lambda} \|(I - T)x - (I - T)y\|^2. \tag{2.1}$$

Proof. Setting $T = (1 - \lambda)I + \lambda S$, one has

$$\begin{aligned}
\|Tx - Ty\|^2 &= \left\|(1 - \lambda)(x - y) + \lambda(Sx - Sy)\right\|^2 \\
&= (1 - \lambda)\|x - y\|^2 + \lambda\|Sx - Sy\|^2 \\
&\quad - \lambda(1 - \lambda)\left\|(I - S)x - (I - S)y\right\|^2 \\
&\le (1 - \lambda)\|x - y\|^2 + \lambda\|x - y\|^2 \\
&\quad - \lambda(1 - \lambda)\left\|(I - S)x - (I - S)y\right\|^2 \\
&= \|x - y\|^2 - \frac{1 - \lambda}{\lambda}\left\|(I - T)x - (I - T)y\right\|^2.
\end{aligned}$$

This completes the proof. □

Using this proposition, we can obtain the following basic conclusion, which is one of the main advantages of averaged nonexpansive mappings.

Proposition 2.1.3. *Let C be a nonempty convex subset of H and let $T : C \to C$ be a nonexpansive mapping with $\mathrm{Fix}(T) \ne \emptyset$. Let $T_\lambda : C \to C$, where $\lambda \in (0, 1)$, be a mapping defined by $T_\lambda := (1 - \lambda)I + \lambda T$. Then*

$$\sum_{n=1}^{\infty} \|T_\lambda^{n+1} x - T_\lambda^n x\|^2 < \infty, \quad \forall x \in C, \tag{2.2}$$

in particular, we have

$$T_\lambda^{n+1}x - T_\lambda^n x \to \theta \quad (n \to \infty). \tag{2.3}$$

Proof. Putting $x_n := T_\lambda^{n-1}x$ for all $n \geq 1$, we have

$$x_{n+1} = T_\lambda^n x = T_\lambda(T_\lambda^{n-1}x) = (1 - \lambda)x_n + \lambda T x_n. \tag{2.4}$$

Fixing $p \in \text{Fix}(T) = \text{Fix}(T_\lambda)$ and using (2.1), one has

$$\|x_{n+1} - p\|^2 = \|T_\lambda^n x - T_\lambda^n p\|^2 \leq \|x_n - p\|^2 - \frac{1-\lambda}{\lambda}\|(I - T_\lambda)x_n\|^2.$$

It follows that

$$\frac{1-\lambda}{\lambda}\|(I - T_\lambda)x_n\|^2 \leq \|x_n - p\|^2 - \|x_{n+1} - p\|^2, \quad \forall n \geq 1. \tag{2.5}$$

This further implies that

$$\frac{1-\lambda}{\lambda}\sum_{n=1}^\infty \|(I - T_\lambda)x_n\|^2 \leq \|x_1 - p\|^2 < \infty.$$

Therefore, (2.2) and (2.3) hold. This completes the proof. $\qquad\square$

Remark 2.1.1. From (3) of Proposition 2.1.1, we know that the complementary mapping of a nonexpansive mapping is $\frac{1}{2}$-inverse strongly monotone. The converse is also true.

Proposition 2.1.4. *Let $T : \text{Dom}(A) \subset H \to H$. Then*
(1) *T is nonexpansive if and only if $A := I - T$ is $\frac{1}{2}$-inverse strongly monotone.*
(2) *If T is v-inverse strongly monotone, then, for all $r > 0$, rT is $\frac{v}{r}$-inverse strongly monotone.*
(3) *For any $\lambda \in (0,1)$, T is λ-averaged nonexpansive if and only if $A := I - T$ is $\frac{1}{2\lambda}$-inverse strongly monotone.*
(4) *A convex combination of finitely many averaged nonexpansive mappings is averaged nonexpansive. Furthermore, if $S : H \to H$ is β-averaged nonexpansive, V is nonexpansive and $\alpha \in (0,1)$, then $T = (1 - \alpha)S + \alpha V$ is λ-averaged nonexpansive, where $\lambda = \alpha + \beta - \alpha\beta$. If $F = \text{Fix}(S) \cap \text{Fix}(V) \neq \emptyset$, then $\text{Fix}(T) = F$.*
(5) *A composition of finitely many averaged nonexpansive mappings is averaged nonexpansive. In particular, if T_i is λ_i-averaged nonexpansive for each $i = 1, 2$, then $T_1 T_2$ and $T_2 T_1$ are λ-averaged nonexpansive, where $\lambda = \lambda_1 + \lambda_2 - \lambda_1\lambda_2$.*
(6) *Let $T_i : H \to H$ be averaged nonexpansive mappings for all $i = 1, \ldots, r$. If $F = \bigcap_{i=1}^r \text{Fix}(T_i) \neq \emptyset$, then*

$$F = \text{Fix}(T_1 T_2 \cdots T_r) = \text{Fix}(T_r T_1 \cdots T_{r-1}) = \cdots = \text{Fix}(T_2 T_3 \cdots T_1).$$

Proof.

(1) The conclusion can be obtained directly from (2) of Proposition 2.1.1.

(2) The conclusion can be verified directly by the definition of an inverse-strongly monotone mapping.

(3) From Proposition 2.1.2, it follows that

$$\frac{1-\lambda}{\lambda}\|Ax - Ay\|^2 \le \|x - y\|^2 - \|Tx - Ty\|^2$$
$$= \|x - y\|^2 - \|Ax - Ay\|^2$$
$$+ 2\langle Ax - Ay, x - y \rangle - \|x - y\|^2$$
$$= 2\langle Ax - Ay, x - y \rangle - \|Ax - Ay\|^2.$$

Hence

$$\langle Ax - Ay, x - y \rangle \ge \frac{1}{2\lambda}\|Ax - Ay\|^2.$$

If $A := I - T$ is $\frac{1}{2\lambda}$-inverse strongly monotone, then it follows from conclusion (2) of this proposition that $\frac{1}{\lambda}A$ is $\frac{1}{2}$-inverse strongly monotone. Define a mapping S by $S = (1 - \frac{1}{\lambda})I + \frac{1}{\lambda}S$. Then $T = (1 - \lambda)I + \lambda S$ and $A = I - T = \lambda(I - S)$. It follows that $\frac{1}{\lambda}A = I - S$ is $\frac{1}{2}$-inverse strongly monotone. It follows from conclusion (1) of this proposition that S is nonexpansive. Hence T is λ-averaged nonexpansive.

(4) We only prove the case of two mappings. Since S is β-averaged nonexpansive, one finds that there exist a constant $\beta \in (0,1)$ and a nonexpansive mapping U such that

$$S = (1 - \beta)I + \beta U.$$

Denote

$$\lambda = \alpha + \beta - \alpha\beta = \alpha + \beta(1 - \alpha) \in (0,1)$$

and

$$W = \frac{(1-\alpha)\beta}{\lambda}U + \frac{\alpha}{\lambda}V.$$

Then W is nonexpansive and

$$T = (1 - \alpha)S + \alpha V$$
$$= (1 - \alpha)[(1 - \beta)I + \beta U] + \alpha V$$
$$= (1 - \alpha - \beta - \alpha\beta)I + (1 - \alpha)\beta U + \alpha V$$
$$= (1 - \lambda)I + \lambda W.$$

Thus T is λ-averaged nonexpansive.

(5) We only prove the case of two mappings. Denote

$$T_1 = (1 - \lambda_1)I + \lambda_1 S_1, \quad T_2 = (1 - \lambda_2)I + \lambda_2 S_2,$$

where λ_1 and λ_2 are two real numbers in $(0, 1)$, and S_1 and S_2 are two nonexpansive mappings. Then

$$\begin{aligned}
T_1 T_2 &= [(1 - \lambda_1)I + \lambda_1 S_1][(1 - \lambda_2)I + \lambda_2 S_2] \\
&= (1 - \lambda_1)[(1 - \lambda_2)I + \lambda_2 S_2] + \lambda_1 S_1 T_2 \\
&= (1 - \lambda_1)(1 - \lambda_2)I + \lambda_2(1 - \lambda_1)S_2 + \lambda_1 S_1 T_2 \\
&= (1 - \lambda)I + \lambda\left[\frac{\lambda_2(1 - \lambda_1)}{\lambda}S_2 + \frac{\lambda_1}{\lambda}S_1 T_2\right].
\end{aligned}$$

Setting

$$B = \frac{\lambda_2(1 - \lambda_1)}{\lambda}S_2 + \frac{\lambda_2}{\lambda}S_1 T_2,$$

one sees that B is nonexpansive. Hence $T_1 T_2$ is λ-averaged nonexpansive. Similarly, we can prove that $T_2 T_1$ is λ-averaged nonexpansive.

(6) It is obvious that $F \subset \mathrm{Fix}(T_1 T_2 \cdots T_r)$. So, we only need to prove the converse. Suppose that $x = T_1 T_2 \cdots T_r x$ and $p \in F$. In view of conclusion (5) of this proposition, one sees that T_r is averaged nonexpansive. Hence, there exist a constant $\alpha \in (0, 1)$ and a nonexpansive mapping S such that $T_r = (1 - \alpha)I + \alpha S$ such that

$$\begin{aligned}
\|x - p\|^2 &= \|T_1 T_2 \cdots T_r x - p\|^2 \\
&\leq \|T_r x - p\|^2 \\
&= \|(1 - \alpha)(x - p) + \alpha(Sx - p)\|^2 \\
&= (1 - \alpha)\|x - p\|^2 + \alpha\|Sx - p\|^2 - \alpha(1 - \alpha)\|x - Sx\|^2 \\
&\leq \|x - p\|^2 - \alpha(1 - \alpha)\|x - Sx\|^2.
\end{aligned}$$

Therefore, $x = Sx \implies T_r x = x$. Similarly, we can prove that $T_{r-1}x = x, \dots, T_1 x = x$, that is, $x \in F$. We can prove other equalities in a similar way. This completes the proof. $\qquad\square$

2.2 Opial conditions and asymptotic centers

In this section, we prove that Hilbert spaces satisfy the Opial condition, which is indeed an important property of Hilbert spaces. The condition plays an important role in proving the existence of fixed points for nonexpansive mappings and convergence of iterative sequences in Hilbert spaces. With the aid of asymptotic centers, we can establish the sufficient and necessary conditions of the weak convergence of Banach–Picard iterative sequences.

Proposition 2.2.1. *Let $\{x_n\}$ be a bounded sequence in H with $x_n \rightharpoonup x$. Then*
(i)

$$\limsup_{n\to\infty} \|x_n - y\|^2 = \limsup_{n\to\infty} \|x_n - x\|^2 + \|x - y\|^2, \quad \forall y \in H; \tag{2.6}$$

(ii)

$$\liminf_{n\to\infty} \|x_n - y\|^2 = \liminf_{n\to\infty} \|x_n - x\|^2 + \|x - y\|^2, \quad \forall y \in H. \tag{2.7}$$

Proof. It follows from the norm and scalar product properties on H that

$$\|x_n - y\|^2 = \|x_n - x + x - y\|^2$$
$$= \|x_n - x\|^2 + 2\langle x_n - x, x - y \rangle + \|x - y\|^2, \quad \forall y \in H. \tag{2.8}$$

From $x_n \rightharpoonup x$, one has $\langle x_n - x, x - y \rangle \to \theta$ as $n \to \infty$. Taking the superior and inferior limits, respectively, we obtain (2.6) and (2.7) immediately. This completes the proof. $\qquad\square$

Theorem 2.2.1 (Opial [60], 1967). *Let $\{x_n\}$ be a sequence in H with $x_n \rightharpoonup x$ as $n \to \infty$. Then*
(1) *For any $y \in H$ with $y \neq x$,*

$$\limsup_{n\to\infty} \|x_n - y\| > \limsup_{n\to\infty} \|x_n - x\|. \tag{2.9}$$

(2) *For any $y \in H$ with $y \neq x$,*

$$\liminf_{n\to\infty} \|x_n - y\| > \liminf_{n\to\infty} \|x_n - x\|. \tag{2.10}$$

Proof. We prove (2.10) only. In a similar way, we can prove (2.9). Letting $a := \liminf_{n\to\infty} \|x_n - x\|$, one has $a \geq 0$. If $a = 0$, then we obtain from the weak lower semicontinuity of the norm that

$$\liminf_{n\to\infty} \|x_n - x\| = 0 < \|x - y\| \leq \liminf_{n\to\infty} \|x_n - y\|, \quad \forall y \neq x.$$

Hence we obtain (2.10). If $a > 0$, then, for all $\varepsilon \in (0, a)$, there exists $n_0 \geq 1$ such that, for all $n \geq n_0$,

$$\|x_n - x\| > (a - \varepsilon) \implies \liminf_{n\to\infty} \|x_n - x\|^2 \geq (a - \varepsilon)^2. \tag{2.11}$$

Using (2.7) and (2.11), one has

$$\liminf_{n\to\infty} \|x_n - y\|^2 = \liminf_{n\to\infty} \|x_n - x\|^2 + \|x - y\|^2$$
$$\geq (a - \varepsilon)^2 + \|x - y\|^2. \tag{2.12}$$

From the arbitrariness of $\varepsilon \in (0, a)$, we find from (2.12) that

$$\liminf_{n\to\infty} \|x_n - y\|^2 \geq a^2 + \|x - y\|^2 > a^2 + \frac{1}{2}\|x - y\|^2, \quad \forall y \neq x. \tag{2.13}$$

For a sufficiently large $n \in \mathbb{N}$, we have

$$\|x_n - y\|^2 > a^2 + \frac{1}{2}\|x - y\|^2, \quad \forall y \neq x$$

$$\Longrightarrow \|x_n - y\| \geq \sqrt{a^2 + \frac{1}{2}\|x - y\|^2}, \quad \forall y \neq x$$

$$> a = \liminf_{n\to\infty} \|x_n - x\|, \quad \forall y \neq x.$$

This completes the proof. □

Remark 2.2.1. In 1967, Opial proved inequality (2.9) and inequality (2.10) in the framework of Hilbert spaces. There are equivalent to each other. Opial conditions (2.9) and (2.10) show that proper, continuous and convex functions

$$f(y) = \limsup_{n\to\infty} \|x_n - y\|$$

and

$$g(y) = \liminf_{n\to\infty} \|x_n - y\|$$

attain their minima at the weak limit point of the sequence $\{x_n\}$.

Definition 2.2.1. Let $\{x_n\}$ be a bounded sequence in H and let C be a nonempty closed convex subset of H. Define a function $r : H \to \mathbb{R}^+$ by

$$r(x) := \limsup_{n\to\infty} \|x_n - x\|, \quad \forall x \in H, \tag{2.14}$$

$$r(C, \{x_n\}) := \inf\{r(y) : y \in C\}, \tag{2.15}$$

and

$$A(C, \{x_n\}) := \{z \in C : r(z) = \min\{r(x) : x \in C\}\}. \tag{2.16}$$

Then $r(C, \{x_n\})$ is called the asymptotic center of $\{x_n\}$ with respect to C and $A(C, \{x_n\})$ is called the set of asymptotic centers of $\{x_n\}$ with respect to C.

Proposition 2.2.2. *Let $\{x_n\}$ be a bounded sequence in H and let C be a nonempty closed convex subset of H. Then $\{x_n\}$ has a unique asymptotic center with respect to C.*

Proof. Since $r : C \to \mathbb{R}^+$ is proper, continuous, and convex, and $r(x) \to \infty$ ($\|x\| \to \infty$), one sees that there exists $z \in C$ such that $r(z) = \min\{r(x) : x \in C\}$. Hence $A(C, \{x_n\}) \neq \emptyset$.

Next, we prove that $A(C, r(x))$ is a singleton. To achieve this, for all $z_1, z_2 \in A(C, \{x_n\})$ and $t \in (0, 1)$, we let $z_t = tz_1 + (1 - t)z_2$. Then

$$r(z_1) \le r(z_t), \quad r(z_2) \le r(z_t). \tag{2.17}$$

It follows that

$$\|x_n - z_t\|^2 = \|t(x_n - z_1) + (1 - t)(x_n - z_2)\|^2$$
$$= t\|x_n - z_1\|^2 + (1 - t)\|x_n - z_2\|^2 - t(1 - t)\|z_1 - z_2\|^2. \tag{2.18}$$

Letting $n \to \infty$ in (2.18), one concludes that

$$\limsup_{n \to \infty} \|x_n - z_t\|^2 \le t\left(\limsup_{n \to \infty} \|x_n - z_1\|\right)^2 + (1 - t)\left(\limsup_{n \to \infty} \|x_n - z_2\|\right)^2$$
$$- t(1 - t)\|z_1 - z_2\|^2$$
$$= tr^2(z_1) + (1 - t)r^2(z_2) - t(1 - t)\|z_1 - z_2\|^2$$
$$\le tr^2(z_t) + (1 - t)r^2(z_t) - t(1 - t)\|z_1 - z_2\|^2$$
$$= r^2(z_t) - t(1 - t)\|z_1 - z_2\|^2. \tag{2.19}$$

If $z_1 \ne z_2$, then $t(1 - t)\|z_1 - z_2\|^2 > \frac{t}{2}(1 - t)\|z_1 - z_2\|^2$. It follows from (2.19) that

$$\limsup_{n \to \infty} \|x_n - z_t\|^2 < r^2(z_t) - \frac{t}{2}(1 - t)\|z_1 - z_2\|^2. \tag{2.20}$$

In view of the property of the $\limsup_{n \to \infty}$, one finds that there exists $n_0 \ge 1$ such that

$$\|x_n - z_t\|^2 < r^2(z_t) - \frac{t}{2}(1 - t)\|z_1 - z_2\|^2, \quad \forall n \ge n_0,$$
$$\implies \|x_n - z_t\| < \sqrt{r^2(z_t) - \frac{t}{2}(1 - t)\|z_1 - z_2\|^2}$$
$$\implies r(z_t) = \limsup_{n \to \infty} \|x_n - z_t\| \le \sqrt{r^2(z_t) - \frac{t}{2}(1 - t)\|z_1 - z_2\|^2} < r(z_t),$$

which is a contradiction. Hence $z_1 = z_2$. Therefore, $A(C, \{x_n\})$ is a singleton. This completes the proof. $\square$

Proposition 2.2.3. *Let C be a closed convex subset of H and let $\{x_n\}$ be a bounded sequence in H. Let $A(C, x_n) = \{z\}$ and $\{y_n\} \subset C$ with $r(y_n) \to r(z)$ as $n \to \infty$. Then $y_n \to z$ as $n \to \infty$.*

Proof. From the definition of $r(\cdot)$, $r(y_n) \ge r(z)$ for each $n \ge 1$, since $r(y_n) \to r(z)$ as $n \to \infty$, we see that, for arbitrary $\varepsilon > 0$, there exists $M \ge 1$ such that

$$r(y_n) < r(z) + \varepsilon, \quad \forall m \ge M. \tag{2.21}$$

Letting

$$u_m = tz + (1-t)y_m, \quad \forall t \in (0,1), \tag{2.22}$$

we arrive at

$$\|x_n - u_m\|^2 = t\|x_n - z\|^2 + (1-t)\|x_n - y_m\|^2 - t(1-t)\|y_m - z\|^2. \tag{2.23}$$

Combining this with (2.21), we obtain that

$$
\begin{aligned}
\limsup_{n\to\infty} \|x_n - u_m\|^2 &\le t\left(\limsup_{n\to\infty} \|x_n - z\|\right)^2 + (1-t)\left(\limsup_{n\to\infty} \|x_n - y_m\|^2\right) \\
&\quad - t(1-t)\|y_m - z\|^2 \\
&= tr^2(z) + (1-t)r^2(y_n) - t(1-t)\|y_m - z\|^2 \\
&\le r^2(y_n) - t(1-t)\|y_m - z\|^2 \\
&< \left(r(z) + \varepsilon\right)^2 - t(1-t)\|y_m - z\|^2, \quad \forall m \ge M.
\end{aligned}\tag{2.24}
$$

Therefore, for all $m \ge M$, there exists $N(m) \ge 1$ such that, for all $N \ge N(m)$,

$$
\begin{aligned}
\|x_n - u_m\| &< \sqrt{\left(r(z) + \varepsilon\right)^2 - t(1-t)\|y_m - z\|^2} \\
\implies r(z) \le r(u_m) &= \limsup_{n\to\infty} \|x_n - u_m\| \le \sqrt{\left(r(z) + \varepsilon\right)^2 - t(1-t)\|y_m - z\|^2} \\
\implies r^2(z) &\le \left(r(z) + \varepsilon\right)^2 - t(1-t)\|y_m - z\|^2, \quad \forall m \ge M, \\
\implies t(1-t)\|y_m - z\|^2 &\le \left(r(z) + \varepsilon\right)^2 - r^2(z) = \varepsilon[2r(z) + \varepsilon], \quad \forall m \ge M.
\end{aligned}\tag{2.25}
$$

Hence $y_n \to z$ as $n \to \infty$. This completes the proof. □

Proposition 2.2.4. *Let C be a closed convex subset of H. Let $\{x_n\}$ be a sequence in C satisfying $x_n \rightharpoonup x$ as $n \to \infty$. Then $A(C, \{x_n\}) = \{x\}$.*

Proof. Since C is weakly closed, we conclude that $x \in C$. In view of Theorem 2.2.1, we find, for all $y \in C$, that

$$
\begin{aligned}
r(x) = \limsup_{n\to\infty} \|x_n - x\| &\le \limsup_{n\to\infty} \|x_n - y\| = r(y) \\
\implies r(x) &\le \inf\{r(y) : y \in C\} \le r(x) \\
\implies r(x) &= \inf\{r(y) : y \in C\}.
\end{aligned}
$$

Hence $x \in A(C, \{x_n\})$. It follows from Proposition 2.2.2 that $A(C, \{x_n\}) = \{x\}$. This completes the proof. □

Proposition 2.2.5. *Let C be a closed convex subset of H and let $T : C \to C$ be a nonexpansive mapping. Let $\{x_n\}$ be a bounded sequence in C defined by $x_n = T^n x$ for all $x \in C$ and $A(C, \{x_n\}) = \{z\}$. If there exists a subsequence $\{x_{n_j}\} \subset \{x_n\}$ such that $x_{n_j} \rightharpoonup x_0$ as $j \to \infty$ and $x_0 \in \mathrm{Fix}(T)$, then $x_0 = z$.*

Proof. Note that

$$r(z) \leq r(x_0) = \limsup_{n \to \infty} \|x_n - x_0\| = \limsup_{n \to \infty} \|T^n x - x_0\|$$

and

$$\begin{aligned}
r(z) \leq r(x_0) &= \limsup_{n \to \infty} \|x_n - x_0\| = \limsup_{n \to \infty} \|T^n x - x_0\| \\
&= \limsup_{n \to \infty} \|T^{n-n_j} T^{n_j} x - T^{n-n_j} x_0\|, \quad \forall n \geq n_j, \\
&\leq \limsup_{j \to \infty} \|T^{n_j} x - x_0\| \leq \limsup_{j \to \infty} \|T^{n_j} x - z\| \\
&\leq \limsup_{n \to \infty} \|T^n x - z\| = r(z).
\end{aligned}$$

Therefore, $r(z) = r(x_0) = r(z)$. In view of $A(C, \{x_n\}) = \{z\}$, one finds that $x_0 = z$. This completes the proof. $\qquad\square$

2.3 The demiclosedness principle and fixed point theorems

In this subsection, we establish the demiclosedness principle and some fixed point theorems for nonexpansive mappings. The demiclosedness principle is not only a technique for the existence of fixed points, but also an essential tool for the convergence of iterative methods.

Theorem 2.3.1 (Demiclosedness principle of nonexpansive mappings). *Let C be a nonempty, closed, and convex subset of H, and let $T : C \to C$ be a nonexpansive mapping. Then $I - T$ is demiclosed at $y \in H$, that is, for any $\{x_n\} \subset C$ with $x_n \rightharpoonup x$ and $x_n - Tx_n \to y \in H$, it follows that $x \in C$ and $x - Tx = y$. In particular, if $y = \theta$, then $I - T$ is said to be demiclosed at the origin.*

Proof. Since C is a closed convex subset of H, we see that C is weakly closed, and so $x \in C$. In view of the (2) of Proposition 2.1.1, we have

$$\langle (I - T)x_n - (I - T)x, x_n - x \rangle \geq \frac{1}{2}\|(I - T)x_n - (I - T)x\|^2, \quad \forall n \geq 1.$$

Letting $n \to \infty$, one has

$$0 \geq \frac{1}{2}\|y - (I - T)x\|^2 \implies (I - T)x = y.$$

This completes the proof. $\qquad\square$

Theorem 2.3.2 (Fixed Point Theorem of Nonexpansive Mappings). *Assume that C is a nonempty, bounded, closed, and convex subset of H, and let $T : C \to C$ be a nonexpansive mapping. Then $\mathrm{Fix}(T)$ is a nonempty, closed, and convex subset of H.*

Proof. Fix $u \in C$. For every $n \geq 1$, we define a mapping $T_n : C \to C$ by

$$T_n x = \frac{1}{n+1} u + \left(1 - \frac{1}{n+1}\right) Tx. \tag{2.26}$$

Then $T_n : C \to C$ is a contraction mapping. Using the Banach's Contraction Principle, there exists a unique $x_n \in C$ such that

$$x_n = T_n x_n = \frac{1}{n+1} u + \left(1 - \frac{1}{n+1}\right) Tx_n, \quad \forall n \geq 1. \tag{2.27}$$

Since C is bounded, $\{x_n\} \subset C$, and $\{Tx_n\} \subset C$, one concludes that both $\{x_n\}$ and $\{Tx_n\}$ are bounded sequences. It follows from (2.27) that

$$x_n - Tx_n = \frac{1}{n+1}(u - Tx_n) \to \theta \quad (n \to \infty).$$

Suppose that $x_n \to x$ as $n \to \infty$. Using the demiclosedness principle of nonexpansive mappings, we have $x = Tx$.

Next, we show that Fix(T) is closed and convex. First, we prove that Fix(T) is closed. Let $\{p_n\}$ be a sequence in Fix(T) with $p_n \to p$ as $n \to \infty$. Then $p_n = Tp_n$ implies $p = Tp$. Hence $p \in$ Fix(T).

Let $p_i \in$ Fix(T) for each $i = 1, 2$ and $t \in (0, 1)$. Let $p_t = (1 - t)p_1 + tp_2$. Since C is convex, we have $p_t \in C$. It follows that

$$\begin{aligned}
\|p_t - Tp_t\|^2 &= \|(1-t)(p_1 - Tp_t) + t(p_2 - Tp_t)\|^2 \\
&= (1-t)\|p_1 - Tp_t\|^2 + t\|p_2 - Tp_t\|^2 - t(1-t)\|p_1 - p_2\|^2 \\
&\leq (1-t)\|p_1 - p_t\|^2 + t\|p_2 - p_t\|^2 - t(1-t)\|p_1 - p_2\|^2 \\
&= [t^2(1-t) + t(1-t)2]\|p_1 - p_2\|^2 - t(1-t)\|p_1 - p_2\|^2. \\
&= 0.
\end{aligned}$$

Thus $p_t = Tp_t$, that is, $p_t \in$ Fix(T). This completes the proof. $\qquad\square$

2.4 Iterative methods of fixed points

Let C be a nonempty, closed, and convex subset of H, and let $T : C \to C$ be a nonexpansive mapping.

Now, we introduce some iterative methods for fixed points of T as follows:

(1) The Banach–Picard iterative method:

$$x_1 \in C, \ x_{n+1} = Tx_n, \quad \forall n \geq 1. \tag{BPIM}$$

(2) The Krasnosel'skiĭ–Man iterative method:

$$x_1 \in C, \ x_{n+1} = (1 - \alpha_n)x_n + \alpha_n Tx_n, \quad \forall n \geq 1, \tag{KMIM}$$

where $\{\alpha_n\}$ is a sequence in $[0, 1]$ satisfying certain conditions.

(3) The Halpern iterative method:

$$x_1, u \in C, \quad x_{n+1} = \alpha_n u + (1 - \alpha_n) T x_n, \quad \forall n \geq 1, \tag{HIM}$$

where $\{\alpha_n\}$ is a sequence in $[0,1]$ satisfying certain conditions;
(4) The Moudafi–Halpern iterative method:

$$x_1 \in C, \quad x_{n+1} = \alpha_n f(x_n) + (1 - \alpha_n) T x_n, \quad \forall n \geq 1, \tag{MHIM}$$

where $f : C \to C$ is a contraction with the contraction coefficient $\rho \in (0,1)$.

2.4.1 Weak convergence theorems

The general methodology to obtain the weak convergence of the iterative sequence $\{x_n\}$ generated by (BPIM) or (KMIM) is to prove that the iterative sequence is a bounded approximate fixed point sequence, that is, $x_n - T x_n \to \theta$ as $n \to \infty$ and then to prove that $\omega_\omega(x_n) \subset \text{Fix}(T)$ with the aid of the demiclosedness principle of nonexpansive mappings. Since $x_n \rightharpoonup x \iff$ all subsequences of $\{x_n\}$ converge weakly to the same point $\iff \omega_\omega(x_n)$ is a singleton set, it is, therefore, essential to find the conditions to guarantee that $\omega_\omega(x_n)$ is a singleton. Thus, we need the following definition.

Definition 2.4.1. Let $\{x_n\}$ be a sequence in H and let S be a subset of H. Then the sequence $\{x_n\}$ is said to be Fejér monotone with respect to S if

$$\|x_{n+1} - p\| \leq \|x_n - p\|, \quad \forall p \in S. \tag{2.28}$$

Example 2.4.1. Let $T : C \to C$ be a nonexpansive mapping with $\text{Fix}(T) \neq \emptyset$. Let $\{x_n\}$ be an iterative sequence generated by (BPIM) or (KMIM). Then $\{x_n\}$ is Fejér monotone with respect to $\text{Fix}(T)$. From (2.28), we see that $\lim_{n \to \infty} \|x_n - p\|$ exists, but is not necessarily zero.

Example 2.4.2. Let $\{e_n\}$ be a standard orthogonal basis in H. Then $\{e_n\}$ is Fejér monotone with respect to the singleton set $S = \{\theta\}$ and $e_n \rightharpoonup \theta$, but it does not converge strongly to θ. However, we have the following proposition.

Proposition 2.4.1. *Let $\{x_n\}$ be a sequence in H. Let C be a nonempty, closed, and convex subset of H, and let $\{x_n\}$ be a Fejér monotone sequence with respect to C. Then $\lim_{n \to \infty} P_C x_n$ exists. Furthermore, if $\omega_\omega(x_n) \subset C$, then $\omega_\omega(x_n)$ is a singleton set and $\{x_n\}$ converges weakly to a point in C.*

Note that Proposition 2.4.1 is called the Browder's convergence theorem.

Proof. It follows from the definition of the metric projection and Fejér monotone sequence that

$$\|x_{n+1} - P_C x_{n+1}\| \leq \|x_{n+1} - P_C x_n\| \leq \|x_n - P_C x_n\|, \quad \forall n \geq 1$$

implies that $\lim_{n\to\infty} \|x_n - P_C x_n\|$ does exist. Using the property of P_C, for all $n, m \geq 1$, we derive

$$\begin{aligned}
\|P_C x_{n+m} - P_C x_n\|^2 &\leq \|x_{n+m} - P_C x_n\|^2 - \|x_{n+m} - P_C x_{n+m}\|^2 \\
&\leq \|x_n - P_C x_n\|^2 - \|x_{n+m} - P_C x_{n+m}\|^2 \\
&\to 0 \quad (n, m \to \infty),
\end{aligned} \tag{2.29}$$

which shows that $\{P_C x_n\}$ is a Cauchy sequence in H. Hence $\lim_{n\to\infty} P_C x_n$ exists. We denote the limit by $\lim_{n\to\infty} P_C x_n = y \in H$. Since C is closed, we see that $y \in C$. If $\omega_\omega(x_n) \subset C$, then, for all $\hat{x} \in \omega_\omega(x_n)$, there is a subsequence $\{x_{n_j}\} \subset \{x_n\}$ such that $x_{n_j} \rightharpoonup \hat{x} \in C$. The properties of P_C imply that

$$\langle x_n - P_C x_n, z - P_C x_n \rangle \leq 0, \quad \forall z \in C. \tag{2.30}$$

In particular, we have

$$\langle x_{n_j} - P_C x_{n_j}, z - P_C x_{n_j} \rangle \leq 0, \quad \forall z \in C. \tag{2.31}$$

Letting $j \to \infty$ in (2.31), we find that

$$\langle \hat{x} - y, z - y \rangle \leq 0, \quad \forall z \in C. \tag{2.32}$$

Letting $z = \hat{x} \in C$ in (2.32), we find that

$$\begin{aligned}
0 \leq \|\hat{x} - y\|^2 &= \langle \hat{x} - y, \hat{x} - y \rangle \leq 0 \\
&\implies \hat{x} = y = \lim_{n\to\infty} P_C x_n,
\end{aligned}$$

which implies that $\omega_\omega(x_n)$ is a singleton set. Hence $\{x_n\}$ converges weakly to a point in C. This completes the proof. □

Theorem 2.4.1. *Let C be a closed convex subset of H and let $T : C \to C$ be a nonexpansive mapping with $\mathrm{Fix}(T) \neq \emptyset$. If, for any $x \in C$, $T^{n+1}x - T^n x \to \theta$ as $n \to \infty$, then $T^n x \rightharpoonup p \in \mathrm{Fix}(T)$ as $n \to \infty$.*

Proof. Let $x_n = T^{n-1}x$ for each $n \geq 1$. Then $x_1 \in C$ and $x_{n+1} = Tx_n$, that is, $\{x_n\}$ is a Banach–Picard iterative sequence. Fixing $p \in \mathrm{Fix}(T)$, we have

$$\|x_{n+1} - p\| = \|Tx_n - p\| \leq \|x_n - p\|.$$

Hence $\{x_n\}$ is a Fejér monotone sequence with respect to $\mathrm{Fix}(T)$. Note that

$$T^{n+1}x - T^n x \to \theta \iff Tx_n - x_n \to \theta.$$

It follows from the demiclosedness principle of nonexpansive mappings that $\omega_\omega(x_n) \subset \mathrm{Fix}(T)$. Therefore, it follows from Proposition 2.4.1 that $x_n \rightharpoonup p \in \mathrm{Fix}(T)$, where $p = \lim_{n\to\infty} P_{\mathrm{Fix}(T)}(x_n)$. This completes the proof. □

Using the technique of the asymptotic center, we can improve Theorem 2.4.1.

Theorem 2.4.2. *Let C be a closed convex subset of H and let $T : C \to C$ be a nonexpansive mapping with $\mathrm{Fix}(T) \neq \emptyset$. Then, for any $x \in C$,*

$$T^n x \rightharpoonup z \in \mathrm{Fix}(T) \iff T^{n+1}x - T^n x \to \theta \quad (n \to \infty).$$

Proof. "$\Longrightarrow$" is obvious. So we only show "$\Longleftarrow$". Suppose that $\{T^{n_j}x\}$ is a subsequence of $\{T^n x\}$. Since $\mathrm{Fix}(T) \neq \emptyset$, we see that $\{T^n x\}$ is a bounded sequence. Assume that $T^{n_j}x \rightharpoonup y$ as $j \to \infty$. Then, for all $m \geq 1$, $T^{n_j + m}x \rightharpoonup y$ as $j \to \infty$. It follows from Proposition 2.2.5 that $A(C, \{T^{n_j+m}x\}) = \{y\}$. For all $m > s \geq 1$, letting $y_s = T^s y$ yields

$$
\begin{aligned}
\|y_s - T^{n_j+m}x\| &= \|T^s y - T^{n_j+m}x\| \\
&= \|T^s y - T^s T^{n_j+m-s}x\| \\
&\leq \|y - T^{n_j+m-s}x\|.
\end{aligned}
\tag{2.33}
$$

Fixing s, n_j and letting $m \to \infty$, we find from (2.33) that

$$r(y) \leq r(y_s) \leq r(y). \tag{2.34}$$

Letting $s \to \infty$ in (2.34), we obtain

$$r(y_s) \to r(y) \quad (s \to \infty).$$

From Proposition 2.2.3, we see that $y_s \to y$ ($s \to \infty$). Since T is continuous, we have $Ty_s \to Ty$ ($s \to \infty$). Note that

$$y_{s+1} = T^{s+1}y = TT^s y = Ty_s \to Ty \quad (s \to \infty).$$

By the uniqueness of the limit, one has $y = Ty$.

Finally, we see from Proposition 2.2.5 that $y = z$ and $\{z\} = A(C, \{T^n x\})$. Hence $T^n x \rightharpoonup z$ as $n \to \infty$. This completes the proof. $\square$

Corollary 2.4.1. *Let C be a closed convex subset of H and let $T : C \to C$ be a nonexpansive mapping with $\mathrm{Fix}(T) \neq \emptyset$. If, for any $\lambda \in (0, 1)$, we define a mapping $T_\lambda : C \to C$ by*

$$T_\lambda x = (1 - \lambda)x + \lambda Tx, \quad \forall x \in C,$$

then, for any $x \in C$, $T_\lambda^n x \rightharpoonup p \in \mathrm{Fix}(T)$ as $n \to \infty$.

Proof. It follows from Proposition 2.1.3 that $T_\lambda^{n+1}x - T_\lambda^n x \to \theta$ as $n \to \infty$. From Theorem 2.4.2, it follows that $T_\lambda^n x \rightharpoonup p \in \mathrm{Fix}(T_\lambda) = \mathrm{Fix}(T)$. This completes the proof. $\square$

Theorem 2.4.3. *Let C be a closed convex subset of H and let $T : C \to C$ be a nonexpansive mapping with $\mathrm{Fix}(T) \neq \emptyset$. Suppose that $\{\alpha_n\}$ is a sequence in $[0, 1]$ satisfying the following conditions:*

$$\sum_{n=1}^{\infty} \alpha_n(1 - \alpha_n) = \infty.$$

If $\{x_n\}$ is the sequence generated by Krasnosel'skiǐ–Man iterative method (KMIM), then the following results hold:
(1) *$\{x_n\}$ is Fejér monotone with respect to $\mathrm{Fix}(T)$;*
(2) *$x_n - Tx_n \to \theta$ as $n \to \infty$;*
(3) *$x_n \rightharpoonup p \in \mathrm{Fix}(T)$ as $n \to \infty$.*

Proof.
(1) Fixing $p \in \mathrm{Fix}(T)$, it follows from the iterative method (KMIM) that

$$\|x_{n+1} - p\| \leq (1 - \alpha_n)\|x_n - p\| + \alpha_n\|Tx_n - p\| \leq \|x_n - p\|.$$

This shows that $\{x_n\}$ is Fejér monotone with respect to $\mathrm{Fix}(T)$. Hence $\lim_{n \to \infty} \|x_n - p\|$ exists. In particular, we have that $\{x_n\}$ is bounded.
(2) By the properties of the norm and scalar product on H, we have

$$\begin{aligned}
\|x_{n+1} - p\|^2 &= \left\|(1 - \alpha_n)(x_n - p) + \alpha_n(Tx_n - p)\right\|^2 \\
&= (1 - \alpha_n)\|x_n - p\|^2 + \alpha_n\left\|(Tx_n - p)\right\|^2 \\
&\quad - \alpha_n(1 - \alpha_n)\|Tx_n - x_n\|^2 \\
&\leq \|x_n - p\|^2 - \alpha_n(1 - \alpha_n)\|Tx_n - x_n\|^2. \tag{2.35}
\end{aligned}$$

This implies that

$$\sum_{n=1}^{\infty} \alpha_n(1 - \alpha_n)\|Tx_n - x_n\|^2 \leq \|x_1 - p\|^2 < \infty. \tag{2.36}$$

Using the assumption that $\sum_{n=1}^{\infty} \alpha_n(1 - \alpha_n) = \infty$, we see that

$$\liminf_{n \to \infty} \|x_n - Tx_n\|^2 = 0.$$

It follows that

$$\liminf_{n \to \infty} \|x_n - Tx_n\| = 0. \tag{2.37}$$

Observe that

$$\begin{aligned}
\|x_{n+1} - Tx_{n+1}\| &= \left\|(1 - \alpha_n)(x_n - Tx_{n+1}) + \alpha_n(Tx_n - Tx_{n+1})\right\| \\
&\leq (1 - \alpha_n)\|x_n - Tx_n\| + \|Tx_n - Tx_{n+1}\| \\
&\leq (1 - \alpha_n)\|x_n - Tx_n\| + \|x_n - x_{n+1}\| \\
&= (1 - \alpha_n)\|x_n - Tx_n\| + \alpha_n\|x_n - Tx_n\| \\
&= \|x_n - Tx_n\|, \quad \forall n \geq 1. \tag{2.38}
\end{aligned}$$

So that $\lim_{n\to\infty} \|x_n - Tx_n\|$ exists. Using (2.37), we obtain

$$\lim_{n\to\infty} \|x_n - Tx_n\| = 0.$$

(3) It follows from the demiclosedness principle of nonexpansive mappings that $\omega_\omega(x_n) \subset \text{Fix}(T)$. Therefore, we conclude from 2.4.1 that $x_n \rightharpoonup p \in \text{Fix}(T)$, where

$$p = \lim_{n\to\infty} P_{\text{Fix}(T)}(x_n).$$

This completes the proof. □

Corollary 2.4.2. *Let λ be a real number in $(0,1)$ and let $T : C \to C$ be a λ-averaged nonexpansive mapping with $\text{Fix}(T) \neq \emptyset$. Suppose that $\{a_n\}$ is a sequence in $[0, \frac{1}{\lambda}]$ satisfying the following condition:*

$$\sum_{n=1}^{\infty} a_n(1 - \lambda a_n) = \infty.$$

If $\{x_n\}$ is the sequence generated by Krasnosel'skiĭ–Man iterative method (KMIM), then the following results hold:
(1) *$\{x_n\}$ is Fejér monotone sequence with respect to $\text{Fix}(T)$;*
(2) *$x_n - Tx_n \to \theta$ as $n \to \infty$;*
(3) *$x_n \rightharpoonup p \in \text{Fix}(T)$ as $n \to \infty$.*

Proof. Let $S = (1 - \frac{1}{\lambda})I + \frac{1}{\lambda}T$ and $\mu_n = \lambda a_n$. Then S is nonexpansive, $\text{Fix}(S) = \text{Fix}(T)$, $\mu_n \in [0, 1]$, $\sum_{n=1}^{\infty} \mu_n(1 - \mu_n) = \infty$, and

$$\begin{aligned}
x_{n+1} &= (1 - a_n)x_n + a_n Tx_n \\
&= x_n - a_n(I - T)x_n \\
&= x_n - \lambda a_n(I - S)x_n \\
&= (1 - \mu_n)x_n + \mu_n Sx_n, \quad \forall n \geq 1.
\end{aligned} \tag{2.39}$$

By Theorem 2.4.3, we obtain the three conclusions immediately. This completes the proof. □

Corollary 2.4.3. *Let $T : C \to C$ be a firmly nonexpansive mapping with $\text{Fix}(T) \neq \emptyset$. Suppose that $\{a_n\}$ is a sequence in $[0, 2]$ satisfying the following condition:*

$$\sum_{n=1}^{\infty} a_n(2 - a_n) = \infty.$$

If $\{x_n\}$ is the sequence generated by Krasnosel'skiĭ–Man iterative method (KMIM), then the following results hold:
(1) *$\{x_n\}$ is Fejér monotone sequence with respect to $\text{Fix}(T)$;*

(2) $x_n - Tx_n \to \theta$ as $n \to \infty$;
(3) $x_n \to p \in \text{Fix}(T)$ as $n \to \infty$.

Proof. Corollary 2.1.1 implies that T is a $\frac{1}{2}$-averaged nonexpansive mapping. Taking $\lambda = \frac{1}{2}$ in Corollary 2.4.2, we obtain the conclusions immediately. This completes the proof. □

2.4.2 Strong convergence theorems

Generally speaking, Banach–Picard iterative method (BPIM) and Krasnoselskiĭ–Man iterative method (KMIM) are only weakly convergent in infinite-dimensional Hilbert spaces. Of course, if $T : C \to C$ is a linear nonexpansive mapping or $T(C)$ is contained in a compact set, then they can be strongly convergent. In order to obtain the strong convergence of iterative sequences, Halpern proposed his iterative method (HIM) in 1967 without the assumption that $T : C \to C$ is a linear nonexpansive mapping or $T(C)$ is contained in a compact set and studied the conditions imposed on control sequences for fixed points of nonexpansive mappings in Hilbert spaces.

One decade later, Lions [42] gave another control condition for the iterative method (HIM), however, his conditions exclude the canonical choice of the control sequences, for example, the sequence $\{\alpha_n\}$ defined by $\alpha_n = \frac{1}{n}$ for each $n \geq 1$. In 1992, Wittmann [97] overcame this flaw with the sequence $\alpha_n = \frac{1}{n}$. In 2002, Xu [99] gave another weak condition. In 2007, Suzuki [84] used new techniques to prove that Halpern iterative method (HIM) for averaged nonexpansive mappings is strongly convergent in case $\{\alpha_n\}$ satisfies the following conditions:
(C1) $\alpha_n \to 0$;
(C2) $\sum_{n=1}^{\infty} \alpha_n = \infty$

In recent two decades, Halpern iterative method (HIM) was extensively investigated by many authors, and many important results were established both in Hilbert and Banach spaces.

Next, we list some recent important and celebrated results.

Theorem 2.4.4 (Browder [7], 1965). *Let C be a nonempty, bounded, closed, and convex subset of H, and let $T : C \to C$ be a nonexpansive mapping. For any $x_0 \in C$ and $n \geq 1$, define a mapping $T_n : C \to C$ by*

$$T_n x = \left(1 - \frac{1}{n+1}\right) Tx + \frac{1}{n+1} x_0, \quad \forall x \in C, n \geq 1. \tag{2.40}$$

Then the following results hold:
(1) *For each $n \geq 1$, T_n has a unique fixed point $u_n \in C$;*
(2) *$\{u_n\}$ converges strongly to $z = P_{\text{Fix}(T)} x_0$.*

Proof.
(1) It is a direct corollary of the Banach's contraction principle.
(2) Note that

$$u_n = \left(1 - \frac{1}{n+1}\right)Tu_n + \frac{1}{n+1}x_0, \quad n \ge 1. \tag{2.41}$$

Since both $\{u_n\}$ and $\{Tu_n\}$ are bounded sequences, we have

$$u_n - Tu_n = \frac{1}{n+1}(x_0 - Tu_n) \to \theta \quad (n \to \infty). \tag{2.42}$$

It follows from Theorem 2.3.2 that $\text{Fix}(T) \neq \emptyset$. Fix $p \in \text{Fix}(T)$. Using (2.41), we have

$$
\begin{aligned}
\|u_n - p\|^2 &= \langle u_n - p, u_n - p \rangle \\
&= \left(1 - \frac{1}{n+1}\right)\langle Tu_n - p, u_n - p \rangle + \frac{1}{n+1}\langle x_0 - p, u_n - p \rangle \\
&\le \left(1 - \frac{1}{n+1}\right)\|u_n - p\|^2 + \frac{1}{n+1}\langle x_0 - p, u_n - p \rangle.
\end{aligned}
\tag{2.43}
$$

This implies that

$$\|u_n - p\|^2 \le \langle x_0 - p, u_n - p \rangle, \quad \forall p \in \text{Fix}(T), \tag{2.44}$$

and

$$\langle u_n - x_0, u_n - p \rangle \le 0, n \ge 1, \quad \forall p \in \text{Fix}(T). \tag{2.45}$$

Since $\{u_n\} \subset C$ is bounded, there exists a subsequence $\{u_{n_j}\} \subset \{u_n\}$ such that $u_{n_j} \rightharpoonup z$ as $j \to \infty$. It follows from the demiclosedness principle of nonexpansive mappings that $z \in \text{Fix}(T)$. In (2.44), taking $p = z \in C$ and $n := n_j$, we obtain

$$
\begin{aligned}
&\|u_{n_j} - z\|^2 \le \langle x_0 - z, u_{n_j} - z \rangle \to 0 \\
&\implies u_{n_j} \to z \quad (j \to \infty).
\end{aligned}
$$

Letting $j \to \infty$ in (2.45), we see that

$$
\begin{aligned}
&\langle z - x_0, z - p \rangle \le 0, \quad \forall p \in \text{Fix}(T) \\
&\implies z = P_{\text{Fix}(T)}x_0, \quad u_n \to z \ (n \to \infty).
\end{aligned}
$$

This completes the proof. $\qquad\square$

We remark that the conclusions of Browder's strong convergence theorem still hold if the boundedness assumption of C is replaced by the assumption that $\text{Fix}(T) \neq \emptyset$.

Theorem 2.4.5 (Halpern [30], 1967). *Let C be a closed convex subset of H and let $T :$ $C \to C$ be a nonexpansive mapping with $\mathrm{Fix}(T) \neq \emptyset$. Let $\alpha_n = \frac{1}{(n+1)^a}$, where $a \in (0,1)$. Let $\{x_n\}$ be a sequence generated by (HIM). Then $\{x_n\}$ converges strongly to a fixed point of T which is closest to u.*

Proof. It follows from Theorem 2.4.4 that there exists a unique $y_n \in C$ such that

$$y_n = \alpha_n u + (1 - \alpha_n) T y_n. \tag{2.46}$$

Using (HIM) and (2.46), we have

$$x_{n+1} - y_n = (1 - \alpha_n)(T x_n - T y_n). \tag{2.47}$$

Since $T : C \to C$ is nonexpansive, it follows that

$$\|x_{n+1} - y_n\| \leq (1 - \alpha_n)\|x_n - y_n\|. \tag{2.48}$$

Hence,

$$\|x_{n+1} - y_n\| \leq (1 - \alpha_n)\|x_n - y_{n-1}\| + \|y_n - y_{n-1}\|. \tag{2.49}$$

Now we estimate $\|y_n - y_{n-1}\|$. It follows from (2.46) that

$$\|y_n - y_{n-1}\| = \|\alpha_n u + (1 - \alpha_n) T y_n - \alpha_n u - (1 - \alpha_{n-1}) T y_{n-1}\|$$
$$\leq |\alpha_n - \alpha_{n-1}|(\|u\| + \|T y_{n-1}\|) + (1 - \alpha_n)\|y_n - y_{n-1}\|.$$

Hence

$$\|y_n - y_{n-1}\| \leq \left|1 - \frac{\alpha_{n-1}}{\alpha_n}\right|(\|u\| + \|T y_{n-1}\|). \tag{2.50}$$

It follows from Theorem 2.4.4 that $y_n \to P_{\mathrm{Fix}(T)}(u)$ as $n \to \infty$. In particular, $\{y_n\}$ is a bounded sequence. So is $\{T y_n\}$. Denote $M = \|u\| + \sup_{n \geq 0}\{\|T y_n\|\}$. Then (2.50) can be rewritten as follows:

$$\|y_n - y_{n-1}\| \leq M\left|1 - \frac{\alpha_{n-1}}{\alpha_n}\right|. \tag{2.51}$$

Since $1 - \frac{\alpha_{n-1}}{\alpha_n} = 1 - \left(\frac{n+1}{n}\right)^a$ and

$$\frac{\alpha_n - \alpha_{n-1}}{\alpha_n} = (n+1)^a\left[1 - \left(\frac{n+1}{n}\right)^a\right] \to 0 \quad (n \to \infty),$$

we have $\|y_n - y_{n-1}\| = o(\alpha_n)$. In view of (2.49), we obtain

$$\|x_{n+1} - y_n\| \leq (1 - \alpha_n)\|x_n - y_{n-1}\| + o(\alpha_n), \quad \forall n \geq 1. \tag{2.52}$$

Using Lemma 1.10.2, we obtain

$$x_{n+1} - y_n \to \theta \quad (n \to \infty).$$

Hence $x_n \to P_{\mathrm{Fix}(T)}(u)$ as $n \to \infty$. This completes the proof. $\qquad\square$

In 1977, Lions [42] further generalized Theorem 2.4.5 as follows.

Theorem 2.4.6 (Lions [42], 1977). *Let C be a closed convex subset of H and let $T : C \to C$ be a nonexpansive mapping with* $\text{Fix}(T) \neq \emptyset$. *Suppose that* $\{\alpha_n\}$ *in* $[0, 1]$ *satisfies the following conditions:*

(C1) $\alpha_n \to 0(n \to \infty)$;

(C2) $\sum_{n=0}^{\infty} \alpha_n = \infty$;

(C3) $\frac{\alpha_{n+1} - \alpha_n}{\alpha_{n+1}^2} \to 0$.

Let $\{x_n\}$ *be a sequence generated by (HIM). Then* $\{x_n\}$ *converges strongly to* $P_{\text{Fix}(T)}u$.

Since the proof is the same as that of Theorem 2.4.5, we omit the proof here.

Remark 2.4.1. Halpern [30] pointed out that conditions (C1) and (C2) are necessary for the convergence of iterative sequences generated in iterative method (KMIM). Indeed, he also put forward the following open question:

Are conditions (C1) and (C2) sufficient to ensure that the sequence generated by (HIM) converges strongly to a fixed point of T in C?

Recently, Suzuki [85] gave an answer to the Halpern's question, however, what are the sufficient and necessary conditions concerning the control sequence $\{\alpha_n\}$ are still not clear.

Remark 2.4.2. Theorems 2.4.5 and 2.4.6 exclude the canonical choice of the control sequence $\{\alpha_n\}$, namely $\alpha_n = \frac{1}{n+1}$ for each $n \geq 0$.

In 1992, Wittmann [97] overcame this flaw and he introduced the following condition:

(C4) $\sum_{n=0}^{\infty} |\alpha_{n+1} - \alpha_n| < \infty$.

With the aid of condition (C4), he established the following strong convergence theorem.

Theorem 2.4.7 (Wittmann [97], 1992). *Let C be a closed convex subset of H and let $T : C \to C$ be a nonexpansive mapping with* $\text{Fix}(T) \neq \emptyset$. *Suppose that* $\{\alpha_n\} \in [0, 1]$ *satisfies the control conditions* (C1), (C2), *and* (C4). *Let* $\{x_n\}$ *be the sequence generated by (HIM). Then* $\{x_n\}$ *converges strongly to* $P_{\text{Fix}(T)}(u)$.

Proof. The proof can be split into five steps as follows:

Step 1. Show that $\{x_n\}$ is bounded.

Fixing $p \in \text{Fix}(T)$, one has

$$\|x_{n+1} - p\| = \alpha_n \|u - p\| + (1 - \alpha_n)\|Tx_n - p\|$$
$$\leq \alpha_n \|u - p\| + (1 - \alpha_n)\|x_n - p\|$$
$$\leq \max\{\|u - p\|, \|x_0 - p\|\} = M, \quad \forall n \geq 0.$$

Step 2. Show that $x_{n+1} - x_n \to \theta$ as $n \to \infty$.

Note that

$$\begin{aligned}
\|x_{n+1} - x_n\| &= \|(\alpha_n - \alpha_{n-1})u + (1 - \alpha_n)Tx_n - (1 - \alpha_{n-1})Tx_{n-1}\| \\
&\leq |\alpha_n - \alpha_{n-1}\|u\|| + |1 - \alpha_n|\|x_n - x_{n-1}\| + |\alpha_n - \alpha_{n-1}|\|Tx_{n-1}\| \\
&\leq (1 - \alpha_n)\|x_n - x_{n-1}\| + M_1|\alpha_n - \alpha_{n-1}|,
\end{aligned}$$

where $M_1 = \|u\| + \sup_{u \geq 0}\{\|Tx_n\|\}$. Using Lemma 1.10.2, we have $x_{n+1} - x_n \to \theta$ as $n \to \infty$.

Step 3. Show that $x_n - Tx_n \to \theta$ as $n \to \infty$.

It follows from (HIM) that $x_{n+1} - Tx_n = \alpha_n(u - Tx_n) \to \theta$ as $n \to \infty$. From Step 2, we see that

$$x_n - Tx_n = (x_n - x_{n+1}) + (x_{n+1} - Tx_n) \to \theta \quad (n \to \infty).$$

Step 4. Show that $\limsup_{n\to\infty}\langle u - P_{\text{Fix}(T)}u, x_n - P_{\text{Fix}(T)}u\rangle \leq 0$.

Let $\{x_{n_k}\}$ be a subsequence of $\{x_n\}$ such that

$$\begin{aligned}
&\limsup_{n\to\infty}\langle u - P_{\text{Fix}(T)}u, x_n - P_{\text{Fix}(T)}u\rangle \\
&= \lim_{k\to\infty}\langle u - P_{\text{Fix}(T)}u, x_{n_k} - P_{\text{Fix}(T)}u\rangle.
\end{aligned}$$

Since $\{x_{n_k}\}$ is bounded, we may, without loss of generality, assume that $x_{n_k} \rightharpoonup y$ as $k \to \infty$. It follows from the Browder's demiclosedness principle 1.9.3 that $y \in \text{Fix}(T)$. By the property of $P_{\text{Fix}(T)}$, we obtain

$$\lim_{k\to\infty}\langle u - P_{\text{Fix}(T)}u, x_{n_k} - P_{\text{Fix}(T)}u\rangle = \langle u - P_{\text{Fix}(T)}u, y - P_{\text{Fix}(T)}u\rangle \leq 0.$$

Step 5. Show that $x_n \to P_{\text{Fix}(T)}u$ as $n \to \infty$.

Note that

$$\begin{aligned}
\|x_{n+1} - P_{\text{Fix}(T)}u\|^2 &= \alpha_n^2\|u - P_{\text{Fix}(T)}u\|^2 + (1 - \alpha_n)^2\|Tx_n - P_{\text{Fix}(T)}u\|^2 \\
&\quad + 2\alpha_n(1 - \alpha_n)\langle u - P_{\text{Fix}(T)}u, Tx_n - P_{\text{Fix}(T)}u\rangle \\
&\leq (1 - \alpha_n)\|x_n - P_{\text{Fix}(T)}u\|^2 + \alpha_n^2\|u - P_{\text{Fix}(T)}u\|^2 \\
&\quad + 2\alpha_n(1 - \alpha_n)\langle u - P_{\text{Fix}(T)}u, Tx_n - x_n\rangle \\
&\quad + 2\alpha_n(1 - \alpha_n)\langle u - P_{\text{Fix}(T)}u, x_n - P_{\text{Fix}(T)}u\rangle \\
&\leq (1 - \alpha_n)\|x_n - P_{\text{Fix}(T)}u\|^2 + 2\alpha_n\|u - P_{\text{Fix}(T)}u\|\|x_n - Tx_n\| \\
&\quad + 2\alpha_n(1 - \alpha_n)\langle u - P_{\text{Fix}(T)}u, x_n - P_{\text{Fix}(T)}u\rangle + \alpha_n^2\|u - P_{\text{Fix}(T)}u\|^2 \\
&\leq (1 - \alpha_n)\|x_n - P_{\text{Fix}(T)}u\|^2 + o(\alpha_n),
\end{aligned}$$

where $o(\alpha_n) = 2\alpha_n\|u - P_{\text{Fix}(T)}u\|\|x_n - Tx_n\| + 2\alpha_n(1 - \alpha_n)\langle u - P_{\text{Fix}(T)}u, x_n - P_{\text{Fix}(T)}u\rangle + \alpha_n^2\|u - P_{\text{Fix}(T)}u\|^2$.

By Lemma 1.10.2, we have $x_n \to P_{\text{Fix}(T)}u$ as $n \to \infty$. This completes the proof. □

In 2002, Xu [100] changed condition (C4) into the following condition in Theorem 2.4.7:

(C5) $\frac{\alpha_{n-1}}{\alpha_n} \to 1$ as $n \to \infty$.

To be more precise, Xu proved the following result.

Theorem 2.4.8 (Xu [100], 2002). *Let C be a closed convex subset of H and let $T : C \to C$ be a nonexpansive mapping with $\text{Fix}(T) \neq \emptyset$. Let $\{x_n\}$ be a sequence generated by (HIM), where $\{\alpha_n\}$ is a sequence in $(0,1)$ satisfying conditions (C1), (C2), and (C5). Then $\{x_n\}$ converges strongly to $P_{\text{Fix}(T)}(u)$.*

Proof. Note that condition (C4) is only used for the second step in Theorem 2.4.7. It guarantees that $x_{n+1} - x_n \to \theta$ as $n \to \infty$. Note that

$$\|x_{n+1} - x_n\| \leq (1 - \alpha_n)\|x_n - x_{n-1}\| + M_1|\alpha_n - \alpha_{n-1}|$$
$$= (1 - \alpha_n)\|x_n - x_{n-1}\| + o(\alpha_n).$$

By Lemma 1.10.2, it follows that $x_{n+1} - x_n \to \theta$ as $n \to \infty$. The rest of the proof is similar to that of Theorem 2.4.7. This completes the proof. □

Remark 2.4.3. Xu pointed out that conditions (C4) and (C5) are mutually independent. Comparing conditions (C3) and (C5), we have

$$(C5) \quad \frac{\alpha_{n-1}}{\alpha_n} \to 1 \quad (n \to \infty) \quad \Longleftrightarrow \quad \frac{\alpha_{n+1} - \alpha_n}{\alpha_{n+1}} \to 0 \quad (n \to \infty).$$

Observe that

$$\frac{|\alpha_{n+1} - \alpha_n|}{\alpha_{n+1}} \leq \frac{|\alpha_{n+1} - \alpha_n|}{\alpha_{n+1}^2}.$$

Therefore, if $\{\alpha_n\}$ satisfies condition (C3), then it satisfies condition (C5). In addition, condition (C5) is satisfied by the canonical choice $\alpha_n = \frac{1}{n+1}$ for each $n \geq 0$, however, condition (C3) is not. Therefore, Theorem 2.4.8 is an improvement of Theorem 2.4.7.

We can further improve Theorems 2.4.7 and 2.4.8 by using the Banach limit technique. By using Suzuki's lemma 1.10.3, we can prove the following more general convergence theorem. The advantage is that it only needs to satisfy conditions (C1) and (C2) for the strong convergence of the iterative sequence $\{x_n\}$.

Theorem 2.4.9. *Let C be a closed convex subset of H and let $T : C \to C$ be a nonexpansive mapping with $\text{Fix}(T) \neq \emptyset$. Suppose that $\{\alpha_n\}$, $\{\beta_n\}$, and $\{\gamma_n\}$ are three sequences in $[0,1]$ such that*

(1) $\alpha_n + \beta_n + \gamma_n = 1$ *for all $n \geq 0$;*

(2) $0 < \liminf_{n \to \infty} \alpha_n \leq \limsup_{n \to \infty} \alpha_n < 1$;

(3) $\beta_n \to 0$ as $n \to \infty$;

(4) $\sum_{n=0}^{\infty} \beta_n = \infty$.

For an arbitrary initial value $x_0 \in C$ and a fixed anchor point $u \in C$, define a sequence $\{x_n\}$ iteratively in C as follows:

$$x_{n+1} = \beta_n u + \alpha_n x_n + \gamma_n T x_n, \quad \forall n \geq 1. \tag{A}$$

Then $\{x_n\}$ converges strongly to $P_{\mathrm{Fix}(T)}(u)$.

Proof. The proof is split into six steps as follows.

Step 1. Show that $\{x_n\}$ is bounded.

Fixing $p \in \mathrm{Fix}(T)$, one finds from the iterative method (A) that

$$
\begin{aligned}
\|x_{n+1} - p\| &= \beta_n \|u - p\| + \alpha_n \|x_n - p\| + \gamma_n \|T x_n - p\| \\
&\leq \beta_n \|u - p\| + (\alpha_n + \gamma_n) \|x_n - p\| \\
&\leq \max\{\|u - p\|, \|x_0 - p\|\} \\
&= M, \quad \forall n \geq 0.
\end{aligned}
$$

Hence $\{x_n\}$ is bounded, so is $\{T x_n\}$.

Step 2. Show that $x_{n+1} - x_n \to \theta$ as $n \to \infty$.

Define a sequence $\{y_n\}$ by

$$y_n = \frac{x_{n+1} - \alpha_n x_n}{1 - \alpha_n}, \quad \forall n \geq 0.$$

Then $x_{n+1} = \alpha_n x_n + (1 - \alpha_n) y_n$ for all $n \geq 0$. Note that

$$
\begin{aligned}
\|y_{n+1} - y_n\| &= \left\| \frac{x_{n+2} - \alpha_{n+1} x_{n+1}}{1 - \alpha_{n+1}} - \frac{x_{n+1} - \alpha_n x_n}{1 - \alpha_n} \right\| \\
&= \left\| \frac{\beta_{n+1} u - y_n + T x_{n+1}}{1 - \alpha_{n+1}} - \frac{\beta_n u - \lambda_n T x_n}{1 - \alpha_n} \right\| \\
&\leq \left\| \left(\frac{\beta_{n+1}}{1 - \alpha_{n+1}} - \frac{\beta_n}{1 - \alpha_n} \right) u \right\| + \|T x_{n+1} - x_n\| \\
&\quad + \left\| \frac{\beta_{n+1} T x_{n+1}}{1 - \alpha_{n+1}} - \frac{\beta_n T x_n}{1 - \alpha_n} \right\| \\
&\leq \left| \frac{\beta_{n+1}}{1 - \alpha_{n+1}} - \frac{\beta_n}{1 - \alpha_n} \right| (\|u\| + \|T x_n\|) \\
&\quad + \left(1 + \frac{\beta_{n+1}}{1 - \alpha_{n+1}} \right) \|x_{n+1} - x_n\| \\
&\leq \left| \frac{\beta_{n+1}}{1 - \alpha_{n+1}} - \frac{\beta_n}{1 - \alpha_n} \right| M_1 + \left(1 + \frac{\beta_{n+1}}{1 - \alpha_{n+1}} \right) \|x_{n+1} - x_n\|,
\end{aligned}
$$

where $M_1 = \|u\| + \sup_{n \geq 0}\{\|T x_n\|\}$. Since $\beta_n \to 0$ as $n \to \infty$, one has

$$\limsup_{n \to \infty} (\|y_{n+1} - y_n\| - \|x_{n+1} - x_n\|) \leq 0.$$

From Suzuki's lemma 1.10.3, we see that $y_n - x_n \to \theta \ (n \to \infty)$. Therefore, we conclude

$$x_{n+1} - x_n = (1 - \alpha_n)(y_n - x_n) \to \theta \quad (n \to \infty).$$

Step 3. Show that $x_n - Tx_n \to \theta \ (n \to \infty)$.

It follows from conditions (1), (2), and (3) that

$$\liminf_{n \to \infty} y_n > 0.$$

Since $x_{n+1} - x_n = \beta_n(u - x_n) + \gamma_n(Tx_n - x_n) \to \theta, \beta_n \to 0$ and $x_{n+1} - x_n \to \theta$ as $n \to \infty$, we find that $x_n - Tx_n \to \theta \ (n \to \infty)$.

Step 4. Show that $\omega_\omega(x_n) \subset \mathrm{Fix}(T)$, where

$$\omega_\omega(x_n) = \{x \in C : \exists \ \{x_{n_j}\} \subset \{x_n\} \text{ such that } x_{n_j} \rightharpoonup x \quad (j \to \infty)\}$$

is a weak-ω limit set of $\{x_n\}$.

For all $x \in \omega_\omega(x_n)$, since $x_{n_j} \rightharpoonup x$ as $j \to \infty$ and $x_{n_j} - Tx_{n_j} \to \theta$ as $j \to \infty$, one concludes from Theorem 1.9.3 that $x = Tx$.

Step 5. Show that $\limsup_{n \to \infty} \langle u - z, x_n - z \rangle \le 0$, where $z = P_{\mathrm{Fix}(T)}u$.

Let $\{x_{n_j}\}$ be a subsequence of $\{x_n\}$ such that

$$\limsup_{n \to \infty} \langle u - z, x_n - z \rangle = \limsup_{j \to \infty} \langle u - z, x_{n_j} - z \rangle.$$

Since $\{x_{n_j}\}$ is bounded, we may, without loss of generality, assume that $x_{n_j} \rightharpoonup x$ as $j \to \infty$. From (4), we see that $x = Tx$. Hence

$$\limsup_{n \to \infty} \langle u - z, x_n - z \rangle = \langle u - z, x - z \rangle \le 0.$$

Letting $\sigma_n = \max\{\langle u - z, x_n - z \rangle, 0\}$, we have $\sigma_n > 0$ and $\sigma_n \to 0$ as $n \to \infty$.

Step 6. Show that $x_n \to z$ as $n \to \infty$.

Note that

$$
\begin{aligned}
\|x_{n+1} - z\|^2 &= \langle x_{n+1} - z, x_{n+1} - z \rangle \\
&= \beta_n \langle u - z, x_{n+1} - z \rangle + \alpha_n \langle x_n - z, x_{n+1} - z \rangle + \gamma_n \langle Tx_n - z, x_{n+1} - z \rangle \\
&\le \beta_n \sigma_{n+1} + \frac{\alpha_n}{2}\|x_n - z\|^2 + \frac{\alpha_n}{2}\|x_{n+1} - z\|^2 \\
&\quad + \frac{\gamma_n}{2}\|x_n - z\|^2 + \frac{\gamma_n}{2}\|x_{n+1} - z\|^2 \\
&= \beta_n \sigma_{n+1} + \frac{1}{2}(1 - \beta_n)\|x_n - z\|^2 + \frac{1}{2}(1 - \beta_n)\|x_{n+1} - z\|^2.
\end{aligned}
$$

This implies that

$$
\begin{aligned}
\|x_{n+1} - z\|^2 &= (1 - \beta_n)\|x_n - z\|^2 + 2\beta_n \sigma_{n+1} \\
&= (1 - \beta_n)\|x_n - z\|^2 + o(\beta_n).
\end{aligned}
$$

It follows from Lemma 1.10.2 that $x_n \to P_{\mathrm{Fix}(T)}u$ as $n \to \infty$. This completes the proof. $\qquad\qquad\square$

Corollary 2.4.4 ([84]). *Let C be a closed convex subset of H and let $T : C \to C$ be a nonexpansive mapping with $\mathrm{Fix}(T) \neq \emptyset$. Suppose that $\{\alpha_n\}$ and $\{\lambda_n\}$ are two sequences in $[0, 1]$ satisfying the following control conditions:*
(C1) $\alpha_n \to 0(n \to \infty)$;
(C2) $\sum_{n=0}^{\infty} \alpha_n = \infty$;
(C3) $0 < \liminf_{n\to\infty} \lambda_n \leq \limsup_{n\to\infty} \lambda_n < 1$.

For an arbitrary initial value $x_0 \in C$ and a fixed anchor $u \in C$, define a sequence $\{x_n\}$ iteratively in C as follows:

$$x_{n+1} = \alpha_n u + (1 - \alpha_n)\big[(1 - \lambda_n)x_n + \lambda_n T x_n\big], \quad \forall n \geq 1. \tag{A1}$$

Then $\{x_n\}$ converges strongly to $P_{\mathrm{Fix}(T)}(u)$.

Proof. It follows from iterative method (A1) that

$$x_{n+1} = \alpha_n u + (1 - \alpha_n)(1 - \lambda_n)x_n + (1 - \alpha_n)\lambda_n T x_n, \quad \forall n \geq 1.$$

Using Theorem 2.4.9, we obtain the desired conclusion immediately. $\qquad\qquad\square$

Corollary 2.4.5. *Let C be a closed convex subset of H and let $\{T_i\}_{i=1}^{r} : C \to C$ be a finite family of r nonexpansive mappings with $F = \bigcap_{i=1}^{r} \mathrm{Fix}(T_i) \neq \emptyset$. Suppose that $\{\alpha_n\}$ and $\{\lambda_n\}$ are two sequences in $[0, 1]$ satisfying the following control conditions:*
(C1) $\alpha_n \to 0$ as $n \to \infty$;
(C2) $\sum_{n=0}^{\infty} \alpha_n = \infty$;
(C3) $0 < \liminf_{n\to\infty} \lambda_n \leq \limsup_{n\to\infty} \lambda_n < 1$.

For an arbitrary initial value $x_0 \in C$ and a fixed anchor $u \in C$, define a sequence $\{x_n\}$ iteratively in C as follows:

$$x_{n+1} = \alpha_n u + (1 - \alpha_n)\left[(1 - \lambda_n)x_n + \lambda_n \sum_{i=1}^{r} \omega_i T_i x_n\right], \quad \forall n \geq 1, \tag{A2}$$

where $\omega_i \in (0, 1)$, $i = 1, \ldots, r$, satisfy $\sum_{i=1}^{r} \omega_i = 1$. Then $\{x_n\}$ converges strongly to $P_F(u)$.

Proof. Let $T = \sum_{i=1}^{r} \omega_i T_i$. Then $T : C \to C$ is a nonexpansive mapping with $F = \mathrm{Fix}(T)$. By Corollary 2.4.4, we obtain the conclusion immediately. This completes the proof. $\qquad\qquad\square$

Theorem 2.4.10 (Bauschke [6], 1996). *Let C be a closed convex subset of H and let $\{T_i\}_{i=1}^{r} : C \to C$ be a finite family of r nonexpansive mappings satisfying the following* *condition:*

$$F = \bigcap_{i=1}^{r} \text{Fix}(T_i) = \text{Fix}(T_r T_{r-1} \cdots T_1)$$

$$= \text{Fix}(T_1 T_r T_{r-1} \cdots T_2) = \cdots = \text{Fix}(T_{r-1} \cdots T_1 T_r)$$

$$\neq \emptyset. \tag{CC}$$

Suppose that $\{\alpha_n\}$ is a real number sequence in $[0,1]$ satisfying the control conditions (C1), (C2), and
(C3) $\sum_{n=1}^{\infty} |\alpha_{n+r} - \alpha_n| < +\infty$ or (C4) $\frac{\alpha_n}{\alpha_{n+r}} \to 1$ as $n \to \infty$.

For an arbitrary initial value $x_0 \in C$ and a fixed element $u \in C$, define a sequence $\{x_n\}$ iteratively in C as follows:

$$x_{n+1} = \alpha_n u + (1 - \alpha_n) T_{n+1} x_n, \quad \forall n \geq 0, \tag{B}$$

where $T_n = T_{n \bmod r}$. Then $x_n \to z$ as $n \to \infty$, where $z = P_F u$ and P_F is the metric projection from C onto F.

Proof. In order to prove $x_n \to P_F u$ as $n \to \infty$, we estimate $\|x_{n+1} - z\|^2$ as follows:

$$\|x_{n+1} - z\|^2 = \left\| (1 - \alpha_n)(T_{n+1} x_n - T_{n+1} z) + \alpha_n (u - z) \right\|^2$$

$$= (1 - \alpha_n)^2 \|T_{n+1} x_n - T_{n+1} z\|^2$$

$$+ 2\alpha_n (1 - \alpha_n) \langle u - z, T_{n+1} x_n - z \rangle + \alpha_n^2 \|u - z\|^2$$

$$\leq (1 - \alpha_n)\|x_n - z\|^2 + 2\alpha_n(1 - \alpha_n)\langle u - z, T_{n+1} x_n - z \rangle + o(\alpha_n). \tag{2.53}$$

Thus we need to prove

$$\limsup_{n \to \infty} \langle u - z, T_{n+1} x_n - z \rangle \leq 0. \tag{2.54}$$

From the assumption $F \neq \emptyset$, it follows that, for any $p \in F$,

$$\|x_{n+1} - p\|^2 = (1 - \alpha_n)\|T_{n+1} x_n - T_{n+1} p\| + \alpha_n \|u - z\|$$

$$\leq (1 - \alpha_n)\|x_n - p\| + \alpha_n \|u - z\|$$

$$\leq \max\{\|x_0 - p\|, \|u - z\|\}$$

$$= M.$$

Hence $\{x_n\}$ is a bounded sequence, so is $\{T_{n+1} x_n\}$. By (B) and (C1), we obtain

$$x_{n+1} - T_{n+1} x_n = \alpha_n(u - T_{n+1} x_n) \to \theta \quad (n \to \infty). \tag{2.55}$$

Thus we only need to prove

$$\limsup_{n \to \infty} \langle x_{n+1} - z, u - z \rangle \leq 0, \tag{2.56}$$

that is,

$$\limsup_{n\to\infty}\langle x_n - z, u - z\rangle \le 0. \tag{2.57}$$

Let

$$\limsup_{n\to\infty}\langle u - z, x_n - z\rangle = \lim_{j\to\infty}\langle u - z, x_{n_j} - z\rangle. \tag{2.58}$$

Without loss of generality, we may assume that $x_{n_j} \to p$ as $j \to \infty$. Then (2.58) can be rewritten as follows:

$$\limsup_{n\to\infty}\langle u - z, x_n - z\rangle = \langle u - z, p - z\rangle. \tag{2.59}$$

If we can prove $p \in F$, then we may conclude from the property of the distance projection P_F that

$$\langle u - z, p - z\rangle \le 0. \tag{2.60}$$

Thus we now need to prove $p \in F$. In fact,

$$\|x_{n+2} - T_{n+2}T_{n+1}x_n\| = \|x_{n+2} - T_{n+2}x_{n+1} + T_{n+2}x_{n+1} - T_{n+2}T_{n+1}x_n\|$$
$$\le \|x_{n+2} - T_{n+2}x_{n+1}\| + \|x_{n+1} - T_{n+1}x_n\|,$$

which implies from (2.55) that $\lim_{n\to\infty}\|x_{n+2} - T_{n+2}x_n\| = 0$. Note that

$$x_{n+r} - T_{n+r}T_{n+r-1}\cdots T_{n+1}x_n \to \theta \quad (n \to \infty). \tag{2.61}$$

Using (B), we have

$$\|x_{n+r+1} - x_{n+1}\| \le (1 - \alpha_n)\|x_n - x_{n+r}\| + M_1|\alpha_n - \alpha_{n+r}|, \tag{2.62}$$

where $M_1 = \sup\{|u| + \|T_{n+1}x_{n+r}\| : n \ge 1\}$. Using assumption (C3) and Lemma 1.10.2, we see that $x_{n+r} - x_n \to \theta$ as $n \to \infty$. In view of (2.61), we obtain

$$x_n - T_{n+r}T_{n+r-1}\cdots T_{n+1}x_n \to \theta \quad (n \to \infty). \tag{2.63}$$

Let $n = k \mod r$, where $k \in \{1, 2, \ldots, r\}$. Hence (2.63) can be rewritten as follows:

$$x_n - T_{k+r}T_{k+r-1}\cdots T_{k+1}x_n \to \theta \quad (n \to \infty).$$

It follows from the demiclosedness principle of nonexpansive mappings that

$$p \in \text{Fix}(T_{k+r}T_{k+r-1}\cdots T_{k+1}) = F.$$

This completes the proof. □

Remark 2.4.4. In fact, in Theorem 2.4.10, we only need to assume

$$F = \bigcap_{i=1}^{r} \text{Fix}(T_i) = \text{Fix}(T_r T_{r-1} \cdots T_1) \neq \emptyset.$$

Then condition (CC) can be obtained.

To weaken condition (C3), we first prove the following proposition.

Proposition 2.4.2. *Let $a \in \mathbb{R}$ and $r \in \mathbb{N}$. Suppose that $\{a_n\} \in l^\infty$ satisfies condition $\mu_n(a_n) \leq a$ for any Banach limit μ_n. If $\limsup_{n\to\infty}(a_{n+r} - a_n) \leq 0$, then $\limsup_{n\to\infty} a_n \leq a$.*

Proof. Set $b_{n+i} = a_{n+ir}$ $(i = 0, 1, 2, \ldots, p-1)$. Then $\mu_n(b_n) = \mu_n(a_n) \leq a$ for any Banach limit μ_n. For any $\varepsilon > 0$, there exist $m \geq 1$ and $p \geq \max\{2, m\}$ such that

$$\frac{b_n + b_{n+1} + \cdots + b_{n+p+1}}{p} < a + \frac{\varepsilon}{2}, \quad \forall n \geq 1. \tag{2.64}$$

Using the assumption $\limsup_{n\to\infty}(a_{n+r} - a_n) \leq 0$, we have

$$a_{n+r} - a_n < \frac{\varepsilon}{p-1}, \quad \forall n \geq m. \tag{2.65}$$

Taking $n \geq m + pr$, one concludes from (2.65) that

$$a_n = a_{n-ir} + (a_{n-(i-1)r} - a_{n+ir}) + \cdots + (a_n - a_{n-r}) \leq \frac{i\varepsilon}{p-1} + a_{n-ir}. \tag{2.66}$$

Taking $i = 0, 1, 2, \ldots, p-1$ in (2.66) and summing up, we arrive at

$$\begin{aligned} a_n &\leq \frac{a_n + a_{n-r} + a_{n-2r} + \cdots + a_{n-(p-1)r}}{p} + \frac{1}{p}\frac{(p-1)p}{2}\frac{\varepsilon}{p-1} \\ &\leq \frac{a_n + a_{n-r} + a_{n-2r} + \cdots + a_{n-(p-1)r}}{p} + \frac{\varepsilon}{2}. \end{aligned} \tag{2.67}$$

Replacing n with $n + (p-1)r$ in (2.67), we have

$$\begin{aligned} a_{n+(p-1)r} &\leq \frac{a_n + a_{n-r} + a_{n-2r} + \cdots + a_{n-(p-1)r}}{p} + \frac{\varepsilon}{2} \\ &= \frac{b_n + b_{n+1} + \cdots + b_{n+p-1}}{p} + \frac{\varepsilon}{2} \\ &\leq a + \frac{\varepsilon}{2} + \frac{\varepsilon}{2} \\ &= a + \varepsilon. \end{aligned} \tag{2.68}$$

Hence $\limsup_{n\to\infty} a_n \leq a$. This completes the proof. □

Theorem 2.4.11. *Let C be a closed convex subset of H and let $\{T_i\}_{i=1}^r : C \to C$ be a finite family of nonexpansive mappings satisfying*

$$
\begin{aligned}
F = \bigcap_{i=1}^{r} \mathrm{Fix}(T_i) &= \mathrm{Fix}(T_r T_{r-1} \cdots T_1) \\
&= \mathrm{Fix}(T_1 T_r T_{r-1} \cdots T_2) = \cdots \\
&= \mathrm{Fix}(T_{r-1} \cdots T_1 T_r) \\
&\neq \emptyset.
\end{aligned}
$$

Suppose that $\{\alpha_n\}$ is a sequence in $[0,1]$ satisfying the control conditions (C1), (C2), *and* (C6) $x_{n+r} - x_n \to \theta$ *as $n \to \infty$.*

Let $\{x_n\}$ be the sequence generated by (B). *Then $x_n \to z$ as $n \to \infty$, where $z = P_F u$ and P_F is the metric projection from C onto F.*

Proof. For all $n \geq 0$, $r \geq 1$ and $t \in (0,1)$, define a contractive mapping by

$$
x \mapsto tu + (1-t)T_{n+r} T_{n+r-1} \cdots T_{n+1} x.
$$

From Banach's contraction principle, one concludes that there exists a unique $x_t^{(n)} \in C$ such that

$$
x_t^{(n)} = tu + (1-t)T_{n+r} T_{n+r-1} \cdots T_{n+1} x_t^{(n)}, \quad \forall n \geq 0,\, r \geq 1,\, t \in (0,1). \tag{2.69}
$$

Note that $T_{2n+r} = T_{n+r}, \ldots, T_{n+r+1} = T_{n+1}$. From the uniqueness of $x_t^{(n)}$, we see that

$$
x_t^{(n+r)} = x_t^{(n)}, \quad \forall n \geq 0,\, r \geq 1,\, t \in (0,1). \tag{2.70}
$$

If $t \to 0$, then we conclude from the Browder's convergence theorem (Theorem 2.4.4) that

$$
x_t^{(n)} \to P_{\mathrm{Fix}(T_{n+r} T_{n+r-1} \cdots T_{n+1})}(u), \quad \forall n \geq 0. \tag{2.71}
$$

It follows that

$$
\mathrm{Fix}(T_{n+r} T_{n+r-1} \cdots T_{n+1}) = F, \quad \forall n \geq 0. \tag{2.72}
$$

Hence

$$
x_t^{(n)} \to P_F(u) \quad (t \to 0), \quad \forall n \geq 0. \tag{2.73}
$$

By the proof of Theorem 2.4.10, we see that

$$
x_{n+r} - T_{n+r} T_{n+r-1} \cdots T_{n+1} x_n \to \theta \quad (n \to \infty). \tag{2.74}
$$

Next, we prove that

$$\mu_n \langle u - P_F u, x_n - P_F u \rangle \le 0. \qquad (2.75)$$

To achieve this, we estimate that

$$\left\| x_t^{(n)} - x_{n+r} \right\|^2$$
$$= \left\| t(u - x_{n+r}) + (1-t)(T_{n+r} T_{n+r-1} \cdots T_{n+1} x_t^{(n)} - x_{n+r}) \right\|^2$$
$$\le (1-t)^2 \left\| x_t^{(n)} - x_n \right\|^2$$
$$\quad + 2 \left\| x_t^{(n)} - x_n \right\| \left\| T_{n+r} T_{n+r-1} \cdots T_{n+1} x_t^{(n)} - x_{n+r} \right\|$$
$$\quad + \left\| T_{n+r} T_{n+r-1} \cdots T_{n+1} x_t^{(n)} - x_{n+r} \right\|^2 + 2t \langle u - x_{n+r}, x_t^{(n)} - x_{n+r} \rangle$$
$$\le (1 + t^2) \left\| x_t^{(n)} - x_n \right\|^2 + 2t \langle u - x_{n+r}, x_t^{(n)} - x_{n+r} \rangle$$
$$\quad + \left\| T_{n+r} T_{n+r-1} \cdots T_{n+1} x_t^{(n)} - x_{n+r} \right\| (2M + \left\| T_{n+r} T_{n+r-1} \cdots T_{n+1} x_t^{(n)} - x_{n+r} \right\|). \qquad (2.76)$$

Hence, we have

$$2t \langle u - x_{n+r}, x_t^{(n)} - x_{n+r} \rangle \le (1 + t^2) \left\| x_t^{(n)} - x_n \right\|^2 - \left\| x_t^{(n)} - x_{n+r} \right\|^2$$
$$+ \left\| T_{n+r} T_{n+r-1} \cdots T_{n+1} x_t^{(n)} - x_{n+r} \right\| (2M$$
$$+ \left\| T_{n+r} T_{n+r-1} \cdots T_{n+1} x_t^{(n)} - x_{n+r} \right\|). \qquad (2.77)$$

Taking the Banach limit μ_n in (2.77), we obtain

$$2t \mu_n \langle u - x_{n+r}, x_t^{(n)} - x_{n+r} \rangle$$
$$\le (1 + t^2) \mu_n \left\| x_t^{(n)} - x_n \right\|^2 - \mu_n \left\| x_t^{(n)} - x_{n+r} \right\|^2$$
$$= (1 + t^2) \mu_n \left\| x_t^{(n)} - x_n \right\|^2 - \mu_n \left\| x_t^{(n+r)} - x_{n+r} \right\|^2$$
$$= (1 + t^2) \mu_n \left\| x_t^{(n)} - x_n \right\|^2 - \mu_n \left\| x_t^{(n)} - x_n \right\|^2$$
$$= t^2 \mu_n \left\| x_t^{(n)} - x_n \right\|^2. \qquad (2.78)$$

Hence, we have

$$\mu_n \langle u - x_{n+r}, x_t^{(n)} - x_{n+r} \rangle \le \frac{t}{2} \mu_n \left\| x_t^{(n)} - x_n \right\|^2.$$

Since $\{x_t^{(n)} - x_n\}$ is bounded, we have

$$\limsup_{t \to 0} \mu_n \langle u - x_{n+r}, x_t^{(n)} - x_{n+r} \rangle \le 0. \qquad (2.79)$$

Since $x_t^{(n)} \to P_F(u)$ as $t \to 0$ uniformly for each $n \ge 0$, for all $\varepsilon > 0$, there exists $\delta > 0$ with $0 < t < \delta$ such that

$$\left\| x_t^{(n)} - P_F u \right\| \le \frac{\varepsilon}{2M}, \quad \forall n \ge 0, \qquad (2.80)$$

where $M \geq \max\{\|u - x_{n+r}\|, \|x_{n+r} - P_F u\|\}$ for all $n \geq 0$ and $t \in (0, 1)$. Note that

$$\langle u - P_F(u), x_{n+r} - P_F(u) \rangle$$
$$= \langle u - x_t^{(n)}, x_{n+r} - x_t^{(n)} \rangle + \langle u - x_t^{(n)}, x_t^{(n)} - P_F(u) \rangle$$
$$+ \langle x_t^{(n)} - P_F(u), x_{n+r} - P_F(u) \rangle$$
$$\leq \langle u - x_t^{(n)}, x_{n+r} - x_t^{(n)} \rangle + 2M \|x_t^{(n)} - P_F(u)\|$$
$$\leq \langle u - x_t^{(n)}, x_{n+r} - x_t^{(n)} \rangle + \varepsilon, \quad n \geq 0, \ \forall t \in (0, \delta). \tag{2.81}$$

Taking the Banach limit μ_n for $n \geq 1$, we obtain

$$\mu_n \langle u - P_F(u), x_{n+r} - P_F(u) \rangle \leq \mu_n \langle u - x_t^{(n)}, x_{n+r} - x_t^{(n)} \rangle + \varepsilon. \tag{2.82}$$

Letting $t \to 0$ in (2.82), we see that

$$\mu_n \langle u - P_F(u), x_{n+r} - P_F(u) \rangle \leq \limsup_{t \to 0} \mu_n \langle u - x_t^{(n)}, x_{n+r} - x_t^{(n)} \rangle + \varepsilon < \varepsilon.$$

From the arbitrariness of $\varepsilon > 0$, we have $\mu_n \langle u - P_F(u), x_{n+r} - P_F(u) \rangle \leq 0$. By the property of the Banach limit μ_n, it follows that

$$\mu_n \langle u - P_F(u), x_n - P_F(u) \rangle = \mu_n \langle u - P_F(u), x_{n+r} - P_F(u) \rangle \leq 0.$$

This proves (2.75). Denote $a_n = \langle u - P_F(u), x_n - P_F(u) \rangle$. Then, for any Banach limit μ_n, one has $\mu_n(a_n) \leq 0$ and

$$\limsup_{n \to \infty}(a_{n+r} - a_n) = \limsup_{n \to \infty} \langle u - P_F(u), x_{n+r} - x_n \rangle$$
$$= \limsup_{j \to \infty} \langle u - P_F(u), x_{n_j+r} - x_{n_j} \rangle.$$

From the assumption that $x_{n_j+r} - x_{n_j} \to \theta$ as $j \to \infty$, one obtains that

$$\limsup_{j \to \infty} \langle u - P_F(u), x_{n_j+r} - x_{n_j} \rangle = 0.$$

Hence $\limsup_{n \to \infty}(a_{n+r} - a_n) \leq 0$. By Proposition 2.4.2, we see that $\limsup_{n \to \infty} a_n \leq 0$. Define

$$r_n = \max\{\langle u - P_F(u), x_n - P_F(u) \rangle, 0\}.$$

Then $r_n \to 0$ as $n \to \infty$. Let $z = P_F u$. Since

$$\|x_{n+1} - z\|^2 = \|(1 - \alpha_n)(T_{n+1}x_n - T_{n+1}z) + \alpha_n(u - z)\|^2$$
$$= (1 - \alpha_n)^2 \|T_{n+1}x_n - T_{n+1}z\|^2 + 2\alpha_n(1 - \alpha_n)\langle u - z, x_n - z \rangle + \alpha_n^2 \|u - z\|^2$$
$$\leq (1 - \alpha_n)\|x_n - z\|^2 + 2\alpha_n r_n + \alpha_n^2 \|u - z\|^2$$
$$= (1 - \alpha_n)\|x_n - z\|^2 + o(\alpha_n),$$

and using Lemma 1.10.2, we obtain $x_n \to z$ as $n \to \infty$. This completes the proof. $\qquad \square$

In order to study common fixed points of a countable family of nonexpansive mappings, we give the following definition of the W-mapping.

Definition 2.4.2 ([80]). Let $\{T_i\}_{i=1}^{\infty} : C \to C$ be a countable family of nonlinear mappings and let $\{y_n\}$ be a sequence in $[0, 1]$. For all $n \geq 1$, define a mapping W_n as follows:

$$U_{n,n+1} = I,$$
$$U_{n,n} = y_n T_n U_{n,n+1} + (1 - y_n)I,$$
$$U_{n,n-1} = y_{n-1} T_{n-1} U_{n,n} + (1 - y_{n-1})I,$$
$$\cdots,$$
$$U_{n,k} = y_k T_k U_{n,k+1} + (1 - y_k)I,$$
$$U_{n,k-1} = y_{k-1} T_{k-1} U_{n,k} + (1 - y_{k-1})I,$$
$$\cdots,$$
$$U_{n,2} = y_2 T_2 U_{n,3} + (1 - y_2)I,$$
$$W_n = U_{n,1} = y_1 T_1 U_{n,2} + (1 - y_1)I.$$

Then W_n is called the W-mapping generated by $T_n, T_{n-1}, \ldots, T_1$ and $y_n, y_{n-1}, \ldots, y_1$.

Next, we list two importance properties of the W-mappings as follows.

Proposition 2.4.3. *Let $\{T_i\}_{i=1}^{n} : C \to C$ be a finite family of nonexpansive mappings with $\bigcap_{i=1}^{n} \text{Fix}(T_i) \neq \emptyset$. Let y_i a real number in $(0,1)$ for each $i = 1, 2, \ldots, n$ and let W_n be the W-mapping generated by $T_n, T_{n-1}, \ldots, T_1$ and $y_n, y_{n-1}, \ldots, y_1$. Then W_n is averaged nonexpansive and*

$$\text{Fix}(W_n) = \bigcap_{i=1}^{n} \text{Fix}(T_i).$$

If $y_i \in (0, b)$, where b is a real number in $0 < b < 1$, then, for all $k \geq 1$ and $x \in C$, the limit $\lim_{n \to \infty} U_{n,k}x$ exists. Furthermore, if D is a bounded subset of C, then the limit $\lim_{n \to \infty} U_{n,k}x$ uniformly exists for all $x \in D$, that is, for all $\varepsilon > 0$, there exists $n_0 \geq k$ (n_0 is independent of x) such that

$$\|U_{n,k}x - U_{m,k}x\| < \varepsilon, \quad \forall m, n \geq n_0, x \in D.$$

Hence, if we define $W : C \to C$ as follows:

$$Wx = \lim_{n \to \infty} W_n x = \lim_{n \to \infty} U_{n,1}x, \quad \forall x \in C, \tag{2.83}$$

then W is called the W-mapping generated by $T_1, T_2, \ldots$ and $y_1, y_2, \ldots$

Proposition 2.4.4. *Let $\{T_i\}_{i=1}^{\infty} : C \to C$ be a countable family of nonexpansive mappings with $F = \bigcap_{i=1}^{\infty} \text{Fix}(T_i) \neq \emptyset$. Let $y_i \in (0, b]$ for each $i = 1, 2, \ldots, n$, where $0 < b < 1$, and let W be the W-mapping generated by $T_1, T_2, \ldots$ and $y_1, y_2, \ldots$ Then $W : C \to C$ is an averaged nonexpansive mapping with $\text{Fix}(W) = F$.*

Since the proof of the above two propositions can be derived from the definition of the W-mapping and some properties of averaged mappings, we here omit the proof.

Theorem 2.4.12. *Let $\{T_i\}_{i=1}^{\infty} : C \to C$ be a countable family of nonexpansive mappings with $F = \bigcap_{i=1}^{\infty} \text{Fix}(T_i) \neq \emptyset$. Let $y_i \in (0, b]$ for each $i = 1, 2, \ldots, n$, where $0 < b < 1$, and let W_n be the W-mapping generated by $T_n, T_{n-1}, \ldots, T_1$ and $y_n, y_{n-1}, \ldots, y_1$. Let $\{\alpha_n\}$ be a real sequence in $[0, 1]$ satisfying the following conditions:*
(C1) $\alpha_n \to 0$ *as* $n \to \infty$;
(C2) $\sum_{n=0}^{\infty} \alpha_n = \infty$.

For an arbitrary initial value $x_0 \in C$ and a fixed anchor $u \in C$, define an iterative sequence $\{x_n\}$ as follows:

$$x_{n+1} = \alpha_n u + (1 - \alpha_n) W_n x_n, \quad \forall n \geq 1. \tag{WNM}$$

Then $\{x_n\}$ converges strongly to $P_F u$.

Proof. For any initial value $y_0 \in C$ and a fixed anchor $u \in C$, define another iterative sequence $\{y_n\}$ as follows:

$$y_{n+1} = \alpha_n u + (1 - \alpha_n) W y_n, \quad \forall n \geq 1, \tag{WM}$$

where W is defined by (2.83). It follows from Proposition 2.4.4 that $W : C \to C$ is an averaged mapping. From (WM) and Corollary 2.4.4, it follows that $y_n \to P_{\text{Fix}(W)} u$ as $n \to \infty$. Using Proposition 2.4.4, we have $\text{Fix}(W) = F$. Hence $\{y_n\}$ converges strongly to $P_F u$.

Now, we only need to prove that $y_n - x_n \to \theta$ as $n \to \infty$. To achieve this, it follows from (WNM) and (WM) that

$$
\begin{aligned}
\|y_{n+1} - x_{n+1}\| &\leq (1 - \alpha_n)\|W y_n - W_n x_n\| \\
&\leq (1 - \alpha_n)\|W y_n - W x_n\| + \|W x_n - W_n x_n\| \\
&\leq (1 - \alpha_n)\|y_n - x_n\| + \sigma_n, \quad \forall n \geq 0, \tag{2.84}
\end{aligned}
$$

where $\sigma_n = \|W x_n - W_n x_n\|$. Now, we estimate σ_n as follows:

$$
\begin{aligned}
\sigma_n &= \lim_{m \to \infty} \|W_m x_n - W_n x_n\| \\
&= \lim_{m \to \infty} \left\| \sum_{j=n}^{m-1} (W_{j+1} x_n - W_j x_n) \right\| \\
&\leq \lim_{m \to \infty} \sum_{j=n}^{m-1} \|W_{j+1} x_n - W_j x_n\| \\
&\leq \sum_{j=n}^{m-1} \left(\prod_{i=1}^{j+1} y_i \right) \|T_{j+1} x_n - x_n\|
\end{aligned}
$$

$$\leq 2M \sum_{j=n}^{\infty} b^{n+1}$$

$$= \frac{2M}{1-b} b^{n+1},$$

where $M \geq \sup\{\|x_n - p\| : n \geq 0\}$ and $p \in \text{Fix}(T)$. Hence we have $\sum_{n=0}^{\infty} \sigma_n < \infty$. Let $a_n = \|y_n - x_n\|$. Then (2.84) can be rewritten as follows:

$$a_n \leq (1 - \alpha_n)a_n + \sigma_n, \quad \forall n \geq 0. \tag{2.85}$$

From Lemma 1.10.2, we obtain $a_n \to 0$, that is,

$$y_n - x_n \to \theta \iff x_n \to P_F u \quad (n \to \infty).$$

This completes the proof. $\square$

Since the convergence of the iterative method (MHIM) can be obtained via the convergence of the iterative method (HIM), we will not discuss the convergence of the iterative method (MHIM) here.

Now, we prove a theorem to show how the convergence result of the iterative method (MHIM) is obtained.

Theorem 2.4.13. *Let C be a closed convex subset of H. Let $T : C \to C$ be a λ-averaged nonexpansive mapping with $\text{Fix}(T) \neq \emptyset$ and let $f : C \to C$ be a ρ-contractive mapping. Suppose that $\{\alpha_n\}$ in $[0,1]$ satisfies the following conditions:*
(C1) $\alpha_n \to 0$ as $n \to \infty$;
(C2) $\sum_{n=0}^{\infty} \alpha_n = \infty$.

Let $\{x_n\}$ be a sequence generated by (MHIM). Then $\{x_n\}$ converges strongly to a fixed point $x^* \in \text{Fix}(T)$ of T. In addition, x^* is also a unique solution of the following variational inequality:

$$\langle (I - f)x^*, x - x^* \rangle \geq 0, \quad \forall x \in \text{Fix}(T). \tag{VIP}$$

Proof. Since $P_{\text{Fix}(T)} : C \to \text{Fix}(T)$ is a nonexpansive mapping and $f : C \to C$ is a ρ-contractive mapping, we find that $P_{\text{Fix}(T)}f : C \to C$ is a ρ-contractive mapping. By the Banach's contraction principle, there exists a unique $x^* \in \text{Fix}(T)$ such that $x^* = P_{\text{Fix}(T)}fx^*$, which is equivalent to the fact that x^* is the unique solution of the variational inequality (VIP). For any $y_1 \in C$, define an iterative sequence $\{y_n\}$ as follows:

$$y_{n+1} = \alpha_n f(x^*) + (1 - \alpha_n)Ty_n, \quad \forall n \geq 1. \tag{2.86}$$

By Corollary 2.4.4, we find that $y_n \to P_{\text{Fix}(T)}fx^* = x^* \in \text{Fix}(T)$. From the iterative methods (MHIM) and (2.86), we see that

$$
\begin{aligned}
\|y_{n+1} - x_{n+1}\| &\leq \alpha_n\|fx^* - fx_n\| + (1 - \alpha_n)\|Ty_n - Tx_n\| \\
&\leq \rho\alpha_n\|y_n - x^*\| + \rho\alpha_n\|y_n - x_n\| + (1 - \alpha_n)\|y_n - x_n\| \\
&= [1 - (1 - \rho)\alpha_n]\|y_n - x_n\| + \rho\alpha_n\|y_n - x^*\| \\
&= [1 - (1 - \rho)\alpha_n]\|y_n - x_n\| + o(\alpha_n), \quad \forall n \geq 1.
\end{aligned}
$$

It follows from Lemma 1.10.2 that $y_n - x_n \to \theta$ as $n \to \infty$. Hence $x_n \to x^* \in \text{Fix}(T)$ as $n \to \infty$. This completes the proof. □

Remark 2.4.5. If $T : C \to C$ is nonexpansive and the sequence $\{\alpha_n\}$ satisfies condition (C4) or (C5) in Theorem 2.4.13, then the conclusion still holds.

In recent years, hybrid projection methods have attracted much attention since they are useful when we investigate the strong convergence of iterative sequences; see [59, 65–69, 86, 81, 91, 95, 116–121]. Next, we give Nakajo–Takahashi's result here.

Theorem 2.4.14. *Let C be a closed convex subset of H and let $T : C \to C$ be a nonexpansive mapping with $\text{Fix}(T) \neq \emptyset$. Define a sequence $\{x_n\}$ as follows:*

$$
\begin{cases}
x_0 \in C, \ C_1 = C, \\
y_n = \alpha_n x_n + (1 - \alpha_n)Tx_n, \\
C_n = \{z \in C : \|y_n - z\| \leq \|x_n - z\|\}, \\
Q_n = \{z \in C : \langle x_n - z, x_0 - x_n \rangle \geq 0\}, \\
x_{n+1} = P_{C_n \cap Q_n}(x_0), \quad \forall n \geq 0,
\end{cases} \tag{HPM}
$$

where $\{\alpha_n\} \subset [0, 1]$ satisfies $0 \leq \alpha_n < a < 1$. Then $\{x_n\}$ converges strongly to $P_{\text{Fix}(T)}(x_0)$.

Proof. The proof is split into five steps as follows:

Step 1. Show that, for all $n \geq 0$, $C_n \cap Q_n$ are closed and convex.

It is obvious that $C_n \cap Q_n$ are closed. Note that

$$
\|y_n - z\| \leq \|x_n - z\| \iff \|x_n - y_n\|^2 + 2\langle y_n - x_n, x_n - z \rangle \leq 0,
$$

which shows that C_n are convex. Hence C_n and Q_n are closed and convex. Thus $C_n \cap Q_n$ are also closed and convex.

Step 2. Show that $\text{Fix}(T) \subset C_n \cap Q_n$ for all $n \geq 0$.

This can be done by induction. It is obvious that $\text{Fix}(T) \subset C_n$. Since $Q_0 = C$, we have $\text{Fix}(T) \subset C_0 \cap Q_0$. Assume that, for some $k \geq 0$, $\text{Fix}(T) \subset C_k \cap Q_k$. It follows from the assumption $\text{Fix}(T) \neq \emptyset$ that $C_k \cap Q_k \neq \emptyset$ is closed and convex. Hence $x_{k+1} = P_{C_k \cap Q_k}(x_0)$ is well-defined. From the property of the metric projection, we see that

$$
\langle x_{k+1} - z, x_0 - x_{n+1} \rangle \geq 0, \quad \forall z \in C_k \cap Q_k. \tag{2.87}
$$

Since $\text{Fix}(T) \subset C_k \cap Q_k$, we obtain

$$\langle x_{k+1} - u, x_0 - x_{n+1} \rangle \geq 0, \quad \forall u \in \text{Fix}(T)$$
$$\implies \text{Fix}(T) \subset Q_{k+1}$$
$$\implies \text{Fix}(T) \subset C_{k+1} \cap Q_{k+1}. \tag{2.88}$$

Hence $\text{Fix}(T) \subset C_n \cap Q_n$ for all $n \geq 0$. So $C_n \cap Q_n \neq \emptyset$ are closed and convex for all $n \geq 0$. This implies that (HPM) is well-defined.

Step 3. Show that $\{x_n\}$ is bounded.

To achieved this, we denote z_0 by $z_0 = P_{\text{Fix}(T)}(x_0)$. Since $x_{n+1} = P_{C_n \cap Q_n}(x_0)$, we have

$$\|x_{n+1} - x_0\| \leq \|z - x_0\|, \quad \forall z \in C_n \cap Q_n, \ n \geq 0. \tag{2.89}$$

In particular, since $z_0 \in \text{Fix}(T) \subset C_n \cap Q_n$, we obtain

$$\|x_{n+1} - x_0\| \leq \|z_0 - x_0\|, \quad \forall n \geq 0, \tag{2.90}$$

which shows that $\{x_n\}$ is a bounded sequence.

Step 4. Show that $x_n - Tx_n \to \theta$ as $n \to \infty$.

Since $x_{n+1} \subset C_n \cap Q_n \subset Q_n$ and $x_n = P_{Q_n}(x_0)$, we see that

$$\|x_n - x_0\| \leq \|x_{n+1} - x_0\|, \quad n \geq 0$$
$$\implies \lim_{n \to \infty} \|x_n - x_0\| \text{ exists}$$
$$\implies \|x_{n+1} - x_0\|^2 - \|x_n - x_0\|^2 \to 0 \quad (n \to \infty). \tag{2.91}$$

Since $x_{n+1} \in Q_n$, we have

$$\langle x_n - x_{n+1}, x_0 - x_n \rangle \geq 0, \quad \forall n \geq 0. \tag{2.92}$$

By the identical equality on H, (2.91) and (2.92), we derive

$$\begin{aligned} \|x_n - x_{n+1}\|^2 &= \left\| (x_n - x_0) - (x_{n+1} - x_0) \right\|^2 \\ &= \|x_n - x_0\|^2 - 2\langle x_n - x_0, x_{n+1} - x_0 \rangle + \|x_{n+1} - x_0\|^2 \\ &= \|x_{n+1} - x_0\|^2 - \|x_n - x_0\|^2 - 2\langle x_n - x_{n+1}, x_0 - x_n \rangle \\ &\leq \|x_{n+1} - x_0\|^2 - \|x_n - x_0\|^2 \to 0 \quad (n \to \infty). \end{aligned}$$

From (HPM) and $x_{n+1} \in C_n$, one arrives at

$$\begin{aligned} \|x_n - Tx_n\| &= \frac{1}{1 - \alpha_n} \|y_n - x_n\| \\ &\leq \frac{1}{1 - \alpha_n} (\|y_n - x_{n+1}\| + \|x_{n+1} - x_n\|) \\ &\leq \frac{1}{1 - \alpha_n} (\|x_n - x_{n+1}\| + \|x_{n+1} - x_n\|) \\ &= \frac{2}{1 - \alpha_n} \|x_{n+1} - x_n\| \to 0 \quad (n \to \infty). \end{aligned}$$

Here we used the condition $0 \le \alpha_n < a < 1$. This proves that $x_n - Tx_n \to \theta$ as $n \to \infty$.

Step 5. We prove that $x_n \to z_0 = P_{\text{Fix}(T)}(x_0)$ as $n \to \infty$.

From Step 4 and the demiclosedness principle for nonexpansive mappings, we see that $\omega_\omega(x_n) \subset \text{Fix}(T)$. It follows from (2.90) that

$$\limsup_{n \to \infty} \|x_n - x_0\| \le \|x_0 - z_0\|. \tag{2.93}$$

Assume that

$$\limsup_{n \to \infty} \langle x_n - z_0, x_0 - z_0 \rangle = \limsup_{j \to \infty} \langle x_{n_j} - z_0, x_0 - z_0 \rangle$$

and $x_{n_j} \to \hat{x}$. Then $\hat{x} \in \text{Fix}(T)$. It follows that

$$\limsup_{n \to \infty} \langle x_n - z_0, x_0 - z_0 \rangle = \langle \hat{x} - z_0, x_0 - z_0 \rangle \le 0. \tag{2.94}$$

By the properties of norm and scalar product on H, we obtain

$$\begin{aligned} \|x_n - z_0\|^2 &= \|x_n - x_0 + x_0 - z_0\|^2 \\ &= \|x_n - x_0\|^2 + 2\langle x_n - x_0, x_0 - z_0 \rangle + \|x_0 - z_0\|^2 \\ &= \|x_n - x_0\|^2 + 2\langle x_n - z_0, x_0 - z_0 \rangle - \|x_0 - z_0\|^2. \end{aligned} \tag{2.95}$$

Letting $n \to \infty$ in (2.95), and combining (2.93) with (2.94), we have

$$\begin{aligned} &\limsup_{n \to \infty} \|x_n - z_0\|^2 \\ &\le \limsup_{n \to \infty} \|x_n - x_0\|^2 + 2\limsup_{n \to \infty} \langle x_n - z_0, x_0 - z_0 \rangle - \|x_0 - z_0\|^2 \\ &\le \left(\limsup_{n \to \infty} \|x_n - x_0\| \right)^2 - \|x_0 - z_0\|^2 \le 0. \end{aligned}$$

Hence $x_n \to z_0 = P_{\text{Fix}(T)}(x_0)$ as $n \to \infty$. This completes the proof. $\qquad\square$

2.5 Fixed points of nonexpansive nonself-mappings

Browder's fixed point theorem (Theorem 1.9.4) claims that a nonexpansive self-mapping T defined on a nonempty, bounded, convex, and closed subset C of a Hilbert space H has fixed points. However, T may have no fixed points if it is a nonself-mapping. To ensure that nonexpansive nonself-mappings have fixed points, it is necessary to add some boundary conditions on the mappings. Recently, some boundary conditions have been proposed in different ways.

Next, we introduce some popular boundary conditions.

(1) Rothe's condition: $T(\partial C) \subset C$, where ∂C denotes the boundary of C.

(2) Inward condition: $Tx \in I_C(x)$ for any $x \in C$, where

$$I_C(x) = \{y \in H : y = x + a(z - x), z \in C, a \geq 1\}.$$

(3) Weak inward condition: $Tx \in \overline{I_C(x)}$ for any $x \in C$.
(4) Generalized weak inward condition: $d(Tx, C) < \|Tx - x\|$ for any $x \in C$ and $x \neq Tx$.
(5) Nowhere normal–outward condition: $Tx \in \wp S_x$ for any $x \in C$, where

$$S_x = \{y \in H : y \neq x, P_C y = x\}$$

and P_C is the metric projection from H onto C.

Proposition 2.5.1. *For the above boundary conditions, the following implication relations hold:*
(1) $\Longrightarrow$ (2) $\Longrightarrow$ (3) $\Longrightarrow$ (4) $\Longrightarrow$ (5).

Proof. (1) $\Longrightarrow$ (2). For any $x \in C$, if $x \in \partial C$, then $Tx \in C$ and $C \subset I_C(x)$. Hence $Tx \in I_C(x)$.
It is obvious that (2) $\Longrightarrow$ (3).
(3) $\Longrightarrow$ (4). Assume that there exists $x \in C$ with $x \neq Tx$ and $d(Tx, C) = \|Tx - x\| > 0$.
Then $\|Tx - P_C(Tx)\| = \|Tx - x\|$. From the existence and the uniqueness of the metric projection $P_C(Tx)$, we see that

$$x = P_C(Tx) \implies \langle Tx - x, y - x \rangle \leq 0, \quad \forall y \in C. \tag{2.96}$$

Since, for any $x \in C$, $Tx \in \overline{I_C(x)}$, there exists $y_n \in T_C(x)$ such that $y_n \to Tx$ as $n \to \infty$, where

$$y_n = x + a_n(z_n - x), \quad a_n \geq 1, \quad z_n \in C. \tag{2.97}$$

Substituting (2.97) into (2.96), we obtain

$$\langle Tx - x, z_n - x \rangle \leq 0$$
$$\implies \langle Tx - x, a_n(z_n - x) \rangle \leq 0$$
$$\implies \langle Tx - x, y_n - x \rangle \leq 0, \quad \forall x \geq 1. \tag{2.98}$$

Letting $n \to \infty$ in (2.98), we have

$$\langle Tx - x, Tx - x \rangle \leq 0 \implies Tx = x,$$

which contradicts the assumption that $x \neq Tx$.
(4) $\Longrightarrow$ (5). Assume that there exists $x \in C$ such that $Tx \notin \wp S_x$. Then $Tx \in S_x$ implies $Tx \neq x$ and $P_C(Tx) = x$. From (4), we have

$$\|Tx - P_C(Tx)\| = d(Tx, C) < \|Tx - x\| = \|Tx - P_C(Tx)\|,$$

which is a contradiction. This completes the proof $\qquad\qquad\square$

Remark 2.5.1. The above relations "$\Longrightarrow$" are not reversible, the readers can illustrate this fact easily.

Proposition 2.5.2. *Let C be a closed convex subset of H and let $T : C \to H$ be a nonexpansive nonself-mapping which satisfies the nowhere normal–outward condition. Then $\mathrm{Fix}(T) = \mathrm{Fix}(P_C T) = \mathrm{Fix}(TP_C)$. In addition, if C is bounded and T is nonexpansive, then $\mathrm{Fix}(T)$ is a nonempty closed convex subset of H.*

Proof. In order to establish the first equality, we only need to prove the inclusion relation $\mathrm{Fix}(P_C T) \subset \mathrm{Fix}(T)$. To achieve this, we assume that $x = P_C(Tx)$. Note that

$$Tx \in \wp S_x \implies Tx \notin S_x \implies Tx = x.$$

To establish the second equality, we assume that $x = TP_C x$. Note that

$$P_C x = P_C T(P_C x) \implies P_C x \in \mathrm{Fix}(P_C T) = \mathrm{Fix}(T),$$

that is, $TP_C x = P_C x$. Since $x = TP_C x$, we see that $x = P_C x \implies x = Tx$. If C is a nonempty, bounded, closed, and convex subset of real Hilbert space H, then $\mathrm{Fix}(T)$ is a nonempty, closed, and convex subset of H. Indeed, since the self-mapping $P_C T : C \to C$ is nonexpansive, it follows from Browder's fixed point theorem (Theorem 1.9.4) that $\mathrm{Fix}(P_C T) \neq \emptyset$ is closed and convex. Noting that $\mathrm{Fix}(P_C T) = \mathrm{Fix}(T)$, we have the desired conclusion. This completes the proof. $\square$

Proposition 2.5.3. *Let C be a closed convex subset of H and let $T : C \to H$ be a nonexpansive nonself-mapping. If $\mathrm{Fix}(T) \neq \emptyset$, then T satisfies the nowhere normal–outward condition.*

Proof. Suppose that there exists $x_0 \in C$ such that $Tx_0 \in S_{x_0} \implies Tx_0 \neq x_0$ and $P_C Tx_0 = x_0$. Taking $p \in \mathrm{Fix}(T)$, we have

$$
\begin{aligned}
\|Tx_0 - p\|^2 &= \|Tx_0 - x_0 + x_0 - p\|^2 \\
&= \|Tx_0 - x_0\|^2 + 2\langle Tx_0 - x_0, x_0 - p \rangle + \|x_0 - p\|^2 \\
&= \|Tx_0 - x_0\|^2 + 2\langle Tx_0 - P_C Tx_0, P_C Tx_0 - p \rangle + \|x_0 - p\|^2 \\
&> \|x_0 - p\|^2 \implies \|Tx_0 - p\| > \|x_0 - p\|,
\end{aligned}
$$

which contradicts $\|Tx_0 - p\| < \|x_0 - p\|$. Hence, $Tx \in \wp S_x$, $\forall x \in C$. This completes the proof. $\square$

Theorem 2.5.1. *Let C be a closed convex subset of H and let $T : C \to H$ be a nonexpansive nonself-mapping satisfying $\mathrm{Fix}(T) \neq \emptyset$. For each $t \in (0,1]$ and $u \in C$, define mappings $T_t : C \to C$ and $S_t : C \to C$ by*

$$T_t(x) = tu + (1 - t)P_C Tx, \quad \forall x \in C, \tag{2.99}$$

and

$$S_t(y) = P_C[tu - (1 - t)Ty], \quad y \in C. \tag{2.100}$$

Then T_t and S_t have unique fixed points $x_t \in C$ and $y_t \in C$, respectively. In addition, both $\{x_t\}$ and $\{y_t\}$ converge strongly to $P_{\mathrm{Fix}(T)}u$.

Proof. The existence and uniqueness of $\{x_t\}$ and $\{y_t\}$ follows from the Banach's contraction principle. From Proposition 2.5.3, we see that $T : C \to H$ satisfies the nowhere normal–outward condition. It follows from Proposition 2.5.2 that $\mathrm{Fix}(T) = \mathrm{Fix}(P_C T)$. Denote $U = P_C T$. Then $x_t = tu + (1 - t)P_C Tx_t$ can be rewritten as follows:

$$x_t = tu + (1 - t)Ux_t, \quad \forall t \in (0, 1]. \tag{2.101}$$

From the Browder's strong convergence theorem for U, it follows that $\{x_t\}$ converges strongly to $P_{\mathrm{Fix}(U)}u = P_{\mathrm{Fix}(T)}u$ as $t \to 0$. Since y_t is the unique fixed point of S_t, we have

$$y_t = P_C[tu - (1 - t)Ty_t], \quad \forall t \in (0, 1]. \tag{2.102}$$

Let $z_t = tu + (1 - t)Ty_t$. Then (2.102) can be rewritten as $y_t = P_C z_t$. So

$$z_t = tu + (1 - t)TP_C z_t, \quad \forall t \in (0, 1]. \tag{2.103}$$

Denote $V = TP_C$. It follows from Proposition 2.5.2 that $\mathrm{Fix}(V) = \mathrm{Fix}(TP_C) = \mathrm{Fix}(T)$. Then (2.103) can be rewritten as follows:

$$z_t = tu + (1 - t)Vz_t, \quad \forall t \in (0, 1]. \tag{2.104}$$

Again, from the Browder's strong convergence theorem for V, it follows that $\{z_t\}$ converges strongly to $P_{\mathrm{Fix}(V)}u = P_{\mathrm{Fix}(T)}u$ as $t \to 0$. Therefore, we have $y_t \to P_{\mathrm{Fix}(T)}u$ as $t \to 0$. This completes the proof. $\qquad\square$

Theorem 2.5.2. *Let C be a closed convex subset of a Hilbert space H and let $T : C \to H$ be a nonexpansive nonself-mapping with $\mathrm{Fix}(T) \neq \emptyset$. Let $\{\alpha_n\}$ be a sequence in $[0, 1]$ satisfying the following conditions:*
(C1) $\alpha_n \to 0$;
(C2) $\sum_{n=1}^{\infty} \alpha_n = \infty$;
(C3) $\sum_{n=1}^{\infty} |\alpha_{n+1} - \alpha_n| < \infty$ *or*
(C4) $\lim_{n \to \infty} \frac{\alpha_n}{\alpha_{n+1}} = 1$.

Define an iterative sequence $\{x_n\}$ as follows:

$$x_1 \in C, \ u \in C, \quad x_{n+1} = \alpha_n u + (1 - \alpha_n)P_C Tx_n, \quad \forall n \geq 1. \tag{2.105}$$

Then $\{x_n\}$ converges strongly to $P_{\mathrm{Fix}(T)}u$.

Proof. Denote $U = P_C T$. It follows from Propositions 2.5.2 and 2.5.3 that

$$\text{Fix}(U) = \text{Fix}(P_C T) = \text{Fix}(T).$$

Note that (2.105) can be rewritten as follows:

$$x_1 \in C, \quad u \in C, \quad x_{n+1} = \alpha_n u + (1 - \alpha_n) U x_n, \quad \forall n \geq 1. \tag{2.106}$$

Thus, by Wittmann's convergence theorem (Theorem 2.4.7) and Xu's convergence theorem (Theorem 2.4.8), we claim that $\{x_n\}$ converges strongly to a point $P_{\text{Fix}(U)} u = P_{\text{Fix}(T)} u$. This completes the proof. $\qquad\square$

Theorem 2.5.3. *Let C be a closed convex subset of a Hilbert space H and let $T : C \to H$ be a nonexpansive nonself-mapping with $\text{Fix}(T) \neq \emptyset$. Let $\{\alpha_n\}$ be a sequence in $[0, 1]$ satisfying the following conditions:*
(C1) $\lim_{n \to \infty} \alpha_n = 0$;
(C2) $\sum_{n=1}^{\infty} \alpha_n = \infty$;
(C3) $\sum_{n=1}^{\infty} |\alpha_{n+1} - \alpha_n| < \infty$ *or*
(C4) $\lim_{n \to \infty} \frac{\alpha_n}{\alpha_{n+1}} = 1$.

Define an iterative sequence $\{x_n\}$ as follows:

$$x_1 \in C, \quad u \in C, \quad x_{n+1} = P_C[\alpha_n u + (1 - \alpha_n) T x_n], \quad \forall n \geq 1. \tag{2.107}$$

Then $\{x_n\}$ converges strongly to $P_{\text{Fix}(T)} u$.

Proof. Let $y_n = \alpha_n u + (1 - \alpha_n) T x_n$ for each ≥ 1. Then (2.107) can be rewritten as $x_{n+1} = P_C y_n$. Hence

$$y_{n+1} = \alpha_{n+1} u + (1 - \alpha_{n+1}) T P_C y_n, \quad \forall n \geq 1. \tag{2.108}$$

Denote $t_n = \alpha_{n+1}$ and $V = T P_C$. Then (2.108) can be rewritten as follows:

$$y_{n+1} = t_n u + (1 - t_n) V y_n, \quad \forall n \geq 1. \tag{2.109}$$

From Propositions 2.5.2 and 2.5.3, we see that

$$\text{Fix}(V) = \text{Fix}(P_C T) = \text{Fix}(T).$$

It is easy to see that $\{t_n\}$ satisfies conditions (C1), (C2) and (C3) or (C4). Thus, from Wittmann's or Xu's convergence theorem, it follows that $\{y_n\}$ converges strongly to a point $P_{\text{Fix}(V)} u = P_{\text{Fix}(T)} u$. Therefore, $\{x_n\}$ converges strongly to $P_{\text{Fix}(T)} u$. This completes the proof. $\qquad\square$

Theorem 2.5.4. *Let C be a closed convex subset of a Hilbert space H and let $T : C \to H$ be a nonexpansive nonself-mapping with $\text{Fix}(T) \neq \emptyset$. Let $\{\alpha_n\}$ be a sequence in $[0, 1]$ satisfying the following conditions:*

(C1) $\lim_{n \to \infty} \alpha_n = 0$;

(C2) $\sum_{n=1}^{\infty} \alpha_n = \infty$

For any $\lambda \in (0,1)$, denote $T_\lambda(x) = \lambda T + (1-\lambda)I$ and define an iterative sequence as follows:

$$x_1 \in C, \ u \in C, \quad x_{n+1} = P_C[\alpha_n u + (1 - \alpha_n)T_\lambda x_n], \quad \forall n \geq 1. \tag{2.110}$$

Then $\{x_n\}$ converges strongly to $P_{\text{Fix}(T)}u$.

Proof. Let $u_n = \alpha_n u + (1 - \alpha_n)T_\lambda x_n$ for each $n \geq 1$. Then (2.110) can be rewritten as $x_{n+1} = P_C u_n$. Thus (2.110) can be written as follows:

$$u_{n+1} = \alpha_{n+1}u + (1 - \alpha_{n+1})T_\lambda P_C u_n, \quad \forall n \geq 1. \tag{2.111}$$

Denote $W = T_\lambda P_C$. Then, from Proposition 2.1.4 (5), we see that W is $\frac{1+\lambda}{2}$-averaged nonexpansive and

$$\text{Fix}(W) = \text{Fix}(T_\lambda) \cap \text{Fix}(P_C) = \text{Fix}(T) \cap C = \text{Fix}(T).$$

Using Suzuki's convergence theorem [84] (see also Corollary 2.4.4) for W, we see that $\{u_n\}$ converges strongly to a point $P_{\text{Fix}(W)}u = P_{\text{Fix}(T)}u$. Therefore, $\{x_n\}$ converges strongly to $P_{\text{Fix}(T)}u$. This completes the proof. $\square$

2.6 Iterative methods for fixed points of quasi-nonexpansive mappings

The class of quasi-nonexpansive mappings is a class of mappings which is much more general than the class of nonexpansive mappings. Due to the word "quasi", this class of mappings loses a number of good properties such as the Lipschitz continuity, the demiclosedness principle, and other important properties. For a long time, we did not know whether the Halpern iterative method (HIM) was valid for the class of quasi-nonexpansive mappings. In 2010, Maingé [45] solved this problem by using new techniques, and he proved that the Halpern-type viscosity iterative method is still valid for quasi-nonexpansive mappings. In this subsection, we discuss this method in a more general framework.

Let us recall the definitions of quasi-nonexpansive and nonspreading mappings as follows.

Definition 2.6.1. Let C be a nonempty subset of a real Hilbert space H and let $T : C \to H$ be a mapping. Then T is said to be

(1) quasi-nonexpansive if $\text{Fix}(T) \neq \emptyset$ and

$$\|Tx - p\| \leq \|x - p\|, \quad \forall x \in C, p \in \text{Fix}(T);$$

(2) nonspreading if

$$2\|Tx - Ty\|^2 \le \|Tx - y\|^2 + \|x - Ty\|^2, \quad \forall x, y \in C.$$

Remark 2.6.1.
(1) Every nonexpansive mapping with a nonempty fixed-point set is quasi-nonexpansive.
(2) Every nonspreading mapping with a nonempty fixed-point set is quasi-nonexpansive.
(3) Every firmly nonexpansive mapping is nonspreading.
(4) Nonexpansive and nonspreading mappings are mutually independent.

Example 2.6.1. Let $E = \{x \in H : \|x\| \le 1\}$. Let $D = \{x \in H : \|x\| \le 2\}$, and $C = \{x \in H : \|x\| \le 3\}$. Define a mapping $T : C \to H$ by

$$Tx = \begin{cases} \theta, & x \in D, \\ P_E x, & x \in C \backslash D. \end{cases}$$

Then T is a nonspreading mapping, but not a nonexpansive mapping.

Example 2.6.2 ([88]). Define a mapping $T : C \to C$, where $C = [-1, 1]$, by $Tx = -x$, $\forall x \in C$. Then T is nonexpansive, but not nonspreading.

Example 2.6.3. Take $C = \mathbb{R}$ and define a mapping $T : C \to C$ by

$$Tx = \begin{cases} \frac{x}{2} \sin \frac{1}{x}, & x \ne 0, \\ 0, & x = 0. \end{cases}$$

Then T is a quasi-nonexpansive mapping but not nonspreading.

Let C be a nonempty, closed, and convex subset of a real Hilbert space H. Let $T : C \to H$ be a nonlinear mapping and let $f : C \to H$ be a ρ-contractive mapping. Let $P_C : H \to C$ be the metric projection. Let $\{\alpha_n\}$, $\{\beta_n\}$, and $\{\gamma_n\}$ be three real number sequences in $[0, 1]$ satisfying the following conditions:
(1) $\alpha_n + \beta_n + \gamma_n = 1$ for all $n \in \mathbb{N}$;
(2) $\alpha_n \to 0 \ (n \to \infty)$;
(3) $\sum_{n=1}^{\infty} \alpha_n = \infty$;
(4) $0 < \liminf_{n \to \infty} \beta_n \le \limsup_{n \to \infty} \beta_n < 1$;

Now, we introduce the following three Halpern-type viscosity iterative methods:

$$x_1 \in C, \quad x_{n+1} = \alpha_n f(x_n) + \beta_n x_n + \gamma_n P_C T x_n, \quad n \ge 1, \tag{I}$$

$$x_1 \in C, \quad x_{n+1} = \beta_n x_n + (1 - \beta_n) P_C [\alpha_n f(x_n) + (1 - \alpha_n) T x_n], \quad n \ge 1, \tag{II}$$

$$x_1 \in C, \quad x_{n+1} = P_C [\alpha_n f(x_n) + \beta_n x_n + \gamma_n T x_n], \quad \forall n \ge 1. \tag{III}$$

To show the convergence of the above three methods, we need the following tools.

Proposition 2.6.1. *Let C be a nonempty, closed, and convex subset of a real Hilbert space H, and let $T : C \to H$ be a quasi-nonexpansive mapping. Then* Fix(T) *is a nonempty, closed, and convex subset of H.*

Proof. Let $\{p_n\} \subset$ Fix(T) such that $p_n \to p$. Then $Tp_n = p_n$ and $p \in C$. It follows from the definition of a quasi-nonexpansive mapping T that $p_n \to Tp$, which implies that $Tp = p$, that is, Fix(T) is closed. Next we show that Fix(T) is also convex. To this end, let $p_i \in$ Fix(T) and $p_t = tp_1 + (1 - t)p_2$, for $t \in (0, 1)$. Then it follows that

$$
\begin{aligned}
\|Tp_t - p_t\|^2 &= \|t(Tp_t - p_1) + (1 - t)(Tp_t - p_2)\|^2 \\
&= t\|Tp_t - p_1\|^2 + (1 - t)\|Tp_t - p_2\|^2 - t(1 - t)\|p_1 - p_2\|^2 \\
&\leq t\|p_t - p_1\|^2 + (1 - t)\|p_t - p_2\|^2 - t(1 - t)\|p_1 - p_2\|^2 \\
&\leq [t(1 - t)^2 + t^2(1 - t)]\|p_1 - p_2\|^2 - t(1 - t)\|p_1 - p_2\|^2 \\
&= 0.
\end{aligned}
$$

This implies that $p_t = Tp_t$. So, Fix(T) is convex. □

Proposition 2.6.2 ([126]). *If $T : C \to H$ is a quasi-nonexpansive mapping, then* Fix($P_C T$) = Fix(T) = Fix(TP_C).

Proof. It is obvious that Fix(T) ⊂ Fix($P_C T$). Next we show the converse inclusion. Assume that $x = P_C Tx$. Then, for all $x \in C$, it follows from the property of P_C that

$$
\begin{aligned}
\|x - p\|^2 &= \|P_C Tx - p\|^2 \\
&\leq \|Tx - p\|^2 - \|x - Tx\|^2 \\
&\leq \|x - p\|^2 - \|x - Tx\|^2.
\end{aligned}
$$

This implies that $x = Tx$. So, Fix($P_C T$) = Fix(T). It is obvious that Fix(T) ⊂ Fix(TP_C). Conversely, we assume $x = TP_C x$, then $P_C x = P_C T(P_C x)$, which shows that $P_C x$ is a fixed point of $P_C T$. It follows that $P_C x = T(P_C x)$, which implies that $P_C x = x$. Therefore, $x = Tx$. Hence the second equality holds. This completes the proof. □

Proposition 2.6.3 ([89]). *Let C be a nonempty, closed, and convex subset of a real Hilbert space H, and let $T : C \to H$ be a nonspreading mapping. Then $I - T$ is demiclosed at the origin.*

Theorem 2.6.1. *Let C be a nonempty, closed, and convex subset of a real Hilbert space H. Let $T : C \to H$ be a quasi-nonexpansive mapping and let $f : C \to C$ be a ρ-contraction mapping, where $\rho \in (0, 1)$. If $I - T$ is demiclosed at the origin, then the iterative sequence $\{x_n\}$ generated by (I) strongly converges to a fixed point $p \in$ Fix(T) of T and p is the unique solution of the following variational inequality:*

$$
\langle (I - f)x, y - x \rangle \geq 0, \quad \forall y \in \text{Fix}(T). \tag{VIP}
$$

Proof. Letting $S = P_C T$, we see that $S : C \to C$ is a quasi-nonexpansive mapping. In fact, from Proposition 2.6.2, we have $\text{Fix}(S) = \text{Fix}(T) \neq \emptyset$. For all $x \in C, p \in \text{Fix}(S)$, it follows from the nonexpansiveness of P_C that

$$\|Sx - p\| = \|P_C Tx - P_C p\| \leq \|Tx - p\| \leq \|x - p\|.$$

This shows that $S : C \to C$ is quasi-nonexpansive.

Next, we prove that $I - S$ is demiclosed at the origin. To achieve this, let $\{x_n\}$ be a sequence in C and assume that $x_n \rightharpoonup x$ and $x_n - Sx_n \to \theta$ as $n \to \infty$. Since C is weakly closed, we have $x \in C$. For all $p \in \text{Fix}(T)$, it follows from the firm nonexpansiveness of P_C and the quasi-nonexpansiveness of T that

$$\begin{aligned} \|Sx_n - p\|^2 &= \|P_C Tx_n - p\|^2 \\ &\leq \|Tx_n - p\|^2 - \|Tx_n - P_C Tx_n\|^2 \\ &\leq \|x_n - p\|^2 - \|Tx_n - Sx_n\|^2. \end{aligned}$$

It follows that

$$\begin{aligned} \|Tx_n - Sx_n\|^2 &\leq \|x_n - p\|^2 - \|Sx_n - p\|^2 \\ &\leq 2\|x_n - Sx_n\| \cdot \|x_n - p\|^2 \\ &\leq M\|x_n - Sx_n\| \to 0, \end{aligned}$$

where $M \geq 2 \sup\{\|x_n - p\| : n \geq 1\}$. Hence $Tx_n \to Sx_n \to \theta$ as $n \to \infty$. Note that

$$\|x_n - Tx_n\| \leq \|x_n - Sx_n\| + \|Sx_n - Tx_n\|,$$

which implies that $x_n - Tx_n \to \theta$ as $n \to \infty$. It follows from the assumption that $I - T$ is demiclosed at the origin that $x = Tx$. Therefore, $x = P_C Tx = Sx$. From [129, Theorem 2.1], we see that $\{x_n\}$ converges to a fixed point of S in norm and $p \in \text{Fix}(S) = \text{Fix}(T)$. This completes the proof. $\qquad\square$

Theorem 2.6.2. *Let C be a nonempty, closed, and convex subset of a real Hilbert space H. Let $T : C \to H$ be a quasi-nonexpansive mapping and let $f : C \to C$ be a ρ-contraction mapping, where $\rho \in (0, 1)$. Suppose that $I - T$ is demiclosed at the origin. Then the iterative sequence $\{x_n\}$ generated by (II) converges to a fixed point $p \in \text{Fix}(T)$ of T in norm and p is the unique solution of variational inequality problem (VIP).*

Proof. Let $y_n = \alpha_n f(x_n) - (1 - \alpha_n)Tx_n$ for each $n \geq 1$. Then it follows from (II) that, for all $p \in \text{Fix}(T)$,

$$\begin{aligned} \|y_n - p\| &\leq \alpha_n \|f(x_n) - p\| + (1 - \alpha_n)\|Tx_n - p\| \\ &\leq \alpha_n \|f(x_n) - f(p)\| + \alpha_n \|f(p) - p\| + (1 - \alpha_n)\|x_n - p\| \\ &\leq [1 - (1 - \rho)\alpha_n]\|x_n - p\| + \alpha_n \|f(p) - p\| \end{aligned}$$

and

$$\|x_n - p\| \leq \beta_n \|x_n - p\| + (1 - \beta_n)\|y_n - p\|$$
$$\leq \alpha_n \|f(x_n) - f(p)\| + \alpha_n \|f(p) - p\| + (1 - \alpha_n)\|x_n - p\|$$
$$\leq \{\beta_n + (1 - \beta_n)[1 - (1 - \rho)\alpha_n]\}\|x_n - p\|$$
$$+ (1 - \beta_n)(1 - \rho)\alpha_n \frac{\|f(p) - p\|}{1 - \rho}$$
$$\leq \max\left\{\|x_1 - p\|, \frac{\|f(p) - p\|}{1 - \rho}\right\} = M_p, \quad \forall n \geq 1.$$

This shows that $\{x_n\}$ is bounded. Hence both $\{Tx_n\}$ and $\{x_n\}$ are bounded. It follows from Proposition 2.6.1 that $\text{Fix}(T) \neq \emptyset$ and it is closed convex. Thus, $P_{\text{Fix}(T)}f$ is a ρ-contractive mapping. By Banach's contraction principle, there exists a unique $p \in \text{Fix}(T)$ such that $p = P_{\text{Fix}(T)}f(p)$, which is equivalent to the fact that the variational inequality problem (VIP) has a unique solution p. From (II), we find that

$$\|x_{n+1} - p\|^2 = \|\beta_n(x_n - p) + (1 - \beta_n)(P_C y_n - p)\|^2$$
$$= \beta_n \|x_n - p\|^2 + (1 - \beta_n)\|P_C y_n - p\|^2 - \beta_n(1 - \beta_n)\|x_n - P_C y_n\|^2$$
$$\leq \beta_n \|x_n - p\|^2 + (1 - \beta_n)\|y_n - p\|^2$$
$$- (1 - \beta_n)\|y_n - P_C y_n\|^2 - \beta_n(1 - \beta_n)\|x_n - P_C y_n\|^2$$
$$= \beta_n \|x_n - p\|^2 + (1 - \beta_n)[\alpha_n^2 \|fx_n - p\|^2$$
$$+ 2\alpha_n(1 - \alpha_n)\langle fx_n - p, Tx_n - p\rangle$$
$$+ (1 - \alpha_n)^2 \|Tx_n - p\|^2] - (1 - \beta_n)\|y_n - P_C y_n\|^2$$
$$- \beta_n(1 - \beta_n)\|x_n - P_C y_n\|^2$$
$$\leq \beta_n \|x_n - p\|^2 + (1 - \beta_n)[1 - (1 - \rho)\alpha_n \|x_n - p\|^2$$
$$+ 2\alpha_n(1 - \alpha_n)\langle fp - p, Tx_n - p\rangle + \alpha_n^2 \|fx_n - p\|^2]$$
$$- (1 - \beta_n)\|y_n - P_C y_n\|^2 - \beta_n(1 - \beta_n)\|x_n - P_C y_n\|^2$$
$$\leq [1 - (1 - \rho)\alpha_n(1 - \beta_n)]\|x_n - p\|^2$$
$$+ 2\alpha_n(1 - \alpha_n)(1 - \beta_n)\langle fp - p, Tx_n - p\rangle$$
$$+ (1 - \beta_n)\alpha_n^2 \|fx_n - p\|^2 - (1 - \beta_n)\|y_n - P_C y_n\|^2$$
$$- \beta_n(1 - \beta_n)\|x_n - P_C y_n\|^2,$$

which implies that

$$\|x_{n+1} - p\|^2 - \|x_n - p\|^2 + (1 - \rho)\alpha_n(1 - \beta_n)\|x_n - p\|^2$$
$$+ (1 - \beta_n)\|y_n - P_C y_n\|^2 + \beta_n(1 - \beta_n)\|x_n - P_C y_n\|^2$$
$$\leq 2\alpha_n(1 - \alpha_n)(1 - \beta_n)\langle fp - p, Tx_n - p\rangle + (1 - \beta_n)\alpha_n^2 \|fx_n - p\|^2, \quad \forall n \geq 1.$$

Denote $\Gamma_n = \|x_n - p\|^2$. The above inequality can be rewritten as

$$\Gamma_{n+1} - \Gamma_n + (1-\rho)\alpha_n(1-\beta_n)\Gamma_n + (1-\beta_n)\|y_n - P_C y_n\|^2$$
$$+ \beta_n(1-\beta_n)\|x_n - P_C y_n\|^2$$
$$\leq 2\alpha_n(1-\alpha_n)(1-\beta_n)\langle fp - p, Tx_n - p\rangle$$
$$+ (1-\beta_n)\alpha_n^2\|fx_n - p\|^2, \quad \forall n \geq 1. \tag{2.112}$$

Now, we discuss (2.112) in two cases:

Case 1. $\Gamma_{n+1} \leq \Gamma_n$ for all $n \geq n_0$. In this case, $\lim_{n\to\infty} \Gamma_n$ exists. Taking the limit in (2.112), we have $x_n - P_C y_n \to \theta$, $y_n - P_C y_n \to \theta$ as $n \to \infty$. Therefore, we have $x_n - y_n \to \theta$ as $n \to \infty$. Since $y_n = \alpha_n fx_n + (1-\alpha_n)Tx_n$ for each $n \geq 1$, we conclude from the boundedness of $\{fx_n\}$ and $\{Tx_n\}$ that $\alpha_n \to 0$ as $n \to \infty$. So

$$y_n - Tx_n = \alpha_n f(x_n - Tx_n) \to \theta \quad (n \to \infty)$$

and

$$x_n - Tx_n = x_n - y_n + y_n - Tx_n \to \theta \quad (n \to \infty).$$

From the assumption that $I - T$ is demiclosed at the origin, we have $\omega_W \subset \text{Fix}(T)$. It follows from the property of P_C that

$$\limsup_{n\to\infty} \langle fp - p, Tx_n - p\rangle = \limsup_{n\to\infty} \langle fp - p, x_n - p\rangle$$
$$= \lim_{j\to\infty} \langle fp - p, x_{n_j} - p\rangle$$
$$= \langle fp - p, \bar{x} - p\rangle \leq 0,$$

where $x_{n_j} \to \bar{x} \in \text{Fix}(T)$. Then (2.112) can be rewritten as follows:

$$\Gamma_{n+1} \leq [1 - (1-\rho)\alpha_n(1-\beta_n)]\Gamma_n + \sigma_n, \quad \forall n \geq 1, \tag{2.113}$$

where $\{\sigma_n\}$ satisfies the condition $\limsup_{n\to\infty} \frac{\sigma_n}{\alpha_n} \leq 0$. Using Lemma 1.10.2, we have $\Gamma_n \to 0$ as $n \to \infty$, that is, $x_n \to p$ as $n \to \infty$.

Case 2. $\Gamma_{n_k} \leq \Gamma_{n_k+1}$ for all $k \geq 1$. Define $\tau : \mathbb{N} \to \mathbb{N}$ by

$$\tau(n) = \max\{k \leq n : \Gamma_k \leq \Gamma_{k+1}\}.$$

Using Lemma 1.10.4, we find that $\tau(n) \to \infty$ as $n \to \infty$ and

$$\Gamma_{\tau(n)} \leq \Gamma_{\tau(n)+1}, \quad \Gamma_n \leq \Gamma_{\tau(n)+1}, \quad \forall n \geq n_0.$$

It follows from (2.112) that $\Gamma_{\tau(n)+1} - \Gamma_n \to 0$, $x_{\tau(n)} - P_C y_{\tau(n)} \to \theta$ and $y_{\tau(n)} - P_C y_{\tau(n)} \to \theta$ as $n \to \infty$. These imply that

$$x_{\tau(n)} - y_{\tau(n)} \to \theta \implies x_{\tau(n)} - Tx_{\tau(n)} \to \theta$$
$$\implies \limsup_{n\to\infty} \langle fp - p, Tx_{\tau(n)} - p\rangle \leq 0. \tag{2.114}$$

Using (2.112) again, we have

$$(1-\rho)\alpha_{\tau(n)}(1-\beta_{\tau(n)})\Gamma_{\tau(n)} \leq 2\alpha_{\tau(n)}(1-\alpha_{\tau(n)})(1-\beta_{\tau(n)})$$
$$+ \langle fp - p, Tx_{\tau(n)} - p \rangle$$
$$+ (1-\beta_{\tau(n)})\alpha_{\tau(n)}^2\|fx_{\tau(n)} - p\|^2.$$

Hence

$$(1-\rho)(1-\beta_{\tau(n)})\Gamma_{\tau(n)} \leq 2(1-\alpha_{\tau(n)})(1-\beta_{\tau(n)})$$
$$+ \langle fp - p, Tx_{\tau(n)} - p \rangle + (1-\beta_{\tau(n)})\alpha_{\tau(n)}\|fx_{\tau(n)} - p\|^2$$
$$\leq 2(1-\alpha_{\tau(n)})(1-\beta_{\tau(n)})\langle fp - p, Tx_{\tau(n)} - p \rangle + \alpha_{\tau(n)}M_1, \tag{2.115}$$

where $M_1 \geq \max\{\|fx_{\tau(n)}\|^2 : n \geq 1\}$. It follows from (2.114) and (2.115) that

$$\Gamma_{\tau(n)} \to 0 \ (n \to \infty) \implies \Gamma_{\tau(n)+1} \to 0 \ (n \to \infty) \implies \Gamma_n \to 0 \ (n \to \infty),$$

that is, $x_n \to p$ as $n \to \infty$. This completes the proof. $\qquad\square$

To obtain the convergence of the iterative method (III), we now establish a more general result.

Theorem 2.6.3. *Let C be a nonempty, closed, and convex subset of a real Hilbert space H. Let $S, T : C \to C$ be two quasi-nonexpansive mappings with $F = \mathrm{Fix}(S) \cap \mathrm{Fix}(T) \neq \emptyset$ and let $f : C \to C$ be a ρ-contractive mapping. Let $\{\alpha_n\}, \{\beta_n\}, \{\gamma_n\},$ and $\{\mu_n\}$ be four real number sequences in $[0,1]$ satisfying the following conditions:*
(1) $\alpha_n + \beta_n + \gamma_n + \mu_n = 1$ for each $n \in N$;
(2) $\alpha_n \to 0$ as $n \to \infty$ and $\sum_{n=1}^{\infty} \alpha_n = \infty$;
(3) $0 < \liminf_{n\to\infty} \gamma_n \leq \limsup_{n\to\infty} \gamma_n < 1$;
(4) $\liminf_{n\to\infty} \gamma_n\mu_n > 0$.

Define a sequence $\{x_n\}$ in C by

$$x_1 \in C, \quad x_{n+1} = \alpha_n fx_n + \beta_n x_n + \gamma_n Sx_n + \mu_n Tx_n, \quad \forall n \geq 1. \tag{IV}$$

If both $I - S$ and $I - T$ are demiclosed at the origin, then the sequence $\{x_n\}$ generated by (IV) converges to a common fixed point $p \in F$ of S and T in norm, and p is the unique solution of the variational inequality problem

$$\langle (I - f)p, y - p \rangle \geq 0, \quad \forall y \in F. \tag{VIP'}$$

Proof. Using the proof method of Theorem 2.6.3, we can establish that $\{x_n\}$ is bounded, so are $\{fx_n\}, \{Sx_n\},$ and $\{Tx_n\}$. Since $\mathrm{Fix}(S)$ and $\mathrm{Fix}(T)$ are (nonempty) closed convex subsets, we find that F is closed and convex. It follows from the assumption that $F \neq \emptyset$ and F is a (nonempty) closed convex subset of H. Therefore, for all $x \in H$, $P_F x$ exists uniquely. Since $P_F f : C \to C$ is a ρ-contractive mapping, there exists a unique $p \in F$ such that $p = P_F fp$. Hence p is the unique solution of (VIP'). Since the rest of the proof is similar to the corresponding part of the proof of Theorem 2.6.2, we omit it here. This completes the proof. $\qquad\square$

Theorem 2.6.4. *Let C be a nonempty, closed, and convex subset of a real Hilbert space H. Let $T : C \to H$ be a quasi-nonexpansive mapping and let $f : C \to H$ be a ρ-contractive mapping. If $I - T$ is demiclosed at the origin, then the sequence $\{x_n\}$ generated by (III) converges to a particular fixed point $p \in \text{Fix}(T)$ of T in norm and p is the unique solution of the variational inequality problem (VIP).*

Proof. Denote $z_n = \alpha_n f x_n + \beta_n x_n + \gamma_n T x_n$ for each $n \geq 1$. Then (III) can be rewritten as $x_n = P_C z_n$ for each $n \geq 1$. So,

$$z_{n+1} = \alpha_{n+1}(fP_C)z_n + \beta_{n+1}P_C z_n + \gamma_{n+1}TP_C z_n$$

$$= \alpha_{n+1}(fP_C)z_n + \frac{1}{2}\beta_{n+1}z_n + \frac{1}{2}\beta_{n+1}Sz_n + \gamma_{n+1}TP_C z_n, \quad \forall n \geq 1, \qquad (2.116)$$

where S is a nonexpansive mapping satisfying $2P_C = I + S$. Denote $\tau_n = \alpha_{n+1}, \eta_n = \frac{1}{2}\beta_{n+1}$, $s_n = \gamma_{n+1}$ for each $n \geq 1$, $g = fP_C$, and $V = TP_C$. Then (2.116) can be rewritten as

$$z_{n+1} = \tau_n g z_n + \eta_n z_n + \eta_n S z_n + s_n V z_n, \quad \forall n \geq 1.$$

From Theorem 2.6.3, we see that $\{x_n\}$ converges to $p = P_{\text{Fix}(S)\cap\text{Fix}(T)}g(p)$ in norm. By Proposition 2.6.2, one has $\text{Fix}(V) = \text{Fix}(T)$ and $\text{Fix}(S) = C$. It follows that

$$\text{Fix}(S) \cap \text{Fix}(V) = C \cap \text{Fix}(T) = \text{Fix}(T),$$

which implies that

$$p = P_{\text{Fix}(T)}g(p) = P_{\text{Fix}(T)}fP_C(p) = P_{\text{Fix}(T)}f(p).$$

This shows that p is the unique solution of the variational inequality problem (VIP). Thus it follows from $x_{n+1} = P_C z_n$ that $\{x_n\}$ converges to $p \in \text{Fix}(T)$ in norm and p is the unique solution of the variational inequality problem (VIP). This completes the proof. $\qquad\square$

Remark 2.6.2. Applying Theorems 2.6.1, 2.6.2, 2.6.3, and 2.6.4 to nonexpansive mappings with fixed points and using Proposition 2.6.3, we can obtain strong convergence results for fixed points of nonspreading mappings.

2.7 Applications

In this subsection, we give some applications of the convergence theorems in the previous subsections to optimization problems, variational inequality problems, and split feasibility problems.

2.7.1 Convex minimization problems

Let C be a nonempty, closed, and convex subset of a real Hilbert space H, and let $\varphi : C \to \mathbb{R}$ be a continuous Fréchet differentiable convex functional.

Consider the following *convex minimization problem*:

Find a point $\tilde{x} \in C$ such that

$$\varphi(\tilde{x}) = \min\{\varphi(x) : x \in C\}. \tag{2.117}$$

We use Ω to denote the solution set of problem (2.117) and use $\nabla\varphi(x)$ to denote the gradient of φ at $x \in C$. Assume that $\Omega \neq \emptyset$. Note that

$$\tilde{x} \in \Omega \iff \langle \nabla\varphi(x), x - \tilde{x} \rangle \geq 0, \quad \forall x \in C \iff \tilde{x} = P_C(I - \gamma\nabla\varphi)\tilde{x}, \quad \forall \gamma > 0.$$

It follows from the Baillon–Haddad's theorem that, if $\nabla\varphi(x)$ is L-Lipschitz continuous, then $\nabla\varphi$ is $\frac{1}{L}$-inverse strongly monotone. Thus, from Proposition 2.1.4 (3), if $\gamma \in (0, \frac{2}{L})$, then $I - \gamma\nabla\varphi$ is $\frac{L\gamma}{2}$-averaged nonexpansive. Denote $T_\gamma := P_C(I - \gamma\nabla\varphi)$. Then, by Proposition 2.1.4 (5), we see that T_γ is $\frac{2+L\gamma}{4}$-averaged nonexpansive. Hence there exists a nonexpansive mapping $S : H \to H$ such that

$$T_\gamma = \left(1 - \frac{2 + L\gamma}{4}\right)I + \frac{2 + L\gamma}{4}S \tag{2.118}$$

and $\text{Fix}(T_\gamma) = \text{Fix}(S) = \Omega$.

Now, we consider the following iterative methods: $x_1 \in C$ is an initial point chosen arbitrarily and

$$x_{n+1} = T_\gamma x_n = T_\gamma^n x_1, \quad \forall n \geq 1; \tag{2.119}$$

$$u \in H, \quad x_{n+1} = P_C[\alpha_n u + (1 - \alpha_n)T_\gamma x_n], \quad \forall n \geq 1; \tag{2.120}$$

$$u \in C, \quad x_{n+1} = \alpha_n u + (1 - \alpha_n)P_C T_\gamma x_n, \quad \forall n \geq 1. \tag{2.121}$$

Theorem 2.7.1. *Assume that $\Omega \neq \emptyset$ and $\gamma \in (0, \frac{2}{L})$. Let $\{\alpha_n\}$ be a real number sequence in $[0, 1]$ satisfying the following conditions:*
(C1) $\alpha_n \to 0$ *as* $n \to \infty$;
(C2) $\sum_{n=1}^\infty \alpha_n = \infty$.

Then the sequence $\{x_n\}$ generated by (2.119) converges weakly to $p = \lim_{n\to\infty} P_\Omega x_n$. In addition, the sequences generated by (2.120) and (2.121) converge strongly to $x^* = P_\Omega u$.

Proof. The first conclusion can be derived directly from Corollary 2.4.1. The second conclusion can be derived directly from Theorem 2.5.4. Note that

$$\text{Fix}(V) = \text{Fix}(P_C) \cap \text{Fix}(T_\gamma) = C \cap \text{Fix}(T_\gamma) = \text{Fix}(T_\gamma) = \Omega.$$

Applying Corollary 2.4.4 to $V := P_C T_\gamma$, we can obtain the third conclusion immediately. This completes the proof. $\qquad \square$

2.7.2 Monotone variational inequality problems

Let C be a nonempty, closed, and convex subset of a real Hilbert space H, and let $A : C \to H$ be a k-Lipschitz continuous and η-strongly monotone mapping.

We consider the following *variational inequality problem:*

Find a $x^* \in C$ such that

$$\langle Ax^*, x - x^* \rangle \geq 0, \quad \forall x \in C. \tag{VIP}$$

Letting $VI(C, A)$ denote the solution set of the variational inequality problem, we see that

$$x^* \in VI(C, A) \iff x^* = P_C(x^* - \mu Ax^*), \quad \forall \mu > 0.$$

If $\mu \in (0, \frac{2\eta}{k})$, then $P_C(I - \mu A)$ is a Banach contractive mapping from C to itself. Indeed, for all $x, y \in C$,

$$\| P_C(I - \mu A)x - P_C(I - \mu A)y \| \leq \sqrt{1 - \mu(2\eta - \mu k^2)} \| x - y \|.$$

From Banach's contraction principle, we see that $P_C(I - \mu A)$ has a unique fixed point $x^* \in C$. Hence $VI(C, A) = \{x^*\}$.

The popular method for solving the variational inequality problem is the projected gradient method:

$$x_1 \in C, \quad x_{n+1} = P_C(I - \mu A)x_n, \quad \forall n \geq 1.$$

Here the projected gradient method is actually the famous Banch–Picard iterative method:

$$x_1 \in C, \quad x_{n+1} = [P_C(I - \mu A)]^n x_1, \quad \forall n \geq 1.$$

It follows from Banach's Contraction Principle that $x_n \to x^*$ as $n \to \infty$.

These methods require us to compute the metric projection at every iteration, however, there are no analytic expressions for the metric projection operator in most cases. Recently, Yamada [108] introduced the following hybrid steepest descent method:

$$x_1 \in C, \quad x_{n+1} = (I - \mu \lambda_n A)Tx_n, \quad \forall n \geq 1, \tag{HSDM}$$

where $T : H \to H$ is a nonexpansive mapping with $\text{Fix}(T) \neq \emptyset$, $\{\lambda_n\} \subset (0, 1)$ and $A : H \to H$ is a k-Lipschitz continuous and η-strongly monotone mapping.

Theorem 2.7.2. *Suppose that $\mu \in (0, \frac{2\eta}{k^2})$ and $\{\lambda_n\}$ is a real number sequence in $[0, 1]$ satisfying the following conditions:*

(C1) $\lambda_n \to 0$ *as* $n \to \infty$;

(C2) $\sum_{n=1}^{\infty} \lambda_n = \infty$;

(C3) $\sum_{n=1}^{\infty} |\lambda_{n+1} - \lambda_n| < \infty$ or

(C4) $\lim_{n \to \infty} \frac{\lambda_n}{\lambda_{n+1}} = 1$.

Then the sequence $\{x_n\}$ generated by (HSDM) converges strongly to x^*, which is the unique solution of the variational inequality problem (VIP).

Proof. Note that (HSDM) can be rewritten as

$$x_{n+1} = \lambda_n(I - \mu A)Tx_n + (1 - \lambda_n)Tx_n, \quad \forall n \geq 1. \tag{2.122}$$

Denote $f(x) = (I - \mu A)Tx$ for all $x \in H$. Then (2.122) can be rewritten as

$$x_{n+1} = \lambda_n f(x_n) + (1 - \lambda_n)Tx_n, \quad \forall n \geq 1, \tag{2.123}$$

where $f : H \to H$ is a ρ-contractive mapping. In fact, for all $x, y \in H$, one has

$$
\begin{aligned}
\|f(x) - f(y)\|^2 &= \|(I - \mu A)Tx - (I - \mu A)Ty\|^2 \\
&= \|Tx - Ty - \mu(ATx - ATy)\|^2 \\
&= \|Tx - Ty\|^2 - 2\mu\langle Tx - Ty, ATx - ATy \rangle + \mu^2\|ATx - ATy\|^2 \\
&\leq \|Tx - Ty\|^2 + 2\eta\mu\|Tx - Ty\|^2 + \mu^2 k^2\|Tx - Ty\|^2 \\
&= [1 - \mu(2\eta - \mu k^2)]\|Tx - Ty\|^2 \\
&\leq [1 - \mu(2\eta - \mu k^2)]\|x - y\|^2.
\end{aligned}
$$

It follows that

$$\|fx - fy\| \leq \rho\|x - y\|, \quad \forall x, y \in H,$$

where $\rho = \sqrt{1 - \mu(2\eta - \mu k^2)} \in (0, 1)$. By Remark 2.4.5, we see that $\{x_n\}$ converges strongly to a fixed point $x^* \in \text{Fix}(T)$, where $x^* = P_{\text{Fix}(T)}fx^*$ of T, which is the unique solution of the following variational inequality:

$$
\begin{aligned}
&\langle fx^* - x^*, y - x^* \rangle \leq 0, \quad \forall y \in \text{Fix}(T) \\
&\iff \langle (I - \mu A)Tx^* - x^*, y - x^* \rangle \leq 0, \quad \forall y \in \text{Fix}(T) \\
&\iff \langle Ax^*, y - x^* \rangle \geq 0, \quad \forall y \in \text{Fix}(T).
\end{aligned}
$$

This completes the proof. □

Theorem 2.7.3. *Let $T : H \to H$ be a nonexpansive mapping with $\text{Fix}(T) \neq \emptyset$ and let $A : C \to H$ be a k-Lipschitz continuous and η-strongly monotone mapping. For any $\lambda \in (0, 1)$, define a mapping $T_\lambda = (1 - \lambda)I + \lambda T$. Suppose that $\mu \in (0, \frac{2\eta}{k^2})$ and $\{\lambda_n\}$ is a real number sequence in $(0, 1)$ satisfying the following conditions:*

(C1) $\lambda_n \to 0$ as $n \to \infty$;

(C2) $\sum_{n=1}^{\infty} \lambda_n = \infty$.

Define an iterative sequence $\{x_n\}$ as follows:

$$x_1 \in C, \quad x_{n+1} = (I - \mu\lambda_n A)T_\lambda x_n, \quad \forall n \geq 1. \tag{HSDM}$$

Then $\{x_n\}$ converges strongly to $x^* \in C$, which is a unique solution of the variational inequality (VIP).

Proof. By a similar proof to that of Theorem 2.6.2, (HSDM) can be rewritten as follows:

$$x_{n+1} = \lambda_n g(x_n) + (1 - \lambda_n)T_\lambda x_n, \quad \forall n \geq 1, \tag{2.124}$$

where $g(x) = (I - \mu A)T_\lambda x$ is a ρ-contractive mapping. By Theorem 2.4.13, it follows that $\{x_n\}$ converges strongly to a fixed point x^* of T_λ, where $x^* \in \mathrm{Fix}(T_\lambda) = \mathrm{Fix}(T)$, and $x^* = P_{\mathrm{Fix}(T_\lambda)}gx^* = P_{\mathrm{Fix}(T)}gx^*$, which is equivalent to

$$\langle gx^* - x^*, y - x^* \rangle \leq 0, \quad \forall y \in \mathrm{Fix}(T)$$
$$\Longleftrightarrow \langle (I - \mu A)Tx^* - x^*, y - x^* \rangle \leq 0, \quad \forall y \in \mathrm{Fix}(T)$$
$$\Longleftrightarrow \langle Ax^*, y - x^* \rangle \geq 0, \quad \forall y \in \mathrm{Fix}(T).$$

This completes the proof. □

2.7.3 Split feasibility problems

Let C and Q be two nonempty, closed, and convex subsets of Hilbert spaces H_1 and H_2, respectively. Let $A : H_1 \rightarrow H_2$ be a bounded linear operator.

Now, we consider the following *split feasibility problem* (SFP):

$$\text{Find a point } \hat{x} \in C \quad \text{such that } A\hat{x} \in Q. \tag{SFP}$$

The split feasibility problem is a special convex feasibility problem. It has very wide applications in signal processing, computer tomography, and radiation therapy treatment planning. For more details, we refer the readers to [15, 16, 18–22, 102].

Denote $A^{-1}(Q) = \{x \in H_1 : Ax \in Q\}$ and $\Gamma = C \cap A^{-1}(Q)$. Let $L := \rho(A^*A)$ denote the spectral radius of A^*A and let A^* be the adjoint operator of A. Let

$$f(x) = \frac{1}{2}\|I - P_Q Ax\|^2, \quad \forall x \in C.$$

Then we have the gradient of f:

$$\nabla f(x) = A^*(I - P_Q)Ax, \quad \forall x \in C, \tag{2.125}$$

and

$$\|\nabla f(x) - \nabla f(y)\| \leq \|A\|^2 \|x - y\| = L\|x - y\|, \quad \forall x, y \in C.$$

Note that

$$x^* \in \operatorname{argmin}_{x \in C} f(x) \iff \langle \nabla f(x^*), y - x^* \rangle \geq 0, \quad \forall y \in C$$
$$\iff x^* \in P_C[(I - \gamma \nabla f)x^*]. \tag{2.126}$$

Under the assumption that $\Gamma \neq \emptyset$, we know that

$$x^* \in \Gamma \iff x^* \in \operatorname{argmin}_{x \in C} f(x). \tag{2.127}$$

In fact, it follows from (2.126) that

$$x^* \in \operatorname{argmin}_{x \in C} f(x) \iff \langle \nabla f(x^*), y - x^* \rangle \geq 0, \quad \forall y \in C.$$

By (2.125), we obtain

$$\langle A^*(I - P_Q)Ax^*, y - x^* \rangle \geq 0, \quad \forall y \in C$$
$$\implies \langle (I - P_Q)Ax^*, Ay - Ax^* \rangle \geq 0, \quad \forall y \in C. \tag{2.128}$$

From the assumption that $\Gamma \neq \emptyset$, we have that if $z \in \Gamma$, then $z \in C$ and, if $Az \in Q$, then $(I - P_Q)Az = \theta$. Taking $y = z \in C$ in (2.128), we conclude from $(I - P_Q)Az = \theta$ that

$$\langle (I - P_Q)Ax^* - (I - P_Q)Az, Az - Ax^* \rangle \geq 0$$
$$\implies \langle (I - P_Q)Ax^* - (I - P_Q)Az, Ax^* - Az \rangle \leq 0. \tag{2.129}$$

Since $I - P_Q$ is firmly nonexpansive, it follows from (2.129) that

$$\|(I - P_Q)Ax^*\|^2 \leq 0 \implies Ax^* = P_Q Ax^* \in Q \implies x^* \in \Gamma.$$

A popular method for solving problem (SFP) is the Byrne's CQ-method:

$$x_1 \in H_1, \quad x_{n+1} = P_C[x_n - \gamma A^*(I - P_Q)Ax_n], \quad \forall n \geq 1. \tag{CQ}$$

Indeed, it is the discretization of fixed point equation (2.126) and belongs to the class of projection gradient methods. In addition, the CQ-method has its specific structure and hence it is necessary to investigate the CQ-method extensively.

Setting $T_\gamma := I - \gamma \nabla f$, for any $\gamma \in (0, \frac{2}{L})$, one sees that T_γ is averaged nonexpansive. So $P_C T_\gamma$ is also averaged nonexpansive and

$$\operatorname{Fix}(P_C T_\gamma) = \operatorname{Fix}(P_C) \cap \operatorname{Fix}(T_\gamma) = C \cap A^{-1}(Q) = \Gamma.$$

Theorem 2.7.4. *Let $\{x_n\}$ be the sequence generated by the CQ-method. Then $\{x_n\}$ converges weakly to a solution $x^* \in \Gamma$ of problem (SFP).*

Proof. Setting $U = P_C T_\gamma$, we see that the CQ-method can be rewritten as follows:

$$x_{n+1} = P_C T_\gamma x_n = U x_n, \quad \forall n \geq 1.$$

From Corollary 2.4.1, we derive the conclusion immediately. This completes the proof. $\square$

Now, we consider the following methods: for any $x_1 \in C$ and for any fixed $u \in C$,
(1) $x_{n+1} = \alpha_n u + (1 - \alpha_n) P_C T_y x_n, \forall n \geq 1;$
(2) $x_{n+1} = P_C[\alpha_n u + (1 - \alpha_n) T_y x_n], \forall n \geq 1.$

If $\{\alpha_n\}$ is a real number sequence in $[0, 1]$ satisfying the following conditions:
(C1) $\alpha_n \to 0$ as $n \to \infty;$
(C2) $\sum_{n=1}^{\infty} \alpha_n = \infty,$

then both methods (1) and (2) converge to a particular solution of the (SFP) in norm, that is, $x^* = P_\Gamma u$. In particular, for method (2), by taking $u = \theta \in H_1$, we can show that $\{x_n\}$ converges strongly to the minimum norm solution of problem (SFP). This is not available in method (1). Further, for more results on problem (SFP), we refer the readers to [15, 58, 62–64, 94, 57, 58, 102, 104, 109–111, 124, 127].

Theorem 2.7.5. *Let C and Q be two nonempty, closed, and convex subsets of Hilbert spaces H_1 and H_2, respectively. Let $A : H_1 \to H_2$ be a bounded linear operator. Suppose that the sequences $\{\alpha_n\}$ and $\{\beta_n\}$ satisfy the following conditions:*
(1) $\alpha_n \to 0$ *and* $\beta_n \to 0$ *as* $n \to \infty;$
(2) $\sum_{n=1}^{\infty} \alpha_n = \infty;$
(3) $\frac{\alpha_n}{\beta_n} \to 0$ *as* $n \to \infty.$

Let $\{x_n\}$ be a sequence generated in the following process: for any $u, x_1 \in H_1$,

$$x_{n+1} = P_C[\alpha_n u + (1 - \alpha_n)(x_n - \beta_n A^*(I - P_Q)Ax_n)], \quad \forall n \geq 1. \tag{2.130}$$

If $\Gamma \neq \emptyset$, then the sequence $\{x_n\}$ generated by (2.130) converges to $x^* = P_\Gamma u$ in norm. In particular, if we take $u = \theta$, then $\{x_n\}$ converges to $x^* = P_\Gamma \theta$ in norm, that is, the minimum norm solution of problem (SFP).

Proof. Since $\Gamma \neq \emptyset$ is closed convex, it follows that $P_\Gamma u$ is uniquely determined; denote it by $x^* = P_\Gamma u$. Denote $v_n = x_n - \beta_n A^*(I - P_Q)Ax_n$ for each $n \geq 1$. Then it follows that, for all $z \in \Gamma$,

$$\|v_n - z\|^2 = \|x_n - z - \beta_n A^*(I - P_Q)Ax_n\|^2$$
$$= \|x_n - z\|^2 - 2\beta_n\langle x_n - z, A^*(I - P_Q)Ax_n\rangle + \beta_n^2\|A^*(I - P_Q)Ax_n\|^2$$
$$= \|x_n - z\|^2 - 2\beta_n\langle Ax_n - Az, (I - P_Q)Ax_n\rangle + \beta_n^2\|A^*(I - P_Q)Ax_n\|^2$$
$$\leq \|x_n - z\|^2 - 2\beta_n\|(I - P_Q)Ax_n\|^2 + \beta_n^2\|A\|^2\|(I - P_Q)Ax_n\|^2$$
$$= \|x_n - z\|^2 - \beta_n(2 - \beta_n\|A\|^2)\|(I - P_Q)Ax_n\|^2$$
$$\leq \|x_n - z\|^2,$$
$$\|x_{n+1} - x^*\| = \|P_C[\alpha_n u + (1 - \alpha_n)v_n] - P_C x^*\|$$
$$\leq \|\alpha_n(u - x^*) + (1 - \alpha_n)(v_n - x^*)\|$$

$$\leq \alpha_n \|u - x^*\| + (1 - \alpha_n)\|v_n - x^*\|$$

$$\leq \alpha_n \|u - x^*\| + (1 - \alpha_n)\|x_n - x^*\|$$

$$\leq \max\{\|u - x^*\|, \|x_1 - x^*\|\}$$

$$= M,$$

and

$$\|x_{n+1} - x^*\|^2 = \|P_C[\alpha_n u + (1 - \alpha_n)v_n] - P_C x^*\|^2$$

$$\leq \|\alpha_n(u - x^*) + (1 - \alpha_n)(v_n - x^*)\|^2$$

$$= (1 - \alpha_n)^2 \|v_n - x^*\|^2$$

$$\quad + 2\alpha_n(1 - \alpha_n)\langle u - x^*, v_n - x^* \rangle + \alpha_n^2 \|u - x^*\|^2$$

$$\leq (1 - \alpha_n)\|x_n - x^*\|^2$$

$$\quad - (1 - \alpha_n)^2 \beta_n(2 - \beta_n \|A\|^2)\|(I - P_Q)Ax_n\|^2$$

$$\quad + 2\alpha_n(1 - \alpha_n)\langle u - x^*, v_n - x^* \rangle + \alpha_n^2 \|u - x^*\|^2. \tag{2.131}$$

Letting $a_n = \|x_n - x^*\|^2$ for each $n \geq 1$, we have

$$r_n = \frac{\beta_n}{\alpha_n}(1 - \alpha_n)^2 (2 - \beta_n \|A\|^2)\|(I - P_Q)Ax_n\|^2$$

$$\quad - 2(1 - \alpha_n)\langle u - x^*, v_n - x^* \rangle - \alpha_n \|u - x^*\|^2.$$

Then (2.131) can be rewritten by

$$a_{n+1} \leq (1 - \alpha_n)a_n + \alpha_n(-r_n), \quad \forall n \geq 1.$$

Since $\{r_n\}$ is bounded below, we see that $\{-r_n\}$ is bounded above. Using Lemma 1.10.5, we have

$$\liminf_{n \to \infty} a_n \leq \limsup_{n \to \infty}(-r_n) = -\liminf_{n \to \infty} r_n < \infty. \tag{2.132}$$

Assume that

$$\liminf_{n \to \infty}(r_n) = \lim_{j \to \infty}(r_{n_j}).$$

Then $\{x_{n_j}\}$ is a subsequence of $\{x_n\}$. So there exists a constant $c_1 > 0$ such that

$$\frac{\beta_{n_k}}{\alpha_{n_k}}(1 - \alpha_{n_k})^2(2 - \beta_{n_k}\|A\|^2)\|(I - P_Q)Ax_{n_k}\|^2 \leq c_1$$

$$\implies \|(I - P_Q)Ax_{n_k}\|^2 \leq c_1 \frac{\alpha_{n_k}}{\beta_{n_k}(1 - \alpha_{n_k})^2(2 - \beta_{n_k}\|A\|^2)}.$$

Since $\frac{\alpha_{n_k}}{\beta_{n_k}} \to 0$ and $\alpha_{n_k} \to 0$ and $\beta_{n_k} \to 0$ as $k \to \infty$, one has

$$\left\| (I - P_Q) A x_{n_k} \right\|^2 \to \theta \quad (k \to \infty).$$

Assume that $x_{n_k} \rightharpoonup \bar{x}$ as $k \to \infty$. Note that $\bar{x} \in C$ and $A x_{n_k} \rightharpoonup A\bar{x}$ as $k \to \infty$. From the demiclosedness principle of $I - P_Q$ it follows that

$$A\bar{x} = P_Q A\bar{x} \in Q \implies \bar{x} \in \Gamma \implies \langle u - x^*, \bar{x} - x^* \rangle \leq 0. \tag{2.133}$$

Since $v_n - x_n = -\beta_n A^*(I - P_Q) A x_n \to \theta$ as $n \to \infty$, we find that $v_{n_k} \rightharpoonup \bar{x}$ as $k \to \infty$. From the definition of $\{r_n\}$ and (2.133), we obtain

$$\liminf_{n \to \infty} r_n = \lim_{k \to \infty} r_{n_k} \geq -2\langle u - x^*, \bar{x} - x^* \rangle \geq 0,$$

which implies from (2.132) that $\limsup_{n \to \infty} a_n \leq 0$. However, we have $\liminf_{n \to \infty} a_n \geq 0$. Therefore, $a_n \to 0$ as $n \to \infty$, that is, $\{x_n\}$ converges to $x^* = P_\Gamma u$ in norm. This completes the proof. $\qquad\square$

Remark 2.7.1. Take $\{\alpha_n\}$ and $\{\beta_n\}$ as follows:

$$\alpha_n = \frac{1}{n^a}, \quad \beta_n = \frac{1}{n^b}, \quad \forall n \geq 1,$$

where $0 < b < a \leq 1$. Then the sequences $\{\alpha_n\}$ and $\{\beta_n\}$ satisfy conditions (C1)–(C3).

2.8 Remark

All results presented in this chapter were obtained by various authors in different periods, however, most of them are reproved and simplified by the authors of the book.

2.9 Exercises

We use H to denote a real Hilbert space.
1. Let $\{T_i\}_{i=1}^r : H \to H$ be a finite family of mappings. Assume that $K = \bigcap_{i=1}^r \mathrm{Fix}(T_i) \neq \emptyset$ and $K = \mathrm{Fix}(T_r T_{r-1} \cdots T_1)$. Show that

$$K = \mathrm{Fix}(T_1 T_r \cdots T_3 T_2) = \cdots = \mathrm{Fix}(T_{r-1} T_{r-2} \cdots T_1 T_r).$$

2. Let C be a nonempty closed convex subset of H and $x_0 \in H$. Prove that (i) If $C = B_r[x_0]$, then

$$P_C x = \begin{cases} x, & \text{if } \|x - x_0\| \leq r, \\ \frac{r(x-x_0)}{\|x-x_0\|}, & \text{if } \|x - x_0\| \geq r; \end{cases}$$

(ii) $P_C : H \to B_r[x_0]$ is firmly nonexpansive.

3. Let $T : B_r[\theta] \to H$ is a nonexpansive mapping. Prove that the following alternative law holds:

   (i) T has a fixed point;

   (ii) There exists some $x \in \partial B_r[\theta]$ and $\lambda \in (0,1)$ such that $x = \lambda Tx$.

4. Let C be a closed subspace of a real Hilbert space H and let P_C be the metric projection of H onto C. Then, show that the following (1), (2), (3), and (4) hold:

   (1) P_C is a linear mapping of H onto C;

   (2) $P_C^2 = P$;

   (3) $\|P_C x\| \le \|x\| \ \forall x \in H$;

   (4) $\|P_C\| = 1$.

5. Let C be a closed subspace of a real Hilbert space H and let $x \in C$. Define

$$N_C(x) = \{z \in H : \langle u - x, z \rangle \le 0, \ \forall u \in C\}.$$

   Show that $N_C(x)$ is a closed convex cone of H.

6. Let C be a nonempty convex and closed subset of a real Hilbert space H. Let A be a mapping of C into H and let P_C be the metric projection of H onto C. Show

$$\text{Fix}(P_C A) = VI(C, I - A),$$

   where $VI(C, I - A)$ denotes the solution set of the variational inequality

$$\langle x - Ax, y - x \rangle \ge 0, \ \forall y \in C.$$

7. Let C be a closed convex subset of a real Hilbert space H and let $T : C \to C$ be a quasi-nonexpansive mapping. Prove that

$$\text{Fix}(T) = \bigcap_{y \in C} \{x \in C : 2\langle y - Ty, x \rangle \le \|y\|^2 - \|Ty\|^2\}.$$

   By means of the above expression, prove that $\text{Fix}(T)$ is a closed convex subset of H.

8. Let $T : H \to H$ be a quasi-nonexpansive mapping with $K = \text{Fix}(T) \ne \emptyset$. Let $A : H \to H$ be k-Lipschitz and η-strongly monotone on H. Let $\{\lambda_n\}$ be a sequence of numbers in (0,1) satisfying (C1) $\lim_{n \to \infty} \lambda_n = 0$, and (C2) $\sum_{n=1}^{\infty} \lambda_n = \infty$. For $\mu \in (0, \frac{2\eta}{k^2})$ and $\sigma \in (0,1)$, define a sequence $\{x_n\}$ iteratively in H by

   (HIM) $\quad x_1 \in H, x_{n+1} = \lambda_n(I - \mu A)x_n + (1 - \lambda_n)[\sigma x_n + (1 - \sigma)Tx_n], \quad n \ge 1.$

   Prove that the sequence $\{x_n\}$ generated by (HIM) converges strongly to the unique solution x^* of the variational inequality problem:

$$\langle Ax, y - x \rangle \ge 0, \ \forall y \in K. \tag{VIP}$$

9. Let $T : H \to H$ be a quasi-nonexpansive mapping with $K = \mathrm{Fix}(T) \neq \emptyset$. Let $A : H \to H$ be k-Lipschitz and η-strongly monotone on H. Let $\{\lambda_n\}$ be a sequence of numbers in $(0,1)$ satisfying (C1) $\lim_{n \to \infty} \lambda_n = 0$, and (C2) $\sum_{n=1}^{\infty} \lambda_n = \infty$. For $\mu \in (0, \frac{2\eta}{k^2})$ and $\sigma \in (0, 1)$, define a sequence $\{x_n\}$ iteratively in H by

$$(\text{HSDIM}) \qquad x_1 \in H, x_{n+1} = \sigma x_n + (1 - \sigma)(Tx_n - \mu \lambda A Tx_n), \quad n \geq 1.$$

Prove that the sequence $\{x_n\}$ generated by (HSDIM) converges strongly to the unique solution x^* of the variational inequality problem:

$$\langle Ax, y - x \rangle \geq 0, \quad \forall y \in K. \qquad (\text{VIP})$$

10. Let C be a nonempty closed convex subset of H. Let $A : C \to H$ be an α-inverse strongly-monotone mapping and let $S : C \to C$ be a nonexpansive mapping such that $K = \mathrm{Fix}(S) \cap VI(C, A) \neq \emptyset$. For arbitrary initial value $x_1 \in C$, define a sequence $\{x_n\}$ iteratively in C by

$$(\text{MIM}) \qquad x_{n+1} = \alpha_n x_n + (1 - \alpha_n)\left[\frac{1}{2}Sx_n + \frac{1}{2}P_C(x_n - \lambda_n A x_n)\right], \quad n \geq 1.$$

where $\lambda_n \in [a, 2\alpha]$, $\alpha_n \in [c, d]$ for $a > 0$, $c > 0$, $d < 1$. Show that the sequence $\{x_n\}$ generated by (MIM) converges weakly to some $x^* \in K$, where $x^* = \lim_{n \to \infty} P_K x_n$.

11. Let C be a nonempty closed convex subset of a real Hilbert space H, let $A : C \to H$ be an α-inverse strongly monotone mapping and $S : C \to C$ be a nonexpansive mapping such that $K = \mathrm{Fix}(S) \cap VI(C, A) \neq \emptyset$. For arbitrary initial value $x_1 \in C$ and a fixed anchor $u \in C$, define a sequence $\{x_n\}$ iteratively in C by

$$(\text{HIM}) \qquad x_{n+1} = \alpha_n u + (1 - \alpha_n)\left[\frac{1}{2}Sx_n + \frac{1}{2}P_C(x_n - \lambda_n A x_n)\right], \quad n \geq 1,$$

where $\{\alpha_n\}$ is a sequence of numbers in $(0, 1)$ satisfying (C1) $\lim_{n \to \infty} \alpha_n = 0$, (C2) $\sum_{n=1}^{\infty} \alpha_n = \infty$, and (C3) $\sum_{n=1}^{\infty} |\alpha_{n+1} - \alpha_n| < \infty$ or (C4) $\frac{\alpha_n}{\alpha_{n+1}} \to 1$; where $\{\lambda_n\}$ is another sequence of numbers in $(0, 1)$ satisfying $\lambda_n \in [a, 2\alpha]$ and $\sum_{n=1}^{\infty} |\lambda_{n+1} - \lambda_n| < \infty$ for some $a > 0$. Prove that the sequence $\{x_n\}$ generated by (HIM) converges strongly to the specific point $z \in K$, where $z = P_K u$.

12. Let $A : H \to H$ be η-strongly monotone and k-Lipschitz mapping. Let $\{T_i\}_{i=1}^{N}$ be nonexpansive self-mappings of H such that $C = \bigcap_{i=1}^{N} \mathrm{Fix}(T_i) \neq \emptyset$. Let $\{\lambda_k\}$ and $\{\beta_k^i\}$ be two sequences of number in $(0,1)$ satisfying conditions: (C1) $\lim_{k \to \infty} \lambda_k = 0$, (C2) $\sum_{k=1}^{\infty} \lambda_k = \infty$, (C3) $\beta_k^i \in (\alpha, \beta)$, for some $\alpha, \beta \in (0, 1)$, and (C4) $\lim_{k \to \infty} |\beta_{k+1}^i - \beta_k^i| = 0$, for $i = 1, 2, \ldots, N$. For any initial value $x_0 \in H$, define a sequence $\{x_n\}$ iteratively in H by

$$x_{n+1} = (I - \lambda_k \mu A)T_N^k T_{N-1}^k \cdots T_1^k x_k, \quad k \geq 0,$$

where $\mu \in (0, \frac{2\eta}{k^2})$ and $T_i^k := (1 - \beta_k^i)I + \beta_k^i T_i$, for $i = 1, 2, \ldots, N$. Prove that the sequence $\{x_k\}$ converges strongly to the unique solution x^* of the variational inequality problem

$$\langle Ax, y - x \rangle \geq 0, \quad \forall y \in C.$$

3 Iterative methods for zeros of monotone mappings and fixed points of pseudocontractive mappings in Hilbert spaces

In this chapter, we systematically discuss the theory of monotone mappings in Hilbert spaces, in particular, the theory of maximal monotone mappings. With the aid of the theory of monotone mappings, we establish the existence and unique theorems of solutions of variational inequalities with strongly monotone hemicontinuous mappings. We also obtain zero theorems of strongly monotone hemicontinuous mappings, from which we derive the existence, uniqueness, and convergence of paths, and some fixed point theorems for hemicontinuous pseudocontractive mappings in Hilbert spaces. Based on these results, we focus on the construction of iterative methods for zeros of monotone mappings, fixed points of pseudocontractive mappings, and solutions of monotone variational inequalities. All iterative methods and convergence theorems discussed in this chapter include some classical results and some of the authors' recent results.

Let H be a real Hilbert space. By $\langle \cdot, \cdot \rangle$ and $\| \cdot \|$ we denote the inner product and norm induced by the inner product in H, respectively. For a multi-valued mapping $A : H \to 2^H$, we denote the effective domain, range, graph, and preimage of A as follows:

(1) $\mathrm{Dom}(A) = \{x \in H : Ax \neq \emptyset\}$;
(2) $\mathrm{Ran}(A) = \{y \in Ax : x \in \mathrm{Dom}(A)\}$;
(3) $\mathrm{Graph}(A) = \{[x, y] : x \in \mathrm{Dom}(A), y \in Ax\}$ and $A^{-1}(y) = \{x \in H : y \in Ax\}$.

For the sake of convenience, we use $A \subset H \times H$ to denote a multi-valued mapping $A : H \to 2^H$, while $A : H \to H$ denotes a single-valued mapping, and we use I to denote the identity mapping in H. Let $\to$ and $\rightharpoonup$ denote the strong and weak convergence, respectively.

3.1 Basic properties of monotone mappings

In this subsection, we discuss some fundamental properties of (maximal) monotone mappings such as the local boundedness, demiclosedness, surjectivity, and so on.

3.1.1 Local boundedness and hemicontinuity

Definition 3.1.1. Let $A \subset H \times H$ be a mapping. Then A is said to be a locally bounded at $x_0 \in \mathrm{Dom}(A)$ if there exists a neighborhood U of x_0 such that $A(U) = \{f : y \in U, f \in Ay\}$ is bounded in H.

https://doi.org/10.1515/9783110667097-003

Definition 3.1.2.
(1) Let $A \subset H \times H$ be a mapping. Then A is said to be hemicontinuous if $\forall x \in \text{Dom}(A)$, $\forall h \in H$, $t_n \to 0$, $A(x + t_n h) \rightharpoonup Ax$ $(n \to \infty)$ provided $x + t_n h \in \text{Dom}(A)$.
(2) $A \subset H \times H$ is said to be demicontinuous if $\forall x \in \text{Dom}(A)$, $\forall \{x_n\} \subset \text{Dom}(A)$, $x_n \to x \Rightarrow Ax_n \rightharpoonup Ax$ $(n \to \infty)$.

Remark 3.1.1. If A is demicontinuous, then A must be hemicontinuous. The converse is not true.

Example 3.1.1. Let $H = \mathbb{R}^2$, and consider the function $\varphi : \mathbb{R}^2 \to \mathbb{R}$ defined by

$$\varphi(x,y) = \begin{cases} xy^2(x^2 + y^4)^{-1}, & (x,y) \in \mathbb{R}^2, \\ 0, & x = y = 0. \end{cases}$$

Then $\varphi : \mathbb{R}^2 \to \mathbb{R}$ is hemicontinuous, but it is not demicontinuous.

Remark 3.1.2. For mappings in finite-dimensional normed linear spaces, demicontinuity and continuity are equivalent, and local boundedness and boundedness are also equivalent. For linear mappings in infinite-dimensional normed linear spaces, demicontinuity and continuity are equivalent, and local boundedness and boundedness are also equivalent.

In order to establish the local boundedness results of monotone mappings, we first prove a useful lemma.

Lemma 3.1.1. *Let $\{x_n\}$ be a sequence in H such that $x_n \to \theta$ and let $\{f_n\}$ be a sequence in H such that $\|f_n\| \to \infty$ as $n \to \infty$. Then, for any $\rho > 0$, there exist $z \in B_\rho[\theta]$ and subsequences $\{x_{n_j}\}$ and $\{f_{n_j}\}$ of $\{x_n\}$ and $\{f_n\}$, respectively, such that*

$$\langle f_{n_j}, x_{n_j} - z \rangle \to \infty \quad (j \to \infty). \tag{3.1}$$

Proof. Assume that the conclusion is not true. Then $\forall z \in B_\rho[\theta]$ there exists a constant $C_z \in R$ such that

$$\langle f_n, x_n - z \rangle \geq C_z, \quad \forall n \geq 1. \tag{3.2}$$

For all $k \geq 1$, let

$$E_k = \{u \in B_\rho[\theta] : \langle f_n, x_n - u \rangle \geq -k, n \geq 1\}.$$

Then E_k is closed and $B_\rho[\theta] = \bigcup_{k \geq 1} E_k$. Since $B_\rho(\theta)$ is a complete metric space, it is of the second category. It follows from the Baire category theorem that there exists $k_0 \in N$ satisfying $\text{int}(E_{k_0}) \neq \emptyset$, that is, there exists $r > 0$, $y \in B_\rho[\theta]$ such that $B_r(y) \subset E_{k_0}$. Note that $-y \in B_\rho[\theta]$. It follows from (3.2) that

$$\langle f_n, x_n + y \rangle \geq C_{-y}. \tag{3.3}$$

For any $u \in B_r(y)$, we have

$$u \in E_{k_0} \implies u \in B_\rho[\theta]$$

and

$$\langle f_n, x_n - u \rangle \geq -k_0,$$

which implies from (3.3) that

$$\langle f_n, 2x_n + y - u \rangle \geq C_{-y} - k_0. \tag{3.4}$$

From the assumption $x_n \to \theta$ as $n \to \infty$, we see that there exists $n_0 \in N$ such that $\|x_n\| < \frac{r}{4}$ for all $n \geq n_0$. For all $z \in H$ with $\|z\| < \frac{r}{2}$, letting $u = 2x_n + y - z$, we obtain

$$\|u - y\| = \|2x_n - z\| \leq \frac{r}{2} + \frac{r}{2} = r, \quad \forall n \geq n_0.$$

This shows that $u = 2x_n + y - z \in B_r(y)$. Using (3.4), we find that

$$\langle f_n, z \rangle \geq C_{-y} - k_0, \quad \forall n \geq n_0, z \in B_{\frac{r}{2}}(\theta).$$

Since $\|-z\| < \frac{r}{2}$, we have $\langle f_n, z \rangle \leq k_0 - C_{-y}$ for all $n \geq n_0$ and $z \in B_{\frac{r}{2}}(\theta)$. Hence

$$|\langle f_n, z \rangle| \leq |k_0 - C_{-y}| < \infty, \quad \forall n \geq n_0, z \in B_{\frac{r}{2}}(\theta).$$

This implies that

$$\|f_n\| \leq \sup_{\|z\| \leq r} |\langle f_n, z \rangle| \leq 2|k_0 - C_{-y}| < \infty,$$

which contradicts the assumption $\|f_n\| \to \infty$, completing the proof. $\qquad \square$

Using Theorem 3.1.1, we find the following important conclusion.

Theorem 3.1.1. *If $A \subset H \times H$ is a monotone mapping, then A is locally bounded in* int(Dom(A)).

Remark 3.1.3. Monotone mappings may be unbounded on the boundary of their effective domains.

Remark 3.1.4. Example 3.1.1 shows that hemicontinuous mappings are not necessarily demicontinuous even in the framework of finite-dimensional spaces. However, for monotone mappings, if they are hemicontinuous in their interiors, then they must be demicontinuous.

Theorem 3.1.2. *Let $A : \text{Dom}(A) \subset H \to H$ be a monotone mapping. If A is hemicontinuous, then A is demicontinuous at* int(Dom(A)).

Proof. Let $\{x_n\}$ be a sequence in int(Dom(A)) such that $x_n \to x \in$ int(Dom(A)). From Theorem 3.1.1, we see that $\{Ax_n\}$ is a bounded sequence in H. Since Hilbert space H

is reflexive, there exists a subsequence $\{Ax_{n_j}\}$ of $\{Ax_n\}$ such that $Ax_{n_j} \rightharpoonup y$ as $j \to \infty$. Since A is monotone, we have

$$\langle Ax_{n_j} - Au, x_{n_j} - u \rangle \geq 0, \quad \forall u \in \text{Dom}(A). \tag{3.5}$$

Letting $j \to \infty$ in (3.5), we obtain

$$\langle y - Au, x - u \rangle \geq 0, \quad \forall u \in \text{Dom}(A). \tag{3.6}$$

Since $x \in \text{int}(\text{Dom}(A))$ and $\text{int}(\text{Dom}(A))$ is an open set of H, for all $h \in H$, one sees that there exists $t_h > 0$ such that, for all $t \in (0, t_h)$, $h_t := x + th \in \text{int}(\text{Dom}(A)) \subset \text{Dom}(A)$. Replacing u with h_t in (3.6), we have

$$\langle y - Ah_t, x - h_t \rangle \geq 0$$
$$\implies \langle y - Ah_t, -th \rangle \geq 0, \quad \forall h \in H$$
$$\implies \langle y - Ah_t, h \rangle \leq 0, \quad \forall h \in H. \tag{3.7}$$

Letting $t \to 0$ in (3.7), one concludes from the hemicontinuity of A at x that

$$\langle y - Ax, h \rangle \leq 0, \quad \forall h \in H$$
$$\implies y = Ax, \quad \text{that is, } Ax_{n_j} \xrightarrow{w} Ax \quad (j \to \infty).$$

Since Ax is unique, we obtain that $Ax_n \rightharpoonup Ax$ as $n \to \infty$. This verifies that A is demicontinuous at any point $x \in \text{Dom}(A)$. Hence A is demicontinuous in $\text{int}(\text{Dom}(A))$. This completes the proof. □

Remark 3.1.5. Theorem 3.1.2 shows that, for single-valued monotone mappings defined on the whole space, demicontinuity is equivalent to hemicontinuity.

Corollary 3.1.1. *Let H be a finite-dimensional space and let $A : H \to H$ be a hemicontinuous monotone mapping. Then A is continuous on H.*

Proof. From Theorem 3.1.2, we find that A is demicontinuous in H. Hence $\forall \{x_n\} \subset H$, $x_n \to x \implies Ax_n \rightharpoonup Ax$ $(n \to \infty)$. In view of $\dim H < \infty$, we obtain that $Ax_n \to Ax$. This completes the proof. □

Remark 3.1.6. Corollary 3.1.1 shows that, for single-valued monotone mappings defined on finite-dimensional spaces, demicontinuity, hemicontinuity, and continuity are equivalent.

Corollary 3.1.2. *If $A : H \to H$ be a linear monotone mapping, then A is continuous monotone. Indeed, it is $\|A\|$-Lipschitz monotone.*

Proof. Since A is linear, we find that A is hemicontinuous. From Theorem 3.1.2, A is hemicontinuous, which implies that A is continuous. Next, we assume that A is discontinuous at the origin θ. Then $\exists \varepsilon_0 > 0$, $\{x_n\} \subset H$, $x_n \to \theta$, but $\|Ax_n\| \geq \varepsilon_0$, $n \in N$. Taking $t_n = \|x_n\|^{-\frac{1}{2}}$, $y_n = t_n x_n$, we find that $y_n \to \theta$. But $\|Ay_n\| = t_n \|Ax_n\| \geq t_n \varepsilon_0 \to \infty$, which contradicts the hemicontinuity of A. Hence A is continuous, therefore, it is bounded. Also $\|Ax\| \le \|A\|\|x\|$, which shows that A is $\|A\|$-Lipschitz monotone. This completes the proof. □

Remark 3.1.7. Corollary 3.1.2 shows that, for linear monotone mappings defined on the whole space, hemicontinuity, demicontinuity, continuity, and Lipschitz continuity are equivalent.

3.1.2 Characteristic description for monotone mappings

Proposition 3.1.1. *Let $A \subset H \times H$ be a mapping. Then A is monotone if and only if, for all $[x,f], [y,g] \in \mathrm{Graph}(A)$,*

$$\|(x + tf) - (y + tg)\| \ge \|x - y\|, \quad \forall t > 0. \tag{3.8}$$

Proof. Using the properties of the norms and the scalar products on H, we have

$$\begin{aligned}
\|(x + tf) - (y + tg)\|^2 &= \|x - y + t(f - g)\|^2 \\
&= \|x - y\|^2 + 2t\langle f - g, x - y\rangle + t^2\|f - g\|^2.
\end{aligned} \tag{3.9}$$

($\Longrightarrow$) If A is monotone, then $\langle f - g, x - y\rangle \ge 0$. Using (3.9), we find that

$$\|(x + tf) - (y + tg)\| \ge \|x - y\|.$$

($\Longleftarrow$) Suppose that (3.8) holds. Then, we find from (3.8) and (3.9) that

$$2\langle f - g, x - y\rangle + t\|f - g\|^2 \ge 0, \quad \forall t > 0.$$

Letting $t \to 0$, we have $\langle f - g, x - y\rangle \ge 0$. This completes the proof. □

Remark 3.1.8. Proposition 3.1.1 describes a characteristic of a monotone mapping A, namely, for all $t > 0$, $(I + tA)$ is an expansive mapping. So, $(I + tA)^{-1}$ always exists on $\mathrm{Ran}(I + tA)$. Hence we can define a resolvent operator by $J_t x = (I + tA)^{-1}x$ for any $x \in \mathrm{Ran}(I + tA)$ and Yosida approximation by $A_t = \frac{1}{t}(I - J_t)$ for any $t > 0$.

Proposition 3.1.2. *Let $A \subset H \times H$ be a monotone mapping. Then A is maximal monotone if and only if $\forall [x,f] \in H \times H$, $[y,g] \in \mathrm{Graph}(A)$, $\langle f - g, x - y\rangle \ge 0 \Longrightarrow [x,f] \in \mathrm{Graph}(A)$.*

Proof. ($\Longrightarrow$) Suppose that $A \subset H \times H$ is maximal monotone. For any $[x,f] \in H \times H$, $[y,g] \in \mathrm{Graph}(A)$, $\langle f - g, x - y\rangle \ge 0$. We need to confirm $[x,f] \in \mathrm{Graph}(A)$. If $[x,f] \notin \mathrm{Graph}(A)$, we construct a set $M = \mathrm{Graph}(A) \bigcup \{[x,f]\}$, and then M is a monotone set and $\mathrm{Graph}(A)$ is a proper subset of M, which contradict the definition of the maximal monotonicity of A.

($\Longleftarrow$) Suppose that M is an arbitrary monotone set containing Graph(A). Then, for any $[x,f] \in H \times H$ and $[y,g] \in$ Graph(A), $\langle f - g, x - y \rangle \geq 0$, which implies that

$$[x,f] \in \text{Graph}(A) \implies M \subseteq \text{Graph}(A) \implies M = \text{Graph}(A).$$

Hence, that A is maximal monotone. This completes the proof. □

Remark 3.1.9. The above proposition is often used to prove that a monotone mapping is maximal monotone.

3.1.3 Demiclosedness principle for monotone mappings

Definition 3.1.3. Let $A \subset H \times H$ be a mapping. Then A is said to be demiclosed if, for any $\{[x_n, f_n]\} \subset \text{Graph}(A)$, $x_n \to x$ and $f_n \rightharpoonup f$ or $x_n \rightharpoonup x$ and $f_n \to f$ imply $[x,f] \in \text{Graph}(A)$.

To establish the demiclosedness principle of monotone hemicontinuous mappings, we first give the famous Minty's lemma.

Lemma 3.1.2 (Minty's lemma). *Let C be a nonempty, closed, and convex subset of a real Hilbert space H. Let $A : C \to H$ be a monotone hemicontinuous mapping and $x_0 \in C$. Then the following assertions are equivalent:*
(1) $\langle Ax, x - x_0 \rangle \geq 0, \forall x \in C$.
(2) $\langle Ax_0, x - x_0 \rangle \geq 0, \forall x \in C$.

Proof. (1) $\implies$ (2). For any $y \in C$, $t \in (0,1)$, let $y_t = (1 - t)x_0 + ty$. Since C is convex, we have $y_t \in C$. Replacing x with y_t, we have

$$\langle Ay_t, y - x_0 \rangle \geq 0, \quad \forall t \in (0,1). \tag{3.10}$$

From the hemicontinuity of A, we obtain $Ay_t \rightharpoonup Ax_0$ as $t \to 0^+$. Letting $t \to 0^+$ in (3.10), we find that $\langle Ax_0, y - x_0 \rangle \geq 0, \forall y \in C$, that is, (2) holds.

(2) $\implies$ (1). For any $x \in C$, we conclude from the monotonicity of A that

$$\langle Ax - Ax_0, x - x_0 \rangle \geq 0, \quad \forall x \in C$$
$$\implies \langle Ax, x - x_0 \rangle \geq \langle Ax_0, x - x_0 \rangle \geq 0, \quad \forall x \in C,$$

that is, (1) holds. This completes the proof. □

Proposition 3.1.3. *Let C be a nonempty, closed, and convex subset of a real Hilbert space H, and let $A : C \to H$ be a monotone hemicontinuous mapping satisfying the "flow-invariance" condition (FIC):*

$$\lim_{h \to 0^+} h^{-1} d((I - hA)x, C) = 0, \quad \forall \in C.$$

Then A is demiclosed.

Proof. Suppose that $\{x_n\}$ is a sequence with $x_n \to x \in H$ and $Ax_n \to \theta$ as $n \to \infty$. Now, we need to prove that $x \in C$ and $Ax = \theta$. Since C is closed and convex, we see that C is w-closed and hence $x \in C$. In order to prove $Ax = \theta$, we write $y_h = x - hAx, \forall h > 0$. It follows from (FIC) that, for any $\varepsilon > 0$, there exists $\delta > 0$ with $0 < h < \delta$ such that

$$d(y_h, C) = d((I - hA)x, C) < h\varepsilon$$
$$\implies \exists u_h \in C \text{ s.t. } \|y_h - u_h\| < h\varepsilon$$
$$\implies \left\| \frac{u_h - x}{h} + Ax \right\| < \varepsilon$$
$$\implies \frac{u_h - x}{h} \to -Ax \quad (h \to 0^+). \tag{3.11}$$

On the other hand, we obtain from the monotonicity of A that

$$\langle Au - Ax_n, u - x_n \rangle \geq 0, \quad \forall u \in C, n \in \mathbb{N}. \tag{3.12}$$

Letting $n \to \infty$ in (3.12), one has

$$\langle Au, u - x \rangle \geq 0, \quad \forall u \in C. \tag{3.13}$$

By Minty's lemma, we derive

$$\langle Ax, u - x \rangle \geq 0, \quad \forall u \in C. \tag{3.14}$$

Taking $u = u_h$ in (3.14), we have

$$\langle Ax, u_h - x \rangle \geq 0, \quad \forall h > 0$$
$$\implies \left\langle Ax, \frac{u_h - x}{h} \right\rangle \geq 0, \quad \forall h > 0. \tag{3.15}$$

Letting $h \to 0^+$ in (3.15) and combining with (3.11), we have

$$-\|Ax\|^2 \geq 0 \implies Ax = \theta.$$

If $x_n \to x$ and $f_n \to f$ as $n \to \infty$, we can prove the conclusion in a similar way. This completes the proof. $\qquad \square$

Proposition 3.1.4. *Let $A \subset H \times H$ be a maximal monotone mapping. Then*
(i) *Ax is a w-closed subset of H, $\forall x \in \text{Dom}(A)$,*
(ii) *A is demiclosed.*

Proof.
(1) First, we prove that, for any $x \in \text{Dom}(A)$, Ax is a convex subset of H. Take $f_1, f_2 \in Ax$ and $t \in (0, 1)$ arbitrarily, and denote $f = (1 - t)f_1 + tf_2$. Then, for any $[y, g] \in \text{Graph}(A)$,

$$\langle f - g, x - y \rangle = (1 - t)\langle f_1 - g, x - y \rangle + t\langle f_2 - g, x - y \rangle \geq 0.$$

By Proposition 3.1.2, we see that $[x,f] \in \text{Graph}(A)$, that is, $f \in Ax$.

Next, we prove that Ax is w-closed. In fact, since

$$Ax = \bigcap_{[y,g] \in \text{Graph}(A)} \{f \in H : \langle f - g, x - y \rangle \geq 0\}$$

and the set $\{f \in H : \langle f - g, x - y \rangle \geq 0\}$ is w-closed, one sees that their intersection Ax is also w-closed.

(2) Let $\{[x_n, f_n]\} \subset \text{Graph}(A)$, $x_n \to x$ and $f_n \to f$ as $n \to \infty$. Then

$$\langle f_n - g, x_n - y \rangle \geq 0, \quad \forall [y,g] \in \text{Graph}(A), \quad n \geq 1.$$

Taking $n \to \infty$, we have

$$\langle f - g, x - y \rangle \geq 0, \quad \forall [y,g] \in \text{Graph}(A).$$

By Proposition 3.1.2, we see that, for any $[x,f] \in \text{Graph}(A)$, if $x_n \to x$ and $f_n \to f$ as $n \to \infty$, then the above conclusion still holds. Hence, Ax is demiclosed. This completes the proof. □

In fact, maximal monotone mappings have more general w-closed property as follows.

Proposition 3.1.5. *Let $A \subset H \times H$ be a maximal monotone mapping. Let $\{[x_n, f_n]\} \subset \text{Graph}(A)$, $x_n \to x$, $f_n \to f$ and*

$$\limsup_{n \to \infty} \langle f_n, x_n \rangle \leq \langle f, x \rangle.$$

Then $[x,f] \in \text{Graph}(A)$ and $\langle f_n, x_n \rangle \to \langle f, x \rangle$

Proof. From the monotonicity of A, we derive that

$$\langle f_n - g, x_n - y \rangle \geq 0, \quad \forall [y,g] \in \text{Graph}(A), \quad n \geq 1. \tag{3.16}$$

From $\limsup_{n \to \infty} \langle f_n, x_n \rangle \leq \langle f, x \rangle$ it follows that

$$\limsup_{n \to \infty} (\langle f_n, x_n \rangle - \langle f, x \rangle) \leq 0.$$

Hence

$$\liminf_{n \to \infty} \langle f_n, x - x_n \rangle \geq 0. \tag{3.17}$$

Using (3.16), we arrive at

$$
\begin{aligned}
\langle f - g, x - y \rangle &= \langle f - f_n, x - y \rangle + \langle f_n - g, x - y \rangle \\
&= \langle f - f_n, x - y \rangle + \langle f_n - g, x - x_n \rangle + \langle f_n - g, x_n - y \rangle \\
&\geq \langle f - f_n, x - y \rangle + \langle f_n, x - x_n \rangle - \langle g, x - x_n \rangle.
\end{aligned}
\tag{3.18}
$$

Taking $n \to \infty$ in (3.18), we obtain

$$\langle f - g, x - y \rangle \geq \liminf_{n \to \infty} \langle f_n, x - x_n \rangle \geq 0, \quad \forall [y, g] \in \mathrm{Graph}(A).$$

It follows from the monotonicity of A that $[x, f] \in \mathrm{Graph}(A)$. Taking $y = x$ and $g = f$ in (3.16), respectively, we have

$$\langle f_n - f, x_n - x \rangle \geq 0, \quad \forall n \geq 1. \tag{3.19}$$

Taking the limit inferior in (3.19), we derive

$$\liminf_{n \to \infty} \langle f_n, x_n \rangle \geq \langle f, x \rangle.$$

Combining with $\limsup_{n \to \infty} \langle f_n, x_n \rangle \leq \langle f, x \rangle$, we have

$$\lim_{n \to \infty} \langle f_n, x_n \rangle = \langle f, x \rangle.$$

This completes the proof. $\qquad\qquad\qquad\qquad\qquad\qquad\qquad\qquad\qquad\square$

Remark 3.1.10. Browder called the mapping with the property in Proposition 3.1.5 a "generalized pseudo-monotone mapping". Compared with the class of (maximal) monotone mappings, this class of mappings has more applications. If we remove the restriction $\langle f_n, x_n \rangle \to \langle f, x \rangle$, then this class of mappings is said to be of M-type.

Remark 3.1.11. If $x_n \to x$ and $f_n \rightharpoonup f$ or $x_n \rightharpoonup x$ and $f_n \to f$ as $n \to \infty$, then $\langle f_n, x_n \rangle \to \langle f, x \rangle$. If $x_n \rightharpoonup x$ and $f_n \rightharpoonup f$ as $n \to \infty$, $\langle f_n, x_n \rangle \to \langle f, x \rangle$ is not necessarily true. Hence, we can not directly take the limit in (3.16) to obtain $\langle f - g, x - y \rangle \geq 0$.

3.1.4 Resolvents and Yosida approximations

In this subsection, we discuss some fundamental properties of resolvent operators and the Yosida approximation of monotone mappings. They play an important role in constructing iterative methods for zeros of monotone mappings. In particular, the asymptotic behavior of the resolvent J_t is the basis of Bruck regularization method.

Suppose that $A \subset H \times H$ is a monotone mapping. Then, for any $t > 0$, the resolvent $J_t = (I + tA)^{-1}$ of A is well defined on $\mathrm{Ran}(I + tA)$ and the Yosida approximation $A_t = t^{-1}(I - J_t)$ of A is also well defined on $\mathrm{Ran}(I + tA)$. In particular, if $A \subset H \times H$ is a maximal monotone mapping, then $\mathrm{Dom}(J_t) = H$, that is, J_t is well defined on Hilbert space H.

Next, we list some fundamental properties.

Proposition 3.1.6. *For any $t > 0$, we have*
(1) $\mathrm{Dom}(J_t) = \mathrm{Ran}(I + tA), \mathrm{Ran}(J_t) = \mathrm{Dom}(A), J_t$ *and A_t are single-valued and monotone;*
(2) $\langle J_t x - J_t y, x - y \rangle \geq \|J_t x - J_t y\|^2, \forall x, y \in \mathrm{Dom}(J_t)$. *In particular,*

$$\|J_t x - J_t y\| \leq \|x - y\|.$$

(3) $\|J_t x - J_t y\|^2 \leq \|x - y\|^2 - \|(I - J_t)x - (I - J_t)y\|^2, \forall x, y \in \text{Dom}(J_t)$.

(4) $[J_t x, A_t x] \in \text{Graph}(A), \forall x \in \text{Dom}(J_t)$.

(5) $\text{Dom}(A_t) = \text{Dom}(J_t)$ and $\langle A_t x - A_t y, x - y \rangle \geq t\|A_t x - A_t y\|^2, \forall x, y \in \text{Dom}(A_t)$.

(6) $\|A_t x\| \leq |Ax|, \forall x \in \text{Dom}(A) \cap \text{Dom}(A_t)$, where $|Ax| = d(\theta, Ax) = \inf\{\|y\| : y \in Ax\}$.

(7) For any $t, s > 0$, $x \in \text{Dom}(J_t)$ and $\frac{s}{t}x + \frac{t-s}{t}J_t x \in \text{Dom}(J_s)$,

$$J_t x = J_s\left(\frac{s}{t}x + \frac{t-s}{t}J_t x\right).$$

(8) If $0 < t < s$, then $\|J_t x - x\| \leq 2\|J_s x - x\|, \forall x \in \text{Dom}(J_t) \cap \text{Dom}(J_s)$.

(9) If $0 < t < s$, then $\|A_s x\| \leq \|A_t x\|, \forall x \in \text{Dom}(A_t) \cap \text{Dom}(A_s)$ and hence $\lim_{t \to 0^+} \|A_t x\|$ exists.

Proof.

(1) From the definition of J_t and A_t, it is easy to find the conclusion.

(2) From the definition of J_t, $\forall x, y \in \text{Dom}(J_t)$, it follows that

$$x \in J_t x + tAJ_t x \implies (I - J_t)x \in tAJ_t x, y \in J_t y + tAJ_t y \implies (I - J_t)y \in tAJ_t y.$$

It follows that

$$\begin{aligned}
\langle J_t x - J_t y, x - y \rangle &= \langle J_t x - J_t y, J_t x - J_t y \rangle \\
&\quad + \langle J_t x - J_t y, (I - J_t)x - (I - J_t)y \rangle \\
&= \|J_t x - J_t y\|^2 + t\langle J_t x - J_t y, x_t - y_t \rangle \\
&\geq \|J_t x - J_t y\|^2,
\end{aligned}$$

where $x_t \in AJ_t x$, $y_t \in AJ_t y$. Thus, we conclude from the monotonicity of A that

$$\langle J_t x - J_t y, x_t - y_t \rangle \geq 0.$$

(3) From (2), we have

$$\begin{aligned}
\|(I - J_t)x - (I - J_t)y\|^2 &= \|x - y\|^2 - 2\langle x - y, J_t x - J_t y \rangle + \|J_t x - J_t y\|^2 \\
&\leq \|x - y\|^2 - \|J_t x - J_t y\|^2,
\end{aligned}$$

that is,

$$\|J_t x - J_t y\|^2 \leq \|x - y\|^2 - \|(I - J_t)x - (I - J_t)y\|^2.$$

(4) From the definition of J_t, we have

$$A_t x = t^{-1}(I - J_t)(x) \in AJ_t(x), \quad \forall x \in \text{Dom}(J_t).$$

(5) It follows from the definition of J_t that $\mathrm{Dom}(A_t) = \mathrm{Dom}(J_t)$, $\forall x, y \in \mathrm{Dom}(A_t)$, $t > 0$. It follows from $A_t x \in AJ_t x$, $A_t y \in AJ_t y$ and the monotonicity of A that

$$\langle A_t x - A_t y, x - y \rangle = \langle A_t x - A_t y, tA_t x + J_t x - tA_t y - J_t y \rangle$$
$$= t\|A_t x - A_t y\|^2 + \langle A_t x - A_t y, J_t x - J_t y \rangle$$
$$\geq t\|A_t x - A_t y\|^2.$$

(6) For all $x \in \mathrm{Dom}(A) \cap \mathrm{Dom}(A_t)$ and $y \in Ax$, we have

$$x + ty \in (I + tA)x \implies J_t(x + ty) = x.$$

Hence

$$\|A_t x\| = \frac{1}{t}\|x - J_t x\| = \frac{1}{t}\|J_t(x + ty) - J_t x\|$$
$$\leq \frac{1}{t}\|x + ty - x\| = \|y\|,$$

which implies that $\|A_t x\| \leq \inf\{\|y\| : y \in Ax\} = |Ax|$.

(7) For all $t, s > 0$ and $x \in \mathrm{Dom}(J_t)$, there exists $[x_1, y_1] \in \mathrm{Graph}(A)$ such that $x = x_1 + ty_1$. Hence

$$J_t x = x_1,$$
$$\frac{s}{t}x + \frac{t-s}{t}J_t x = \frac{s}{t}(x_1 + ty_1) + \frac{t-s}{t}x_1 = x_1 + sy_1 \in \mathrm{Ran}(I + sA) = \mathrm{Dom}(J_s)$$

and

$$J_s\left(\frac{s}{t}x + \frac{t-s}{t}J_t x\right) = J_s(x_1 + sy_1) = x_1 = J_t x.$$

(8) For all $x \in \mathfrak{D}(J_t) \cap \mathrm{Dom}(J_s)$ and $0 < t < s$, it follows from (2) and (7) that

$$|J_s x - J_t x| = \left\|J_t\left(\frac{t}{s}x + \frac{s-t}{s}J_s x\right) - J_t x\right\|$$
$$\leq \left(1 - \frac{t}{s}\right)\|x - J_s x\|$$
$$\leq \|x - J_s x\|.$$

Hence

$$\|J_t x - x\| \leq \|J_t x - J_s x\| + \|J_s x - x\| \leq 2\|x - J_s x\|.$$

(9) For all $0 < t < s$ and $x \in \mathrm{Dom}(A_t) \cap \mathrm{Dom}(A_s)$, it follows from (2) and (7) that

$$\|J_s x - x\| = \left\|J_t\left(\frac{t}{s}x + \left(1 - \frac{t}{s}\right)J_s x\right) - x\right\|$$
$$\leq \left\|J_t\left(\frac{t}{s}x + \left(1 - \frac{t}{s}\right)J_s x\right) - J_t x\right\| + \|J_t x - x\|$$
$$\leq \left(1 - \frac{t}{s}\right)\|J_s x - x\| + \|J_t x - x\|.$$

It follows that $t\|J_sx - x\| \le s\|J_tx - x\|$, that is, $s^{-1}\|J_sx - x\| \le t^{-1}\|J_tx - x\|$. Therefore, $\|A_sx\| \le \|A_tx\|$. Hence $\lim_{t\to 0^+} \|A_tx\|$ exists. This completes the proof. $\qquad\square$

For maximal monotone mappings, we have the following better conclusions.

Proposition 3.1.7. *Let $A : H \times H$ be a maximal monotone mapping. Then the following conclusions hold:*

(1) *For any $t > 0$, $A_t : H \to H$ is a maximal monotone mapping and*

$$A_t = \left(A^{-1} + tI\right)^{-1}, \quad (A_t)_s = A_{t+s}.$$

(2) *For any $x \in H$, $\lim_{t\to 0^+} J_tx = P_{\overline{co}\,\mathrm{Dom}(A)}x$.*

(3) *$\overline{\mathrm{Dom}(A)}$ and $\overline{\mathrm{Ran}(A)}$ are closed convex subsets of H.*

(4) *$A^{-1}(\theta) = \mathrm{Fix}(J_t)$ and so the zero set $A^{-1}(\theta)$ of A is closed and convex.*

(5) *For any $x \in \mathrm{Dom}(A)$ and $t > 0$, $\lim_{t\to 0^+} A_t(x) = A^0(x)$, where A^0 is the minimum section of A defined by*

$$A^0x = \{y \in Ax : |Ax| = \|y\|\}.$$

(6) *If $\theta \in \overline{\mathrm{Ran}(A)}$, then, for all $x \in H$, $\lim_{t\to\infty} J_tx$ exists and belongs to the zero set $A^{-1}(\theta)$ of A, more precisely, $\lim_{t\to\infty} J_tx = P_{A^{-1}(\theta)}x$, where $P_{A^{-1}(\theta)}x$ is the metric projection from H onto the zero set $A^{-1}(\theta)$ of A.*

Proof.

(1) The conclusion can be directly obtained from the definition and properties of monotone mappings. So we omit it here.

(2) For any $x \in H$, $t > 0$, since $A \subset H \times H$ is maximal monotone, we have $\mathrm{Ran}(I + tA) = H$. Hence $\mathrm{Dom}(J_t) = H$ and $x_t = J_tx$, which imply that

$$[x_t, t^{-1}(x - x_t)] \in \mathrm{Graph}(A).$$

From the monotonicity of A, we see that, $\forall [x', y'] \in \mathrm{Graph}(A)$,

$$\langle t^{-1}(x - x_t) - y', x_t - x' \rangle \ge 0$$
$$\implies \langle x - x_t - ty', x_t - x' \rangle \ge 0$$
$$\implies \langle x_t - x + ty', x_t - x' \rangle \le 0, \quad \forall t > 0.$$

This implies that $\{x_t\}$ is bounded. If $t_n \to 0$ and $x_n := x_{t_n} \to x_0$ as $n \to \infty$, then

$$\|x_n\|^2 \le \langle x - t_ny', x_n - x' \rangle + \langle x_n, x' \rangle$$
$$\implies \|x_0\|^2 = \liminf_{n\to\infty} \|x_n\|^2 \le \langle x, x_0 - x' \rangle + \langle x_0, x \rangle$$
$$\implies \langle x_0 - x, x_0 - x' \rangle \le 0, \quad \forall x' \in \mathrm{Dom}(A)$$
$$\implies \langle x_0 - x, x_0 - x' \rangle \le 0, \quad \forall x' \in \overline{co}\,\mathrm{Dom}(A).$$

Hence $x_0 = P_{\overline{co}\,\mathrm{Dom}(A)}$, which shows that $\{x_t\}$ converges to x_0 weakly. On the other hand, for any $x' \in \overline{co}\,\mathrm{Dom}(A)$, one has

$$\limsup_{t\to 0^+} \|x_t\|^2 \le \langle x, x_0 - x' \rangle + \langle x_0, x' \rangle.$$

Taking $x' = x_0$ in the inequality above, we have

$$\limsup_{t\to 0^+} \|x_t\|^2 \le \|x_0\|^2 \implies \limsup_{t\to 0^+} \|x_t\| \le \|x_0\|.$$

Since $x_{t_n} \xrightarrow{w} x_0$ as $n \to \infty$, we conclude from the weak lower semicontinuity of the norm that

$$\|x_0\| \le \liminf_{t\to 0^+} \|x_t\| \le \limsup_{t\to 0^+} \|x_t\| \le \|x_0\| \implies \|x_t\| \to \|x_0\| \quad (t \to 0^+).$$

Hence $x_t \to x_0$ as $t \to 0^+$.

(3) For any $x \in \overline{co}\,\mathrm{Dom}(A)$, it follows from (1) that $x_t \to x_0$ as $t \to 0^+$. Hence $x \in \mathrm{Dom}(A)$ and $x \in \overline{\mathrm{Dom}(A)}$. Thus

$$\overline{co}\,\mathrm{Dom}(A) \subseteq \overline{\mathrm{Dom}(A)}.$$

On the contrary, it is obvious that

$$\overline{\mathrm{Dom}(A)} \subseteq \overline{co}\,\mathrm{Dom}(A).$$

Hence $\overline{\mathrm{Dom}(A)} = \overline{co}\,\mathrm{Dom}(A)$ is closed and convex. Since A is maximal monotone, one finds that A^{-1} is also maximal monotone. So, $\overline{\mathrm{Ran}(A)} = \overline{\mathrm{Dom}(A^{-1})}$ is closed and convex.

(4) Note that

$$x \in A^{-1}(\theta) \iff \theta \in Ax \iff x = J_t x, \quad \forall t > 0.$$

Since J_t is nonexpansive, we find that $\mathrm{Fix}(J_t)$ is closed and convex.

(5) For any $x \in \mathrm{Dom}(A)$ and $t > 0$, since A is monotone, we have

$$\langle A^0 x - A_t x, x - J_t x \rangle \ge 0$$

and

$$\|A_t x\|^2 \le \langle A^0 x, A_t x \rangle \le \|A^0 x\| \|A_t x\|,$$

which imply that $\|A_t x\| \le \|A^0 x\|$. This shows that $\{A_t x\}$ is bounded. Take $t_n \to 0$ arbitrarily and denote $x_n := A_{t_n} x$. Suppose that $x_n \to y$ as $n \to \infty$. It follows from (2) that $J_{t_n} x \to x$ as $n \to \infty$, $x_n \in A_{t_n} x$ and $x_n \to y$ as $n \to \infty$. Thus, we conclude from the demiclosedness of A that $y \in Ax$. Using the weak lower semicontinuity of the norm, we have

$$\|A^0x\| \le \|y\| \le \liminf_{n\to\infty} \|x_n\| \le \limsup_{n\to\infty} \|x_n\| \le \|A^0x\|$$

$$\implies \lim_{n\to\infty} \|x_n\| = \|A^0x\| = \|y\|$$

$$\implies y = A^0x.$$

Thus $\lim_{t\to 0^+} \|A_tx\| = \|A^0x\|$. Note that

$$\|A_tx - A^0x\|^2 = \|A_tx\|^2 - 2\langle A_tx, A^0x\rangle + \|A^0x\|^2$$

$$\le \|A_tx\|^2 - 2\|A_tx\|^2 + \|A^0x\|^2$$

$$= \|A^0x\|^2 - \|A_tx\|^2 \to 0 \quad (t \to 0^+).$$

So, we obtain that $\lim_{t\to 0^+} A_tx = A^0x$.

(6) Since $A \subset H \times H$ is maximal monotone, we see that $\text{Ran}(I + tA) = H$, $\forall t > 0$. Hence, for any $x \in H$ and $t > 0$, J_tx is well defined. For any $x \in H$, there exists a unique $x_t \in \text{Dom}(A)$ such that

$$x \in x_t + tAx_t, \tag{R1}$$

that is, $x_t = J_tx$, $\forall t > 0$.

For the sake of simplicity, we suppose that $x = \theta$. If not, we consider a maximal monotone mapping $\hat{A}(\cdot) = A(\cdot - x)$.

Now, we are in a position to show that $\lim_{t\to\infty} J_t\theta = z$, where $z = P_{A^{-1}(\theta)}(\theta)$. First, we prove that $P_{A^{-1}(\theta)}(\theta)$ is well defined. By the assumption $\theta \in \text{Ran}(A)$, we see that $A^{-1}(\theta) \ne \emptyset$. It follows from (4) that $A^{-1}(\theta)$ is closed and convex. Hence $P_{A^{-1}(\theta)}(\theta)$ is uniquely determined. For any $x^* \in A^{-1}(\theta)$ and $y_t \in Ax_t$, we obtain from (R1) that

$$\theta = x_t + ty_t, \quad \forall t > 0.$$

This implies that

$$0 = \langle x_t, x_t - x^*\rangle + t\langle y_t, x_t - x^*\rangle$$

$$= \|x_t - x^*\|^2 + \langle x^*, x_t - x^*\rangle + t\langle y_t - \theta, x_t - x^*\rangle$$

$$\ge \|x_t - x^*\|^2 + \langle x^*, x_t - x^*\rangle.$$

It follows that

$$\|x_t - x^*\|^2 \le \langle x^*, x^* - x_t\rangle, \quad \forall x^* \in A^{-1}(\theta), t > 0, \tag{R2}$$

and

$$\langle x_t, x_t - x^*\rangle \le 0, \quad \forall x^* \in A^{-1}(\theta), t > 0. \tag{R3}$$

It follows from (R2) that $\|x_t - x^*\| \leq \|x^*\|$. Hence $\{x_t\}$ is bounded.

For any $t_n > 0$ with $t_n \to \infty$ as $n \to \infty$, one denotes $x_n := x_{t_n}$ and $y_n := y_{t_n}$. Since sequence $\{x_n\}$ is bounded and space H is reflexive, we may assume, without loss of generality, that $x_n \to z$ as $n \to \infty$. Then, it follows from $\theta = x_n + t_n y_n$ that $y_n = -\frac{1}{t_n} x_n \to \theta$ as $n \to \infty$. Using the demiclosedness of Graph(A), we have $\theta \in Az$. Taking $x^* = z$ in (R2), we obtain $x_n \to z$ as $n \to \infty$.

From (R3), we see that $\langle x_n, x_n - x^* \rangle \leq 0$, $\forall x^* \in A^{-1}(\theta)$ and $n \geq 1$. Taking $n \to \infty$, we have

$$\langle z, z - x^* \rangle \leq 0, \quad \forall x^* \in A^{-1}(\theta).$$

It follows from some properties of the metric projection that

$$z = P_{A^{-1}(\theta)} \theta.$$

This proves that $\lim_{t \to \infty} J_t \theta = z$. This completes the proof. □

Remark 3.1.12. One knows that some properties of maximal monotone mappings are better than those of general monotone mappings. Hence, we expect to study general monotone mappings via maximal monotone mappings. Indeed, this is possible since every monotone mapping $A \subset H \times H$ has a maximal monotone extension $\tilde{A} \subset H \times H$ satisfying $\text{Dom}(\tilde{A}) \subset \overline{\text{co}}\,\text{Dom}(A)$ and $\text{Graph}(A) \subset G(\tilde{A})$ when $\text{Dom}(A) = C$ is closed and convex, $\text{Dom}(\tilde{A}) = \text{Dom}(A) = C$ and $\text{Ran}(A) \subset \text{Ran}(\tilde{A})$. For example, in order to study monotone operator equations

$$\theta \in Ax,$$

one first studies the case when A is maximal monotone. Based on this, one further considers its maximal monotone extension $\tilde{A}$. With the aid of properties of A, we can obtain some results, which are valid for $\tilde{A}$, also for A.

3.2 Criteria of maximal monotone mappings

It is an important question whether a monotone mapping is maximal monotone.

Theorem 3.2.1. *If $A : \text{Dom}(A) = H \to H$ is a monotone hemicontinuous mapping, then A is maximal monotone.*

Proof. For any $[x, f] \in H \times H$,

$$\langle f - Ay, x - y \rangle \geq 0, \quad \forall y \in H. \tag{3.20}$$

Next, we prove $Ax = f$. If $Ax \neq f$, then there exists $z \in H$ such that $\langle f - Ax, z \rangle \neq 0$. Assume that $\langle f - Ax, z \rangle > 0$. Letting $y_t := x + tz$, $t > 0$ and taking $y = y_t$ in (3.20), we have

$$\langle f - Ay_t, -tz \rangle \geq 0, \quad \forall t > 0.$$

It follows that

$$\langle f - Ay_t, z \rangle \le 0. \tag{3.21}$$

Note that $A : H \to H$ is hemicontinuous. Letting $t \to 0$ in (3.21), we arrive at

$$\langle f - Ax, z \rangle \le 0,$$

which contradicts $\langle f - Ax, z \rangle > 0$. This completes the proof. □

Theorem 3.2.2. *Let $A : \text{Dom}(A) = H \to H$ be a maximal monotone mapping. Then A is demicontinuous. Equivalently, if A is not maximal monotone, then A is not demicontinuous.*

Proof. For any $\{x_n\} \subset H$ with $x_n \to x$ as $n \to \infty$, we need to prove $Ax_n \to Ax$ as $n \to \infty$. In fact, we see from Theorem 3.2.1 that $\{Ax_n\}$ is a bounded sequence in H. Since H is reflexive, $\{Ax_n\}$ has a subsequence $\{Ax_{n_j}\}$ such that $Ax_{n_j} \rightharpoonup y$ as $j \to \infty$. It follows that

$$\langle Ax_{n_j} - g, x_{n_j} - y \rangle \ge 0, \quad \forall [y, g] \in \text{Graph}(A), \ n \ge 1.$$

Letting $j \to \infty$ in the above inequality, we have

$$\langle y - g, x - y \rangle \ge 0, \quad \forall [y, g] \in \text{Graph}(A).$$

It follows from the maximal monotonicity of A that $y = Ax$. Therefore, it follows that $Ax_{n_j} \xrightarrow{w} Ax$ as $j \to \infty$. Since Ax is unique, we must have $Ax_n \xrightarrow{w} Ax$ as $n \to \infty$. This completes the proof. □

Remark 3.2.1. The above theorem gives another important property of a maximal monotone mapping.
(1) If $A : H \to H$ is not demicontinuous, then $A : H \to H$ is not maximal monotone.
(2) We know that the sum of finitely many monotone mappings is still monotone, however, the sum of two maximal monotone mappings is not necessarily maximal monotone. The following theorems give the necessary and sufficient conditions for the sum of two maximal monotone mappings still be maximal monotone.

Theorem 3.2.3. *Let $A \subset H \times H$ be a set-valued maximal monotone mapping and let $B : H \to H$ be single-valued monotone mapping. If $T = A + B \subset H \times H$ is maximal monotone, then A must be maximal monotone.*

Proof. Suppose that, for any $[x, f] \in H \times H$,

$$\langle f - g, x - y \rangle \ge 0, \quad \forall [y, g] \in \text{Graph}(A).$$

Denote $u = Bx$ and $v = By$. Since $g \in Ay$ is arbitrary, the elements in Ty take the form as $w = g + v$. So, $\forall [y, g] \in \text{Graph}(T)$, we obtain

$$\langle w - u - f, y - x \rangle = \langle g - f, y - x \rangle + \langle v - u, y - x \rangle \ge 0.$$

Since T is maximal monotone, we have $[x, f + u] \in \text{Graph}(T)$, that is, $x \in \text{Dom}(T)$ and

$$f + u \in Tx = Ax + Bx \implies f \in Ax.$$

So A is maximal monotone. This completes the proof. □

Theorem 3.2.4. *Let $A \subset H \times H$ be a monotone mapping and $\lambda > 0$. If $\text{Ran}(A + \lambda I) = H$, then $(A+\lambda I)^{-1} : H \to H$ is single-valued, $\frac{1}{\lambda}$-Lipschitz continuous and maximal monotone.*

Proof. By Proposition 3.1.1, for any $[x, f], [y, g] \in \text{Graph}(A)$, one has

$$\|(\lambda x + f) - (\lambda y + g)\| \geq \lambda \|x - y\|.$$

This implies that $(A + \lambda I)^{-1} : H \to H$ is single-valued, $\text{Dom}((A + \lambda I)^{-1}) = H$ and $\frac{1}{\lambda}$-Lipschitz continuous. By Theorem 3.2.1, we see that $(A + \lambda I)^{-1} : H \to H$ is maximal monotone. This completes the proof. □

Thus we have the following criterion.

Theorem 3.2.5. *Let $A \subset H \times H$ be a monotone mapping. If there exists $\lambda > 0$ such that $\text{Ran}(A + \lambda I) = H$, then $A \subset H \times H$ is a maximal monotone mapping.*

Proof. By Theorem 3.2.4, we see that $(A+\lambda I)^{-1} : H \to H$ is maximal monotone, so is $(A+\lambda I) \subset H \times H$. In view of $B = \lambda I$ is monotone mapping, one concludes from Theorem 3.2.3 that $A \subset H \times H$ must be maximal monotone. This completes the proof. □

Indeed, the converse of the above theorem is also true.

Theorem 3.2.6. *Let $A \subset H \times H$ be a monotone mapping. If A is maximal monotone, then there exists $\lambda_0 > 0$ such that*

$$R(A + \lambda_0 I) = H. \tag{3.22}$$

Proof. Suppose that $A \subset H \times H$ is a monotone mapping. In order to prove (3.22), we prove, without loss of generality, that $\text{Ran}(A + I) = H$. To this end, for any $y \in H$, we need to prove that there exists $x \in \text{Dom}(A)$ such that $y \in (I + A)x$. By the maximal monotonicity of A, we only need to prove that there exists $x \in H$ such that

$$\langle v - (y - x), u - y \rangle \geq 0, \quad \forall [u, v] \in \text{Graph}(A). \tag{3.23}$$

Fixing $[u, v] \in \text{Graph}(A)$, we let

$$M(u, v) = \{x \in H : \langle v + x, u - x \rangle \geq \langle y, u - x \rangle\}.$$

Now, we are in a position to show that $M(u, v)$ is a nonempty, bounded, closed, and convex subset of H. Since $u \in M(u, v)$, we have $M(u, v) \neq \emptyset$. By the continuity of the inner product $\langle \cdot, \cdot \rangle$, we see that $M(u, v)$ is closed. Then the construction of $M(u, v)$ implies that

$$\|x\|^2 - \langle y, x \rangle + \langle v, x \rangle - \langle u, x \rangle \le \langle u, v \rangle, \quad \forall x \in M(u, v)$$

$$\implies \|x\|^2 \le \|u\|\|v\| + (\|y\| + \|v\| + \|u\|)\|x\|. \tag{3.24}$$

Thus $M(u, v)$ is bounded. Note that $\|\cdot\|^2$ is convex and $\langle y, \cdot \rangle, \langle v, \cdot \rangle, \langle u, \cdot \rangle$ are linear. From (3.24), we see that $M(u, v)$ is convex and then $M(u, v)$ is w-compact. Thus, in order to prove (3.23), we only need to prove that

$$\bigcap \{M(u, v) : [u, v] \in \text{Graph}(A)\} \ne \emptyset.$$

From the Heine–Borel theorem on compact sets, we only need to prove, for any finite number of $[u_i, v_i] \in \text{Graph}(A)$ $(i = 1, 2, \ldots, m)$, that

$$\bigcap_{i=1}^{m} M(u_i, v_i) \ne \emptyset.$$

To achieve this, one considers the bounded convex subset K of $\mathbb{R}^m$ given by

$$K = \left\{ \lambda = (\lambda_1, \lambda_2, \ldots, \lambda_m)^T : \lambda_i \ge 0, \sum_{i=1}^{m} \lambda_i = 1 \right\}$$

and the functional $f : K \times K \to \mathbb{R}$. Then

$$f(\lambda, \mu) = \sum_{i=1}^{m} \mu_i \langle x(\lambda) + v_i - y, x(\lambda) - u_i \rangle, \quad \forall \lambda, \mu \in K,$$

where $\lambda = (\lambda_1, \lambda_2, \ldots, \lambda_m)^T$, $\mu = (\mu_1, \mu_2, \ldots, \mu_m)^T$ and $x(\lambda) = \sum_{i=1}^{m} \lambda_i \mu_i$.

It is not hard to see that f is continuous and convex in the first variable λ and is linear in the second variable μ. It satisfies all the conditions of the Minimax Theorem [87]. Using the Minimax Theorem, one sees that there exists $\lambda_0 \in K$ such that

$$f(\lambda_0, \mu) \le \max_{\lambda \in K} f(\lambda, \lambda), \quad \forall \mu \in K.$$

Since $\sum_{i=1}^{m} \sum_{j=1}^{m} a_{ij} = \sum_{j=1}^{m} \sum_{i=1}^{m} a_{ij}$ and due to the monotonicity of A, we have

$$f(\lambda, \lambda) = \sum_{i=1}^{m} \sum_{j=1}^{m} \lambda_i \lambda_j \langle v_i, u_j - u_i \rangle = \frac{1}{2} \sum_{i=1}^{m} \sum_{j=1}^{m} \lambda_i \lambda_j \langle v_i - v_j, u_j - u_i \rangle \le 0.$$

It follows that

$$\sum_{i=1}^{m} \mu_i \langle x(\lambda_0) + v_i - y, x(\lambda_0) - u_i \rangle \le 0, \quad \forall \mu \in K.$$

Hence,

$$\langle v_i + x(\lambda_0) - y, u_i - x(\lambda_0)_i \rangle \ge 0, \quad \forall i = 1, 2, \ldots, m.$$

Therefore, $x(\lambda_0) \in \bigcap_{i=1}^{m} M(u_i, v_i) \ne \emptyset$ and $\text{Ran}(A + I) = H$. This completes the proof. $\square$

Combining Theorems 3.2.5 and 3.2.6, we can obtain the following criterion.

Theorem 3.2.7. *Let $A \subset H \times H$ be a monotone mapping. Then the following statements are equivalent:*

(1) *A is maximal monotone,*

(2) $\mathrm{Ran}(A + \lambda I) = H, \forall \lambda > 0,$

(3) *there exists $\lambda_0 > 0$ such that $\mathrm{Ran}(A + \lambda_0 I) = H$.*

Proof. If $A \subset H \times H$ is maximal monotone, then $\frac{1}{\lambda} A, \forall \lambda > 0$, is also maximal monotone. By Theorem 3.2.6, we see that $\mathrm{Ran}(\frac{1}{\lambda} A + I) = H$. So, $\mathrm{Ran}(A + \lambda I) = H, \forall \lambda > 0$. In particular, there exists $\lambda_0 > 0$ such that $\mathrm{Ran}(A + \lambda_0 I) = H$.

If there exists $\lambda_0 > 0$ such that $\mathrm{Ran}(A + \lambda_0 I) = H$, then we can conclude from Theorem 3.2.5 that $A \subset H \times H$ is maximal monotone. This completes the proof. $\square$

As an application of Theorem 3.2.7, we give the following very useful conclusion.

Theorem 3.2.8. *Let $f : H \to \overline{\mathbb{R}}$ be a proper, lower semicontinuous, and convex functional. Then the subdifferential $\partial f \subset H \times H$ of f is a maximal monotone mapping.*

Proof. From Example 1.9.8, we see that $\partial f \subset H \times H$ is monotone. From Theorem 3.2.7, we only need to prove that

$$\mathrm{Ran}(I + \partial f) = H, \tag{3.25}$$

that is, for any $y_0 \in H$, the equation $y_0 \in x + \partial f(x)$ has at least one solution $x_0 \in \mathrm{Dom}(\partial f)$. To this end, we consider the functional $\varphi : H \to \overline{\mathbb{R}}$ defined by

$$\varphi(x) = \frac{1}{2}\|x\|^2 + f(x) - \langle x, y_0 \rangle, \quad \forall x \in H. \tag{3.26}$$

It is easy to verify that $\varphi : H \to \overline{\mathbb{R}}$ is proper, lower semicontinuous, and convex. Since $\mathrm{Dom}(\partial f) \neq \emptyset$, there exist $z \in \mathrm{Dom}(f)$ and $x^* \in H$ such that

$$f(x) \geq f(z) + \langle x^*, x - z \rangle, \quad \forall x \in H. \tag{3.27}$$

Substituting (3.27) into (3.26), we obtain

$$\varphi(x) \geq \frac{1}{2}\|x\|^2 + f(z) + \langle x^*, x - z \rangle - \langle x, y_0 \rangle$$

$$\geq \frac{1}{2}\|x\|^2 + f(z) + \|x^*\|\|x - z\| - \|x\|\|y_0\|.$$

It follows that

$$\lim_{\|x\| \to \infty} \varphi(x) = \infty.$$

Using the existence of the functional minimum theorem, there exists $x_0 \in \text{Dom}(f)$ such that

$$\varphi(x_0) \le \varphi(x), \quad \forall x \in H. \tag{3.28}$$

By some properties of the subdifferential, we have $\theta \in \partial\varphi(x_0)$. Note that

$$\begin{aligned}
\partial\varphi(x) &= \partial\frac{1}{2}\|x\|^2 + \partial f(x) + \partial(\langle x, -y_0\rangle) \\
&= \{x\} + \partial f(x) - \{y_0\} \\
&= \{x - y_0\} + \partial f(x).
\end{aligned}$$

Hence, $y_0 - x_0 \in \partial f(x_0)$ and so $y_0 \in x_0 + \partial f(x_0)$. This completes the proof. □

Remark 3.2.2. Using Theorem 3.2.8, one sees that the convex optimization problem

$$\min_{x \in H} f(x)$$

can be transformed into a zero point problem of a maximal monotone mapping $A = \partial f$:

$$\theta \in Ax = \partial f(x).$$

This shows that the theory of maximal monotone mappings is a powerful tool for solving convex optimization problems.

3.3 Acute angle principle for monotone mappings

The acute angle principle of monotone mappings is an important constituent of monotone mappings. We can establish a series of important results, such as the surjectivity of maximal monotone mappings, from the acute angle principle.

Theorem 3.3.1. *Let $A \subset H \times H$ be a maximal monotone mapping. Suppose that there exists a fixed constant $\alpha > 0$ such that*

$$\langle x^*, x \rangle \ge 0, \quad \forall x \in \text{Dom}(A) \text{ with } \|x\| > \alpha,\, x^* \in Ax. \tag{3.29}$$

Then there exists $x \in B_\alpha[\theta] \cap \text{Dom}(A)$ such that

$$\theta \in Ax. \tag{3.30}$$

Proof. From Theorem 3.2.7, we see that, for any $\varepsilon > 0$, $\text{Ran}(A + \frac{\varepsilon}{\alpha}I) = H$. So, there exists a unique $x_\varepsilon \in \text{Dom}(A)$ such that

$$\theta \in \left(A + \frac{\varepsilon}{\alpha}I\right)x_\varepsilon.$$

Letting $x_\varepsilon^* = -\frac{\varepsilon}{\alpha} x_\varepsilon$, one has $x_\varepsilon^* \in A x_\varepsilon$ and

$$0 = \langle x_\varepsilon^*, x \rangle + \frac{\varepsilon}{\alpha} \langle x_\varepsilon, x_\varepsilon \rangle = \langle x_\varepsilon^*, x \rangle + \frac{\varepsilon}{\alpha} \| x_\varepsilon \|^2.$$

Therefore,

$$\langle x_\varepsilon^*, x \rangle = -\frac{\varepsilon}{\alpha} \| x_\varepsilon \|^2. \tag{3.31}$$

It follows from (3.29) that $\| x_\varepsilon \| \leq \alpha$ and

$$\| x^* \| = \frac{\varepsilon}{\alpha} \| x_\varepsilon \| \leq \varepsilon, \quad \forall \varepsilon > 0. \tag{3.32}$$

Taking $\varepsilon_n > 0$ with $\varepsilon_n \to 0$ as $n \to \infty$, and setting $x_n := x_{\varepsilon_n}$ and $x_n^* := x_{\varepsilon_n}^*$, we obtain

$$\| x_n \| \leq \alpha, \| x_n^* \| \leq \varepsilon_n, \quad \forall n \geq 1.$$

Since $\{x_n\}$ is a bounded sequence in H and H is reflexive, one sees that there exists a subsequence $\{x_{n_j}\} \subset \{x_n\}$ such that $x_{n_j} \rightharpoonup x$ as $j \to \infty$. By the w-lower semicontinuity of the norm, we see that

$$\| x \| \leq \liminf_{j \to \infty} \| x_{n_j} \| \leq \alpha.$$

Hence $x \in B_\alpha[\theta]$. Since $x_{n_j} \rightharpoonup x$ as $j \to \infty$, $x_{n_j}^* \to x$ as $j \to \infty$ and $A \subset H \times H$ is maximal monotone, we conclude from Proposition 3.1.4 that $x \in \text{Dom}(A)$ and $\theta \in Ax$. Hence, (3.30) holds. This completes the proof. $\qquad \square$

Proposition 3.3.1. *Let $A : \text{Dom}(A) = H \to H$ be a hemicontinuous mapping. Suppose there exists a constant $\alpha > 0$ such that*

$$\langle Ax, x \rangle \geq 0, \quad \forall x \in H \text{ with } \| x \| > \alpha.$$

Then the equation $Ax = \theta$ has at least one solution in $B_\alpha[\theta]$.

Proof. From Theorem 3.2.1, we see that $A : \text{Dom}(A) = H \to H$ is maximal monotone. From Theorem 3.3.1, we can obtain the desired conclusion immediately. This completes the proof. $\qquad \square$

Next, we establish the surjectivity results for maximal monotone mappings. First, let us discuss the surjectivity under some coercitivity conditions.

Definition 3.3.1. A mapping $A \subset H \times H$ is said to be coercive if there exists a function $c : \mathbb{R}^+ \to \mathbb{R}$ such that $c(r) \to \infty$ as $r \to \infty$ and

$$\langle f, x \rangle \geq c(\| x \|) \| x \|, \quad \forall [x, f] \in \text{Graph}(A). \tag{3.33}$$

It is obvious that if $A \subset H \times H$ is an η-strongly monotone mapping, then A is coercive.

Theorem 3.3.2 ([10]). *Let $A \subset H \times H$ be a maximal monotone mapping satisfying the coercitivity condition (3.33). Then $\text{Ran}(A) = H$.*

Proof. Since translation does not change the maximal monotonicity and coercitivity of A, we only need to prove

$$\theta \in \text{Ran}(A).$$

Using coercitivity condition (3.33), we see that there exists a constant $\alpha > 0$ such that

$$\langle f, x \rangle \geq c(\|x\|)\|x\| \geq 0$$

as soon as $\|x\| > \alpha$. Using Theorem 3.3.1, we see that there exists $x \in B_\alpha(\theta) \cap \text{Dom}(A)$ such that $\theta \in A(x)$. This completes the proof. $\qquad \square$

Corollary 3.3.1 ([5]). *Let $A : \text{Dom}(A) = H \to H$ be hemicontinuous, monotone, and coercive. Then $\text{Ran}(A) = H$.*

Proof. Using Theorem 3.2.1, we see that $A : \text{Dom}(A) = H \to H$ is maximal monotone. Using Theorem 3.3.2, we can derive the desired conclusion immediately. This completes the proof. $\qquad \square$

We know that local boundedness is a fundamental property of monotone mappings. Next, we prove that local boundedness is a characteristic property related to the surjectivity of a maximal monotone mapping.

Definition 3.3.2. Let $A \subset H \times H$ be a mapping. Then A^{-1} is said to be locally bounded if for any $y \in H$, there exists $r > 0$ such that the set $\{x \in H : B_r(y) \cap Ax \neq \emptyset\}$ is bounded in H.

It is easy to see that A^{-1} is locally bounded if and only if, for any $\{x_n\} \subset H$ and $\{[x_n, f_n]\} \subset \text{Graph}(A)$, if $f_n \to f$ as $n \to \infty$, then $\{x_n\}$ is bounded.

Remark 3.3.1. For any coercive mapping A, A^{-1} is locally bounded. Indeed, if $\langle f_n, x_n \rangle \geq c(\|x_n\|)\|x_n\|$, then $c(\|x_n\|) \leq \|f_n\|$. If $f_n \to f$ $(n \to \infty)$, then $\|f_n\| \to \|f\|$ $(n \to \infty)$. This shows that $\{c(\|x_n\|)\}$ is a bounded sequence, so is $\{x_n\}$. Therefore, A^{-1} is locally bounded.

Theorem 3.3.3. *Let $A \subset H \times H$ be a maximal monotone mapping. Then the following are equivalent:*
(1) $\text{Ran}(A) = H$.
(2) A^{-1} *is locally bounded on H.*

Proof. (1) $\Longrightarrow$ (2). Since $A \subset H \times H$ is maximal monotone, one sees that A^{-1} is also maximal monotone. Since $\mathrm{Ran}(A) = H$, one has $\mathrm{Dom}(A^{-1}) = \mathrm{Ran}(A) = H$. Using Theorem 3.1.1, one sees that A^{-1} is locally bounded on H.

(2) $\Longrightarrow$ (1). Since H is connected, we only need to prove that $\mathrm{Ran}(A)$ is open and closed in H. First, we shows that $\mathrm{Ran}(A)$ is closed. For any $y \in \overline{\mathrm{Ran}(A)}$, there exists $\{y_n\} \subset \mathrm{Ran}(A)$ such that $y_n \to y$ as $n \to \infty$. Since A^{-1} is locally bounded on H, there exists $r > 0$ such that $\|x_n\| \le r$ for all $n \ge 1$ and $x_n \in A^{-1}(y_n)$, that is, $y_n \in Ax_n$.

Since H is reflexive, there exists a subsequence $\{x_{n_j}\} \subset \{x_n\}$ such that $x_{n_j} \rightharpoonup x$ as $j \to \infty$. Using the monotonicity of A, we obtain

$$\langle x_{n_j} - v, y_{n_j} - u \rangle \ge 0, \quad \forall [u, v] \in \mathrm{Graph}(A^{-1}).$$

Taking $j \to \infty$ in the above inequality, we have

$$\langle x - v, y - u \rangle \ge 0, \quad \forall [u, v] \in \mathrm{Graph}(A^{-1}).$$

Using the maximal monotonicity of A^{-1}, we see that $[y, x] \in \mathrm{Graph}(A^{-1})$, that is, $x \in A^{-1}(y)$. It follows that $y \in Ax \subset \mathrm{Ran}(A)$.

Next, we prove that $\mathrm{Ran}(A)$ is open. Suppose that $y_0 \in \mathrm{Ran}(A)$. Then there exists $x_0 \in \mathrm{Dom}(A)$, such that $x_0 \in A^{-1}(y_0)$, because A^{-1} is locally bounded, we see that there exist $r > 0$ and an open ball $B(y_0, r) \subset H$ such that $A^{-1}(B(y_0, r) \cap \mathrm{Ran}(A))$ is bounded. For all $y \in B(y_0, r)$, $\varepsilon > 0$ and $x_0 + \frac{1}{\varepsilon} y \in H$, it follows from $\mathrm{Ran}(I + \frac{1}{\varepsilon} A) = H$ that there exists $x_\varepsilon \in \mathrm{Dom}(A)$ such that

$$x_0 + \frac{1}{\varepsilon} y \in x_\varepsilon + \frac{1}{\varepsilon} Ax_\varepsilon.$$

It follows then that

$$\varepsilon x_0 + y \in \varepsilon x_\varepsilon + Ax_\varepsilon.$$

Denote $z_\varepsilon := y + \varepsilon(x_0 - x_\varepsilon)$. It follows from the monotonicity of A that

$$\langle y_0 - z_\varepsilon, x_0 - x_\varepsilon \rangle \ge 0 \Longrightarrow \langle y_0 - z_\varepsilon, z_\varepsilon - y \rangle \ge 0$$
$$\Longrightarrow \|y_0 - z_\varepsilon\|^2 \le \langle y_0 - z_\varepsilon, y_0 - y \rangle \le \|y_0 - z_\varepsilon\| \|y_0 - y\|$$
$$\Longrightarrow \|y_0 - z_\varepsilon\| \le \|y_0 - y\| < r,$$

which implies $z_\varepsilon \in B(y_0, r)$. Since $x_\varepsilon \in A^{-1}(z_\varepsilon)$ and $A^{-1}(B(y_0, r) \cap \mathrm{Ran}(A))$ is bounded, we see that $\{x_\varepsilon\}$ is bounded. Take $\varepsilon_n > 0$ with $\varepsilon_n \to 0$ as $n \to \infty$. Setting $x_n := x_{\varepsilon_n}$ and $z_n := z_{\varepsilon_n}$ and letting $x_n \rightharpoonup x$ as $n \to \infty$, we obtain from $z_n = y + \varepsilon_n(x_0 - x_n)$ that $z_n \to y$ as $n \to \infty$. Since A^{-1} is maximal monotone, we conclude from the demiclosedness that $y \in \mathrm{Dom}(A^{-1})$ and $x \in A^{-1}(y)$, that is, $y \in Ax \subset \mathrm{Ran}(A)$. It follows that

$$B(y_0, r) \subset \mathrm{Ran}(A),$$

which shows that $y_0 \in \mathrm{int}(\mathrm{Ran}(A))$. Hence $\mathrm{Ran}(A)$ is an open set of H. This completes the proof. $\qquad\square$

Corollary 3.3.2. *Let $A \subset H \times H$ be a maximal monotone mapping. If A^{-1} is bounded, then* $\mathrm{Ran}(A) = H$. *In particular, if* $\mathrm{Dom}(A)$ *is bounded, then* $\mathrm{Ran}(A) = H$.

In the following, we discuss sufficient conditions under which the sum of two maximal monotone mappings is maximal monotone.

Theorem 3.3.4 (Rockafellar [77], 1970). *Let $A, B \subset H \times H$ be two maximal monotone mappings. If they satisfy one of the following conditions:*
(1) $\mathrm{Dom}(A) \cap \mathrm{int}(\mathrm{Dom}(B)) \neq \emptyset$.
(2) *There exists $x \in \overline{\mathrm{Dom}(A)} \cap \overline{\mathrm{Dom}(B)}$ such that B is locally bounded at x,*

then $A + B \subset H \times H$ *is maximal monotone.*

Remark 3.3.2. By Rockafellar's Theorem [77], we see that conditions (1) and (2) are equivalent.

Proof. By Remark 3.3.2, we only consider condition (1). Since the translation does not change the (maximal) monotonicity, we may assume, without loss of generality, that $\theta \in A\theta$ and $\theta \in \mathrm{int}(\mathrm{Dom}(B))$.

Now, consider the following two cases.
Case I. $\mathrm{Dom}(B) \subset H$ is bounded.

To prove that $A + B$ is maximal monotone, we find from Theorem 3.2.7 that we only need to prove $\mathrm{Ran}(A + B + I) = H$. That is, for any $y \in H$, we only need to prove $y \in (A + B + I)$. Since the translation does not change the maximal monotonicity, we may assume $y = \theta$. Then we only need to prove that there exists $x \in H$ such that

$$\theta \in (A + B + I)(x). \tag{3.34}$$

It is obvious that x satisfies (3.34) if and only if there exists $x^* \in H$ such that

$$-x^* \in \left(A + \frac{1}{2}I\right)(x), \quad x^* \in \left(B + \frac{1}{2}I\right)(x). \tag{3.35}$$

Define two mappings S_1 and $S_2 : H \to H$ by

$$S_1(y) = -\left(A + \frac{1}{2}I\right)^{-1}(-y) \tag{3.36}$$

and

$$S_2(y) = \left(B + \frac{1}{2}I\right)^{-1}(y). \tag{3.37}$$

Then

$$x^* \text{ satisfies } (3.43) \iff \theta = S_1(x^*) + S_2(x^*). \tag{3.38}$$

To prove (3.34), we only need to prove

$$\theta \in \text{Ran}(S_1 + S_2). \tag{3.39}$$

By Theorem 3.2.4, we see that S_1 and S_2 are two single-valued, 2-Lipschitz continuous, maximal monotone, and $\text{Dom}(S_1) = H = \text{Dom}(S_2)$. Hence $S_1 + S_2 : H \to H$ is single-valued, 2-Lipschitz continuous maximal monotone, and $\text{Dom}(S_1 + S_2) = H$. By Theorem 3.2.1, we see that $S_1 + S_2$ is maximal monotone. Using the assumption that $\theta \in A\theta$, we have $S_1(\theta) = \theta$. By the monotonicity of S_1, we obtain that

$$\langle S_1(y), y \rangle \geq 0, \quad \forall y \in H. \tag{3.40}$$

Since $\text{Ran}(S_2) = \text{Dom}(B + \frac{1}{2}I) = \text{Dom}(B)$, we see that $\text{Ran}(S_2)$ is bounded and $\theta \in \text{int}(\text{Ran}(S_2))$. Hence we claim that there exists $\alpha > 0$ such that

$$\langle S_2(y), y \rangle \geq 0, \quad \forall y \in H \text{ with } \|y\| > \alpha. \tag{3.41}$$

If (3.41) holds, then we derive from (3.40) that

$$\langle (S_1 + S_2)y, y \rangle \geq 0, \quad \forall y \in H \text{ with } \|y\| > \alpha.$$

By Theorem 3.3.1, we see that $\theta \in \text{Ran}(S_1 + S_2)$, that is, (3.39) holds.

Next, we prove that (3.41) is true. For all $y, z \in H$, we conclude from the monotonicity of S_2 that

$$\langle S_2(y) - S_2(z), y - z \rangle \geq 0$$
$$\implies \langle S_2(y), y \rangle \geq \langle S_2(z), y \rangle + \langle S_2(y) - S_2(z), z \rangle. \tag{3.42}$$

Since $\text{Ran}(S_2)$ is bounded, there exists $\alpha_1 > 0$ such that

$$|\langle S_2(y) - S_2(z), z \rangle| \leq 2\alpha_1 \|z\|. \tag{3.43}$$

Note that $\text{Dom}(S_2^{-1}) = \text{Ran}(S_2)$. In view of (3.40), one sees that

$$\theta \in \text{int}(\text{Dom}(S_2^{-1})).$$

By Theorem 3.1.1, S_2^{-1} is locally bounded at the origin θ and thus there exist $\varepsilon > 0$ and $\alpha_2 > 0$ such that

$$\{S_2(z) : \|z\| \leq \alpha_2\} \supset \{u \in H : \|u\| \leq \varepsilon\}. \tag{3.44}$$

By (3.42) and (3.43), it follows that, for all $z \in H$, $\|z\| \leq \alpha_2$,

$$\langle S_2(y), y \rangle \geq \langle S_2(z), y \rangle - 2\alpha_1\alpha_2. \tag{3.45}$$

By (3.44) and (3.45), we see that

$$\langle S_2(y), y \rangle \geq \sup_{\|u\| \leq \varepsilon} \{\langle u, y \rangle - 2\alpha_1\alpha_2\} = \varepsilon\|y\| - 2\alpha_1\alpha_2. \tag{3.46}$$

Take $\alpha = \frac{2\alpha_1\alpha_2}{\varepsilon}$. Then, if $\|y\| > \alpha$, then we see from (3.46) that $\langle S_2(y), y \rangle \geq 0$, which implies that (3.41) holds.

Case II. $\mathrm{Dom}(B) \subset H$ is unbounded.

In this case, using the perturbation techniques of monotone mappings and Case I, we can derive the desired conclusion.

To complete the proof, let us collect some necessary and known conclusions.

Let C be a nonempty, closed, and convex subset of a Hilbert space H. Recall that the indicator function of C is defined as follows:

$$i_C(x) = \begin{cases} 0, & x \in C, \\ \infty, & x \notin C. \end{cases}$$

It is well known that $i_C : H \to \overline{\mathbb{R}}$ is proper, lower semicontinuous, and convex. Further, the subdifferential $\partial i_C \subset H \times H$ of i_C is maximal monotone. Recall also that the normal cone of C at x is defined as follows:

$$N_C(x) = \{z \in H : \langle u - x, z \rangle \leq 0, \quad \forall y \in C\}. \tag{3.47}$$

It is clear that $N_C(x)$ is a closed cone and

$$\partial i_C(x) = N_C(x). \tag{3.48}$$

Let $C = B_\alpha[\theta] = \{x \in H : \|x\| \leq \alpha\}, \forall \alpha > 0$. Then

$$N_C(x) = \begin{cases} \{\theta\}, & \|x\| < \alpha, \\ \{\lambda x : \lambda > 0\}, & \|x\| = \alpha. \end{cases}$$

Thus, we have that

$$\partial i_{B_\alpha[\theta]}(x) = \begin{cases} \{\theta\}, & \|x\| < \alpha, \\ \{\lambda x : \lambda > 0\}, & \|x\| = \alpha, \\ \emptyset, & \|x\| > \alpha. \end{cases}$$

Denote $\varphi_\alpha(x) = \partial i_{B_\alpha[\theta]}(x), \forall x \in H$ and $\alpha > 0$. If $T \subset H \times H$ is monotone, then $T + \varphi_\alpha$ is monotone and

$$(T + \varphi_\alpha)(x) = Tx, \quad \forall x \in B_\alpha[\theta], \tag{3.49}$$

$$\mathrm{Dom}(T + \varphi_\alpha) = \mathrm{Dom}(T) \cap B_\alpha[\theta]. \tag{3.50}$$

Now, we in a position to prove Theorem 3.3.4 under Case II. Assume that

$$\theta \in \mathrm{Dom}(A) \cap \mathrm{int}(\mathrm{Dom}(B))$$

Since $\mathrm{Dom}(\varphi_\alpha) = B_\alpha[\theta]$, we have

$$\mathrm{int}(\mathrm{Dom}(\varphi_\alpha)) = B_\alpha[\theta].$$

It follows that

$$\theta \in \text{Dom}(B) \cap \text{Dom}(\varphi_\alpha) \neq \emptyset.$$

Furthermore, $\text{Dom}(\varphi_\alpha)$ is bounded. Since $\text{Dom}(B+\varphi_\alpha) = \text{Dom}(B) \cap B_\alpha[\theta]$, we derive that

$$\theta \in \text{Dom}(A) \cap \text{int}(\text{Dom}(B + \varphi_\alpha)) \neq \emptyset,$$

where $\text{Dom}(B + \varphi_\alpha)$ is bounded. Letting $T = B$ in (3.49), we have $(B + \varphi_\alpha)x = Bx$, $\forall x \in B_\alpha[\theta]$. If $\|x\| = \alpha$, then

$$(B + \varphi_\alpha)x = Bx + \varphi_\alpha(x) = Bx + \lambda x = (B + \lambda I)x.$$

Note that B and φ_α are maximal monotone. From Case I, we can prove that $B + \varphi_\alpha$ is maximal monotone. Since $A + (B + \varphi_\alpha) = (A + B) + \varphi_\alpha$, we see that $A + (B + \varphi_\alpha)$ is maximal monotone. Hence $(A + B) + \varphi_\alpha$ is maximal monotone. By Theorem 3.2.3, we see that $A + B$ is maximal monotone. This completes the proof. $\square$

Theorem 3.3.5 (Rockafellar [77], 1970). *Let C be a nonempty convex and closed subset of a real Hilbert space H and let $A : C \to H$ be a monotone and hemicontinuous mapping. Define a mapping $T : H \to 2^H$ by*

$$Tx = \begin{cases} Ax + N_C(x), & x \in C, \\ \emptyset, & x \notin C, \end{cases}$$

where $N_C(x)$ is the normal cone of C at x. Then $T \subset H \times H$ is maximal monotone.

Proof. Note that the effective domain of T is $\text{Dom}(T) = C$. If $x \in C$, then $N_C(x) = \partial i_C(x)$. If $x \in \text{Dom}(T) = C$, then $Tx = Ax + \partial i_C(x)$ and A and ∂i_C are monotone. Thus T is also monotone.

Next, we prove that T is maximal monotone. From Proposition 3.1.2, we only need to prove, for any $[x, f] \in H \times H$, that

$$\langle y - x, g - f \rangle \geq 0, \quad \forall [y, g] \in \text{Graph}(A) \implies [x, f] \in \text{Graph}(A).$$

Since $[y, g] \in \text{Graph}(A)$, $g \in Ty = Ay + N_C(y)$, we have $g = Ay + z$ and $z \in N_C(y)$. It follows that

$$\langle y - x, z \rangle + \langle y - x, Ay - f \rangle \geq 0, \quad \forall y \in C, z \in N_C(y). \tag{3.51}$$

Since $N_C(y)$ is a closed cone, we see that $z \in N_C(y)$. So, for any $\lambda \geq 0$, one has $\lambda z \in N_C(y)$. By (3.51), we obtain

$$\langle y - x, \lambda z \rangle + \langle y - x, Ax - f \rangle \geq 0, \quad \forall y \in C, \tag{3.52}$$

which implies from (3.52) that

$$\langle y - x, z \rangle \geq 0, \quad \forall z \in N_C(y). \tag{3.53}$$

If (3.53) does not hold, then there exists $z \in N_C(y)$ such that

$$\langle y - x, z \rangle < 0.$$

For any $\lambda > 0$, letting $\lambda \to \infty$, one has

$$\lambda \langle y - x, z \rangle = \langle y - x, \lambda z \rangle \to -\infty, \quad \forall y \in C,$$

which contradicts (3.52). Since $N_C(y) = \partial i_C(y)$, we see from (3.53) that

$$\langle y - x, z - \theta \rangle \geq 0, \quad \forall z \in \partial i_C(y).$$

Note that i_C is maximal monotone. From Proposition 3.1.2, we have $\theta \in \partial i_C(x)$. Hence $x \in C$. Letting $x_t = tu + (1 - t)x$, $u \in C$, $\forall t \in (0, 1)$, we have $x_t \in C$. Taking $y = x_t$ and $z = \theta$ in (3.51), we obtain

$$\langle x_t - x, \theta \rangle + \langle x_t - x, Ax_t - f \rangle \geq 0,$$

that is,

$$\langle x_t - x, Ax_t - f \rangle \geq 0 \implies t \langle u - x, Ax_t - f \rangle \geq 0, \quad \forall t \in (0, 1)$$
$$\implies \langle u - x, Ax_t - f \rangle \geq 0, \quad \forall t \in (0, 1).$$

Letting $t \to 0^+$, we find from the hemicontinuity of A that

$$\langle u - x, Ax - f \rangle \geq 0, \quad \forall u \in C \implies f - Ax \in N_C(x) \implies Ax + N_C(x) = Tx.$$

Therefore, $T \subset H \times H$ is maximal monotone. This completes the proof. $\qquad \square$

3.4 Monotone variational inequalities

In this subsection, we study the existence and uniqueness of solutions for a class of monotone variational inequalities via the monotone mapping theory. First, we establish existence and uniqueness of solutions for variational inequality problems (in short, (VIP)) involving η-strongly monotone hemicontinuous mappings. Then we drive some zero point theorems of monotone hemicontinuous mappings. Using the existence theorems of solutions of variational inequalities, we introduce some iterative methods for finding the solutions of the monotone variational inequality problems.

Theorem 3.4.1. *Let C be a nonempty closed convex subset of a real Hilbert space H and let $A : C \to H$ be an η-strongly monotone hemicontinuous mapping. Then*

$$\langle Ax, y - x \rangle \geq 0, \quad \forall y \in C, \tag{3.54}$$

has a unique solution $x^* \in C$.

Proof. Denote the solution set of (3.54) by $VI(C, A)$ and define $T \subset H \times H$ as in Theorem 3.3.5 by

$$Tx = \begin{cases} Ax + N_C(x), & x \in C, \\ \emptyset, & x \notin C, \end{cases}$$

where $N_C(x)$ is the normal cone of C at x. By Theorem 3.3.5, $T \subset H \times H$ is maximal monotone and

$$VI(C, A) = T^{-1}(\theta) = \{z \in C : \theta \in Tz\}. \tag{3.55}$$

In fact, we have

$$z \in T^{-1}(\theta) \iff \theta \in Az + N_C(z)$$
$$\iff \langle u - z, -Az \rangle \leq 0, \quad \forall u \in C$$
$$\iff \langle u - z, Az \rangle \geq 0, \quad \forall u \in C$$
$$\iff z \in VI(C, A).$$

To prove that (3.54) has a unique solution, we only need to prove that T has a unique zero point. From Theorem 3.3.1, we just need to prove that there exists $\alpha > 0$ such that

$$\langle x^*, x \rangle \geq 0, \quad \forall x \in C \text{ with } \|x\| > \alpha, \ x^* \in Tx. \tag{3.56}$$

Now, we consider the following two cases.
1. $\theta \in C$.
    Since $A : C \to H$ is η-strongly monotone, one has

    $$\langle Ax - A\theta, x \rangle \geq \eta\|x\|^2, \quad \forall x \in C \tag{3.57}$$
    $$\implies \langle Ax, x \rangle \geq \eta\|x\|^2 - \|A\theta\|\|x\|, \quad \forall x \in C. \tag{3.58}$$

On the other hand, we have $x^* \in Tx = Ax + N_C(x)$ and

$$x^* = Ax + z, \quad \forall z \in N_C(x), \tag{3.59}$$
$$z \in N_C(x) \implies \langle u - x, z \rangle \leq 0, \quad \forall u \in C.$$

Letting $u = \theta \in C$ yields

$$\langle -x, z \rangle \leq 0 \implies \langle x, z \rangle \geq 0. \tag{3.60}$$

From (3.58)–(3.60), we see that

$$\langle x^*, x \rangle = \langle Ax + z, x \rangle = \langle Ax, x \rangle + \langle x, z \rangle$$
$$\geq \eta \|x\|^2 - \|A\theta\| \|x\|$$
$$= \|x\|(\eta\|x\| - \|A\theta\|).$$

Set $\alpha = \frac{\|A\theta\| + 1}{\eta} > 0$. If $\|x\| > \alpha$, then

$$\langle x^*, x \rangle \geq 0, \quad \forall x \in C, \, x^* \in Tx.$$

2. $\theta \notin C$.

Fixing $x_0 \in C$, we obtain a shift $\tilde{C} = C - x_0$. This shows $\theta \in \tilde{C}$. Define a mapping $\tilde{T}(x) = T(x + x_0), \forall x \in \tilde{C}$ (that is, $x + x_0 \in C$). This implies that $\tilde{T} \subset H \times H$ is still maximal monotone. In fact, we define $\tilde{A} = A(x + x_0), \forall x \in \tilde{C}$. From $N_{\tilde{C}}(x) = N_C(x + x_0), \forall x \in \tilde{C}$, one has

$$\tilde{T}(x) = T(x + x_0) = A(x + x_0) + N_C(x + x_0) = \tilde{A}x + N_{\tilde{C}}(x), \quad \forall x \in \tilde{C}.$$

It is easy to verify that $\tilde{A} : \tilde{C} \to H$ is η-strongly monotone and hemicontinuous. So, it follows from Theorem 3.3.5 that $\tilde{T} \subset H \times H$ is maximal monotone. From Case I, one sees that there exists $\tilde{x} \in B_\alpha[\theta] \cap \tilde{C}$ such that

$$\theta \in \tilde{T}(\tilde{x}),$$

where $\alpha > 0$ is a fixed constant, and $B_\alpha[\theta]$ is a closed ball with the center origin and the radius α. Hence we obtain $\theta \in T(\tilde{x} + x_0)$. Setting $z = \tilde{x} + x_0$, we have $z \in C$ and $\theta \in Tz$. Hence $T^{-1}\theta \neq \emptyset$. The uniqueness is guaranteed by the η-strong monotonicity of A. This completes the proof. $\square$

Corollary 3.4.1. *Let C be a nonempty bounded closed convex subset of a real Hilbert space H and let $A : C \to H$ be a monotone hemicontinuous mapping. Then $VI(C, A)$ is nonempty, closed, and convex.*

Proof. Define a mapping $A_n : C \to H$ by

$$A_n(x) = Ax + \frac{1}{n}x, \quad \forall x \in C, n \geq 1.$$

Then $A_n : C \to H$ is $\frac{1}{n}$-strongly monotone and hemicontinuous. From Theorem 3.4.1, there exists a unique $x_n \in C$ such that $x_n \in VI(C, A)$, that is,

$$\langle A_n x_n, y - x_n \rangle \geq 0, \quad \forall y \in C, n \geq 1.$$

From Minty's lemma, it follows that $\langle A_n y, y - x_n \rangle \geq 0, \forall y \in C$ and $\forall n \geq 1$, that is,

$$\left\langle Ay + \frac{1}{n}y, y - x_n \right\rangle \geq 0, \quad \forall y \in C, n \geq 1. \tag{3.61}$$

Since C is bounded, one sees that $\{x_n\}$ is a bounded sequence. We may assume that $x_n \rightharpoonup x$ as $n \to \infty$. Letting $n \to \infty$ in (3.61), we arrive at

$$\langle Ay, y - x \rangle \geq 0, \quad \forall y \in C. \tag{3.62}$$

In view of Minty's lemma, we obtain

$$\langle Ax, y - x \rangle \geq 0, \quad \forall y \in C,$$

that is, $VI(C, A) \neq \emptyset$. It follows from (3.62) that $VI(C, A)$ is closed and convex. This completes the proof. $\qquad\square$

Corollary 3.4.2. *Let C be a nonempty, closed, and convex subset of a real Hilbert space H, and let $A : C \to H$ be an η-strongly monotone hemicontinuous mapping satisfying the "flow-invariance" condition (FIC):*

$$\lim_{t \to 0^+} t^{-1} d((I - tA)x, C) = 0, \quad \forall x \in C.$$

Then there exists a unique $x^* \in C$ such that $Ax^* = \theta$.

Proof. From Theorem 3.4.1, we see that there exists a unique $x^* \in C$ such that

$$\langle Ax^*, x - x^* \rangle \geq 0, \quad \forall x \in C. \tag{3.63}$$

By using the "flow-invariance" condition, we see that, for any $\varepsilon > 0$, there exists $\delta > 0$ with $0 < t < \delta$ such that

$$d((I - tA)x^*, C) < t\varepsilon \implies \exists x_t \in C \text{ such that } \|(I - tA)x^* - x_t\| < t\varepsilon$$

$$\implies \left\| \frac{x_t - x^*}{t} + Ax^* \right\| < \varepsilon$$

$$\implies \frac{x_t - x^*}{t} \to -Ax^* \quad (t \to 0^+).$$

Letting $x = x_t \in C$ in (3.63), we obtain

$$\langle Ax^*, x_t - x^* \rangle \geq 0, \quad \forall t > 0 \implies \left\langle Ax^*, \frac{x_t - x^*}{t} \right\rangle \geq 0, \quad \forall t > 0$$

$$\implies \langle Ax^*, -Ax^*, \rangle \geq 0$$

$$\implies Ax^* = \theta.$$

This completes the proof. $\qquad\square$

Corollary 3.4.3. *Let C be a nonempty, bounded, closed, and convex subset of a real Hilbert space H, and let $A : C \to H$ be a monotone hemicontinuous mapping satisfying the "flow-invariance" condition (FIC). Then there exists some $x^* \in C$ such that $Ax^* = \theta$.*

Proof. From Corollary 3.4.1, we see that $VI(C, A) \neq \emptyset$, that is, there exists $x^* \in C$ such that

$$\langle Ax^*, x - x^* \rangle \geq 0, \quad \forall x \in C.$$

The rest of the proof is similar to the proof of Corollary 3.4.2, so we omit it here. This completes the proof. □

Let C be a nonempty, closed, and convex subset of a Hilbert space H. Let $A : C \to H$ be a monotone hemicontinuous mapping and let $R : H \to H$ be an η-strongly monotone hemicontinuous mapping. Take $u \in H$ and a sequence $\{r_n\}$, where $r_n > 0$ and $r_n \to 0$ as $n \to \infty$. Then, for any $n \in \mathbb{N}$, $r_n(R - u) + A : C \to H$ is ηr_n-strongly monotone and hemicontinuous. From Theorem 3.4.1, we have that the following variational inequality

$$\langle r_n(Ry - u) + Ay, x - y \rangle \geq 0, \quad \forall x \in C, \tag{3.64}$$

has a unique solution $y_n \in C$, $\forall n \in \mathbb{N}$.

Letting $r_n = \frac{\alpha_n}{\beta_n}$ in (3.64), we see that (3.64) can be rewritten as

$$\langle \alpha_n(Ry_n - u) + \beta_n Ay_n, x - y_n \rangle \geq 0, \quad \forall x \in C, \tag{3.65}$$

which is equivalent to

$$\langle y_n - \alpha_n(Ry_n - u) - \beta_n Ay_n - y_n, x - y_n \rangle \leq 0, \quad \forall x \in C. \tag{3.66}$$

From the property of P_C and (3.66), we have

$$y_n = P_C[(I - \alpha_n R)y_n + \alpha_n u - \beta_n Ay_n], \quad \forall n \geq 1. \tag{3.67}$$

Now, we are in a position to prove that if $\frac{\alpha_n}{\beta_n} \to 0$ as $n \to \infty$ and $VI(C, A) \neq \emptyset$, then $\{y_n\}$ converges in norm to a special solution of variational inequality (3.54).

Theorem 3.4.2. *Let C be a nonempty, closed, and convex subset of a real Hilbert space H. Let $A : C \to H$ be a monotone hemicontinuous mapping and let $R : H \to H$ be an η-strongly monotone hemicontinuous mapping. Assume that $\frac{\alpha_n}{\beta_n} \to 0$ as $n \to \infty$ and $VI(C, A) \neq \emptyset$. Let $\{y_n\}$ be a sequence generated by (3.67). Then $\{y_n\}$ converges in norm to a special solution $x^* = P_{VI(C,A)}[(I - R)x^* + u]$ of the variational inequality (3.54). In particular, if $R = I$ and $u = \theta$, then $\{y_n\}$ converges in norm to the minimum norm solution $x^* = P_{VI(C,A)}\theta$ of variational inequality (3.54).*

Proof. From the assumption $VI(C, A) \neq \emptyset$, it follows from Minty's lemma that $VI(C, A)$ is closed and convex. Hence $VI(C, A)$ is a nonempty, closed, and convex subset of H. Therefore, for any $h \in H$, $P_{VI(C,A)}h$ is well defined. Since (3.67) and (3.65) are equivalent, we find from Minty's lemma that

$$\langle \alpha_n Rx - \alpha_n u + \beta_n Ax, x - y_n \rangle \geq 0, \quad \forall x \in C. \tag{3.68}$$

For all $x^* \in VI(C, A)$, we derive

$$\langle Ax^*, x - x^* \rangle \geq 0, \quad \forall x \in C. \tag{3.69}$$

In view of Minty's lemma, we obtain

$$\langle Ax, x - x^* \rangle \geq 0, \quad \forall x \in C. \tag{3.70}$$

Taking $x = y_n$ in (3.70), we see that

$$\langle Ay_n, y_n - x^* \rangle \geq 0, \quad \forall n \geq 1. \tag{3.71}$$

Taking $x = x^*$ in (3.65), we have

$$\langle \alpha_n Ry_n - \alpha_n u + \beta_n Ay_n, y_n - x^* \rangle \leq 0, \quad \forall n \geq 1. \tag{3.72}$$

It follows from (3.71), (3.72) and the strong monotonicity of R that

$$\begin{aligned}
0 &\geq \langle \alpha_n Ry_n - \alpha_n u + \beta_n Ay_n, y_n - x^* \rangle \\
&= \alpha_n \langle Ry_n - Rx^*, y_n - x^* \rangle + \alpha_n \langle Rx^* - u, y_n - x^* \rangle + \beta_n \langle Ay_n, y_n - x^* \rangle \\
&\geq \eta \alpha_n \| y_n - x^* \|^2 + \alpha_n \langle Rx^* - u, y_n - x^* \rangle,
\end{aligned}$$

which implies that

$$\| y_n - x^* \|^2 \leq \frac{1}{\eta} \langle Rx^* - u, x^* - y_n \rangle, \quad \forall x^* \in VI(C, A). \tag{3.73}$$

In particular, we have

$$\| y_n - x^* \|^2 \leq \frac{1}{\eta} \| Rx^* - u \|^2, \quad \forall x^* \in VI(C, A), \, n \geq 1, \tag{3.74}$$

which shows that $\{y_n\}$ is bounded. We may assume, without loss of generality, that $y_n \rightharpoonup \bar{y}$ as $n \to \infty$. Then $\bar{y} \in VI(C, A)$. In fact, it follows from (3.68) that

$$\left\langle \frac{\alpha_n}{\beta_n}(Rx - u) + Ax, x - y_n \right\rangle \geq 0, \quad \forall x \in C. \tag{3.75}$$

Letting $n \to \infty$ in (3.75), we obtain

$$\langle Ax, x - \bar{y} \rangle \geq 0, \quad \forall x \in C. \tag{3.76}$$

It follows from Minty's lemma that

$$\langle A\bar{y}, x - \bar{y} \rangle \geq 0, \quad \forall x \in C,$$

that is, $\bar{y} \in VI(C,A)$. Letting $x^* = \bar{y}$, we have

$$\|y_n - \bar{y}\|^2 \leq \frac{1}{\eta} \langle R\bar{y} - u, \bar{y} - y_n \rangle, \quad \forall n \geq 1. \tag{3.77}$$

Since $y_n \rightharpoonup \bar{y}$, it follows from (3.77) that $y_n \to \bar{y}$ as $n \to \infty$. From (3.73), we arrive at

$$\langle Rx^* - u, x^* - y_n \rangle \geq 0, \quad \forall x^* \in VI(C,A).$$

Since $y_n \rightharpoonup \bar{y}$ as $n \to \infty$, we obtain

$$\langle Rx^* - u, x^* - \bar{y} \rangle \geq 0, \quad \forall x^* \in VI(C,A).$$

By Minty's lemma, we have

$$\langle R\bar{y} - u, x^* - \bar{y} \rangle \geq 0, \quad \forall x^* \in VI(C,A). \tag{3.78}$$

It follows from Theorem 3.4.1 that $\bar{y}$ is the unique solution of variational inequality (3.78). Hence $\{y_n\}$ converges in norm to $\bar{y} \in VI(C,A)$. From (3.78), we see that

$$\bar{y} = P_{VI(C,A)}[(I - R)\bar{y} + u].$$

This completes the proof. □

In view of the convergence of iterative method (3.67), it is natural to consider the following explicit form:

$$x_1 \in C, \quad x_{n+1} = P_C[(I - \alpha_n R)x_n + \alpha_n u - \beta_n Ax_n], \quad n \geq 1. \tag{3.79}$$

Recall that a mapping $T : C \to H$ is said to be generalized Lipschitz continuous if there exists a constant $L > 0$ such that

$$\|Tx - Ty\| \leq L(1 + \|x - y\|), \quad \forall x, y \in C.$$

It is obvious that a Lipschitz continuous mapping must be a generalized Lipschitz continuous mapping, but the converse is not true. Generalized Lipschitz continuous mappings may not be continuous, for example, sign function is a simple example.

Theorem 3.4.3. *Let C be a nonempty, closed, and convex subset of a real Hilbert space H. Let $A : C \to H$ be a hemicontinuous and generalized Lipschitz continuous monotone mapping, and let $R : H \to H$ be a hemicontinuous and generalized Lipschitz continuous, η-strongly monotone mapping. Let $\{\alpha_n\}$ and $\{\beta_n\}$ be two sequences in $(0,1)$ satisfying the following conditions:*
(1) $\frac{\alpha_n}{\beta_n} \to 0$ and $\frac{\beta_n^2}{\alpha_n} \to 0$ as $n \to \infty$;
(2) $\alpha_n \to 0$ as $n \to \infty$ and $\sum_{n=1}^{\infty} \alpha_n = \infty$;
(3) $\frac{|\alpha_{n+1} - \alpha_n| + |\beta_{n+1} - \beta_n|}{\alpha_n^2} \to 0$ as $n \to \infty$.

If $VI(C, A) \neq \emptyset$, then the sequence $\{x_n\}$ generated by (3.79) converges to $x^* = P_{VI(C,A)}[(I - R)x^* + u]$ in norm.

Proof. Assume that $\{y_n\}$ is the sequence defined by (3.67). It follows from Theorem 3.4.2 that $y_n \to x^* = P_{VI(C,A)}[(I - R)x^* + u]$. Hence, we only need to prove that $x_{n+1} - y_n \to \theta$ as $n \to \infty$.

First, we prove that $\{x_n\}$ is bounded. In fact, for any $p \in VI(C, A)$, one has

$$p = P_C[(1 - \alpha_n)p + \alpha_n p - \beta_n Ap].\tag{3.80}$$

From (3.69), (3.80) and the property of P_C, we have

$$\begin{aligned}
\|x_{n+1} - p\|^2 &= \left\|P_C[(I - \alpha_n R)x_n + \alpha_n u - \beta_n Ax_n] - P_C[(1 - \alpha_n)p + \alpha_n p - \beta_n Ap]\right\|^2 \\
&\leq \left\|x_n - p - \alpha_n(Rx_n - p) + \alpha_n(u - p) - \beta_n(Ax_n - Ap)\right\|^2 \\
&= \|x_n - p\|^2 - 2\alpha_n\langle Rx_n - p, x_n - p\rangle - 2\alpha_n\langle Ax_n - Ap, x_n - p\rangle \\
&\quad + 2\alpha_n\langle u - p, x_n - p\rangle - 2\beta_n\langle Ax_n - Ap, x_n - p\rangle \\
&\quad + \left\|\alpha_n(Rx_n - u) + \beta_n(Ax_n - Ap)\right\|^2 \\
&\leq (1 - 2\eta\alpha_n)\|x_n - p\|^2 - 2\alpha_n\langle Rp - u, x_n - p\rangle \\
&\quad + 2\alpha_n^2\|Rx_n - u\|^2 + 2\beta_n^2\|Ax_n - Ap\|^2.
\end{aligned}\tag{3.81}$$

Since A and R are generalized Lipschitz continuous and

$$\begin{aligned}
2\left|\langle Rp - u, x_n - p\rangle\right| &\leq 2\eta\frac{\|Rp - u\|}{\eta}\|x_n - p\| \\
&\leq \eta\left(\frac{\|Rp - u\|^2}{\eta^2} + \|x_n - p\|^2\right),
\end{aligned}\tag{3.82}$$

we have

$$\|Ax_n - Ap\|^2 \leq L^2(1 + \|x_n - p\|^2) \leq 2L^2(1 + \|x_n - p\|^2),\tag{3.83}$$

$$\|Rx_n - Rp\|^2 \leq L^2(1 + \|x_n - p\|^2) \leq 2L^2(1 + \|x_n - p\|^2),\tag{3.84}$$

and

$$\|Rx_n - u\|^2 \leq 2(\|Rx_n - Rp\|^2 + \|Rp - u\|^2).\tag{3.85}$$

From condition (1) and (2), we may assume, without loss of generality, that

$$8L^2\alpha_n^2 + 4L^2\beta_n^2 \leq \frac{1}{2}\eta\alpha_n, \quad \forall n \geq 1.\tag{3.86}$$

Substituting (3.82)–(3.86) into (3.81), we obtain

$$\|x_{n+1} - p\|^2 \leq (1 - \eta\alpha_n + 8L^2\alpha_n^2 + 4L^2\beta_n^2)\|x_n - p\|^2$$

$$+ \frac{1}{2}\eta\alpha_n\left(\frac{2}{\eta^2} + 4 + \frac{L^2}{\eta}\right)\|Rp - u\|^2$$

$$\leq \left(1 - \frac{1}{2}\eta\alpha_n\right)\|x_n - p\|^2 + \frac{1}{2}\eta\alpha_n\left(\frac{2}{\eta^2} + 4 + \frac{L^2}{\eta}\right)\|Rp - u\|^2$$

$$\leq \max\left\{\|x_1 - p\|^2, \left(\frac{2}{\eta^2} + 4 + \frac{L^2}{\eta}\right)\|Rp - u\|^2\right\}$$

$$= M,$$

which shows that $\{x_n\}$ is bounded. Since R is η-strongly monotone and A is monotone, one has

$$\langle Rx_n - Ry_n, x_n - y_n \rangle \geq \eta\|x_n - y_n\|^2, \quad \forall n \geq 1, \tag{3.87}$$

and

$$\langle Ax_n - Ay_n, x_n - y_n \rangle \geq \eta\|x_n - y_n\|^2, \quad \forall n \geq 1. \tag{3.88}$$

Since R and A are both generalized Lipschitz continuous, we have

$$\|Rx_n - Ry_n\|^2 \leq 2L^2(1 + \|x_n - y_n\|^2), \quad \forall n \geq 1, \tag{3.89}$$

and

$$\|Ax_n - Ay_n\|^2 \leq 2L^2(1 + \|x_n - y_n\|^2), \quad \forall n \geq 1. \tag{3.90}$$

Further, we conclude from (3.67), the property of P_C, monotonicity of A and strongly monotonicity of R that

$$\|y_{n+1} - y_n\|^2 \leq \langle y_{n+1} - y_n, (I - \alpha_n R)y_{n+1} - (I - \alpha_n R)y_n + (\alpha_{n+1} - \alpha_n)u \rangle$$

$$- \beta_{n+1}Ay_{n+1} - \beta_n Ay_n$$

$$= \|y_{n+1} - y_n\|^2 - \alpha_{n+1}\langle Ry_{n+1} - Ry_n, y_{n+1} - y_n \rangle$$

$$- (\alpha_{n+1} - \alpha_n)\langle Ry_n - u, y_{n+1} - y_n \rangle - \beta_{n+1}\langle Ay_{n+1} - Ay_n, y_{n+1} - y_n \rangle$$

$$- (\beta_{n+1} - \beta_n)\langle Ay_n, y_{n+1} - y_n \rangle$$

$$\leq (1 - \eta\alpha_{n+1})\|y_{n+1} - y_n\|^2 + |\alpha_{n+1} - \alpha_n|\|Ry_n\|\|y_{n+1} - y_n\|$$

$$+ |\alpha_{n+1} - \alpha_n|\|u\|\|y_{n+1} - y_n\| + |\beta_{n+1} - \beta_n|\|Ay_n\|\|y_{n+1} - y_n\|. \tag{3.91}$$

It follows that

$$\|y_{n+1} - y_n\| \leq \frac{M}{\eta}\frac{|\alpha_{n+1} - \alpha_n| + |\beta_{n+1} - \beta_n|}{\alpha_{n+1}}, \quad \forall n \geq 1, \tag{3.92}$$

where $M \geq \max\{\|Ry_n\| + \|u\|, \|Ay_n\|\}$. Using conditions (1) and (2), we may assume that

$$\alpha_n^2 + \beta_n^2 \leq \frac{\eta}{4L^2}\alpha_n, \quad \forall n \geq 1. \tag{3.93}$$

It follows from the property of P_C, and (3.87)–(3.93) that

$$\begin{aligned}
\|x_{n+2} - y_{n+1}\|^2 &\leq \|x_{n+1} - y_{n+1} - \alpha_{n+1}(Rx_{n+1} - Ry_{n+1}) - \beta_n(Ax_{n+1} - Ay_{n+1})\|^2 \\
&= \|x_{n+1} - y_{n+1}\|^2 - 2\alpha_{n+1}\langle Rx_{n+1} - Ry_{n+1}, x_{n+1} - y_{n+1}\rangle \\
&\quad - 2\beta_{n+1}\langle Ax_{n+1} - Ay_{n+1}, x_{n+1} - y_{n+1}\rangle \\
&\quad + \|\alpha_{n+1}(Rx_{n+1} - Ry_{n+1}) + \beta_{n+1}(Ax_{n+1} - Ay_{n+1})\|^2 \\
&\leq (1 - 2\eta\alpha_{n+1})\|x_{n+1} - y_{n+1}\|^2 + 2\alpha_{n+1}^2\|Rx_{n+1} - Ry_{n+1}\|^2 \\
&\quad + 2\beta^2\|Ax_{n+1} - Ay_{n+1}\|^2 \\
&\leq (1 - 2\eta\alpha_{n+1})\|x_{n+1} - y_{n+1}\|^2 + 4L^2\alpha_{n+1}^2(1 + \|x_{n+1} - y_{n+1}\|^2) \\
&\quad + 4L^2\beta_{n+1}^2(1 + \|x_{n+1} - y_{n+1}\|^2) \\
&= [1 - 2\eta\alpha_{n+1} + 4L^2(\alpha_{n+1}^2 + \beta_{n+1}^2)]\|x_{n+1} - y_{n+1}\|^2 + 4L^2(\alpha_{n+1}^2 + \beta_{n+1}^2) \\
&\leq (1 - \eta\alpha_{n+1})\|x_{n+1} - y_{n+1}\|^2 + 4L^2(\alpha_{n+1}^2 + \beta_{n+1}^2) \\
&\leq (1 - \eta\alpha_{n+1})[\|x_{n+1} - y_n\|^2 + 2\|x_{n+1} - y_n\|\|y_{n+1} - y_n\| \\
&\quad + \|y_{n+1} - y_n\|^2] + 4L^2(\alpha_{n+1}^2 + \beta_{n+1}^2) \\
&\leq (1 - \eta\alpha_{n+1})\|x_{n+1} - y_n\|^2 + M_1\frac{|\alpha_{n+1} - \alpha_n| + |\beta_{n+1} - \beta_n|}{\alpha_{n+1}} \\
&\quad + 4L^2(\alpha_{n+1}^2 + \beta_{n+1}^2) \\
&\leq (1 - \eta\alpha_{n+1})\|x_{n+1} - y_n\|^2 + o(\eta\alpha_{n+1}).
\end{aligned}$$

In view of Lemma 1.10.2, we have

$$x_{n+1} - y_n \to \theta \quad (n \to \infty).$$

This completes the proof. $\qquad\qquad\qquad\qquad\qquad\qquad\qquad\qquad\qquad\qquad\qquad\square$

Remark 3.4.1. Taking $\alpha_n = \frac{1}{n^a}$ and $\beta_n = \frac{1}{n^b}$ for all $n \geq 1$, where $a < \frac{b+1}{2}$ and $0 < b < a < 2b$, we see that $\{\alpha_n\}$ and $\{\beta_n\}$ satisfy conditions (1)–(3) in Theorem 3.4.3.

Let A be a mapping on a Hilbert space H. Recall that A is said to be pseudomonotone if

$$0 \leq \langle Ax, y - x\rangle \Longrightarrow 0 \leq \langle Ay, y - x\rangle, \quad \forall x, y \in H.$$

Also A is said to be γ-strongly pseudomonotone iff there exists $\gamma > 0$ such that

$$0 \leq \langle Ax, y - x\rangle \Rightarrow \gamma\|x - y\|^2 \leq \langle Ay, y - x\rangle, \quad \forall x, y \in H.$$

If A is L-Lipschitz continuous and monotone, Korpelevich [40] proposed the following well known extragradient method (with double projections) that reduces the monotonicity of operator A:

$$\begin{cases} x_0 \in H, \\ y_n = P_C(x_n - \tau A x_n), \\ x_{n+1} = P_C(x_n - \tau A y_n), \quad \forall n \geq 0, \end{cases}$$

where τ is a real number in $(0, \frac{1}{L})$. It was proved that sequence $\{x_n\}$ generated in the above extragradient algorithm converges weakly to a solution of the variational inequality. However, the price is that one needs to calculate two projections from H onto the feasibility set C. In most situations, there are no analytic expressions for the metric projection. Hence, the extragradient method is not very convenient and efficient in practical calculations. In 2011, Censor, Gibali, and Reich [22] studied a subgradient–extragradient method for solving the following variational inequality in a Hilbert space

$$\text{find } x^* \in C \text{ such that } \langle Ax^*, x - x^* \rangle \geq 0, \quad \forall x \in C, \tag{3.94}$$

where A is a monotone and Lipschitz continuous mapping.

Algorithm 3.4.7. Step 0. Select a starting point $x_0 \in H$ and $\tau > 0$, and set $n = 0$.
  Step 1. Given the current iterate x_n, compute

$$y_n = P_C(x_n - \tau A x_n),$$

construct the half-space T_n, the bounding hyperplane of which supports C at y_n, as

$$T_n := \{w \in H : \langle (x_n - \tau A x_n) - y_n, w - y_n \rangle \leq 0\},$$

and calculate the next iterate

$$x_{n+1} = P_{T_n}(x_n - \tau A x_n).$$

Step 2. If $x_n = y_n$, then stop. Otherwise, set $k \leftarrow (k+1)$ and return to Step 1.

Theorem 3.4.4 (Censor, Gibali, and Reich [22], 2011). *Let C be a nonempty, closed, and convex subset of a Hilbert space H. Assume A is L-Lipschitz continuous and monotone on C and let $\tau \leq \frac{1}{L}$. If the solution set of (3.94) is not empty, then the sequences $\{x_n\}$ and $\{y_n\}$ generated by Algorithm 3.4.7 weakly converge to the same point in the variational inequality (3.94).*

Recently, convergence rate problems received much attention. One of acceleration methods is the inertial method which is a two-step iterative method. Its feature is that the next iterate is defined by making use of the previous two iterates.

Next, we discuss an inertial subgradient–extragradient algorithm to solve variational inequality problem (3.94) with A being strongly pseudomonotone; see [28] for more details.

Algorithm 3.4.8. Step 0. Choose $x_0, x_1 \in H$ arbitrarily.

Step 1. Given the current iterate x_n, compute

$$\begin{cases} w_n = x_n + \alpha_n(x_n - x_{n-1}), \\ y_n = P_C(w_n - \tau_n A w_n), \end{cases}$$

construct the half-space T_n, the bounding hyperplane of which supports C at y_n, as

$$T_n = \{x \in H \mid \langle w_n - \tau_n A w_n - y_n, x - y_n \rangle \leq 0\},$$

and calculate the next iterate

$$x_{n+1} = \lambda_n P_{T_n}(w_n - \tau_n A y_n) + (1 - \lambda_n) w_n,$$

where $\{\alpha_n\}$ is nondecreasing with $\alpha_1 = 0$ and $0 \leq \alpha_n \leq \alpha < 1$ (α is a fixed real number in $(0,1)$), $\{\tau_n\}$ is a real sequence such that $\sum_{n=1}^{\infty} \tau_n = \infty$ and $\lim_{n \to \infty} \tau_n = 0$, and λ, σ, and δ are three positive real numbers such that

$$\delta > \frac{4\alpha[\sigma + \alpha(\alpha + 1)]}{1 - \alpha^2}$$

and

$$0 < \lambda \leq \lambda_n \leq \frac{\delta - 4\alpha[\sigma + \alpha(\alpha + 1) + \frac{1}{4}\alpha\delta]}{4\delta[\sigma + \alpha(\alpha + 1) + \frac{1}{4}\alpha\delta]}.$$

Step 2. If $x_n = y_n$, then stop. Otherwise, set $k \leftarrow (k + 1)$ and return to Step 1.

Theorem 3.4.5. *Let C be a nonempty, convex, and closed subset of a real Hilbert space H, and let $A : H \to H$ be L-Lipschitz continuous and strongly γ-pseudomonotone. Let $\{x_n\}$ be a sequence generated by Algorithm 3.4.8. If $VI(C, A)$, the solution set, is not empty, then the sequence $\{x_n\}$ converges strongly to a solution of variational inequality* (3.94).

Proof. Let $z_n = w_n - \tau_n A y_n$. Fixing $p \in VI(C, A)$, we have

$$\|x_{n+1} - p\|^2 = (1 - \lambda_n)\|w_n - p\|^2 + \lambda_n(\|z_n - p\|^2 + \|P_{T_n}(z_n) - z_n\|^2 \\ + 2\langle P_{T_n}(z_n) - z_n, z_n - p \rangle).$$

In view of $p \in C \subset T_n$, we conclude

$$2\|P_{T_n}(z_n) - z_n\|^2 + 2\langle z_n - p, P_{T_n}(z_n) - z_n \rangle \\ = 2\langle P_{T_n}(z_n) - z_n, P_{T_n}(z_n) - p \rangle \\ \leq 0.$$

Hence

$$-\left\|P_{T_n}(z_n) - z_n\right\|^2 \geq \left\|P_{T_n}(z_n) - z_n\right\|^2 + 2\langle P_{T_n}(z_n) - z_n, z_n - p\rangle.$$

It follows that

$$\|x_{n+1} - p\|^2 \leq (1 - \lambda_n)\|w_n - p\|^2 + \lambda_n(\|w_n - p\|^2 - \left\|P_{T_n}(z_n) - w_n\right\|^2$$
$$+ 2\tau_n\langle Ay_n, p - P_{T_n}(z_n)\rangle).$$

From the fact that p is in VI(C, A), one concludes $0 \leq \langle Ap, x - p\rangle$, $\forall x \in C$. Since A is strongly pseudomonotone, one obtains $-\gamma\|x - p\|^2 \leq \langle Ax, x - p\rangle$, $\forall x \in C$. By setting $x = y_n \in C$, one has $\gamma\|y_n - p\|^2 \leq \langle Ay_n, y_n - p\rangle$. It follows that

$$\langle Ay_n, p - P_{T_n}(z_n)\rangle \leq \langle Ay_n, y_n - P_{T_n}(z_n)\rangle - \gamma\|y_n - p\|^2.$$

Using the above two inequalities, we obtain

$$\|x_{n+1} - p\|^2 \leq \lambda_n\|w_n - p\|^2 + (1 - \lambda_n)\|w_n - p\|^2 - \lambda_n\left\|P_{T_n}(z_n) - w_n\right\|^2$$
$$- 2\lambda_n\tau_n\gamma\|y_n - p\|^2 + 2\lambda_n\tau_n\langle Ay_n, y_n - P_{T_n}(z_n)\rangle$$
$$= \|w_n - p\|^2 - \lambda_n\left\|P_{T_n}(z_n) - y_n\right\|^2 - \lambda_n\|y_n - w_n\|^2$$
$$- 2\lambda_n\tau_n\gamma\|y_n - p\|^2 + 2\lambda_n\langle w_n - y_n - \tau_n Ay_n, P_{T_n}(z_n) - y_n\rangle.$$

From the fact that $y_n = P_C(w_n - \tau_n Aw_n)$, one finds that

$$2\lambda_n\langle P_{T_n}(z_n) - y_n, w_n - y_n - \tau_n Ay_n\rangle$$
$$\leq \lambda_n\tau_n L\left\|P_{T_n}(z_n) - y_n\right\|^2 + \lambda_n\tau_n L\|w_n - y_n\|^2.$$

This further yields

$$\|x_{n+1} - p\|^2 \leq \|w_n - p\|^2 - \lambda_n\|y_n - w_n\|^2 - \lambda_n\left\|P_{T_n}(z_n) - y_n\right\|^2$$
$$+ \lambda_n\tau_n L\left\|P_{T_n}(z_n) - y_n\right\|^2 - 2\lambda_n\tau_n\gamma\|y_n - p\|^2 + \lambda_n\tau_n L\|w_n - y_n\|^2$$
$$\leq \|w_n - p\|^2 - \frac{\lambda_n(1 - \tau_n L)}{2}\left\|P_{T_n}(z_n) - w_n\right\|^2.$$

From the restriction that $\tau_n \to 0$ as $n \to \infty$, we find that there exists $n_0 \in \mathbb{N}$ with $\tau_n \leq \frac{1}{2L}$ for all $n \geq n_0$. Hence, $\frac{1}{4} \leq \frac{1 - \tau_n L}{2}$. For all $n \geq n_0$, one has

$$\|x_{n+1} - p\|^2 \leq \|w_n - p\|^2 - \frac{\lambda_n}{4}\left\|P_{T_n}(z_n) - w_n\right\|^2.$$

This further implies

$$\left\|P_{T_n}(z_n) - w_n\right\| = \frac{1}{\lambda_n}\|w_n - x_{n+1}\|.$$

Hence,

$$\|x_{n+1} - p\|^2 \le \|w_n - p\|^2 - \frac{1}{4\lambda_n}\|w_n - x_{n+1}\|^2.$$

One also has

$$\|w_n - p\|^2 = \alpha_n(1 + \alpha_n)\|x_n - x_{n-1}\|^2 + (1 + \alpha_n)\|x_n - p\|^2 - \alpha_n\|x_{n-1} - p\|^2$$

and

$$\|x_{n+1} - w_n\|^2 \ge \left(\alpha_n^2 - \frac{\alpha_n}{\rho_n}\right)\|x_n - x_{n-1}\|^2 + (1 - \alpha_n\rho_n)\|x_n - x_{n+1}\|^2,$$

where $\rho_n = \frac{1}{\delta\lambda_n + \alpha_n}$. This shows that

$$\|x_{n+1} - p\|^2 \le (1 + \alpha_n)\|x_n - p\|^2 - \alpha_n\|x_{n-1} - p\|^2$$
$$- \frac{1}{4\lambda_n}(1 - \alpha_n\rho_n)\|x_n - x_{n+1}\|^2 + \gamma_n\|x_n - x_{n-1}\|^2,$$

where

$$\gamma_n = \alpha_n(\alpha_n + 1) + \frac{1}{4\lambda_n}\alpha_n\left(\frac{1}{\rho_n} - \alpha_n\right) > 0.$$

In view of $\delta = \frac{1 - \rho_n\alpha_n}{\lambda_n\rho_n}$, one has

$$\gamma_n = \frac{1}{4\lambda_n}\alpha_n\left(\frac{1}{\rho_n} - \alpha_n\right) + \alpha_n(\alpha_n + 1) \le \frac{1}{4}\alpha\delta + \alpha(\alpha + 1).$$

Putting $\Gamma_n = \|x_n - p\|^2 + \gamma_n\|x_n - x_{n-1}\|^2 - \alpha_n\|x_{n-1} - p\|^2$, we can obtain

$$\Gamma_{n+1} - \Gamma_n \le \|x_{n+1} - p\|^2 - (\alpha_n + 1)\|x_n - p\|^2$$
$$+ \alpha_n\|x_{n-1} - p\|^2 + \gamma_{n+1}\|x_{n+1} - x_n\|^2 - \gamma_n\|x_n - x_{n-1}\|^2$$
$$\le \left(\gamma_{n+1} + \frac{1}{4\lambda_n}(\alpha_n\rho_n - 1)\right)\|x_{n+1} - x_n\|^2.$$

We can prove that $\gamma_{n+1} + \frac{1}{4\lambda_n}(\alpha_n\rho_n - 1) \le -\sigma$. Since

$$\gamma_{n+1} + \frac{1}{4\lambda_n}(\alpha_n\rho_n - 1) \le -\sigma$$
$$\iff (\alpha_n\rho_n - 1) + 4\lambda_n(\gamma_{n+1} + \sigma) \le 0$$
$$\iff 4(\alpha_n + \delta\lambda_n)(\gamma_{n+1} + \sigma) \le \delta,$$

one has

$$4(\alpha_n + \delta\lambda_n)(\gamma_{n+1} + \sigma) \le 4\left[\sigma + \alpha(\alpha + 1) + \frac{1}{4}\alpha\delta\right](\alpha + \delta\lambda_n) \le \delta.$$

Hence, $\{\Gamma_n\}$ is a nonincreasing sequence. In view of

$$-\alpha\|x_{n-1} - p\|^2 \le \|x_n - p\|^2 - \alpha\|x_{n-1} - p\|^2 \le \Gamma_n \le \Gamma_1,$$

one has

$$\|x_n - p\|^2 \le \alpha^n\|x_0 - p\|^2 + \Gamma_1\sum_{k=1}^{n-1}\alpha^k \le \alpha^n\|x_0 - p\|^2 + \frac{\Gamma_1}{1-\alpha},$$

where $\Gamma_1 = \|x_1 - p\|^2 \ge 0$. In view of the above three inequalities, we obtain that

$$\sigma\sum_{k=1}^{n}\|x_{k+1} - x_k\|^2 \le \Gamma_1 - \Gamma_{n+1}$$

$$\le \Gamma_1 + \alpha\|x_n - p\|^2$$

$$\le \frac{\Gamma_1}{1-\alpha} + \alpha^{n+1}\|x_0 - p\|^2.$$

This further yields that $\lim_{n\to\infty}\|x_n - x_{n+1}\| = 0$. In view of $\alpha_n \le \alpha$, one arrives at

$$\|w_n - x_{n+1}\|^2 = \|x_n - x_{n+1}\|^2 + 2\alpha_n\langle x_n - x_{n+1}, x_n - x_{n-1}\rangle + \alpha_n^2\|x_n - x_{n-1}\|^2 \to 0.$$

As in Alvarez and Attouch [4], one derives that $\lim_{n\to\infty}\|x_n - p\|^2 = l$. Hence, $\lim_{n\to\infty}\|w_n - p\|^2 = l$. This further implies $0 \le \|x_n - w_n\| \le \|x_n - x_{n+1}\| + \|x_{n+1} - w_n\| \to 0$ as $n \to \infty$. It follows that $\lim_{n\to\infty}\|w_n - y_n\| = 0$ and $\lim_{n\to\infty}\|y_n - x_n\| = 0$. Finally, one will show that $\{x_n\}$ converges strongly to p.

Note that

$$\sum_{n=1}^{k}2\lambda_n\tau_n\gamma\|y_n - p\|^2 \le \alpha\|x_k - p\|^2 + \|x_1 - p\|^2 + 2\alpha\sum_{n-1}^{k}\|x_n - x_{n-1}\|^2 \le M,$$

where $M > 0$ is some constant. It is not hard to find that $\liminf_{n\to\infty}\|y_n - p\| = 0$. Since $\{y_n\}$ is bounded, there exists a subsequence $\{y_{n_k}\}$ of $\{y_n\}$ such that $\|y_{n_k} - p\| \to 0$ as $k \to \infty$. So, one also has $\lim_{k\to\infty}\|x_{n_k} - p\| = 0$. Using $\lim_{n\to\infty}\|x_n - p\|^2 = l$, we obtain $\lim_{n\to\infty}\|x_n - p\| = 0$, that is, $x_n \to p$ as $n \to \infty$. This completes the proof. □

3.5 Fixed point theory of pseudocontractive mappings

In this subsection, using the conclusions of the previous sections, we prove some fixed point theorems for hemicontinuous pseudocontractive mappings. In the following, we always assume that C is a nonempty, closed, and convex subset of a Hilbert space H.

Setting $A := I - T$, we have the following result.

Proposition 3.5.1. *Let* $T : C \to H$ *be a mapping. Then the following conclusions hold:*
(1) *T is pseudocontractive if and only if A is monotone;*

(2) *T is η-strongly pseudocontractive if and only if A is a $(1 - \eta)$-strongly monotone.*
(3) *T is k-strictly pseudocontractive if and only if A is k-inverse strongly monotone.*

Proof.

(1)

$$T \text{ is pseudocontractive} \iff \langle Tx - Ty, x - y \rangle \leq \|x - y\|^2$$
$$\iff \|x - y\|^2 - \langle Tx - Ty, x - y \rangle \geq 0$$
$$\iff \langle (I - T)x - (I - T)y, x - y \rangle \geq 0$$
$$\iff \langle Ax - Ay, x - y \rangle \geq 0, \quad \forall x, y \in C.$$

(2)

$$T \text{ is } \eta\text{-strongly pseudocontractive}$$
$$\iff \langle Tx - Ty, x - y \rangle \leq \eta \|x - y\|^2$$
$$\iff \langle Tx - Ty, x - y \rangle \leq \|x - y\|^2 - (1 - \eta)\|x - y\|^2$$
$$\iff \|x - y\|^2 - \langle Tx - Ty, x - y \rangle \geq (1 - \eta)\|x - y\|^2$$
$$\iff \langle (I - T)x - (I - T)y, x - y \rangle \geq (1 - \eta)\|x - y\|^2$$
$$\iff \langle Ax - Ay, x - y \rangle \geq (1 - \eta)\|x - y\|^2, \quad \forall x, y \in C.$$

(3) The claim follows from the definitions of strictly pseudocontractive mappings and inverse strongly monotone mappings. This completes the proof. □

Remark 3.5.1. The class of k-strictly pseudocontractive mappings, which is an important class of mappings between the class of nonexpansive mappings and the class of pseudocontractive mappings, is a subclass of Lipschitz continuous mappings. In general, pseudocontractive and strongly pseudocontractive mappings are not necessarily continuous. The above three types of pseudocontractive mappings are independent of each other.

Example 3.5.1. Let $C = [0, 1]$ and define a mapping $T : C \to \mathbb{R}$ by $Tx = -2x, \forall x \in C$. Then $T : C \to \mathbb{R}$ is $\frac{1}{3}$-strictly pseudocontractive, but not nonexpansive.

Example 3.5.2 (Chidume and Mutangadura [24], 2001). Let $H = \mathbb{R}^2, C = \{x \in \mathbb{R}^2 : \|x\| \leq 1\}, C_1 = \{x \in C : \|x\| \leq \frac{1}{2}\}$, and

$$C_2 = \left\{ x \in C : \frac{1}{2} \leq \|x\| \leq 1 \right\}, \quad \forall x = (a, b) \in \mathbb{R}^2.$$

Set $x^\perp = (b, -a)$ and define a mapping $T : C \to C$ by

$$Tx = \begin{cases} x + x^\perp, & x \in C_1, \\ \frac{x}{\|x\|} - x + x^\perp, & x \in C_2. \end{cases}$$

Then T is 5-Lipschitz pseudocontractive, but not strictly pseudocontractive.

Example 3.5.3. Let $C = (0, \infty)$ and define a mapping $T : C \to C$ by $Tx = \frac{x^2}{1+x}$, $\forall x \in C$. Then T is strictly pseudocontractive, but not strongly pseudocontractive.

Example 3.5.4. Let $H = \mathbb{R}$ and define a mapping $T : H \to H$ by

$$
Tx = \begin{cases}
1, & x \in (-\infty, -1), \\
\sqrt{1 - (x + 1)^2}, & x \in [-1, 0), \\
-\sqrt{1 - (x - 1)^2}, & x \in [0, 1], \\
-1, & x \in (1, \infty).
\end{cases}
$$

Then T is strongly pseudocontractive, but not strictly pseudocontractive.

Based on the results on hemicontinuous monotone mappings, we establish some fixed point theorems for hemicontinuous pseudocontractive mappings. They act as the theoretical basis for iterative methods when finding fixed points of such mappings. To this end, we recall the concepts of weak inward conditions.

Definition 3.5.1.
(1) A mapping $T : C \to H$ is said to satisfy the inward condition (IC) if

$$Tx \in I_C(x), \quad \forall x \in C.$$

(2) A mapping $T : C \to H$ is said to satisfy the weak inward condition (WIC) if

$$Tx \in \overline{I_C(x)}, \quad \forall x \in C,$$

where $I_C(x) = \{x + \lambda(u - x) : \lambda \geq 0, u \in C\}$ and C is called an inward set at x.

It is easy to verify the following basic properties:
(1) $C \subset I_C(x)$, $\forall x \in C$;
(2) if $x \in \text{int}(C) \neq \emptyset$, then $I_C(x) = H$;
(3) if C is a convex set, then $I_C(x)$ is also convex and $I_C(x) = \{x + \lambda(u - x) : \lambda \geq 1, u \in C\}$.

Let $A := I - T$. The following proposition provides a close relation between weak inward condition and flow-invariance condition:

Proposition 3.5.2 (Caristi [17], 1976). *Let C be a nonempty, closed, and convex subset of a Hilbert space H, and let $T : C \to H$ be a mapping. Then T satisfies the weak inward condition if and only if A satisfies the flow-invariance condition (FIC):* $\lim_{h \to 0^+} h^{-1} d((I - hA)x, C) = 0$, $\forall x \in C$.

Proof. ($\Longrightarrow$) Suppose that $x \in C$ and $Tx \in \overline{I_C(x)}$. Then, for any $\varepsilon > 0$, there exists $y \in I_C(x)$ such that $\|y - Tx\| < \varepsilon$. Since $y \in I_C(x)$, we see that there exist $u \in C$ and $\lambda_0 \geq 1$

such that

$$y = x + h^{-1}(u - x), \quad \forall h \in (0, \lambda_0]$$
$$\implies x + h(y - x) = u \in C, \quad \forall h \in (0, \lambda_0]$$
$$\implies h^{-1}d((I - hA)x, C) \leq h^{-1}\|(1 - h)x + hTx - [x + h(y - x)]\| = \|y - Tx\| < \varepsilon,$$

that is, (FIC) holds.

($\Longleftarrow$) For any $x \in C$ and $\varepsilon > 0$, we find $h \in (0, 1)$ and $y \in C$ such that

$$\|x - h(Tx - x) - y\| \leq d(x + h(Tx - x), C) + h\varepsilon$$
$$\implies \|Tx - [(1 - h^{-1})x + h^{-1}y]\| \leq h^{-1}d(x + h(Tx - x), C) + \varepsilon$$
$$\implies Tx \in \overline{I_C(x)}.$$

This completes the proof. □

Theorem 3.5.1. *Let C be a nonempty, convex, and closed subset of a real Hilbert space H. Let $T : C \to H$ be a hemicontinuous pseudocontractive mapping satisfying the weak inward condition (WIC). Then $I - T$ is demiclosed at the origin, that is, if, for any $\{x_n\} \subset C$ with $x_n \rightharpoonup x$, $x_n - Tx \to \theta$ as $n \to \infty$, then $x \in C$ and $x = Tx$.*

Proof. Letting $A := I - T$, we see that $A : C \to H$ is hemicontinuous and monotone. From Proposition 3.5.2, we see that A satisfies the flow-invariance condition (FIC). Thus, by Proposition 3.1.3, we see that A is demiclosed. Hence $x \in C$ and $Ax = \theta \implies x = Tx$. This completes the proof. □

Theorem 3.5.2. *Let C be a nonempty, convex, and closed subset of a real Hilbert space H. Let $T : C \to H$ be a hemicontinuous η-strongly pseudocontractive mapping satisfying the weak inward condition (WIC). Then T has a unique fixed point in C.*

Proof. Letting $A := I - T$, we see that $A : C \to H$ is hemicontinuous and $(1 - \eta)$-strongly monotone. By Proposition 3.5.2, we see that A satisfies the flow-invariance condition (FIC). Thus it follows from Corollary 3.4.2 that there exists a unique $x^* \in C$ such that $Ax^* = \theta$, that is, T has a unique fixed point in C. This completes the proof. □

Theorem 3.5.3. *Let C be a nonempty, convex, and closed subset of a real Hilbert space H. Let $T : C \to H$ be a hemicontinuous pseudocontractive mapping satisfying the weak inward condition (WIC). Then $\mathrm{Fix}(T) \neq \emptyset$.*

Proof. Letting $A := I - T$, we see that $A : C \to H$ is hemicontinuous and monotone. By Proposition 3.5.2, we see that A satisfies the flow-invariance condition (FIC). It follows from Corollary 3.4.3 that there exists a unique $x^* \in C$ such that $Ax^* = \theta$, that is, $x^* \in \mathrm{Fix}(T) \neq \emptyset$. This completes the proof. □

Theorem 3.5.4. *Let C be a nonempty, convex, and closed subset of a real Hilbert space H. Let $T : C \to H$ be a mapping. Let $P_C : H \to C$ be the metric projection from H onto C and let $A := I - T$. Then the following conclusions hold:*

(1) $\text{Fix}(P_C T) = VI(C, A)$;

(2) *if T satisfies the weak inward invariant condition (WIC), then* $\text{Fix}(P_C T) = \text{Fix}(T) = \text{Fix}(TP_C)$.

Proof.

(1)

$$u \in \text{Fix}(P_C T) \Longrightarrow u = P_C Tu$$
$$\Longleftrightarrow \langle Tu - u, y - u \rangle \leq 0, \quad \forall y \in C$$
$$\Longleftrightarrow \langle Au, y - u \rangle \geq 0, \quad \forall y \in C$$
$$\Longleftrightarrow u \in VI(C, A).$$

(2) We first prove $\text{Fix}(P_C T) = \text{Fix}(T)$. Letting $x = P_C Tx$, we find from (1) that

$$x \in VI(C, A) \implies \langle Ax, y - x \rangle \geq 0, \quad \forall y \in C. \tag{3.95}$$

Since $Tx \in \overline{I_C(x)}$, there exists $\{y_n\} \subset I_C(x)$ such that $y_n \to Tx$ as $n \to \infty$. From the definition of $I_C(x)$, we see that there exist $u_n \in C$ and $\lambda_n \geq 1$ such that $y_n = x + \lambda_n(u_n - x)$ for all $n \geq 1$. Taking $y := u_n \in C$ in (3.95), we have

$$\langle Ax, u_n - x \rangle \geq 0, \quad \forall n \geq 1$$
$$\implies \langle Ax, \lambda_n(u_n - x) \rangle \geq 0, \quad \forall n \geq 1$$
$$\implies \langle Ax, y_n - x \rangle \geq 0, \quad \forall n \geq 1. \tag{3.96}$$

Letting $n \to \infty$ in (3.96), we obtain

$$\langle Ax, Tx - x \rangle \geq 0 \implies -\|Ax\|^2 \geq 0 \implies Ax = \theta,$$

that is, $x = Tx$. Hence $\text{Fix}(P_C T) \subseteq \text{Fix}(TP_C)$. It is obvious to see that the inverse is also true. Therefore, we have $\text{Fix}(P_C T) = \text{Fix}(T)$.

Next, we prove $\text{Fix}(T) = \text{Fix}(TP_C)$. Letting $x = TP_C x$, we have $P_C x = (P_C T)P_C x$, which shows that $P_C x \in \text{Fix}(P_C T)$. It follows that

$$TP_C x = P_C x \implies P_C x = x \implies x = Tx.$$

This completes the proof. $\qquad\qquad\qquad\qquad\qquad\qquad\qquad\qquad\qquad\quad\square$

Corollary 3.5.1. *Let C be a nonempty, convex, and closed subset of a real Hilbert space H. If $T : C \to H$ is a hemicontinuous pseudocontractive mapping, then $\text{Fix}(P_C T)$ is closed and convex in H. Furthermore, if $T : C \to H$ satisfies the weak inward condition (WIC), then $\text{Fix}(T)$ is closed and convex in H.*

Theorem 3.5.5. *Let C be a nonempty, convex, and closed subset of a real Hilbert space H. Let $T : C \to H$ be a hemicontinuous pseudocontractive mapping satisfying the weak* inward condition (WIC) and $\text{Fix}(T) \neq \emptyset$. Then, for any $u \in C$, there exists a continuous path $\{x_t\}_{t>0}$ such that $x_t \to z$ as $t \to 0^+$, where $z = P_{\text{Fix}(T)}u$ and $\{x_t\}$ satisfies the following equation:

$$tu + (1-t)Tx_t = x_t, \quad \forall t \in (0,1). \tag{3.97}$$

Proof. For any $t \in (0,1)$ and $u \in C$, we define a mapping $T_t : C \to H$ by

$$T_t(x) = tu + (1-t)Tx, \quad \forall x \in C. \tag{3.98}$$

Then $T_t : C \to H$ is hemicontinuous and $(1-t)$-strongly pseudocontractive, and satisfies the weak inward condition (WIC). So, we conclude from Theorem 3.5.2 that there exists a unique $x_t \in C$ such that

$$x_t = T_t x_t = tu + (1-t)Tx_t, \quad \forall t \in (0,1).$$

Hence (3.97) holds.

Next, we prove that the net $\{x_t\}_{t>0}$ is continuous. To this end, we see from (3.97) that

$$\begin{aligned}
\|x_t - x_{t_0}\|^2 &= \langle x_t - x_{t_0}, x_t - x_{t_0} \rangle \\
&= \langle tu + (1-t)Tx_t - t_0 u - (1-t_0)Tx_{t_0}, x_t - x_{t_0} \rangle \\
&= (t - t_0)\langle u - Tx_{t_0}, x_t - x_{t_0} \rangle + (1-t)\langle Tx_t - Tx_{t_0}, x_t - x_{t_0} \rangle \\
&\le |t - t_0|\,\|u - Tx_{t_0}\|\,\|x_t - x_{t_0}\| + (1-t)\|x_t - x_{t_0}\|^2, \quad \forall t \in (0,1).
\end{aligned}$$

This implies that

$$\|x_t - x_{t_0}\| \le \frac{|t - t_0|}{t}\|u - Tx_{t_0}\|.$$

If $t \to t_0$, then $x_t \to x_{t_0}$, which verifies the continuity of the path $\{x_t\}_{t>0}$.

Next, we prove the convergence of $\{x_t\}$. For any $p \in \text{Fix}(T)$, it follows from (3.97) and the definition of pseudocontractive mappings that

$$\begin{aligned}
\|x_t - p\|^2 &= \langle x_t - p, x_t - p \rangle \\
&= \langle tu + (1-t)Tx_t - p, x_t - p \rangle \\
&= t\langle u - p, x_t - p \rangle + (1-t)\langle Tx_t - p, x_t - p \rangle \\
&\le t\langle u - p, x_t - p \rangle + (1-t)\|x_t - p\|^2 \\
&\Rightarrow \|x_t - p\|^2 \le \langle u - p, x_t - p \rangle, \quad \forall t \in (0,1),\ p \in \text{Fix}(T). \tag{3.99}
\end{aligned}$$

In particular, one has $\|x_t - p\| \le \|u - p\|$, $\forall t \in (0,1)$, which shows that the net $\{x_t\}_{t>0}$ is bounded. From (3.97), we see that $\{Tx_t\}$ is also bounded. Hence

$$x_t - Tx_t = t(u - Tx_t) \to \theta \quad (t \to 0). \tag{3.100}$$

It follows from (3.99) that

$$\langle x_t - p - u + p, x_t - p \rangle \le 0$$
$$\Rightarrow \langle x_t - u, x_t - p \rangle \le 0, \quad \forall t \in (0,1), \, p \in \text{Fix}(T). \tag{3.101}$$

Take a sequence $\{t_n\}$ with $t_n \to \infty$ arbitrarily and put $x_n := x_{t_n}$. Since H is reflexive and sequence $\{x_n\}$ is bounded in H, we see that there exists a subsequence $\{x_{n_i}\} \subset \{x_n\}$ such that $x_{n_k} \to z$ as $k \to \infty$. It follows from (3.100) that $x_{n_k} - Tx_{n_k} \to \theta$ as $k \to \infty$. From Theorem 3.5.1, we see that $z \in C$ and $z = Tz$. Using (3.99), we derive

$$\|x_{n_k} - z\|^2 \le \langle u - z, x_{n_k} - z \rangle. \tag{3.102}$$

Letting $k \to \infty$ in (3.102), we have $x_{n_k} \to z$ as $k \to \infty$. It follows from (3.101) that

$$\langle x_{n_k} - u, x_{n_k} - p \rangle \le 0, \quad \forall p \in \text{Fix}(T). \tag{3.103}$$

Letting $k \to \infty$ in (3.103), we have

$$\langle z - u, z - p \rangle \le 0, \quad \forall p \in \text{Fix}(T),$$

which shows that $z = P_{\text{Fix}(T)}u$. Hence $x_n \to z$ as $n \to \infty$. Therefore, it follows that $x_t \to z$ as $t \to 0$. This completes the proof. $\square$

3.6 Iterative methods of fixed points for pseudocontractive mappings and zeros for monotone mappings

This section is devoted a discussion of the iterative methods of fixed points of pseudo-contractive mappings and zeros of monotone mappings, including the normal Mann's iterative method, Ishikawa's iterative method, Bruck regularization iterative method, hybrid projection iterative methods, and Ishikawa–Halpern-type iterative methods. By the transformation $A := I - T$ or $T := I - A$, every convergence result of fixed points for pseudocontractive mappings implies a convergence result of zeros for monotone mappings.

3.6.1 Normal Mann iterative method

For η-strongly pseudocontractive mappings and k-strictly pseudocontractive mappings defined on a Hilbert space H, we can find their fixed points via the normal Mann iterative method.

Theorem 3.6.1. *Let C be a nonempty, bounded, closed, and convex subset of a real Hilbert space H, and let $T : C \to C$ be a hemicontinuous η-strongly pseudocontractive mapping. Let $\{\alpha_n\}$ be a sequence in $[0,1]$ satisfying the following conditions: $\alpha_n \to 0$*

as $n \to \infty$ and $\sum_{n=1}^{\infty} \alpha_n = \infty$. Let $\{x_n\}$ be a sequence generated in the following normal Mann iterative method (NMIM): $x_1 \in C$ and

$$x_{n+1} = (1 - \alpha_n)x_n + \alpha_n T x_n, \quad \forall n \geq 1.$$

Then $\{x_n\}$ converges strongly to the unique fixed point of T.

Proof. From Theorem 3.5.2, we see that T has a unique fixed point in C; denote it by q. Note that

$$
\begin{aligned}
\|x_{n+1} - q\|^2 &= \left\| (1 - \alpha_n)(x_n - q) + \alpha_n(Tx_n - q) \right\|^2 \\
&= (1 - \alpha_n)^2 \|x_n - q\|^2 + 2\alpha_n(1 - \alpha_n)\langle Tx_n - q, x_n - q \rangle + \alpha_n^2 \|Tx_n - q\|^2 \\
&\leq (1 - \alpha_n)^2 \|x_n - q\|^2 + 2\eta\alpha_n(1 - \alpha_n)\|x_n - q\|^2 + \alpha_n^2 M^2 \\
&\leq \left[1 - (1 - \eta)\alpha_n \right] \|x_n - q\|^2 + \alpha_n^2 M^2,
\end{aligned}
$$

where $M = \mathrm{diam}(C)$. By Lemma 1.10.2, we have $x_n \to q$ as $n \to \infty$. This completes the proof. $\qquad\square$

Corollary 3.6.1. *Let C be a nonempty, bounded, closed, and convex subset of a real Hilbert space H, and let $A : C \to H$ be a hemicontinuous σ-strongly monotone mapping satisfying the condition that $T := I - A : C \to C$ is a self-mapping. Let $\{t_n\}$ be a sequence in $(0, 1)$ satisfying the following conditions: $t_n \to \infty$ as $n \to \infty$ and $\sum_{n=1}^{\infty} t_n = \infty$. Let $\{x_n\}$ be a sequence generated in the steepest descent method (SDM): $x_1 \in C$ and*

$$x_{n+1} = x_n - t_n A x_n, \quad \forall n \geq 1.$$

Then $\{x_n\}$ converges strongly to the unique zero point x^* of A.

Proof. Since $T : C \to C$ is hemicontinuous and $(1 - \sigma)$-strongly pseudocontractive, one sees that T has a unique fixed point x^*. So, A has a unique zero point x^*. One can rewrite (SDM) as

$$x_{n+1} = x_n - t_n(I - T)x_n = (1 - t_n)x_n + t_n T x_n, \quad \forall n \geq 1.$$

From Theorem 3.6.1, we can derive the conclusion immediately. This completes the proof. $\qquad\square$

Theorem 3.6.2. *Let C be a nonempty, closed, and convex subset of a real Hilbert space H, and let $T : C \to C$ be a k-strictly pseudocontractive mapping with $\mathrm{Fix}(T) \neq \emptyset$. Let $\{\alpha_n\}$ be a sequence satisfying the following conditions: $\alpha_n \in (k, 1)$ and $\sum_{n=1}^{\infty}(\alpha_n - k)(1 - \alpha_n) = \infty$. Let $\{x_n\}$ be a sequence generated in the normal Mann iterative method (NMIM): $x_1 \in C$ and $x_{n+1} = (1 - \alpha_n)x_n + \alpha_n T x_n$, $\forall n \geq 1$. Then $\{x_n\}$ converges weakly to a fixed point $p = \lim_{n \to \infty} P_{\mathrm{Fix}(T)}(x_n)$ of T.*

Proof. Letting $\beta_n = \frac{\alpha_n - k}{1-k}$, we have $\alpha_n = \beta_n + k(1 - \beta_n)$ and

$$1 - \alpha_n = (1 - k)(1 - \beta_n).$$

It follows that

$$
\begin{aligned}
x_{n+1} &= \alpha_n x_n + (1 - \alpha_n) T x_n \\
&= [\beta_n + k(1 - \beta_n)] x_n + [(1 - k)(1 - \beta_n)] T x_n \\
&= \beta_n x_n + (1 - \beta_n)[k x_n + (1 - k) T x_n], \quad \forall n \geq 1.
\end{aligned}
\tag{3.104}
$$

Define a mapping $S : C \to C$ by $Sx = kx + (1-k)Tx$, $\forall x \in C$. Then (3.104) can be rewritten as

$$x_{n+1} = \beta_n x_n + (1 - \beta_n) S x_n, \quad \forall n \geq 1. \tag{3.105}$$

We know that $S : C \to C$ is a nonexpansive mapping. Indeed,

$$
\begin{aligned}
\|Sx - Sy\|^2 &= \|k(x - y) + (1 - k)(Tx - Ty)\|^2 \\
&= k\|x - y\|^2 + (1 - k)\|Tx - Ty\|^2 - k(1 - k)\|(I - T)x - (I - T)y\|^2 \\
&\leq k\|x - y\|^2 + (1 - k)\|x - y\|^2 + k(1 - k)\|(I - T)x - (I - T)y\|^2 \\
&\quad - k(1 - k)\|(I - T)x - (I - T)y\|^2 \\
&= \|x - y\|^2.
\end{aligned}
$$

It is easy to check that Fix(S) = Fix(T). Hence,

$$\sum_{n=1}^{\infty} \beta_n (1 - \beta_n) = \frac{1}{(1 - k)^2} \sum_{n=1}^{\infty} (\alpha_n - k)(1 - \alpha_n) = \infty.$$

From Reich's weak convergence theorem, we see that $\{x_n\}$ weakly converges to a fixed point p of S, where $p = \lim_{n \to \infty} P_{\text{Fix}(T)}(x_n)$. Furthermore, Fix($S$) = Fix($T$). Hence $\{x_n\}$ weakly converges to a fixed point p of T. This completes the proof. □

Theorem 3.6.3. *Let C be a nonempty, closed, and convex subset of a real Hilbert space H. Let $A : C \to H$ be a v-inverse strongly monotone mapping such that $A^{-1}(\theta) \neq \emptyset$ and $T := I - A : C \to C$ is a self-mapping. Let $\{t_n\}$ be a sequence in $(0, 1)$ satisfying the following conditions: $t_n \in (k, 1)$ and $\sum_{n=1}^{\infty}(t_n - k)(1 - t_n) = \infty$, where $k = 1 - 2v$. Let $\{x_n\}$ be a sequence generated in the steepest descent method (SDM): $x_1 \in C$ and $x_{n+1} = x_n - t_n A x_n$, $\forall n \geq 1$. Then $\{x_n\}$ converges weakly to a zero point x^* of A, where*

$$x^* = \lim_{n \to \infty} P_{A^{-1}(\theta)}(x_n).$$

Proof. It follows from Proposition 3.5.2 (3) that $T : C \to C$ is hemicontinuous and k-strictly pseudocontractive, where $k = 1 - 2v$, and Fix(T) $\neq \emptyset$. Note that the (SDM) method can be rewritten as $x_{n+1} = (1 - t_n)x_n + t_n T x_n$, $\forall n \geq 1$. We can obtain from Theorem 3.6.2 the desired conclusion immediately. This completes the proof. □

Remark 3.6.1. The above theorems show that, in the framework of Hilbert spaces, we can transform k-strictly pseudocontractive mappings into nonexpansive mappings via an appropriate convex combination of the identity mapping and a nonexpansive mapping.

Since the normal Mann iterative method is valid for nonexpansive mappings, we see that it is also valid for k-strictly pseudocontractive mappings. However, for more general pseudocontractive mappings, even Lipschitz continuous, the normal Mann iterative method is not valid. In 2001, Chidume and Mutangadura gave a counterexample (see [24]).

3.6.2 Ishikawa iterative method

In 1974, Ishkikawa introduced a two-step iterative method for fixed points of Lipschitz pseudocontractive mappings.

Lemma 3.6.1 (Ishikawa [32], 1974). *Let C be a convex subset of a real Hilbert space H and let $T : C \to C$ be an L-Lipschitz pseudocontractive mapping with $\mathrm{Fix}(T) \neq \emptyset$. Let $\{\alpha_n\}$ and $\{\beta_n\}$ be two sequences in $[0,1]$ satisfying the following condition: $\alpha_n \leq \beta_n$, $\forall n \geq 1$. Let $\{x_n\}$ be a sequence generated in the following Ishikawa iterative method (IIM): $x_1 \in C$ and*

$$\begin{cases} y_n = (1 - \beta_n)x_n + \beta_n Tx_n, \\ x_{n+1} = (1 - \alpha_n)x_n + \alpha_n Ty_n, \quad \forall n \geq 1. \end{cases}$$

Then,

$$\|x_{n+1} - p\|^2 \leq \|x_n - p\|^2 - \alpha_n\beta_n(1 - 2\beta_n - L^2\beta_n^2)\|x_n - Tx_n\|^2, \quad \forall p \in \mathrm{Fix}(T). \tag{3.106}$$

Proof. Fixing $p \in \mathrm{Fix}(T)$, we have

$$\|x_{n+1} - p\|^2 = \left\|(1 - \alpha_n)(x_n - p) + \alpha_n(Ty_n - p)\right\|^2$$
$$= (1 - \alpha_n)\|x_n - p\|^2 + \alpha_n\|Ty_n - p\|^2 - \alpha_n(1 - \alpha_n)\|x_n - Ty_n\|^2, \tag{3.107}$$
$$\|Ty_n - p\|^2 \leq \|y_n - p\|^2 + \|y_n - Ty_n\|^2$$
$$= \beta_n\|Ty_n - p\|^2 + (1 - \beta_n)\|x_n - p\|^2 - \beta_n(1 - \beta_n)\|x_n - Tx_n\|^2,$$
$$+ \|y_n - Ty_n\|^2, \tag{3.108}$$
$$\|y_n - Ty_n\|^2 = \beta_n\|Tx_n - Ty_n\|^2 + (1 - \beta_n)\|x_n - Ty_n\|^2 - \beta_n(1 - \beta_n)\|x_n - Tx_n\|^2$$
$$\leq [L^2\beta_n^2 - \beta_n(1 - \beta_n)]\|x_n - Tx_n\|^2 + (1 - \beta_n)\|x_n - Tx_n\|^2, \tag{3.109}$$

and

$$\|Tx_n - p\|^2 \leq \|x_n - p\|^2 + \|x_n - Tx_n\|^2. \tag{3.110}$$

Substituting (3.108)–(3.110) into (3.107), we find from condition $\alpha_n \leq \beta_n$ that

$$\|x_{n+1} - p\|^2 \leq \|x_n - p\|^2 - \alpha_n\beta_n(1 - 2\beta_n - L^2\beta_n^2)\|x_n - Tx_n\|^2.$$

This completes the proof. □

Theorem 3.6.4 (Zhou [116], 2008). *Let C be a closed convex subset of a real Hilbert space H and let $T : C \to C$ be an L-Lipschitz pseudocontractive mapping with $\mathrm{Fix}(T) \neq \emptyset$. Let $\{\alpha_n\}$ and $\{\beta_n\}$ be two real sequence satisfying the following conditions:*
(1) $\liminf_{n \to \infty} \alpha_n > 0$;
(2) $\alpha_n \leq \beta_n, \forall n \geq 1$;
(3) $\beta_n \leq \beta < \frac{\sqrt{1+L^2}-1}{L^2}$ *for all $n \geq 1$.*

Let $\{x_n\}$ be a sequence generated by (IIM) in Lemma 3.6.1. Then $x_n \to p$ as $n \to \infty$, where

$$p = \lim_{n \to \infty} P_{\mathrm{Fix}(T)}(x_n).$$

Proof. From Lemma 3.6.1, we see that, for all $p \in \mathrm{Fix}(T)$,

$$\|x_{n+1} - p\|^2 \leq \|x_n - p\|^2 - \alpha_n\beta_n(1 - 2\beta_n - L^2\beta_n^2)\|x_n - Tx_n\|^2. \tag{3.111}$$

Conditions (1)–(3) guarantee the existence of constant $c > 0$ such that

$$\alpha_n\beta_n(1 - 2\beta_n - L^2\beta_n^2) \geq c.$$

It follows from (3.111) that

$$c\|x_n - Tx_n\|^2 \leq \|x_n - p\|^2 - \|x_{n+1} - p\|^2, \forall p \in \mathrm{Fix}(T),$$
$$\implies \|x_{n+1} - p\| \leq \|x_n - p\|, \quad \forall p \in \mathrm{Fix}(T). \tag{3.112}$$

This shows that $\{x_n\}$ is Fejér monotone. Hence $\lim_{n \to \infty} \|x_n - p\|$ exists. This implies that $\{x_n\}$ is bounded. Letting $n \to \infty$ in (3.111), we obtain $x_n - Tx_n \to \theta$. By Theorem 3.5.1, we have $\omega_\omega(x_n) \subset \mathrm{Fix}(T)$. It follows from Browder's convergence principle that $x_n \to p$ as $n \to \infty$, where $p = \lim_{n \to \infty} P_{\mathrm{Fix}(T)}(x_n)$. This completes the proof. □

Corollary 3.6.2. *Let C be a closed convex subset of a real Hilbert space H. Let $A : C \to H$ be an L-Lipschitz monotone mapping such that $A^{-1}(\theta) \neq \emptyset$ and $T := I - A$ is a self-mapping on C. Let $\{x_n\}$ be a sequence generated in the following generalized steepest descent method (GSDM): $x_1 \in C$ and*

$$\begin{cases} y_n = x_n - \beta_n Ax_n, \\ x_{n+1} = x_n - \alpha_n(x_n - Ty_n), \quad \forall n \geq 1, \end{cases}$$

where $\{\alpha_n\}$ and $\{\beta_n\}$ are two sequences in $(0,1)$ satisfying the following conditions:
(1) $\liminf_{n \to \infty} \alpha_n > 0$;

(2) $\alpha_n \leq \beta_n$ for all $n \geq 1$;

(3) $\beta_n \leq \beta < \frac{\sqrt{1+(1+L)^2}-1}{(1+L)^2}$ for all $n \geq 1$.

Then $\{x_n\}$ weakly converges to a zero point x^* of A, where $x^* = \lim_{n \to \infty} P_{A^{-1}(\theta)} x_n$.

Proof. Note that $T : C \to C$ is an $(L + 1)$-Lipschitz pseudocontractive mapping with $\text{Fix}(T) = A^{-1}(\theta) \neq \emptyset$ and the (GSMD) can be rewritten as the (IIM) in Lemma 3.6.1. From Theorem 3.6.4, we can obtain the desired conclusion immediately. This completes the proof. $\qquad\square$

Theorem 3.6.5 (Ishikawa [32], 1974). *Let C be a compact convex subset of a real Hilbert space H and let $T : C \to C$ be a Lipschitz pseudocontractive mapping. Let $\{\alpha_n\}$ and $\{\beta_n\}$ be two sequences in $(0, 1)$ satisfying the following conditions:*

(1) $\alpha_n \leq \beta_n$ *for all* $n \geq 1$;

(2) $\beta_n \to 0$ *as* $n \to \infty$;

(3) $\sum_{n=1}^{\infty} \alpha_n \beta_n = \infty$.

Let $\{x_n\}$ be a sequence generated by the (IIM) in Lemma 3.6.1. Then $\{x_n\}$ converges strongly to a fixed point of T.

Proof. From Schauder's fixed point theorem, we see that $\text{Fix}(T) \neq \emptyset$. Using (3.111) and conditions (2) and (3), we derive that

$$\sum_{n=1}^{\infty} \alpha_n \beta_n (1 - 2\beta_n - L^2 \beta_n^2) \|x_n - Tx_n\|^2 < \infty$$

$$\implies \liminf_{n \to \infty} \|x_n - Tx_n\| = 0$$

$$\implies \exists \{x_{n_j}\} \text{ such that } x_{n_j} \to z \quad (j \to \infty)$$

$$\implies Tx_{n_j} \to Tz \quad (j \to \infty)$$

$$\implies z = Tz.$$

This shows that $z \in \text{Fix}(T)$ and, for all $p \in \text{Fix}(T)$, the limit $\lim_{n \to \infty} \|x_n - p\|$ exists. Hence $x_n \to z$ as $n \to \infty$. This completes the proof. $\qquad\square$

Corollary 3.6.3. *Let C be a compact convex subset of a real Hilbert space. Let $A : C \to H$ be an L-Lipschitz monotone mapping such that $T := I - A : C \to C$ is a self-mapping on C. Let $\{x_n\}$ be a sequence generated by the algorithm given in Corollary 3.6.2, where $\{\alpha_n\}$ and $\{\beta_n\}$ are two sequences satisfying conditions (1)–(3) in Theorem 3.6.5. Then $\{x_n\}$ converges strongly to a zero point of A.*

Let C be a nonempty closed convex subset of a real Hilbert space H and $T : C \to C$ a mapping. For a fixed anchor $u \in C$ and arbitrary initial value $x_1 \in C$, define a sequence

$\{x_n\}$ iteratively in C by the following manner:

$$\begin{cases} y_n = (1 - \beta_n)x_n + \beta_n Tx_n, \\ z_n = (1 - \alpha_n)x_n + \alpha_n Ty_n, \\ x_{n+1} = \lambda_n u + (1 - \lambda_n)z_n, \quad n \geq 1, \end{cases} \tag{HIIM}$$

where $\{\alpha_n\}$, $\{\beta_n\}$, and $\{\lambda_n\}$ are three real sequences in $(0, 1)$ satisfying certain restrictions. We call above iterative method Halpern–Ishikawa iterative method. Next we study strong convergence of the Halpern–Ishikawa iterative method for Lipschitz continuous pseudocontractive mappings.

By means of the methods and techniques used in Theorem 3.6.4, we can prove the following strong convergence results. Because the proof lines are similar, we omit the proofs.

Theorem 3.6.6. *Let C be a closed convex subset of a real Hilbert space H and let $T : C \to C$ be an L-Lipschitz pseudocontractive mapping with $\mathrm{Fix}(T) \neq \emptyset$. Let $\{\alpha_n\}$, $\{\beta_n\}$, and $\{\lambda_n\}$ be three real sequence in $(0, 1)$ satisfying the following conditions:*
(1) $\liminf_{n\to\infty} \alpha_n > 0$;
(2) $\alpha_n \leq \beta_n, \forall n \geq 1$;
(3) $\beta_n \leq \beta < \frac{\sqrt{1+L^2}-1}{L^2}$ *for all $n \geq 1$;*
(4) $\lambda_n \to 0$ *as $n \to \infty$;*
(5) $\sum_{n=1}^{\infty} \lambda_n = \infty$.

Let $\{x_n\}$ be a sequence generated by (HIIM). Then $x_n \to p$ as $n \to \infty$, where

$$p = P_{\mathrm{Fix}(T)}u.$$

Let C be a nonempty closed convex subset of a real Hilbert space H. Let $T : C \to C$ be a mapping and $f : C \to C$ a p-strict contraction. Consider

$$\begin{cases} x_1 \in C \text{ chosen arbitrarily,} \\ y_n = (1 - \beta_n)x_n + \beta_n Tx_n, \\ z_n = (1 - \alpha_n)x_n + \alpha_n Ty_n, \\ x_{n+1} = \lambda_n f(x_n) + (1 - \lambda_n)z_n, \quad n \geq 1, \end{cases} \tag{MHIIM}$$

where $\{\alpha_n\}$, $\{\beta_n\}$, and $\{\lambda_n\}$ are three real sequences in $(0, 1)$ satisfying certain restrictions. We call above iterative method Moudafi–Halpern–Ishikawa iterative method.

Theorem 3.6.7. *Let C be a closed convex subset of a real Hilbert space H and let $T : C \to C$ be an L-Lipschitz pseudocontractive mapping with $\mathrm{Fix}(T) \neq \emptyset$. Let $f : C \to C$ be a p-strict contraction. Let $\{\alpha_n\}$, $\{\beta_n\}$, and $\{\lambda_n\}$ be three real sequence in $(0, 1)$ satisfying the following conditions:*
(1) $\liminf_{n\to\infty} \alpha_n > 0$;

(2) $\alpha_n \le \beta_n, \forall n \ge 1$;

(3) $\beta_n \le \beta < \frac{\sqrt{1+L^2}-1}{L^2}$ for all $n \ge 1$;

(4) $\lambda_n \to 0$ as $n \to \infty$;

(5) $\sum_{n=1}^{\infty} \lambda_n = \infty$.

Let $\{x_n\}$ be a sequence generated by (MHIIM). Then $x_n \to p$ as $n \to \infty$, where

$$p = P_{\mathrm{Fix}(T)} f(p),$$

and p is the unique solution of the variational inequality:

$$\langle (I - f)p, y - p \rangle \ge 0, \quad \forall y \in \mathrm{Fix}(T).$$

As a consequence of the above theorem, we have the following interesting result.

Theorem 3.6.8. *Let H be a real Hilbert space and let $T : H \to H$ be an L-Lipschitz pseudocontractive mapping with $\mathrm{Fix}(T) \ne \emptyset$. Let $A : H \to H$ be a k-Lipschitz continuous and η-strongly monotone operator. Let $\{\alpha_n\}$, $\{\beta_n\}$, and $\{\lambda_n\}$ be three real sequence in $(0, 1)$ satisfying the following conditions:*

(1) $\liminf_{n\to\infty} \alpha_n > 0$;

(2) $\alpha_n \le \beta_n, \forall n \ge 1$;

(3) $\beta_n \le \beta < \frac{\sqrt{1+L^2}-1}{L^2}$ for all $n \ge 1$;

(4) $\lambda_n \to 0$ as $n \to \infty$;

(5) $\sum_{n=1}^{\infty} \lambda_n = \infty$.

For $\mu \in (0, \frac{2\eta}{k^2})$ and initial value $x_1 \in H$ chosen arbitrarily, define a sequence $\{x_n\}$ iteratively in H by (MHIIM) with $f(x) = (I - \mu A)x$. Then $x_n \to x^*$ as $n \to \infty$, where

$$x^* = P_{\mathrm{Fix}(T)} f(x^*),$$

and x^* is the unique solution of the monotone variational inequality:

$$\langle Ax^*, x - x^* \rangle \ge 0, \quad \forall x \in \mathrm{Fix}(T).$$

3.6.3 Bruck regularization iterative method

Note that there is a strong restriction that the domain of the mapping is compact in Ishikawa convergence theorem (Theorem 3.6.5). Is there an iterative method that does not require the compactness so that the sequence generated by the Ishikawa iteration method is still convergent? The answer is positive. In 1974, Bruck [11] introduced the following regularization iteration method.

Let $A \subset H \times H$ be a mapping. Let x_1 be an arbitrary initial point in $\text{Dom}(A)$ and let z be a fixed element in H. Let $\{x_n\}$ be a sequence generated by

$$x_{n+1} = x_n - \lambda_n(v_n + \theta_n(x_n - z)), \quad n \geq 1, v_n \in Ax_n, \tag{BRIM-1}$$

where $\{\lambda_n\}$ and $\{\theta_n\}$ are two nonnegative real number sequences satisfying certain conditions. Then (BRIM-1) is called the Bruck regularization iterative method.

Let C be a nonempty, closed, and convex subset of a Hilbert space H. Let $T : C \rightarrow C$ be a mapping. Let x_1 be an arbitrary initial point in C and let z be a fixed element in H. Let $\{x_n\}$ be a sequence generated by

$$x_{n+1} = [1 - \lambda_n(1 + \theta_n)]x_n + \lambda_n Tx_n + \lambda_n \theta_n z, \quad n \geq 1, \tag{BRIM-2}$$

where $\{\lambda_n\}$ and $\{\theta_n\}$ are two nonnegative real number sequences satisfying some certain conditions.

Bruck regularization iterative methods are motivated at least by the following two aspects. One is the steepest descent method, which is for zeros of η-strongly monotone mappings, and the other is the Halpern iterative method, which is for fixed points of nonexpansive mappings.

For η-strongly monotone mappings, steepest descent methods are effective, however, they are not available for general monotone mappings. Thanks to the fact that the perturbations $A + \theta_n I$ of A are η_n-strongly monotone, it is natural to replace A in the steepest descent method with $A + \theta_n I$. This leads to (BRIM-1).

On the other hand, for a maximal monotone mapping $A \subset H \times H$ if $\theta \in \text{Ran}(A)$, then the resolvent J_t of A has the following asymptotic property: $\lim_{t\rightarrow\infty} J_t x = P_{A^{-1}(\theta)}x$, $\forall x \in H$. Letting $t_n \rightarrow \infty$ and $\theta_n = t_n^{-1} \rightarrow 0$, we find that there exists a unique $y_n \in \text{Dom}(A)$ such that

$$\theta \in \theta_n y_n + Ay_n, \quad n \geq 1 \tag{3.113}$$

and $y_n \rightarrow x^* = P_{A^{-1}(\theta)}\theta \ (n \rightarrow \infty)$.

If we can find a subsequence $\{x_{n_j}\}$ of $\{x_n\}$ such that $x_{n_j} \rightarrow x^*$ and prove that $x_n - y_{n_j} \rightarrow \theta \ (j \rightarrow \infty)$, then we can prove $x_n \rightarrow x^* \ (n \rightarrow \infty)$. Therefore, it is essential to choose $\{\lambda_n\}$ and $\{\theta_n\}$ to guarantee that sequence $\{x_n\}$ has a subsequence $\{x_{n_j}\}$ which strongly converges to x^*. To this end, Bruck [11] introduced the definition of admissible pairs.

Definition 3.6.1. A pair of nonnegative sequences $\{\lambda_n\}$ and $\{\theta_n\}$ is said to be an admissible pair if the following conditions hold:
(1) $\theta_n > 0$ and $\{\theta_n\}$ monotonically and decreasingly converges to 0;
(2) There exists a strictly increasing subsequence $\{n(i)\}_{i=1}^{\infty}$ satisfying:
$$\liminf_{i\rightarrow\infty} \theta_{n(i)} \sum_{j=n(i)}^{n(i+1)} \lambda_j > 0, \quad \lim_{i\rightarrow\infty}[\theta_{n(i)} - \theta_{n(i+1)}]\sum_{j=n(i)}^{n(i+1)} \lambda_j = 0, \text{ and}$$
$$\lim_{i\rightarrow0} \sum_{j=n(i)}^{n(i+1)} \lambda_j^2 = 0.$$

Examples of admissible pairs are as follows:

(1) $\lambda_n = \frac{1}{n}, \theta_n = \frac{1}{\log\log n}, n \geq 2, n(i) = i^i$;

(2) $\lambda_n = n^{-\frac{1}{2}}, \theta_n = n^{-\frac{1}{4}}, n \geq 1, n(i) = i^6$.

Bruck [11] first established the convergence of method (BRIM-1). As a direct consequence, the convergence of method (BRIM-2) can be derived.

Next, we first prove the convergence of method (BRIM-2). As a corollary, the convergence of method (BRIM-1) can be easily obtained.

Theorem 3.6.9. *Let C be a nonempty, bounded, closed, and convex subset of a Hilbert space H. Let $T : C \to C$ be a single-valued hemicontinuous pseudocontractive mapping. Then T has at least one fixed point in C. Let $\{\lambda_n\}$ and $\{\theta_n\}$ be a admissible pair and $\lambda_n(1 + \theta_n) \leq 1, \forall n \in N$. Let $\{x_n\}$ be a sequence generated by (BRIM-2). Then $\{x_n\}$ converges strongly to $p = P_{\text{Fix}(T)}z$.*

Proof. From Theorem 3.5.3, we see that $\text{Fix}(T) \neq \emptyset$. It follows from Corollary 3.5.1 that $\text{Fix}(t)$ is a closed and convex subset of H. Hence $P_{\text{Fix}(T)}z$ is well defined. Let $t_i = \frac{\theta_i}{1+\theta_i} \to 0$ as $i \to \infty$. Using Theorem 3.5.5, we see that there exists a unique $y_i \in C$ such that

$$y_i = (1 - t_i)Ty_i + t_i z, \quad i \geq 1, \tag{3.114}$$

and $y_i \to x^* = P_{\text{Fix}(T)}z$ as $i \to \infty$. Letting $A := I - T$, we find from (3.114) that

$$\theta_i(z - y_i) = Ay_i, \quad i \geq 1. \tag{3.115}$$

Note that (BRIM-2) can be rewritten as (BRIM-1):

$$x_{n+1} = x_n - \lambda_n[Ax_n + \theta_n(x_n - z)], \quad i \geq 1. \tag{3.116}$$

For $n \geq i \geq 2$, one has

$$x_n - y_i = x_{n-1} - y_i - \lambda_{n-1}[Ax_{n-1} + \theta_{n-1}(x_{n-1} - z)], \tag{3.117}$$

which implies that

$$\begin{aligned}
\|x_n - y_i\|^2 &= \|x_{n-1} - y_i\|^2 - 2\lambda_{n-1}\langle Ax_{n-1} + \theta_{n-1}(x_{n-1} - z), x_{n-1} - y_i \rangle \\
&\quad + \lambda_{n-1}^2 \|Ax_{n-1} + \theta_{n-1}(x_{n-1} - z)\|^2 \\
&= \|x_{n-1} - y_i\|^2 + 2\lambda_{n-1}(\theta_i - \theta_{n-1})\langle x_{n-1} - z, x_{n-1} - y_i \rangle \\
&\quad - 2\lambda_{n-1}\langle Ax_{n-1} + \theta_i(x_{n-1} - z), x_{n-1} - y_i \rangle \\
&\quad + \lambda_{n-1}^2 \|Ax_{n-1} + \theta_{n-1}(x_{n-1} - z)\|^2.
\end{aligned} \tag{3.118}$$

From (3.115) and the monotonicity of A, we find that

$$\langle Ax_{n-1} + \theta_i(z - y_i), x_{n-1} - y_i \rangle \geq 0. \tag{3.119}$$

Hence,

$$
\begin{aligned}
&\langle Ax_{n-1} + \theta_i(x_{n-1} - z), x_{n-1} - y_i \rangle \\
&= \langle Ax_{n-1} + \theta_i(z - y_i), x_{n-1} - y_i \rangle + \langle \theta_i(z - y_i) + \theta_i(x_{n-1} - z), x_{n-1} - y_i \rangle \\
&\geq \theta_i \| x_{n-1} - y_i \|^2.
\end{aligned} \tag{3.120}
$$

Substituting (3.120) into (3.118), we obtain

$$
\begin{aligned}
\| x_n - y_i \|^2 &\leq (1 - 2\lambda_{n-1}\theta_i)\| x_{n-1} - y_i \|^2 \\
&\quad + 2\lambda_{n-1}(\theta_i - \theta_{n-1})\langle x_{n-1} - z, x_{n-1} - y_i \rangle \\
&\quad + \lambda_{n-1}^2 \| Ax_{n-1} + \theta_{n-1}(x_{n-1} - z) \|^2.
\end{aligned} \tag{3.121}
$$

Since C is bounded, we find that $\{x_n\}$ is bounded, and so is $\{Ax_n\}$. There exists a constant $M > 0$ such that $\forall n \geq i \geq 2$,

$$
2\langle x_{n-1} - z, x_{n-1} - y_i \rangle \leq M, \| Ax_{n-1} + \theta_{n-1}(x_{n-1} - z) \|^2 \leq M.
$$

In view of $1 - 2\lambda_{n-1}\theta_i \leq \exp\{-2\lambda_{n-1}\theta_i\}$, we find from (3.121) that

$$
\| x_n - y_i \|^2 \leq \exp(-2\lambda_{n-1}\theta_i)\| x_{n-1} - y_i \|^2 + M\lambda_{n-1}(\theta_i - \theta_{n-1}) + M\lambda_{n-1}^2. \tag{3.122}
$$

It follows that

$$
\| x_n - y_i \|^2 \leq \exp\left(-2\theta_i \sum_{j=i}^{n-1} \lambda_j \right)\| x_i - y_i \|^2 + M \sum_{j=i}^{n-1} (\theta_i - \theta_j)\lambda_j + M \sum_{j=i}^{n-1} \lambda_j^2.
$$

If $j \leq n$, $\theta_i - \theta_j \leq \theta_i - \theta_n$, then

$$
\| x_n - y_i \|^2 \leq \exp\left(-2\theta_i \sum_{j=i}^{n-1} \lambda_j \right)\| x_i - y_i \|^2 + M(\theta_i - \theta_n) \sum_{j=i}^{n} \lambda_j + M \sum_{j=i}^{n} \lambda_j^2. \tag{3.123}
$$

From (3.123), we first prove that $x_{n(k)} \to x^*$ as $k \to \infty$. In fact, letting $i = n(k)$ and $n = n(k + 1)$ in (3.123), one has

$$
\begin{aligned}
\| x_{n(k+1)} - y_{n(k)} \|^2 &\leq \exp\left(-2\theta_{n(k)} \sum_{j=n(k)}^{n(k+1)} \lambda_j \right) \exp(2\theta_{n(k)}\lambda_{n(k+1)})\| x_{n(k)} - y_{n(k)} \|^2 \\
&\quad + M(\theta_{n(k)} - \theta_{n(k+1)}) \sum_{j=n(k)}^{n(k+1)} \lambda_j + M \sum_{j=n(k)}^{n(k+1)} \lambda_j^2.
\end{aligned} \tag{3.124}
$$

From the definition of admissible pairs, one concludes that $\lambda_n \to 0$. Thus

$$
\exp(\theta_{n(k)}\lambda_{n(k)}) \to 1.
$$

Let $r \in (0, 1)$ such that

$$\limsup_{k \to \infty} \|x_{n(k+1)} - y_{n(k)}\|^2 \leq r \limsup_{k \to \infty} \|x_{n(k)} - y_{n(k)}\|^2. \tag{3.125}$$

In view of $\lim_{k \to \infty} y_{n(k)} = x^*$, one has

$$\limsup_{k \to \infty} \|x_{n(k+1)} - y_{n(k)}\|^2 = \limsup_{k \to \infty} \|x_{n(k+1)} - x^*\|^2$$

$$= \limsup_{k \to \infty} \|x_{n(k)} - y_{n(k)}\|^2. \tag{3.126}$$

Combining (3.125) and (3.126), and using the fact that $r \in (0, 1)$, we claim that $\limsup_{k \to \infty} \|x_{n(k)} - x^*\|^2 = 0$, that is, $x_{n(k)} \to x^*$ $(k \to \infty)$.

Now, we are in a position to show that $x_n - y_{n(k)} \to \theta$. To this end, $\forall n > n(1)$, pick a k such that $n(k) \leq n \leq n(k+1)$ and let $i = k$ in (3.123). It follows that

$$\|x_n - y_{n(k)}\|^2 \leq \|x_{n(k)} - y_{n(k)}\|^2 + M(\theta_{n(k)} - \theta_n) \sum_{j=n(k)}^{n} \lambda_j + M \sum_{j=n(k)}^{n} \lambda_j^2$$

$$\leq \|x_{n(k)} - y_{n(k)}\|^2 + M(\theta_{n(k)} - \theta_{n(k+1)}) \sum_{j=n(k)}^{n(k+1)} \lambda_j + M \sum_{j=n(k)}^{n(k+1)} \lambda_j^2. \tag{3.127}$$

From the definition of admissible pairs, one obtains $x_{n(k)} \to x^*$ as $k \to \infty$. Note that $k \to \infty$ if $n \to \infty$. Letting $n \to \infty$ in (3.127), we obtain $x_n - y_{n(k)} \to \theta$. Hence, $y_{n(k)} \to x^*$. It follows that $x_n \to x^*$ as $n \to \infty$. This completes the proof. $\qquad \square$

Theorem 3.6.10. *Let $A \subset H \times H$ be a maximal monotone mapping and $\theta \in \operatorname{Ran}(A)$. Let $\{\lambda_n\}$ and $\{\theta_n\}$ be an admissible pair and $z \in H$. Let $\{x_n\} \subset \operatorname{Dom}(A)$ be a sequence generated by (BRIM-1). If $\{x_n\}$ and $\{v_n\}$ are bounded, then $\{x_n\}$ converges strongly to $x^* \in A^{-1}(\theta)$ and $x^* = P_{A^{-1}(\theta)}(z)$.*

Proof. Letting $t_i = \theta_i^{-1}$, $x = z$ in Proposition 3.1.4 (2), we have $\lim_{i \to \infty} J_{t_i} z = P_{A^{-1}(\theta)}(z)$. Without loss of generality, we may assume $z = \theta$ (otherwise, we consider $A'(\cdot) = A'(\cdot - z)$). Putting $y_i = J_{t_i}\theta$, we have $y_i \subset \operatorname{Dom}(A)$, $y_i \to P_{A^{-1}(\theta)}$ as $i \to \infty$. $\{y_i\}$ is the unique solution satisfying equation $\theta = \theta_i y_i + A y_i$ $(i = 1, 2, \dots)$. From the proof of Theorem 3.6.9, we can obtain the desired conclusion immediately. This completes the proof. $\qquad \square$

From the examples of admissible pairs given by Bruck, we see that his choice is difficult. As mentioned above, $n(i) = i^i$. Is it possible to give control sequences that are relatively easier to choose? The answer is positive! The following theorem gives another choice that guarantees the convergence of (BRIM-2).

Theorem 3.6.11. *Let C be a nonempty, bounded, closed, and convex subset of a Hilbert space H, and let $T : C \to C$ be a single-valued hemicontinuous pseudocontractive mapping. Then T has at least one fixed point. Suppose that the nonnegative real number sequences $\{\lambda_n\}$ and $\{\theta_n\}$ satisfy the following conditions:*

(1) $\lambda_n(1 + \theta_n) \le 1$ *for all* $n \ge 1$;
(2) $\theta_n \to 0$ *as* $n \to \infty$ (*not necessarily monotonically decreasing*);
(3) $\sum_{n=1}^{\infty} \lambda_n \theta_n = \infty$;
(4) $\frac{\theta_{n-1}}{\theta_n} - 1 = o(\lambda_n \theta_n)$;
(5) $\lambda_n = o(\theta_n)$.

Let $\{x_n\}$ *be the sequence generated by* (BRIM-2). *Then* $\{x_n\}$ *converges strongly to* $p = P_{\text{Fix}(T)}(z)$.

Proof. From Theorem 3.5.3, we see that $\text{Fix}(T) \ne \emptyset$. From Corollary 3.5.1, we see that $\text{Fix}(T)$ is closed and convex. Hence $P_{\text{Fix}(T)}(z)$ is well defined. Letting $t_n = \frac{\theta_n}{1+\theta_n}$, one has $t_n \to 0$ as $n \to \infty$. From Theorem 3.5.5, there exists a unique $\{y_n\} \subset C$ such that

$$y_n = t_n z + (1 - t_n) T y_n, \quad \forall n \ge 1, \tag{3.128}$$

and $y_n \to p = P_{\text{Fix}(T)}(z)$ as $n \to \infty$. From (3.128), we have

$$(1 + \theta_n) y_n = \theta_n z + T y_n, \quad \forall n \ge 1. \tag{3.129}$$

It follows that

$$\lambda_n (1 + \theta_n) y_n = \lambda_n \theta_n z + \lambda_n T y_n, \quad \forall n \ge 1. \tag{3.130}$$

If we can prove $x_n - y_n \to \theta$ as $n \to \infty$, then $x_n \to p$ as $n \to \infty$, which yields the desired conclusion. Therefore, we only need to prove $x_n - y_n \to \theta$ as $n \to \infty$. First, we estimate $\|y_n - y_{n-1}\|$. From (3.129) and the definition of pseudocontractive mappings, we derive

$$\begin{aligned}
\|y_n - y_{n-1}\|^2 &= \langle y_n - y_{n-1}, y_n - y_{n-1} \rangle \\
&= \langle \theta_n(z - y_n) - \theta_{n-1}(z - y_{n-1}) + T y_n - T y_{n-1}, y_n - y_{n-1} \rangle \\
&= \langle \theta_n(z - y_{n-1}) + \theta_n(y_n - y_{n-1}) - \theta_{n-1}(z - y_{n-1}), y_n - y_{n-1} \rangle \\
&\quad + \langle T y_n - T y_{n-1}, y_n - y_{n-1} \rangle \\
&= (\theta_n - \theta_{n-1}) \langle z - y_{n-1}, y_n - y_{n-1} \rangle \\
&\quad - \theta_n \|y_n - y_{n-1}\|^2 + \langle T y_n - T y_{n-1}, y_n - y_{n-1} \rangle \\
&\le (\theta_n - \theta_{n-1}) \langle z - y_{n-1}, y_n - y_{n-1} \rangle + (1 - \theta_n) \|y_n - y_{n-1}\|^2 \\
&\le |\theta_n - \theta_{n-1}| \|z - y_{n-1}\| \|y_n - y_{n-1}\| + (1 - \theta_n) \|y_n - y_{n-1}\|^2.
\end{aligned}$$

Then

$$\|y_n - y_{n-1}\|^2 \le \frac{|\theta_n - \theta_{n-1}|}{\theta_n} \|z - y_{n-1}\| \|y_n - y_{n-1}\|.$$

This implies that

$$\|y_n - y_{n-1}\| \le \frac{|\theta_n - \theta_{n-1}|}{\theta_n} \|z - y_{n-1}\| \le M \frac{|\theta_n - \theta_{n-1}|}{\theta_n}, \tag{3.131}$$

where $M = \sup\{\|z - y_{n-1}\| : n \geq 1\}$. For (BRIM-2) and (3.130), we have

$$x_{n+1} - y_n = [1 - \lambda_n(1 + \theta_n)](x_n - y_n) + \lambda_n(Tx_n - Ty_n). \tag{3.132}$$

Therefore,

$$
\begin{aligned}
\|x_{n+1} - y_n\|^2 &\leq [1 - \lambda_n(1 + \theta_n)]^2 + 2\lambda_n[1 - \lambda_n(1 + \theta_n)] \\
&\quad + \langle x_n - y_n, Tx_n - Ty_n \rangle + \lambda_n^2\|Tx_n - Ty_n\|^2 \\
&\leq \{[1 - \lambda_n(1 + \theta_n)]^2 + 2\lambda_n[1 - \lambda_n(1 + \theta_n)]\}\|x_n - y_n\|^2 + D\lambda^2 \\
&\leq (1 - 2\lambda_n\theta_n)\|x_n - y_n\|^2 + D\lambda^2, \quad \forall n \geq 1,
\end{aligned}
\tag{3.133}
$$

where $D = \sup\{\|Tx_n - Ty_n\|^2\}$. Since

$$\|x_n - y_n\| \leq \|x_n - y_{n-1}\| + \|y_n - y_{n-1}\|,$$

we find from (3.131) that

$$
\begin{aligned}
\|x_n - y_n\|^2 &\leq \|x_n - y_{n-1}\|^2 + 2\|x_n - y_{n-1}\|\|y_n - y_{n-1}\| + \|y_n - y_{n-1}\|^2 \\
&\leq \|x_n - y_{n-1}\|^2 + M_1\frac{|\theta_n - \theta_{n-1}|}{\theta_n}.
\end{aligned}
\tag{3.134}
$$

Substituting (3.134) into (3.133), we obtain

$$\|x_{n+1} - y_n\|^2 \leq (1 - 2\lambda_n\theta_n)\|x_n - y_{n-1}\|^2 + M_1\frac{|\theta_n - \theta_{n-1}|}{\theta_n} + D\lambda_n^2. \tag{3.135}$$

From conditions (2)–(5), we see that (3.135) can be rewritten as

$$\|x_{n+1} - y_n\|^2 \leq (1 - 2\lambda_n\theta_n)\|x_n - y_{n-1}\|^2 + o(\lambda_n\theta_n). \tag{3.136}$$

Setting $a_n = \|x_n - y_{n-1}\|^2$ and $t_n = 2\lambda_n\theta_n$, we see that (3.136) can be rewritten as

$$a_{n+1} \leq (1 - t_n)a_n + o(t_n),$$

where $\{t_n\}$ is a sequence in $[0, 1]$ satisfying $\sum_{n=1}^{\infty} t_n = \infty$. It follows from Lemma 1.10.2 that $a_n \to 0$ as $n \to \infty$, that is, $x_n \to p = P_{\text{Fix}(T)}(z)$ as $n \to \infty$. This completes the proof. □

The control sequences satisfying the above theorem are:

$$\lambda_n = \frac{1}{(n + 1)^a}, \qquad \theta_n = \frac{1}{(n + 1)^b},$$

where $0 < b < a$ and $a + b < 1$. It is obvious that the above two sequences $\{\lambda_n\}$ and $\{\theta_n\}$ are relatively simple.

Corollary 3.6.4. *Let C be a nonempty, bounded, closed, and convex subset of a Hilbert space H. Let $A : C \to H$ be a monotone hemicontinuous mapping such that $T = I - A : C \to C$ is a self-mapping. Suppose that nonnegative real number sequences $\{\lambda_n\}$ and $\{\theta_n\}$ satisfy the conditions of Theorem 3.6.11. Let $\{x_n\}$ be a sequence generated in (BRIM-1). Then $\{x_n\}$ converges strongly to a zero point $x^* = P_{A^{-1}(\theta)}(z)$ of A.*

Proof. It is clear that $T : C \to C$ is a hemicontinuous pseudocontractive mapping. From Theorem 3.5.3 we see that $\mathrm{Fix}(T) \neq \emptyset$. It follows that $A^{-1}(\theta) \neq \emptyset$. Using Corollary 3.5.1, we find that $\mathrm{Fix}(T)$ is closed and convex. Hence $A^{-1}(\theta)$ is nonempty, closed, and convex. This shows that $P_{A^{-1}(\theta)}(z)$ is well defined. From $T = I - A$, one sees that (BRIM-1) can be rewritten as (BRIM-2). Thus the assumptions of Theorem 3.6.11 all are satisfied. From Theorem 3.6.11, we see that $\{x_n\}$ converges strongly to a fixed point $x^* = P_{\mathrm{Fix}(T)}z = P_{A^{-1}(\theta)}z$, that is, $x^* \in A^{-1}(\theta)$. This completes the proof. $\square$

The following theorem removes the boundedness requirement of set C, however, the price is the continuity of T.

Theorem 3.6.12. *Let C be a nonempty, closed, and convex subset of a Hilbert space H. Let $T : C \to C$ be an L-Lipschitz pseudocontractive mapping such that $\mathrm{Fix}(T) \neq \emptyset$. Let $\{x_n\}$ be the sequence generated by (BRIM-2), where $\{\lambda_n\}$ and $\{\theta_n\}$ satisfy conditions (1)–(5) of Theorem 3.6.11. Then $\{x_n\}$ converges strongly to a fixed point $x^* = P_{\mathrm{Fix}(T)}(z)$ of T.*

Proof. It follows from (BRIM-2) and (3.130) that

$$
\begin{aligned}
\|x_{n+1} - y_n\|^2 &= \left\| [1 - \lambda_n(1 + \theta_n)](x_n - y_n) + \lambda_n(Tx_n - Ty_n) \right\|^2 \\
&= [1 - \lambda_n(1 + \theta_n)]^2 \|x_n - y_n\|^2 + \lambda_n^2 \|Tx_n - Ty_n\|^2 \\
&\quad + 2\lambda_n[1 - \lambda_n(1 + \theta_n)]\langle x_n - y_n, Tx_n - Ty_n \rangle \\
&\leq \{[1 - \lambda_n(1 + \theta_n)]^2 + 2\lambda_n[1 - \lambda_n(1 + \theta_n)] + L^2\lambda_n^2\}\|x_n - y_n\|^2 \\
&\leq [1 - 2\lambda_n\theta_n + (4 + L^2)\lambda_n^2]\|x_n - y_n\|^2 \\
&\leq (1 - \lambda_n\theta_n)\|x_n - y_n\|^2.
\end{aligned}
$$

For large enough $n \geq 1$, we have

$$
\begin{aligned}
\|x_{n+1} - y_n\| &\leq \sqrt{1 - \lambda_n\theta_n}\|x_n - y_n\| \\
&\leq \left(1 - \frac{1}{2}\lambda_n\theta_n\right)\|x_n - y_n\|.
\end{aligned}
$$

Using (3.131), we obtain

$$
\begin{aligned}
\|x_{n+1} - y_n\| &\leq \left(1 - \frac{1}{2}\lambda_n\theta_n\right)\|x_n - y_{n-1}\| + \|y_n - y_{n-1}\| \\
&\leq \left(1 - \frac{1}{2}\lambda_n\theta_n\right)\|x_n - y_{n-1}\| + M\left|\frac{\theta_{n-1}}{\theta_n} - 1\right| \\
&= \left(1 - \frac{1}{2}\lambda_n\theta_n\right)\|x_n - y_{n-1}\| + o(\lambda_n\theta_n).
\end{aligned}
$$

By Lemma 1.10.2, we derive that $x_{x+1} - y_n \to \theta$ as $n \to \infty$. Hence $x_n \to P_{\text{Fix}(T)}(z)$ as $n \to \infty$. This completes the proof. □

Corollary 3.6.5. *Let C be a nonempty, closed, and convex subset of a Hilbert space H. Let $A : C \to H$ be an L-Lipschitz monotone mapping such that $A^{-1}(\theta) \neq \emptyset$ and $T = I - A : C \to C$ is a self-mapping. Let $\{x_n\}$ be a sequence generated by (BRIM-1), where $\{\lambda_n\}$ and $\{\theta_n\}$ satisfy conditions (1)–(5) of Theorem 3.6.11. Then $\{x_n\}$ converges strongly to a zero point $x^* = P_{A^{-1}(\theta)}(z)$ of A.*

Proof. Note that $T : C \to C$ is $(L+1)$-Lipschitz pseudocontractive with $\text{Fix}(T) = A^{-1}(\theta) \neq \emptyset$ and $\{x_n\}$ can be rewritten as (BRIM-2). All the conditions of Theorem 3.6.12 are satisfied. Hence $\{x_n\}$ converges strongly to a fixed point $x^* = P_{\text{Fix}(T)}(z) = P_{A^{-1}(\theta)}(z)$, which is a zero point of A. This completes the proof. □

3.7 Remarks

(1) Theorem 3.3.5 is due to Rochafellar [77]. It establishes the connection between variational inequalities and maximal monotone operators, and it is a powerful tool for the solvability of monotone variational inequalities.
(2) Theorem 3.4.1 is due to the authors [127]. It establishes the solvability of a class of variational inequalities with hemicontinuous strongly monotone operators and it is a powerful tool for iterative solutions of monotone variational inequalities.
(3) Theorems 3.4.2 and 3.4.3 are both due to the authors [127]. Theorem 3.6.8 is due to the author of [116]. It improves the convergence theorem established by Ishikawa [32] in 1974.
(4) Theorem 3.6.9 improves the convergence theorem established by Bruck [11] in 1974. It weakens the continuity assumption. Theorem 3.6.11 seems to be a new result compared with the results of Chidume and Zegeye [25]. And the requirement on the continuity of mappings is also reduced.

3.8 Exercises

We use H to denote a real Hilbert space.
1. Show that $T : [0, 1] \to \mathbb{R}$ defined by $Tx = 1 - x^{\frac{2}{3}}$ is a continuous pseudocontractive mapping, which is not nonexpansive.
2. Let K be a closed, bounded, and convex subset of H, and $T : K \to K$ a k-strictly pseudocontractive mapping. Show that there exist values of $\lambda \in (0, 1)$ such that the averaged mapping

$$T_\lambda(x) = (1 - \lambda)x + \lambda Tx, \quad x \in K,$$

is nonexpansive.

3. Let K be a closed, bounded, and convex subset of H, $T : K \to K$ a k-strictly pseudocontractive mapping, and let $\{\alpha_n\}$ be a sequence of real numbers satisfying the
   following conditions:
   (i) $\alpha_n \in (0, 1)$ for all $n \geq 1$;
   (ii) $\sum_{n=1}^{\infty} \alpha_n = \infty$, and
   (iii) $\lim_{n \to \infty} \alpha_n = \alpha < 1 - k$.
   Show that the Mann iteration method generated from an arbitrary $x_1 \in K$ by

   $$x_{n+1} = (1 - \alpha_n)x_n + \alpha_n T x_n, \quad n \geq 1,$$

   converges weakly to a fixed point of T.

4. Let C be a nonempty, closed, and convex subset of a real Hilbert space H, and
   $T : C \to C$ be k-Lipschitz continuous and quasi-pseudocontractive. Show that
   $\text{Fix}(T)$ is a nonempty, closed, and convex subset of C.

5. Let C be a nonempty, closed, and convex subset of a real Hilbert space H, and $T :
   C \to C$ be a demicontinuous pseudocontraction. Show that $\text{Fix}(T)$ is a nonempty,
   closed, and convex subset of C, and $I - T$ is demiclosed at zero.

6. Let C be a nonempty, closed, and convex subset of a real Hilbert space H, and
   $T : C \to C$ be a hemicontinuous pseudocontraction. Show that $F(T)$ is a nonempty,
   closed, and convex subset of C, and $I - T$ is demiclosed at zero.

7. Let $T : H \to H$ be mapping. Write $A := I - T$. Prove that T is pseudocontractive
   $\Longleftrightarrow A$ is monotone $\Longleftrightarrow \|Tx - Ty\|^2 \leq \|x - y\|^2 + \|Ax - Ay\|^2$, $\forall x, y \in H$.

8. Let $T : H \to H$ be mapping. Prove that if T is k-strictly pseudocontractive, then T
   is Lipschitz continuous and pseudocontractive.

9. Let C be a nonempty, closed, and convex subset of H. Then $T : C \to C$ is said to be
   hemicontractive if $\|Tx - p\|^2 \leq \|x - p\|^2 + \|x - Tx\|^2$ for all $x \in H$ and all $p \in \text{Fix}(T)$.
   Prove that if $T : C \to C$ is hemicontinuous and hemicontractive, then $\text{Fix}(T)$ is a
   nonempty, closed, and convex subset of H.

10. Let C be a nonempty, closed, and convex subset of H, and $T : C \to C$ be an
    L-Lipschitz pseudocontraction such that $\text{Fix}(T) \neq \emptyset$. Let $\{\alpha_n\}$, $\{\beta_n\}$, and $\{\lambda_n\}$ be
    three sequences of numbers in $(0,1)$ satisfying the following conditions:
    (1) $0 < a \leq \alpha_n \leq \beta_n < b < \frac{\sqrt{1+L^2}-1}{L}$, for all $n \geq 1$;
    (2) $\lambda_n \to 0$ $(n \to \infty)$ and $\sum_{n=1}^{\infty} \lambda_n = \infty$.
    For an arbitrary initial value $x_1 \in C$ and a fixed anchor $u \in C$, define a sequence
    $\{x_n\}$ iteratively in C by

    $$\begin{cases} x_1, u \in C, \\ y_n = (1 - \beta_n)x_n + \beta_n T x_n, \\ u_n = (1 - \alpha_n)x_n + \alpha_n T y_n, \\ x_{n+1} = \lambda_n u + (1 - \lambda_n)u_n, \quad n \geq 1. \end{cases} \qquad \text{(IHIM)}$$

    Prove that the sequence $\{x_n\}$ defined by (IHIM) converges strongly to a fixed point
    of T, $x^* = P_{\text{Fix}(T)}u$.

11. In the above method (IHIM), if u is replaced by a contraction $f : C \rightarrow C$, that is, u is replaced by $f(x_n)$, then the corresponding sequence converges strongly to a fixed point of T, $x^* = P_{\text{Fix}(T)}f(x^*)$. Try to prove it.

12. Let $A : H \rightarrow H$ be an η-strongly monotone and k-Lipschitz mapping. Let $T : H \rightarrow H$ be an L-Lipschitz continuous pseudocontractive mapping such that $\text{Fix}(T) \neq \emptyset$. Let $\{\alpha_n\}$, $\{\beta_n\}$, and $\{\lambda_n\}$ be the same as in Exercise 10. For $\mu \in (0, \frac{2\eta}{k^2})$, define a sequence $\{x_n\}$ iteratively in H by

$$\begin{cases} x_1 \in H, \\ y_n = (1 - \beta_n)x_n + \beta_n Tx_n, \\ u_n = (1 - \alpha_n)x_n + \alpha_n Ty_n, \\ x_{n+1} = \lambda_n(I - \mu A)x_n + (1 - \lambda_n)u_n, \quad n \geq 1. \end{cases} \quad \text{(VIHIM)}$$

Prove that the sequence $\{x_n\}$ defined by (VIHIM) converges strongly to the unique solution x^* of the variational inequality:

$$\langle Ax, y - x \rangle \geq 0, \quad \forall y \in \text{Fix}(T).$$

13. Let A, T, $\{\alpha_n\}$, $\{\beta_n\}$, $\{\lambda_n\}$, and μ be the same as in Exercise 12. Define a sequence $\{x_n\}$ iteratively by

$$\begin{cases} x_1 \in H, \\ y_n = (1 - \beta_n)x_n + \beta_n Tx_n, \\ u_n = (1 - \alpha_n)x_n + \alpha_n Ty_n, \\ x_{n+1} = (I - \mu A)u_n, \quad n \geq 1. \end{cases} \quad \text{(HSDM)}$$

Prove that the sequence $\{x_n\}$ defined by (HSDM) converges strongly to the unique solution x^* of the variational inequality:

$$\langle Ax, y - x \rangle \geq 0, \quad \forall y \in \text{Fix}(T).$$

14. Try to extend the result of Exercise 13 to a finite family of L-Lipschitz continuous pseudocontractive mappings such that $K = \bigcap_{i=1}^r \text{Fix}(T_i) \neq \emptyset$.

15. Let C be a nonempty, closed, and convex subset of a real Hilbert space H, and $A : C \rightarrow H$ be an L-Lipschitz and monotone mapping such that the solution $\text{SOL}(C, A)$ of the variational inequality $VI(C, A)$ is nonempty. Define a sequence $\{x_n\} \subset C$ by

$$\begin{cases} x_1 \in C, \\ y_n = P_C(x_n - \lambda Ax_n), \\ x_{n+1} = P_C(x_n - \lambda Ay_n), \quad n \geq 1, \end{cases} \quad \text{(KIM)}$$

where $\lambda \in (0, 1/L)$ and P_C denotes the metric projection from H onto C. Prove that the sequence $\{x_n\}$ defined by (KIM) converges weakly to some point

$$x^* = \lim_{n \to \infty} P_{\text{SOL}(C,A)}x_n.$$

16. Let C be a nonempty, closed, and convex subset of a real Hilbert space H, and $A : C \to H$ be an L-Lipschitz continuous and monotone mapping such that the solution $\mathrm{SOL}(C, A) \neq \emptyset$. Also $\forall x_1, u \in C$, define a sequence $\{x_n\} \subset C$ by

$$
\begin{cases}
x_1, u \in C, \\
y_n = P_C(x_n - \tau A x_n), \\
z_n = P_C(x_n - \tau A y_n), \\
x_{n+1} = \lambda_n u + (1 - \lambda_n) z_n, \quad n \geq 1,
\end{cases}
\tag{MEGIM}_1
$$

where $\tau \in (0, 1/L)$ and $\{\lambda_n\}$ satisfies the following conditions: (C1) $\lambda_n \to 0$ $(n \to \infty)$, and (C2) $\sum_{n=1}^{\infty} \lambda_n = \infty$. Prove that the sequence $\{x_n\}$ defined by (MEGIM)$_1$ converges strongly to some point $x^* = P_{\mathrm{SOL}(C,A)}(u)$.

17. Let C be a nonempty, closed, and convex subset of a real Hilbert space H, and $A : C \to H$ be an L-Lipschitz continuous and monotone mapping such that the solution $\mathrm{SOL}(C, A) \neq \emptyset$. Let $S : C \to C$ be a nonexpansive mapping such that $K = \mathrm{Fix}(S) \cap \mathrm{SOL}(C, A) \neq \emptyset$. Also $\forall x_1 \in C$, define a sequence $\{x_n\}$ iteratively in C by

$$
\begin{cases}
x_1 \in C, \\
y_n = P_C(x_n - \tau A x_n), \\
z_n = P_C(x_n - \tau A y_n), \\
x_{n+1} = \alpha_n x_n + (1 - \alpha_n) S z_n, \quad n \geq 1,
\end{cases}
\tag{MEGIM}_2
$$

where $\tau \in (0, 1/L)$ and $\{\alpha_n\}$ is a sequence of numbers satisfying the following condition: (C1) $\alpha_n \in [a, b]$ for $a, b \in (0, 1)$. Prove that the sequence $\{x_n\}$ defined by (MEGIM)$_2$ converges weakly to some point

$$
x^* = \lim_{n \to \infty} P_K(x_n).
$$

18. Let C be a nonempty, closed, and convex subset of a real Hilbert space H, and $A : C \to H$ be an L-Lipschitz continuous and monotone mapping such that the solution $\mathrm{SOL}(C, A) \neq \emptyset$. Let $S : C \to C$ be a nonexpansive mapping such that $K = \mathrm{Fix}(S) \cap \mathrm{SOL}(C, A) \neq \emptyset$. Also $\forall x_1, u \in C$, define a sequence $\{x_n\}$ iteratively in C by

$$
\begin{cases}
x_1, u \in C, \\
y_n = P_C(x_n - \tau A x_n), \\
z_n = P_C(x_n - \tau A y_n), \\
x_{n+1} = \alpha_n u + (1 - \alpha_n) S z_n, \quad n \geq 1,
\end{cases}
\tag{MEGIM}_1
$$

where $\tau \in (0, 1/L)$ and $\{\alpha_n\}$ is a sequence of numbers satisfying the following conditions: (C1) $\alpha_n \to 0$ $(n \to \infty)$ and (C2) $\sum_{n=1}^{\infty} \alpha_n = \infty$. Prove that the sequence $\{x_n\}$ defined by (MEGIM)$_1$ converges strongly to a specific point $x^* = P_K u$.

4 Fixed point theory and iterative methods for fixed points of nonexpansive mappings in Banach spaces

In this chapter, we present some known fixed point theorems in the framework of general Banach spaces. We focus on the generalized Browder fixed point theorem proposed by Caristi in 1976. We also discuss the normal Mann iterative method and the Halpern iterative method for fixed points of nonexpansive mappings.

4.1 Several celebrated fixed point theorems

Theorem 4.1.1 (Banach–Caristi [17], 1976). *Let E be a real Banach space and let C be a nonempty, closed, and convex subset of E. Let $T : C \to E$ be a contractive mapping satisfying the weak inward condition (WIC). Then T has a unique fixed point in C. If, in addition, T is a self-mapping on C, then, for any initial point $x_1 \in C$, the Banach–Picard iterative sequence $\{T^n x_1\}$ converges to the fixed point.*

In 1969, Meir and Keeler improved the above result and established the following theorem.

Theorem 4.1.2 (Meir–Keeler [50], 1969). *Let E be a real Banach space and let C be a nonempty, closed, and convex subset of E. Let $T : C \to C$ be an MK contraction, that is, for any $\epsilon > 0$, there exists $\delta > 0$ such that*

$$\epsilon \le \|x - y\| < \epsilon + \delta \implies \|Tx - Ty\| < \epsilon, \quad \forall x, y \in C. \tag{4.1}$$

Then T has a unique fixed point in C.

Remark 4.1.1.

(i) It is easy to check that (4.1) is equivalent to the following fact: for any $\epsilon > 0$, there exists $\delta > 0$ such that

$$\|x - y\| < \epsilon + \delta \implies \|Tx - Ty\| < \epsilon, \quad \forall x, y \in C. \tag{4.2}$$

(ii) An MK mapping is a mapping in-between contractions and nonexpansive mappings.

(iii) Theorem 4.1.2 may be still valid for the MK type nonself-mappings that satisfy the weak inward condition (WIC).

In 1965, Browder and Göhde obtained the following known fixed point theorem independently. In the same year, Kirk also proved a fixed point theorem in reflexive Banach spaces with normal structures. In 1976, Caristi extended Browder–Göhde fixed point theorem from self-mappings to nonself-mappings.

https://doi.org/10.1515/9783110667097-004

Theorem 4.1.3 (Browder–Göhde–Caristi [17], 1976). *Let E be a real uniformly convex Banach space and let C be a nonempty, bounded, closed, and convex subset of E. Let $T : C \to E$ be a nonexpansive mapping satisfying the weak inward condition (WIC). Then T has a fixed point in C. Moreover, $\mathrm{Fix}(T) = \{x \in C : Tx = x\}$ is nonempty, closed, and convex.*

Proof. Fixing $u \in C$, for any $n \geq 1$, we define a mapping $T_n : C \to E$ by

$$T_n x = \frac{1}{1+n} u + \frac{n}{1+n} Tx, \quad \forall x \in C. \tag{4.3}$$

Then, for any $n \geq 1$, $T_n : C \to E$ is a contractive mapping satisfying the weak inward condition (WIC). Using Theorem 4.1.1, we see that, for any $n \geq 1$, T_n has a unique fixed point x_n in C, that is,

$$x_n = T_n x_n = \frac{1}{1+n} u + \frac{n}{1+n} Tx_n, \quad n \geq 1. \tag{4.4}$$

Due to $\{Tx_n\} \subset C$ and the boundedness of C, we derive that $\{Tx_n\}$ is bounded. From (4.4), we see that

$$x_n - Tx_n \to \theta \quad (n \to \infty). \tag{4.5}$$

The uniform convexity of E asserts that E is reflexive. From the boundedness of $\{x_n\}$, we assume that $x_n \rightharpoonup x$. By applying Theorem 1.9.3, we find that $x \in \mathrm{Fix}(T)$.

Next, we show that $\mathrm{Fix}(T)$ is closed and convex. We first show the closedness of $\mathrm{Fix}(T)$. To this end, we fix $\{p_n\} \subset \mathrm{Fix}(T)$ with $p_n \to p$ as $n \to \infty$. Since $Tp_n = p_n$ and T is continuous, we find that $Tp = p$, that is, $p \in \mathrm{Fix}(T)$.

Now, we are in a position to show that $\mathrm{Fix}(T)$ is convex. Let $p_1, p_2 \in \mathrm{Fix}(T)$ and $t \in (0, 1)$ and put $p_t = tp_1 + (1 - t)p_2$. It follows that $p_t \in C$ and

$$\|p_1 - Tp_t\| = \|Tp_1 - Tp_t\| \leq \|p_1 - p_t\| = (1 - t)\|p_1 - p_2\| \tag{4.6}$$

and

$$\|p_2 - Tp_t\| = \|Tp_2 - Tp_t\| \leq \|p_2 - p_t\| = t\|p_1 - p_2\|. \tag{4.7}$$

Combining (4.6) with (4.7), we find that

$$\|p_1 - p_2\| \leq \|p_1 - Tp_t\| + \|Tp_t - p_2\| \leq \|p_1 - p_t\| + \|p_t - p_2\| = \|p_1 - p_2\|.$$

This implies that

$$\|p_1 - p_2\| = \|p_1 - Tp_t\| + \|Tp_t - p_2\|.$$

Set $x = p_1 - Tp_t, y = Tp_t - p_2$. It follows that

$$\|x + y\| = \|x\| + \|y\|. \tag{4.8}$$

Since E is uniformly convex, we see that it is also strictly convex. Using (4.8), there exists $\lambda > 0$ such that $x = \lambda y$. Hence

$$Tp_t = \frac{1}{1+\lambda}p_1 + \frac{\lambda}{1+\lambda}p_2. \tag{4.9}$$

It follows that

$$\frac{\lambda}{1+\lambda}\|p_1 - p_2\| = \|Tp_t - p_1\| \le \|p_t - p_1\| = (1-t)\|p_1 - p_2\|$$

and

$$\frac{1}{1+\lambda}\|p_1 - p_2\| = \|Tp_t - p_2\| \le \|p_t - p_2\| = t\|p_1 - p_2\|. \tag{4.10}$$

Hence $t = \frac{1}{1+\lambda}$, that is, $Tp_t = p_t$. This completes the proof. $\qquad\square$

4.2 Normal Mann iterative method and Reich's weak convergence theorem

Let E be a real Banach space and let C be a nonempty, closed, and convex subset of E. Let $T : C \to C$ be a mapping and let $\{\alpha_n\}$ be a sequence in $(0,1)$. The normal Mann iterative method (NMIM) generates a sequence in the following manner:

$$x_1 \in C, \quad x_{n+1} = (1 - \alpha_n)x_n + \alpha_n Tx_n.$$

If $\{\alpha_n\}$ is a fixed constant in $(0,1)$, then the (NMIM) is also called the Krasnosel'skiĭ–Mann iterative method.

Theorem 4.2.1 (Reich [73], 1979). *Let E be a real uniformly convex Banach space such that the norm of E is Fréchet differentiable. Let C be a nonempty, closed, and convex subset of E, and let $T : C \to C$ be a nonexpansive mapping with a nonempty fixed point set $\mathrm{Fix}(T) \ne \emptyset$. Let $\{x_n\}$ be a sequence generated by (NMIM), where $\{\alpha_n\}$ is a real sequence such that $\sum_{n=1}^{\infty} \alpha_n(1 - \alpha_n) = \infty$. Then $\{x_n\}$ converges weakly to a fixed point of T.*

Proof. Define a mapping $T_n : C \to C$ by

$$T_n x = (1 - \alpha_n)x + \alpha_n Tx, \quad \forall x \in C, n \ge 1. \tag{4.11}$$

Then $\{T_n\}_{n\ge 1} : C \to C$ is a family of nonexpansive mappings satisfying the condition $\mathrm{Fix}(T) \subset \bigcap_{i=1}^{\infty} \mathrm{Fix}(T_n)$. It follows from (4.11) that

$$x_{n+1} = T_n x_n, \quad \forall n \ge 1. \tag{4.12}$$

Using Theorem 1.9.8, we need to prove $\omega_w(x_n) \subset \mathrm{Fix}(T)$. To this end, we first show that $\lim_{n\to\infty}\|x_n - p\|$ exists for all $p \in \mathrm{Fix}(T)$. Indeed, using either (NMIA) or (4.12), we find that

$$\|x_{n+1} - p\| = \|T_n x_n - Tp\| \le \|x_n - p\|, \quad \forall p \in \mathrm{Fix}(T).$$

This shows that $\{\|x_n - p\|\}$ is decreasing. This further implies that $\lim_{n\to\infty} \|x_n - p\|$ exists for all $p \in \text{Fix}(T)$. In particular, we see that sequence $\{x_n\}$ is bounded. Fix $p \in \text{Fix}(T)$ and let $r > 0$ be a sufficiently large number such that $\|x_n - p\| \leq r$. Hence, $\|Tx_n - p\| \leq r$. Using Theorem 1.8.9, we find that

$$\|x_{n+1} - p\|^2 \leq (1 - \alpha_n)\|x_n - p\|^2 + \alpha_n\|Tx_n - p\|^2 - (1 - \alpha_n)\alpha_n g(\|Tx_n - x_n\|)$$
$$\leq \|x_n - p\|^2 - (1 - \alpha_n)\alpha_n g(\|Tx_n - x_n\|). \tag{4.13}$$

It follows that

$$(1 - \alpha_n)\alpha_n g(\|Tx_n - x_n\|) \leq \|x_n - p\|^2 - \|x_{n+1} - p\|^2. \tag{4.14}$$

Hence,

$$\sum_{n=1}^{\infty} \alpha_n(1 - \alpha_n)g(\|Tx_n - x_n\|) < \infty. \tag{4.15}$$

In view of $\sum_{n=1}^{\infty} \alpha_n(1 - \alpha_n) = \infty$ and (4.15), we find that $\liminf_{n\to\infty} g(\|Tx_n - x_n\|) = 0$. Hence, $\liminf_{n\to\infty} \|Tx_n - x_n\| = 0$. From (NMIM), we also have $\|x_{n+1} - Tx_{n+1}\| \leq \|x_n - Tx_n\|$. This shows that $\lim_{n\to\infty} \|x_n - Tx_n\|$ exists. It follows that $\lim_{n\to\infty} \|x_n - Tx_n\| = 0$. Using Theorem 1.9.3, we see that $\omega_w(x_n) \subset \text{Fix}(T)$. With the aid of Theorem 1.9.8, we find the desired conclusion immediately. $\square$

Remark 4.2.1. If E is a Hilbert space, the sequence $\{x_n\}$ defined by (NMIM) in Theorem 4.2.1 converges weakly to $z \in \text{Fix}(T)$ and z is uniquely defined by $\lim_{n\to\infty} P_{\text{Fix}(T)}x_n = z$. If E is not Hilbert, we are not sure whether sequence $\{x_n\}$ converges weakly to $z \in \text{Fix}(T)$. Such a z can be defined by the limit of a generalized projection, which is an interesting problem. Recently, Takahashi and Yao investigated this problem; see [90] and the references therein.

Remark 4.2.2. Generally speaking, the sequence $\{x_n\}$ generated by Theorem 4.2.1 is weakly convergent. Under an additional condition, it is also strongly convergent.

To illustrate this, we introduce the following definitions.

Definition 4.2.1. Let E be a Banach space and let C be a nonempty, closed, and convex subset of E. Let $T : C \to C$ be a mapping. Then T is said to satisfy Condition (A) if $\text{Fix}(T) \neq \emptyset$ and there exits a nondecreasing function $f : \mathbb{R}^+ \to \mathbb{R}^+$ with $f(0) = 0$, $f(t) > 0$, $\forall t > 0$ such that

$$\|x - Tx\| \geq f(d(x, \text{Fix}(T))), \quad \forall x \in C,$$

where $d(x, \text{Fix}(T)) = \inf\{\|x - z\| : z \in \text{Fix}(T)\}$.

Also $T : C \to C$ is said to satisfy Condition (B) if $\text{Fix}(T) \neq \emptyset$ and there exits a constant $r > 0$ such that

$$\|x - Tx\| \geq rd(x, \text{Fix}(T)), \quad \forall x \in C.$$

It is obvious mappings, which satisfy Condition (B), also satisfy Condition (A).

Example 4.2.1. Let $E = \mathbb{R}^2$ and $\|x\| = \sqrt{\langle x, x \rangle}$, $\forall x \in E$. Let

$$C = \left\{ (r, \theta) : 0 \le r \le 1, -\frac{\pi}{2} \le \theta \le -\frac{\pi}{4} \right\}.$$

Define a mapping $T : C \to C$ by $T(r, \theta) = (r, -\frac{\pi}{2})$, $(r, \theta) \in C$. Then $\text{Fix}(T) = \{(r, -\frac{\pi}{2}) : 0 \le r \le 1\} \ne \emptyset$ and

$$\|x - Tx\| \ge d(x, \text{Fix}(T)), \quad \forall x \in C.$$

This shows that T satisfies Condition (B), which further implies that T also satisfies Condition (A).

Definition 4.2.2. A mapping T is said to be semicompact if any bounded sequence $\{x_n\} \subset \text{Dom}(T)$, such that $\{x_n - Tx_n\}$ converges, has a convergent subsequence.

Remark 4.2.3.
(i) Let E be a Banach space and let C be a nonempty, closed, and convex subset of E. Let $T : C \to C$ be a mapping with fixed points. If $I - T$ maps bounded closed subsets of C to closed subsets of E, then T satisfies Condition (A).
(ii) If $T : C \to C$ is continuous semicompact, then T satisfies Condition (A).

Theorem 4.2.2 (Senter and Dotson [78], 1974). *Let E be a real uniformly convex Banach space and let C be a nonempty, closed, and convex subset of E. Let $T : C \to C$ be a nonexpansive mapping with a nonempty fixed point set $\text{Fix}(T) \ne \emptyset$. Let $\{x_n\}$ be a sequence generated by (NMIM). If T satisfies Condition (A) and control sequence $\{\alpha_n\}$ satisfies $\sum_{n=1}^{\infty} \alpha_n (1 - \alpha_n) = \infty$, then $\{x_n\}$ converges strongly to a fixed point of T in norm.*

Proof. From the proof of Theorem 4.2.1, we find that $x_n - Tx_n \to \theta$ as $n \to \infty$. Using Condition (A), we conclude that $f(d(x_n, \text{Fix}(T))) \to 0$ as $n \to \infty$, which implies that $d(x_n, \text{Fix}(T)) \to 0$ as $n \to \infty$. Thus, a standard argument yields that $x_n \to p \in \text{Fix}(T)$ as $n \to \infty$. $\square$

In 1976, Ishikawa [33] obtained the following remarkable result.

Theorem 4.2.3 (Ishikawa [33], 1976). *Let E be a Banach space and let C be a nonempty, closed, and convex subset of E. Let $T : C \to C$ be a nonexpansive mapping. Let $\{x_n\}$ be generated by the iterative method (NMIM), where $\{\alpha_n\}$ is such that $0 \le \alpha_n \le c < 1$ and $\sum_{n=1}^{\infty} \alpha_n = \infty$. If, in addition, $\{x_n\}$ is bounded, then $x_n - Tx_n \to \theta$ $(n \to \infty)$.*

Using Theorem 4.2.3 and Condition (A), we establish the following strong convergence result.

Theorem 4.2.4 (Ishikawa [33], 1976). *Let E be a Banach space and let C be a nonempty, closed, and convex subset of E. Let $T : C \to C$ be a nonexpansive mapping with a* *nonempty fixed-point set* Fix(T) $\neq \emptyset$. *Let* $\{x_n\}$ *be the sequence generated by (NMIM). If T satisfies Condition (A), then* $\{x_n\}$ *converges strongly to a fixed point of T.*

Proof. It follows from Theorem 4.2.3 that $x_n - Tx_n \to \theta$ as $n \to \infty$. Using Condition (A), we find that $f(d(x_n, \mathrm{Fix}(T))) \to 0$ as $n \to \infty$, which implies that $d(x_n, \mathrm{Fix}(T)) \to 0$ as $n \to \infty$. Thus, a standard argument yields that $x_n \to p \in \mathrm{Fix}(T)$ as $n \to \infty$. $\qquad\square$

Theorem 4.2.5. *Let E be a reflexive Banach space. Let* $T : E \to E$ *be a linear nonexpansive mapping, that is,* $\|Tx\| \le \|x\|$, $\forall x \in E$. *Fix* $\lambda \in (0,1)$ *and define a mapping* $S_\lambda : E \to E$ *by* $S_\lambda = \lambda I + (1-\lambda)T$, $\forall \lambda \in (0,1)$. *Then, for any* $x_1 \in E$, $S_\lambda^n x_1 \to y \in \mathrm{Fix}(T)$ *as* $n \to \infty$.

Proof. Letting $x_n = S_\lambda^n x_1$, one has $\|x_n\| = \|S_\lambda^n x_1\| \le \|x_1\|$ and

$$x_{n+1} = S_\lambda^{n+1} x_1 = S^\lambda x_n = \lambda x_n + (1-\lambda)Tx_n, \quad n \ge 1. \tag{4.16}$$

From Theorem 4.2.3, we obtain that $x_n - Tx_n \to \theta$ as $n \to \infty$, that is,

$$(T - I)S_\lambda^n x_1 \to \theta \text{ as } n \to \infty. \tag{4.17}$$

In view of $S_\lambda^n x_1 = [\lambda I + (1-\lambda)T]^n x_1 = \sum_{j=0}^n C_n^j \lambda^j (1-\lambda)^{n-j} T^{n-j} x_1$, we assert that

$$(T - I)S_\lambda^n x_1 = S_\lambda^n (T - I)x_1 \to \theta \quad \text{as } n \to \infty. \tag{4.18}$$

Next, we prove that

$$\overline{R(T-I)} = \left\{ x \in E : \lim_{n\to\infty} S_\lambda^n x = 0 \right\}. \tag{4.19}$$

Let $z \in \overline{R(T-I)}$. For any $\varepsilon > 0$, there exists $\omega \in R(I - T)$ such that

$$\| z - \omega \| < \varepsilon \Longrightarrow \| S_\lambda^n (z - \omega) \| \le \| z - \omega \| < \varepsilon.$$

It follows that

$$S_\lambda^n(z - \omega) = S_\lambda^n z - S_\lambda^n \omega \to \theta \quad \text{as } n \to \infty.$$

However, we find from (4.18) that $S_\lambda^n \omega \to \theta$ as $n \to \infty$, which in turn implies that $S_\lambda^n z \to \theta$ as $n \to \infty$. This shows that

$$z \in \left\{ x \in E : \lim_{n\to\infty} S_\lambda^n x_1 = \theta \right\}.$$

Conversely, assume that $\lim_{n\to\infty} S_\lambda^n x = \theta$. Then there exists $N \ge 1$ such that, for any $\varepsilon > 0$,

$$\left\| x - (x - S_\lambda^n x) \right\| = \left\| S_\lambda^n x \right\| < \varepsilon, \quad \forall n \ge N. \tag{4.20}$$

In view of the equality

$$x - S_\lambda^n x = \left[I - (\lambda I + (1 - \lambda)T)^n\right]x = (1 - \lambda)(I - T)(I + S_\lambda + S_\lambda^2 + \cdots + S_\lambda^{n-1})x \in R(I - T),$$

we assert that $x \in \overline{R(T - I)}$. This establishes (4.19).

Due to the facts that E is reflexive and $\{S_\lambda^n x_1\}$ is bounded, we may assume, without loss of generality, that $S_\lambda^n x_1 \rightharpoonup y$ $(n \to \infty)$. Since $x_1 - S_\lambda^n x_1 \in R(I - T)$ and $\overline{R(I - T)}$ is weakly closed, we find that $x_1 - S_\lambda^n x_1 \rightharpoonup x_1 - y \in \overline{R(I - T)}$. It follows from (4.19) that $\lim_{n \to \infty} S_\lambda^n(x_1 - y) = \theta$, which further implies that

$$S_\lambda^n x_1 - S_\lambda^n y \to \theta \quad (n \to \infty). \tag{4.21}$$

On the other hand, from $S_\lambda^n x_1 \rightharpoonup y$ and $(I - T)S_\lambda^n x_1 \to \theta$, we see that $y = Ty$. So, $S_\lambda^n y = y$, $\forall n \geq 1$. It follows from (4.21) that $S_\lambda^n x_1 \to y$ $(n \to \infty)$. This completes the proof. □

Remark 4.2.4. Theorem 4.2.5 shows that the Krasnosel'skiĭ–Mann iterative method is strongly convergent for linear nonexpansive mappings. It is an interesting problem whether normal Mann iterative method (NMIM) is also strongly convergent. Recently, Takahashi and Yao [90] investigated this problem in a real uniformly convex Banach space; see [90] for more details and the references therein.

4.3 Halpern iterative method and its strong convergence theorems

Theorem 4.3.1. *Let E be a real reflexive Banach space whose norm is uniformly Gâteaux differentiable. Let C be a nonempty, closed, and convex subset of E such that every nonempty, bounded, closed, and convex subset of C has the f. p. p. for nonexpansive self-mappings. Let $T : C \to E$ be a nonexpansive mapping satisfying the weak inward condition (WIC) and $\mathrm{Fix}(T) \neq \emptyset$. Then, for any fixed $u \in C$, for any $t \in (0, 1]$, there exists a unique bounded continuous path $\{x_t\} \subset C$ such that*

$$x_t = tu + (1 - t)Tx_t, \quad t \in (0, 1] \tag{4.22}$$

and $x_t \to p = Q_{\mathrm{Fix}(T)}u = \lim_{t \to 0} x_t$, where $Q_{\mathrm{Fix}(T)} : C \to \mathrm{Fix}(T)$ is a unique sunny nonexpansive retraction mapping from C onto $\mathrm{Fix}(T)$.

Proof. For any fixed $u \in C$ and any $t \in (0, 1]$, we define $T_t : C \to E$ by

$$T_t x = tu + (1 - t)Tx, \quad \forall x \in C. \tag{4.23}$$

Then, for any $t \in (0, 1]$, one sees that $T_t : C \to E$ is contractive. Applying Banach–Caristi theorem (Theorem 4.1.1), one concludes that T_t has a unique fixed point $x_t \in C$, that is, $x_t = T_t x_t$, $t \in (0, 1]$. From (4.23), one has

$$x_t = tu + (1 - t)Tx_t, \quad \forall t \in (0, 1]. \tag{4.24}$$

This shows the existence and uniqueness of the path $\{x_t\}$, which satisfies (4.22). From the assumption that $\text{Fix}(T) \neq \emptyset$ and (4.22), we find that

$$\|x_t - p\|^2 = \langle x_t - p, j(x_t - p) \rangle$$
$$= t\langle u - p, j(x_t - p) \rangle + (1 - t)\langle Tx_t - p, j(x_t - p) \rangle$$
$$\leq t\langle u - p, j(x_t - p) \rangle + (1 - t)\|x_t - p\|^2, \quad \forall p \in \text{Fix}(T).$$

It follows that

$$\|x_t - p\|^2 \leq \langle u - p, j(x_t - p) \rangle, \quad \forall p \in \text{Fix}(T). \tag{4.25}$$

Hence $\|x_t - p\| \leq \|u - p\|$, $\forall t \in (0,1)$, which implies that $\{x_t\}$ is bounded. This yields that $\{Tx_t\}$ is also bounded.

Next, we prove that $\{x_t\}$ is continuous. To this end, $\forall t, s \in (0,1)$, we see that there exist unique paths $\{x_t\}$ and $\{x_s\}$, respectively, satisfying (4.22). Thus,

$$x_t - x_s = (t - s)u + (1 - t)Tx_t - (1 - s)Tx_s. \tag{4.26}$$

It follows that

$$\|x_t - x_s\| \leq |t - s|\|u\| + (1 - t)\|Tx_t - Tx_s\| + |t - s|\|Tx_s\|$$
$$\leq |t - s|(\|u\| + \|Tx_s\|) + (1 - t)\|x_t - x_s\|.$$

This implies that

$$\|x_t - x_s\| \leq \frac{|t - s|}{t}(\|u\| + \|Tx_s\|) \tag{4.27}$$

Fix $s \in (0,1)$ and let $t \to s$. Since $\{Tx_s\}$ is bounded, we find that $x_t \to x_s$ as $t \to s$. This shows that the path $\{x_t\}$ is continuous.

Next, we show that $x_t \to p \in \text{Fix}(T)$ as $t \to 0$. Denote the Banach limit by μ_t and define a function $\varphi : C \to \mathbb{R}^+$ by

$$\varphi(y) = \mu_t \|x_t - y\|, \quad \forall y \in C. \tag{4.28}$$

Then $\varphi : C \to \mathbb{R}^+$ is continuous convex and $\varphi(y) \to +\infty$ ($\|y\| \to +\infty$). From the assumption that E is reflexive, while C is closed and convex in E, we see that there exists $z \in C$ such that

$$\varphi(z) = \min_{y \in C} \varphi(y). \tag{4.29}$$

Let

$$K = \left\{ z \in C : \varphi(z) = \min_{y \in C} \varphi(y) \right\}.$$

Then $K \neq \emptyset$. Further, K is bounded and closed convex. Indeed,

$$K = \bigcap_{y \in C} \{z \in C : \varphi(z) \leq \varphi(y)\}.$$

From the above equality, one easily verifies that K is bounded, closed, and convex by means of the continuity, convexity, and the property $\varphi(z) \to \infty$ as $\|z\| \to \infty$.

It follows from Banach contraction principle that $C \subset (2I - T)(C)$. Write $S = (2I - T)^{-1}$. Then $S : C \to C$ is nonexpansive such that $\text{Fix}(S) = \text{Fix}(T)$. We claim that $S(K) \subseteq K$. To see this, in view of (4.22), we have $x_t - Tx_t \to 0$ as $t \to \theta$, which yields that $x_t - Sx_t \to 0$ as $t \to \theta$.

By virtue of the definition of φ, we have for any $z \in K$ that

$$\begin{aligned}
\varphi(Sz) &= \mu_t \|x_t - Sz\| \\
&\leq \mu_t \|x_t - z\| \\
&\leq \varphi(y)
\end{aligned}$$

for all $y \in C$. This shows that $S : K \to K$ is a self-mapping. From the assumption that there exits $z_0 \in K$ such that $z_0 = Sz_0$, that is, $z_0 \in \text{Fix}(T)$, and from Takahashi's result, one has $\mu_t(\langle x - z_0, j(x_t - z_0)\rangle) \leq 0$, $\forall x \in C$. In particular, letting $x = u \in C$, we arrive at

$$\mu_t(\langle u - z_0, j(x_t - z_0)\rangle) \leq 0, \quad \forall t \in (0, 1]. \tag{4.30}$$

On the other hand, one concludes from (4.25) that

$$\langle x_t - u, j(x_t - z_0)\rangle \leq 0, \quad \forall t \in (0, 1].$$

Taking the Banach limit in the above inequality, we arrive at

$$\mu_t(\langle x_t - u, j(x_t - z_0)\rangle) \leq 0. \tag{4.31}$$

Adding (4.30) and (4.31), we obtain $\mu_t(\langle x_t - z_0, j(x_t - z_0)\rangle) \leq 0$. So, $\mu_t(\|x_t - z_0\|^2) = 0$. Thus, for any $t_n \to 0$, there exits a subsequence of $\{x_{t_n}\}$ such that $x_{t_n} \to z_0$ as $n \to \infty$. If $s_n \to 0$ and $x_{s_n} \to p \in \text{Fix}(T)$ as $n \to \infty$, then $p = z_0$. It follows from (4.25) that

$$\langle z_0 - u, j(z_0 - p)\rangle \leq 0 \tag{4.32}$$

and

$$\langle p - u, j(p - z_0)\rangle \leq 0. \tag{4.33}$$

Adding (4.32) and (4.33), we find that

$$\|z_0 - p\|^2 = \langle z_0 - p, j(z_0 - p)\rangle \leq 0,$$

that is, $p = z_0$. Hence, $\{x_t\}$ strongly converges to $p \in \text{Fix}(T)$. Defining $Q_{\text{Fix}(T)} : C \to \text{Fix}(T)$ by $Q_{\text{Fix}(T)}u = \lim_{t \to 0} x_t$, one easily verifies that $Q_{\text{Fix}(T)} : C \to \text{Fix}(T)$ is a unique sunny nonexpansive retraction from C onto $\text{Fix}(T)$. This completes the proof. $\qquad\square$

Corollary 4.3.1 (Reich [73], 1979). *Let E be a uniformly smooth Banach space and let C be a closed convex subset of E. Let T : C → E be a nonexpansive mapping satisfying the weak inward condition (WIC) and* Fix(T) ≠ ∅. *Then, for any fixed u ∈ C and for any t ∈ (0, 1], there exists a unique bounded continuous path* {x_t} ⊂ *C satisfying (4.22) and the path* {x_t} *strongly converges to* $p = Q_{\text{Fix}(T)}u$ *as t → 0.*

Corollary 4.3.2. *Let E be a uniformly convex Banach space whose norm is uniformly Gâteaux differentiable. Let C be a closed and convex subset of E. Let T : C → E be a nonexpansive mapping satisfying the weak inward condition (WIC) and* Fix(T) ≠ ∅. *Then, for any fixed u ∈ C and for any t ∈ (0, 1], there exists a unique bounded continuous path* {x_t} ⊂ *C satisfying (4.22) and the path* {x_t} *strongly converges to* $p = Q_{\text{Fix}(T)}u$, *as t → 0.*

In order to establish strong convergence theorems of Halpern iterative methods, we first give the following result.

Lemma 4.3.1. *Let E be a real reflexive Banach space whose norm is uniformly Gâteaux differentiable. Let C be a nonempty, closed, and convex subset of E such that every nonempty, bounded, closed, and convex subset of C has the f. p. p. for every nonexpansive self-mapping. Let T : C → E be a nonexpansive mapping satisfying the weak inward condition (WIC) and* Fix(T) ≠ ∅. *Let* {x_n} ⊂ *C be a bounded approximate fixed point sequence of T, that is,* {x_n} *is bounded and* $x_n - Tx_n \to \theta$ *as n → ∞. Then*

$$\limsup_{n\to\infty}\langle u - p, j(x_n - p)\rangle \le 0$$

where $p = Q_{\text{Fix}(T)}u$, $Q_{\text{Fix}(T)} : C \to F(T)$ *is the unique sunny nonexpansive retraction from C onto* Fix(T).

Proof. From Theorem 4.3.1, one sees that there exists a unique sunny nonexpansive retraction $Q_{\text{Fix}(T)} : C \to F(T)$, where $Q_{\text{Fix}(T)}u = \lim_{t\to 0} x_t = p$. On the other hand, {$x_t$} satisfies (4.22), that is,

$$x_t = tu + (1 - t)Tx_t, \quad \forall t \in (0, 1]. \tag{4.34}$$

Let $M = \sup\{\|x_n - x_t\|, t \in (0, 1), n \ge 1\}$. It follows that

$$\begin{aligned}
\|x_t - x_n\|^2 &= \langle x_t - x_n, j(x_t - x_n)\rangle \\
&= t\langle u - x_n, j(x_t - x_n)\rangle + (1 - t)\langle Tx_t - x_n, j(x_t - x_n)\rangle \\
&= t\langle u - x_t, j(x_t - x_n)\rangle + t\|x_t - x_n\|^2 \\
&\quad + (1 - t)\langle Tx_t - Tx_n, j(x_t - x_n)\rangle + (1 - t)\langle Tx_n - x_n, j(x_t - x_n)\rangle \\
&\le t\langle u - x_t, j(x_t - x_n)\rangle + \|x_t - x_n\|^2 + (1 - t)M\|x_n - Tx_n\|.
\end{aligned}$$

It follows that

$$\langle u - x_t, j(x_n - x_t)\rangle \le \frac{M}{t}\|x_n - Tx_n\|. \tag{4.35}$$

Fixing $t \in (0,1)$ and letting $n \to \infty$, it follows from (4.35) that

$$\limsup_{n\to\infty}\langle u - x_t, j(x_n - x_t)\rangle \le 0. \tag{4.36}$$

Note that

$$\langle u-x_t, j(x_n-x_t)\rangle = \langle u-p, j(x_n-p)\rangle + \langle u-p, j(x_n-x_t)-j(x_n-p)\rangle + \langle p-x_t, j(x_n-x_t)\rangle. \tag{4.37}$$

Due to the fact that $x_t \to p$ as $t \to 0$, we find from the boundedness of $\{x_n\}$ and $\{x_t\}$ that

$$\langle p - x_t, j(x_n - x_t)\rangle \to \theta \quad \text{as } t \to \theta. \tag{4.38}$$

Since the norm of E is uniformly Gâteaux differentiable, we have that $j : E \to E^*$ is s-w uniformly continuous on any bounded subset of E. Hence,

$$\langle u - p, j(x_n - x_t) - j(x_n - p)\rangle \to \theta \quad \text{as } t \to \theta. \tag{4.39}$$

From (4.38) and (4.39), for any $\varepsilon > 0$, we see that there exists $\delta = \delta(\varepsilon) > 0$ such that, for any $n \ge 1$,

$$\left|\langle p - x_t, j(x_n - x_t)\rangle\right| < \frac{\varepsilon}{2}$$

and

$$\left|\langle u - p, j(x_n - x_t) - j(x_n - p)\rangle\right| < \frac{\varepsilon}{2}$$

as $0 < t < \delta$. Taking the limits on both sides of (4.37), we find that

$$\limsup_{n\to\infty}\langle u - p, j(x_n - p)\rangle \le \frac{\varepsilon}{2} + \frac{\varepsilon}{2} = \varepsilon.$$

Since $\varepsilon > 0$ is chosen arbitrarily, we find that

$$\limsup_{n\to\infty}\langle u - p, j(x_n - p)\rangle \le 0.$$

This completes the proof. □

Remark 4.3.1.
(i) If path $\{x_t\}$ exists and strongly converges to p, then the assumptions that E is reflexive and C has the f. p. p. can be removed.
(ii) If E is strictly convex, then the assumption that C has the f. p. p. can be removed.
(iii) If E satisfies Opial condition and the duality mapping $J : E \to E^*$ is a weakly sequentially continuous at zero, then the assumptions that the norm of E is uniformly Gâteaux differentiable and C has the f. p. p. can be removed.

Theorem 4.3.2. *Let E be a real reflexive Banach space whose norm is uniformly Gâteaux differentiable. Let C be a nonempty, closed, and convex subset of E such that every nonempty, bounded, closed, and convex subset of C has the f. p. p. for every nonexpansive self-mapping. Let $T : C \to C$ be a nonexpansive mapping such that $\mathrm{Fix}(T) \neq \emptyset$. Let $\{\alpha_n\}$ be a sequence in $[0, 1]$ satisfying the control conditions:*
(i) $\lim_{n\to\infty} \alpha_n = 0$;
(ii) $\sum_{n=1}^{\infty} \alpha_n = \infty$;
(iii) $\sum_{n=1}^{\infty} |\alpha_{n+1} - \alpha_n| < \infty$ *or*
(iv) $\lim_{n\to\infty} \frac{\alpha_{n+1}}{\alpha_n} = 1$.

Let $u \in C$ be a fixed element. For any initial value $x_1 \in C$, define a sequence $\{x_n\}$ in following manner:

$$x_{n+1} = \alpha_n u + (1 - \alpha_n) T x_n, \quad n \geq 1. \tag{HIM}$$

Then the sequence $\{x_n\}$ converges strongly to some point $p = Q_{\mathrm{Fix}(T)} u$, where $Q_{\mathrm{Fix}(T)} : C \to \mathrm{Fix}(T)$ is a unique sunny nonexpansive retraction mapping from C onto $\mathrm{Fix}(T)$.

Proof. The proof is split into three steps.
Step 1. Show that $\{x_n\}$ is a bounded sequence.
Fixing $p \in \mathrm{Fix}(T)$, we have

$$\begin{aligned}
\|x_{n+1} - p\| &= \|\alpha_n(u - p) + (1 - \alpha_n)(T x_n - p)\| \\
&\leq \alpha_n \|u - p\| + (1 - \alpha_n)\|x_n - p\| \\
&\leq \max\{\|u - p\|, \|x_1 - p\|\} \triangleq M.
\end{aligned}$$

Then $\{x_n\}$ is bounded, so is $\{T x_n\}$. Since $\alpha_n \to 0$ as $n \to \infty$, one has

$$x_{n+1} - T x_n = \alpha_n(u - T x_n) \to \theta \quad \text{as } n \to \infty. \tag{4.40}$$

Step 2. Show that $x_n - T x_n \to \theta$ as $n \to \infty$.
Note that

$$\begin{aligned}
x_{n+1} - x_n &= (\alpha_n - \alpha_{n-1})u + (1 - \alpha_n)T x_n - (1 - \alpha_{n-1})T x_{n-1} \\
&= (\alpha_n - \alpha_{n-1})u + (1 - \alpha_n)T x_n - (1 - \alpha_n)T x_{n-1} \\
&\quad + (1 - \alpha_n)T x_{n-1} - (1 - \alpha_{n-1})T x_{n-1} \\
&= (\alpha_n - \alpha_{n-1})(u - T x_{n-1}) + (1 - \alpha_n)(T x_n - T x_{n-1}).
\end{aligned}$$

It follows that

$$\begin{aligned}
\|x_{n+1} - x_n\| &\leq (1 - \alpha_n)\|x_n - x_{n-1}\| + |\alpha_n - \alpha_{n-1}|(\|u\| + \|T x_{n-1}\|) \\
&\leq (1 - \alpha_n)\|x_n - x_{n-1}\| + M|\alpha_n - \alpha_{n-1}|, \tag{4.43}
\end{aligned}$$

where $M = \sup\{\|u\| + \|T x_n\|, \ n \geq 1\}$. Using Lemma 1.10.2, we have $x_{n+1} - x_n \to \theta$ as $n \to \infty$. In view of (4.40), we find that $x_n - T x_n = x_n - x_{n+1} + x_{n+1} - T x_n \to \theta$ as $n \to \infty$.

Step 3. Show that $x_n \to p = Q_{\text{Fix}(T)}u \ (n \to \infty)$.
From Lemma 4.3.1, we find that

$$\limsup_{n\to\infty}\langle u - p, j(x_n - p)\rangle \leq 0. \tag{4.41}$$

It follows that

$$\begin{aligned}
\|x_{n+1} - p\|^2 &= \alpha_n\langle u - p, j(x_{n+1} - p)\rangle + (1 - \alpha_n)\langle Tx_n - p, j(x_{n+1} - p)\rangle \\
&\leq (1 - \alpha_n)\|x_n - p\|\|x_{n+1} - p\| + \alpha_n\langle u - p, j(x_{n+1} - p)\rangle \\
&\leq \frac{1 - \alpha_n}{2}\|x_n - p\|^2 + \frac{1 - \alpha_n}{2}\|x_{n+1} - p\|^2 + \alpha_n\langle u - p, j(x_{n+1} - p)\rangle.
\end{aligned}$$

This yields that

$$\|x_{n+1} - p\|^2 \leq (1 - \alpha_n)\|x_n - p\|^2 + 2\alpha_n\langle u - p, j(x_{n+1} - p)\rangle. \tag{4.42}$$

Setting $a_n = \|x_n - p\|^2$, $t_n = \alpha_n$, and $b_n = 2\alpha_n\langle u - p, j(x_{n+1} - p)\rangle$, we obtain from (4.42) that

$$a_{n+1} \leq (1 - t_n)a_n + b_n,$$

where $\{t_n\}$ and $\{b_n\}$ satisfy the following conditions:
(1) $\sum_{n=1}^{\infty}t_n = \sum_{n=1}^{\infty}\alpha_n = \infty$;
(2) $\limsup_{n\to\infty}\frac{b_n}{t_n} = 2\limsup_{n\to\infty}\langle u - p, j(x_{n+1} - p)\rangle \leq 0$.
By virtue of Lemma 1.10.2, we find that $a_n \to 0$ as $n \to \infty$, that is, $x_n \to p = Q_{\text{Fix}(T)}u$ as $n \to \infty$. This completes the proof. $\qquad\square$

Corollary 4.3.3. *Let E be a real uniformly smooth Banach space. Let C be a nonempty, closed, and convex subset of E. Let $T : C \to C$ be a nonexpansive mapping such that $\text{Fix}(T) \neq \emptyset$. Let $\{x_n\}$ be a sequence defined in Theorem 4.3.2. Then $x_n \to p = Q_{\text{Fix}(T)}u$ as $n \to \infty$.*

Corollary 4.3.4. *Let E be a real uniformly convex Banach space whose norm is uniformly Gâteaux differentiable. Let C be a nonempty, closed, and convex subset of E. Let $T : C \to C$ be a nonexpansive mapping such that $\text{Fix}(T) \neq \emptyset$. Let $\{x_n\}$ be a sequence defined in Theorem 4.3.2. Then $x_n \to p = Q_{\text{Fix}(T)}u$ as $n \to \infty$.*

In 2009, Suzuki [85] pointed out that conditions (i) and (ii) of Theorem 4.3.2 are not sufficient to guarantee the strong convergence of (HIA).

Example 4.3.1. Let $E = \mathbb{R}$, $C = [-1, 1]$, and $u = 1$. Define a nonexpansive mapping $T : C \to C$ by $Tx = -x$, $\forall x \in C$. Then $\text{Fix}(T) = \{0\}$. Let $\{\alpha_n\}$ be a real sequence in $[0, 1]$ defined by

$$\alpha_n = \begin{cases} 0, & n \text{ is an odd number}, \\ \frac{1}{n}, & n \text{ is an even number}. \end{cases}$$

Then $\{\alpha_n\}$ satisfies conditions (i) and (ii) of Theorem 4.3.2. However, $x_n = (-1)^{(n+1)}$ is not convergent for the initial value $x_1 = 1$. This further implies that $\{x_n\}$ does not converge to the fixed point of T.

In 2007, Suzuki [84] introduced the following iterative method:

$$x_1, u \in C, \quad x_{n+1} = \alpha_n u + (1 - \alpha_n)[\lambda T x_n + (1 - \lambda)x_n], \tag{HSIM}$$

where λ is a real number in $(0, 1)$. This method is now called the Halpern–Suzuki iterative method. Suzuki [84], based on new analysis techniques, proved that the sequence generated by (HSIM) converges to a fixed point of T under conditions (i) and (ii) of Theorem 4.3.2.

Theorem 4.3.3 (Suzuki [84], 2007). *Let E be a real Banach space whose norm is uniformly Gâteaux differentiable. Let C be a nonempty, closed, and convex subset of E. Let $T : C \to C$ be a nonexpansive mapping such that $\mathrm{Fix}(T) \neq \emptyset$. Let $\{\alpha_n\}$ be a real sequence satisfying conditions (i) and (ii) of Theorem 4.3.2. Let $\{x_n\}$ be a sequence defined by (HSIM) and let $\{x_t\}$ be a path defined by (4.22). If $\{x_t\}$ converges strongly to some $p \in \mathrm{Fix}(T)$ as $t \to 0$, then $x_n \to p = Q_{\mathrm{Fix}(T)} u \ (n \to \infty)$.*

Proof. The key step is to prove that $\{x_n\}$ is a bounded approximate fixed point sequence of T.

(i) $\{x_n\}$ is bounded.

Let $T_\lambda = \lambda T + (1 - \lambda)T$, $\lambda \in (0, 1)$. For any $P \in \mathrm{Fix}(T)$, one has

$$\begin{aligned}
\|x_{n+1} - p\| &= \|\alpha_n(u - p) + (1 - \alpha_n)(T_\lambda x_n - p)\| \\
&\leq \alpha_n \|u - p\| + (1 - \alpha_n)\|x_n - p\| \\
&\leq \max\{\|u - p\|, \|x_1 - p\|\} \triangleq M,
\end{aligned}$$

which shows that $\{x_n\}$ is bounded, so is $\{T_\lambda x_n\}$. Since $\alpha_n \to 0$ as $n \to \infty$, we have that

$$x_{n+1} - T_\lambda x_n = \alpha_n(u - T_\lambda x_n) \to \theta \quad (n \to \infty). \tag{4.43}$$

(ii) $\lim_{n \to \infty} \|x_n - T x_n\| = 0$.

Write $\lambda_n = (1 - \lambda)(1 - \alpha_n)$ and define a sequence $\{y_n\}$ by

$$y_n = \frac{\alpha_n}{1 - \lambda_n} u + \frac{\lambda(1 - \alpha_n)}{1 - \lambda_n} T x_n.$$

Then (HSIM) reduces to

$$x_{n+1} = \lambda_n x_n + (1 - \lambda_n)y_n, \quad \forall n \geq 1. \tag{4.44}$$

It follows that

$$\|y_{n+1} - y_n\| \leq \left|\frac{\alpha_{n+1}}{1 - \lambda_{n+1}} - \frac{\alpha_n}{1 - \lambda_n}\right|\|u\| + \frac{\lambda(1 - \alpha_{n+1})}{1 - \lambda_{n+1}}\|x_{n+1} - x_n\|$$
$$+ \left|\frac{\lambda(1 - \alpha_{n+1})}{1 - \lambda_{n+1}} - \frac{\lambda(1 - \alpha_n)}{1 - \lambda_n}\right|\|Tx_n\|.$$

Since $\alpha_n \to 0$, we see that $1 - \lambda_n \to \lambda$. This, together with the boundedness of $\{Tx_n\}$, implies that

$$\limsup_{n\to\infty}(\|y_{n+1} - y_n\| - \|x_{n+1} - x_n\|) \leq 0.$$

From Lemma 1.10.3, we find that $y_n - x_n \to \theta$ as $n \to \infty$. Hence $x_{n+1} - x_n = (1 - \lambda_n)(y_n - x_n) \to \theta$ as $n \to \infty$. It follows from (4.43) that $x_n - T_\lambda x_n \to \theta$ as $n \to \infty$, which yields $x_n - Tx_n \to \theta$ as $n \to \infty$. From Lemma 4.3.1 and Remark 4.3.1, we have $\limsup_{n\to\infty}\langle u - p, j(x_n - p)\rangle \leq 0$.

(iii) $x_n \to p = Q_{\text{Fix}(T)}u$ $(n \to \infty)$.

From (HSIM), we have

$$\|x_{n+1} - p\|^2 = \alpha_n\langle u - p, j(x_n - p)\rangle + (1 - \alpha_n)\langle T_\lambda x_n - p, j(x_{n+1} - p)\rangle$$
$$\leq (1 - \alpha_n)\|x_n - p\|\|x_{n+1} - p\| + \alpha_n\langle u - p, j(x_{n+1} - p)\rangle$$
$$\leq \frac{1 - \alpha_n}{2}\|x_n - p\|^2 + \frac{1 - \alpha_n}{2}\|x_{n+1} - p\|^2$$
$$+ \alpha_n\langle u - p, j(x_{n+1} - p)\rangle,$$

which yields

$$\|x_{n+1} - p\|^2 \leq (1 - \alpha_n)\|x_n - p\|^2 + 2\alpha_n\langle u - p, j(x_{n+1} - p)\rangle. \tag{4.45}$$

Putting $a_n = \|x_{n+1} - p\|^2$, $\alpha_n = t_n$, and

$$b_n = 2\alpha_n\langle u - p, j(x_{n+1} - p)\rangle,$$

we conclude from (4.45) that

$$a_{n+1} \leq (1 - t_n)a_n + b_n, \quad \forall n \geq 1.$$

It follows from Lemma 1.10.2 that $a_n \to 0$ as $n \to \infty$, that is, $x_n \to p = Q_{\text{Fix}(T)}u$ $(n \to \infty)$. □

Corollary 4.3.5. *Let E be a real uniformly smooth Banach space. Let C be a nonempty, closed, and convex subset of E. Let $T : C \to C$ be a nonexpansive mapping such that $\text{Fix}(T) \neq \emptyset$. Let $\{\alpha_n\}$ be a real sequence satisfying conditions (i) and (ii) of Theorem 4.3.2. Let $\{x_n\}$ be generated by (HSIM). Then $x_n \to p = Q_{\text{Fix}(T)}u$ as $n \to \infty$.*

Corollary 4.3.6. *Let E be a real uniformly convex Banach space whose norm is uniformly Gâteaux differentiable. Let C be a nonempty, closed, and convex subset of E. Let $T : C \to C$ be a nonexpansive mapping such that $\mathrm{Fix}(T) \neq \emptyset$. Let $\{\alpha_n\}$ be a real sequence satisfying conditions (i) and (ii) of Theorem 4.3.2. Let $\{x_n\}$ be defined by (HSIM). Then $x_n \to p = Q_{\mathrm{Fix}(T)}u(n \to \infty)$ as $n \to \infty$.*

Recently, several authors also investigated the following so-called general Halpern iterative method:

$$x_{n+1} = \alpha_n u + \beta_n x_n + \gamma_n T x_n, \quad n \geq 1,$$

where u is a fixed element in set C, and $\{\alpha_n\}$, $\{\beta_n\}$, and $\{\gamma_n\}$ are real sequences in $(0, 1)$ such that $\alpha_n + \beta_n + \gamma_n = 1$. With some conditions imposed on the control sequences $\{\alpha_n\}$, $\{\beta_n\}$, and $\{\gamma_n\}$, we can show that the sequence generated by the above iterative method is also strongly convergent. Setting $\lambda := \lambda_n$ in (HSIM), the above general Halpern iterative method is equivalent to the (HSIM).

In 2010, Xu [103] investigated Reich-type implicit iterative methods and Halpern-type explicit iterative method.

Theorem 4.3.4 (Xu [103], 2010). *Let E be a real uniformly smooth Banach space. Let C be a closed convex subset of E. Let $T : C \to E$ be a nonexpansive mapping satisfying the weak inward condition (WIC). Then, for any $u \in C$ and $t \in (0, 1]$, there exists a unique bounded continuous path $\{z_t\} \subset C$ satisfying*

$$z_t = T(tu + (1 - t)z_t), \quad t \in (0, 1]. \tag{4.46}$$

Further, $\{z_t\}$ is bounded as $t \to 0$ if and only if $\mathrm{Fix}(T) \neq \emptyset$. If, in addition, $\mathrm{Fix}(T) \neq \emptyset$, then $\{z_t\}$ strongly converges to $z = Q_{\mathrm{Fix}(T)}u$ as $t \to 0$, where $Q_{\mathrm{Fix}(T)}u : C \to \mathrm{Fix}(T)$ is a unique sunny nonexpansive retraction from C onto $\mathrm{Fix}(T)$.

Proof. Fix $u \in C$. For any $t \in (0, 1]$, we define the mapping $T_t : C \to C$ by

$$T_t x = T(tu + (1 - t)x), \quad \forall x \in C. \tag{4.47}$$

Then, for any $t \in (0, 1]$, $T_t : C \to E$ is a contractive mapping satisfying the weak inward condition (WIC). From Banach–Caristi theorem (Theorem 4.1.1), we find that for any $t \in (0, 1]$, there exists a unique continuous path $\{z_t\} \subset C$ satisfying (4.46).

Let $\{x_t\}$ be defined by (4.22). Thus

$$
\begin{aligned}
\|z_t - T x_t\| &= \|T(tu + (1 - t)z_t) - T x_t\| \\
&\leq \|tu + (1 - t)z_t - x_t\| \\
&= \|tu + (1 - t)z_t - tu - (1 - t)T x_t\| \\
&= \|(1 - t)(z_t - T x_t)\| \\
&= (1 - t)\|z_t - T x_t\|, \quad \forall t \in (0, 1).
\end{aligned}
$$

It follows that

$$z_t = Tx_t, \quad \forall t \in (0, 1]. \tag{4.48}$$

From Theorem 4.3.1, $\{x_t\}$ is bounded as $t \to 0$ if and only if $\text{Fix}(T) \neq \emptyset$. From the fact that $\{x_t\}$ is bounded if and only if $\{Tx_t\}$ is bounded, we find from (4.48) that $\{z_t\}$ is bounded as $t \to 0$ if and only if $\text{Fix}(T) \neq \emptyset$. If $\text{Fix}(T) \neq \emptyset$, it follows from Theorem 4.3.1 that $x_t \to Q_{\text{Fix}(T)}u$ as $t \to 0$. Hence $Tx_t \to Q_{\text{Fix}(T)}u$ as $t \to 0$, that is, $z_t \to Q_{\text{Fix}(T)}u$ as $t \to 0$. This completes the proof. $\qquad\square$

Remark 4.3.2. In fact, the paths defined by (4.45) and (4.22) are equivalent to each other. If $z_t \to Q_{\text{Fix}(T)}u$ $(t \to 0)$, it follows from (4.48) that $Tx_t \to Q_{\text{Fix}(T)}u$ $(t \to 0)$. Hence $x_t = tu + (1 - t)Tx_t \to Q_{\text{Fix}(T)}u$ $(t \to 0)$.

Theorem 4.3.5 (Xu [103], 2010). *Let E be a real uniformly smooth Banach space. Let C be a nonempty, closed, and convex subset of E. Let $T : C \to C$ be a nonexpansive mapping such that $\text{Fix}(T) \neq \emptyset$. Let $\{\alpha_n\}$ be a sequence satisfying conditions (i)–(iv) of Theorem 4.3.2. For any initial value $x_1 \in C$ and any fixed element $u \in C$, we define a sequence $\{x_n\}$ iteratively by*

$$x_{n+1} = T[\alpha_n u + (1 - \alpha_n)x_n], \quad n \geq 1. \tag{XIM-1}$$

Then $x_n \to z = Q_{\text{Fix}(T)}u$ $(n \to \infty)$, where $Q_{\text{Fix}(T)} : C \to \text{Fix}(T)$ is a unique sunny nonexpansive retraction mapping from C onto $\text{Fix}(T)$.

Proof. For any initial value $y_1 \in C$ and given anchor $u \in C$, we define another sequence $\{y_n\}$ iteratively by

$$y_{n+1} = \alpha_n u + (1 - \alpha_n)Ty_n, \quad n \geq 1. \tag{4.49}$$

By applying Theorem 4.3.2, we have $y_n \to Q_{\text{Fix}(T)}u$ as $n \to \infty$, so that $Ty_{n+1} \to Q_{\text{Fix}(T)}u$ as $n \to \infty$, since T is continuous.

It follows from (4.49) and (XIM-1) that

$$\|x_{n+1} - Ty_{n+1}\| \leq (1 - \alpha_n)\|x_n - Ty_n\|.$$

By using condition $\sum_{n=1}^{\infty} \alpha_n = \infty$, we conclude that $x_n - Ty_n \to \theta$ as $n \to \infty$, that is, $x_n \to Q_{\text{Fix}(T)}u$ as $n \to \infty$. This completes the proof. $\qquad\square$

Remark 4.3.3. In fact, the convergence of sequence $\{x_n\}$ defined by (XIM-1) and sequence $\{y_n\}$ defined by (4.49) implies one another. If $x_n \to Q_{\text{Fix}(T)}u$ $(n \to \infty)$, then $Ty_n \to Q_{\text{Fix}(T)}u$ $(n \to \infty)$. From (4.49), one has

$$y_{n+1} = \alpha_n u + (1 - \alpha_n)Ty_n \to Q_{\text{Fix}(T)}u \quad (n \to \infty).$$

Theorem 4.3.6 (Xu [103], 2010). *Let E be a real uniformly smooth Banach space. Let C be a nonempty, closed, and convex subset of E. Let $T : C \to C$ be a nonexpansive mapping such that $\mathrm{Fix}(T) \neq \emptyset$. Let $\{\alpha_n\}$ be a real sequence satisfying the control conditions (i) and (ii) of Theorem 4.3.2. Fix $u \in C$. For any initial value $x_1 \in C$, we define a sequence $\{x_n\}$ iteratively by*

$$x_{n+1} = \lambda x_n + (1 - \lambda)T[\alpha_n u + (1 - \alpha_n)x_n], \quad \lambda \in (0,1), n \geq 1. \qquad \text{(XIM-2)}$$

Then $x_n \to z = Q_{\mathrm{Fix}(T)}u$ as $n \to \infty$, where $Q_{\mathrm{Fix}(T)} : C \to \mathrm{Fix}(T)$ is a unique sunny nonexpansive retraction mapping from C onto $\mathrm{Fix}(T)$.

Proof. The key is to prove $x_n - Tx_n \to \theta$ as $n \to \infty$.

Step 1. Show that $\{x_n\}$ is bounded.

For any $p \in \mathrm{Fix}(T)$, it follows from (XIM-2) that

$$
\begin{aligned}
\|x_{n+1} - p\| &\leq \lambda\|x_n - p\| + (1 - \lambda)(\alpha_n\|u - p\| + (1 - \alpha_n)\|x_{n+1} - p\|) \\
&= (1 - (1 - \lambda)\alpha_n)\|x_{n+1} - p\| + (1 - \lambda)\alpha_n\|u - p\| \\
&\leq \{\|u - p\|, \|x_1 - p\|\} \triangleq M, \quad \forall n \geq 1,
\end{aligned}
$$

which shows that $\{x_n\}$ is bounded, and so is $\{Tx_n\}$.

Step 2. Show that $x_n - Tx_n \to \theta$ as $n \to \infty$.

Letting $y_n = T[\alpha_n u + (1 - \alpha_n]x_n)$, one sees that (XIM-2) reduces to

$$x_{n+1} = \lambda x_n + (1 - \lambda)y_n, \quad n \geq 1. \qquad (4.50)$$

Letting $M = \sup\{\|u\| + \|x_n\| : n \geq 1\}$, one has

$$
\begin{aligned}
\|y_{n+1} - y_n\| &= \|T[\alpha_{n+1}u + (1 - \alpha_{n+1})x_{n+1}] - T[\alpha_n u + (1 - \alpha_n)x_n]\| \\
&\leq |\alpha_{n+1} - \alpha_n|(\|u\| + \|x_n\|) + (1 - \alpha_{n+1})\|x_{n+1} - x_n\| \\
&\leq M|\alpha_{n+1} - \alpha_n| + \|x_{n+1} - x_n\|.
\end{aligned}
$$

This implies that $\limsup_{n\to\infty}(\|y_{n+1} - y_n\| - \|x_{n+1} - x_n\|) \leq 0$. Using (1.10.3), we find that $y_n - x_n \to \theta$ as $n \to \infty$. Observe that

$$\|x_n - Tx_n\| \leq \|x_n - y_n\| + \|y_n - Tx_n\| \leq \|x_n - y_n\| + \alpha_n\|u - x_n\|,$$

we deduce that $x_n - Tx_n \to \theta$ as $n \to \infty$.

Step 3. Show that $x_n \to z = Q_{\mathrm{Fix}(T)}u$ as $n \to \infty$.

Lemma 4.3.1 ensures that $\limsup_{n\to\infty}\langle u - z, j(x_n - z)\rangle \leq 0$. From (XIM-2), the convexity of $\|\cdot\|^2$, and Reich inequality (RI), for some fixed positive constant M, we find that

$$
\begin{aligned}
\|x_{n+1} - z\|^2 &\le \lambda\|x_n - z\|^2 + (1 - \lambda)\|T(\alpha_n u + (1 - \alpha_n)x_n) - z\|^2 \\
&\le \lambda\|x_n - z\|^2 + (1 - \lambda)\|\alpha_n(u - z) + (1 - \alpha_n)(x_n - z)\|^2 \\
&\le \lambda\|x_n - z\|^2 + (1 - \lambda)(1 - \alpha_n)\|x_n - z\|^2 \\
&\quad + 2(1 - \lambda)\alpha_n(1 - \alpha_n)\langle u - z, j(x_n - z)\rangle \\
&\quad + (1 - \lambda)\max\{\|x_n - z\|, 1\}\alpha_n\|u - z\|\beta(\alpha_n\|u - z\|) \\
&\le [1 - (1 - \lambda)\alpha_n]\|(x_n - z)\|^2 \\
&\quad + 2(1 - \lambda)\alpha_n(1 - \alpha_n)\langle u - z, j(x_n - z)\rangle + M\alpha_n\beta(\alpha_n). \qquad (4.51)
\end{aligned}
$$

Let $t_n = (1 - \lambda)\alpha_n$,

$$
b_n = 2(1 - \lambda)\alpha_n(1 - \alpha_n)\langle u - z, j(x_n - z)\rangle + M\alpha_n\beta(\alpha_n),
$$

and $a_n = \|(x_n - z)\|^2$. Then (4.51) yields

$$
a_n \le (1 - t_n)a_n + b_n, \quad n \ge 1,
$$

where $\{t_n\}$ and $\{b_n\}$ satisfy conditions (i) and (ii) of Lemma 1.10.2. From Lemma 1.10.2, one see that $a_n \to 0 \ (n \to \infty)$, that is $x_n \to z = Q_{\text{Fix}(T)}u$ as $n \to \infty$. This completes the proof. □

Remark 4.3.4. It is not clear whether the convergence of (XIM-2) can be deduced from the convergence of (HSIM). However, (XIM-2) can be modified so that its convergence can be deduced from the convergence of the (HSIM).

Theorem 4.3.7. *Let E be a real uniformly smooth Banach space. Let C be a nonempty, closed, and convex subset of E. Let $T : C \to C$ be a nonexpansive mapping with $\text{Fix}(T) \ne \emptyset$. Let $\{\alpha_n\}$ be a sequence satisfying the control conditions (i) and (ii) of Theorem 4.3.2. Let λ be a real number in $(0, 1)$ and define $T_\lambda = \lambda I + (1 - \lambda)T$. Fix $u \in C$. For any initial value $x_1 \in C$, we define a sequence $\{x_n\}$ iteratively by*

$$
x_{n+1} = T_\lambda[\alpha_n u + (1 - \alpha_n)x_n], \quad n \ge 1. \qquad \text{(MXIM)}
$$

Then $x_n \to z = Q_{\text{Fix}(T)}u$ as $n \to \infty$, where $Q_{\text{Fix}(T)} : C \to \text{Fix}(T)$ is a unique sunny nonexpansive retraction mapping from C onto $\text{Fix}(T)$.

Proof. Letting $y_n = \alpha_n u + (1 - \alpha_n)x_n$, we find from (MXIM) that $x_{n+1} = T_\lambda y_n$. Hence

$$
y_{n+1} = \alpha_{n+1}u + (1 - \alpha_{n+1})T_\lambda y_n, \quad n \ge 1.
$$

Using Theorem 4.3.4, we conclude that $y_n \to z = Q_{\text{Fix}(T)}u$ as $n \to \infty$, that is, $x_n \to z = Q_{\text{Fix}(T)}u$ as $n \to \infty$. □

Remark 4.3.5. Indeed, the convergence of (MXIM) is equivalent to the convergence of (HSIM). Let $\{y_n\}$ be a sequence generated by (HSIM), that is,

$$y_{n+1} = \alpha_n u + (1 - \alpha_n)T_\lambda y_n, \quad n \geq 1. \tag{4.52}$$

Then

$$T_\lambda y_{n+1} = T_\lambda(\alpha_n u + (1 - \alpha_n)T_\lambda y_n), \quad n \geq 1. \tag{4.53}$$

Combining (MXIM) with (4.52), we find that

$$\|x_{n+1} - T_\lambda y_{n+1}\| \leq (1 - \alpha_n)\|x_n - T_\lambda y_n\|, \quad n \geq 1. \tag{4.54}$$

It follows from condition $\sum_{n=1}^{\infty}\alpha_n = \infty$ that $x_n - T_\lambda y_n \to \theta$ as $n \to \infty$. If $x_n \to z = Q_{\text{Fix}(T)}u$ as $n \to \infty$, we have $T_\lambda y_n \to z$ as $n \to \infty$. Using (4.52), we conclude that $y_n \to z = Q_{\text{Fix}(T)}u$ as $n \to \infty$.

4.4 Moudafi's viscosity iterative method

In 2000, Moudafi [56] first introduced a viscosity iterative method with respect to non-expansive mappings in a Hilbert space and established strong convergent theorems for this kind of iterative method. In 2004, Xu [100] revisited Moudafi's viscosity iterative method in the framework of uniformly smooth Banach spaces. He established strong convergent theorems of fixed points, which are solutions of some monotone variational inequality. In 2007, Suzuki [83] extended the Moudafi's viscosity iterative method to the case with Meir–Keeler contractive mappings, which is called the Moudafi-like viscosity iterative method, and established the convergence criteria. In view of Suzuki's results, it is trivial to extend the Halpern-like iterative methods to the Moudafi-like viscosity iterative methods.

In this section, we first introduce Suzuki's results. Then, we derive Xu's results based on Suzuki's results.

It is obvious that a contractive mapping $f : C \to C$ is a Meir–Keeler contractive mapping, however, a Meir–Keeler contractive mapping is only a conditional contraction. To be more clear, it is contractive if we put some restriction on the domain of f.

Proposition 4.4.1. *Let E be a real Banach space. Let C be a nonempty, closed, and convex subset of E. Let $\Phi : C \to C$ be a Meir–Keeler contractive mapping. Then, for any $\varepsilon > 0$, there exists $r \in (0, 1)$ such that*

$$\|x - y\| \geq \varepsilon \implies \|\Phi x - \Phi y\| \leq r\|x - y\|, \quad \forall x, y \in C. \tag{4.55}$$

This implies that the restriction of Φ on the set of $C \cap \{(x, y) \mid \|x - y\| \geq \varepsilon\}$ is contractive.

Proof. Fixing $\varepsilon > 0$, one sees that there exists $\delta \in (0, \varepsilon)$ satisfying

$$\|u - v\| < \frac{\varepsilon}{4} + \delta \implies \|\Phi u - \Phi v\| < \frac{\varepsilon}{4}, \quad \forall u, v \in C.$$

Let $r := \frac{4\varepsilon - \delta}{4\varepsilon} \in (0, 1)$ and denote $[\frac{a}{\varepsilon}]$ by the largest integer not exceeding $\frac{a}{\varepsilon}$. Fix $x, y \in C$ such that $\|x - y\| \geq \varepsilon$. Let $a = \|x - y\|$. From the nonexpansivity of Φ, we have

$$\|\Phi x - \Phi y\| \leq \sum_{j=0}^{[\frac{a}{\varepsilon}]} \left\| \Phi\left(1 - \frac{\varepsilon + \delta}{4a} j\right) x + \frac{\varepsilon + \delta}{4a} jy - \Phi\left(1 - \frac{\varepsilon + \delta}{4a}(j + 1)\right) x + \frac{\varepsilon + \delta}{4a}(j + 1)y \right\|$$

and

$$\|\Phi x - \Phi y\| \leq \sum_{j=0}^{[\frac{a}{\varepsilon}]} \left\| \Phi\left(1 - \frac{\varepsilon + \delta}{4a} j\right) x + \frac{\varepsilon + \delta}{4a} jy - \Phi\left(1 - \frac{\varepsilon + \delta}{4a}(j + 1)\right) x + \frac{\varepsilon + \delta}{4a}(j + 1)y \right\|$$

$$+ \left\| \Phi\left(1 - \frac{\varepsilon + \delta}{4a}\left(\left[\frac{a}{\varepsilon}\right] + 1\right)\right) x + \frac{\varepsilon + \delta}{4a}\left(\left[\frac{a}{\varepsilon}\right] + 1\right) y - \Phi y \right\|$$

$$\leq \left(\left[\frac{a}{\varepsilon}\right] + 1\right) \frac{\varepsilon}{4} + \left(1 - \frac{\varepsilon + \delta}{4a}\left(\left[\frac{a}{\varepsilon}\right] + 1\right) a\right)$$

$$\leq a - \frac{\delta}{4}\left(\left[\frac{a}{\varepsilon}\right] + 1\right) \leq r\|x - y\|. \qquad \square$$

Proposition 4.4.2. *Let E be a real Banach space. Let C be a nonempty, closed, and convex subset of E. Let $T : C \to C$ be a nonexpansive mapping. Let $\Phi : C \to C$ be a Meir–Keeler contractive mapping. Then, we have the following conclusions:*
(1) *$T \circ \Phi : C \to C$ is a Meir–Keeler contractive mapping.*
(2) *$\forall t \in (0, 1)$, the mapping $T_t : C \to C$, defined by*

$$T_t x = (1 - t)Tx + t\Phi x, \quad \forall x \in C,$$

is Meir–Keeler contractive.

Proof. Claim (i) can be derived by the definition of Meir–Keeler contractive mappings.

(ii) Fixing $t \in (0, 1]$, one concludes from Proposition 4.4.1 that, for any $\varepsilon > 0$, there exists $r \in (0, 1)$ satisfying (4.55). Set

$$\delta := \frac{t\varepsilon(1 - r)}{1 - t + tr} > 0$$

and fix $x, y \in C$ such that $\|x - y\| < \varepsilon + \delta$.

Next, we divide the proof into two cases.
(1) If $\|x - y\| \geq \varepsilon$, then we conclude from Proposition 4.4.1 that

$$\|T_t x - T_t y\| \leq (1 - t)\|Tx - Ty\| + t\|\Phi x - \Phi y\|$$

$$\leq (1 - t)\|x - y\| + tr\|x - y\|$$

$$= (1 - t + tr)\|x - y\|$$

$$< (1 - t + tr)(\varepsilon + \delta) = \varepsilon.$$

(2) If $\|x - y\| < \varepsilon$, then

$$\|T_t x - T_t y\| \leq (1 - t)\|Tx - Ty\| + t\|\Phi x - \Phi y\|$$
$$\leq (1 - t)\|x - y\| + t\|x - y\|$$
$$= \|x - y\| < \varepsilon.$$

So, for any $\varepsilon > 0$, there exists $\delta > 0$ such that

$$\|T_t x - T_t y\| < \varepsilon,$$

provided that $\|x - y\| < \varepsilon + \delta$. Therefore, $T_t : C \rightarrow C$ is Meir–Keeler contractive. This completes the proof. □

Proposition 4.4.3. *Let E be a real smooth Banach space. Let C be a nonempty, closed, and convex subset of E, and let K be a subset of C. Let $P : C \rightarrow K$ be the unique sunny nonexpansive retraction and let $\Phi : C \rightarrow C$ be a mapping. For any $z \in K$, we have the following assertions, which are equivalent to each other:*
(i) $z = P \circ \Phi z$;
(ii) $\langle \Phi z - z, j(z - y) \rangle \geq 0, \forall y \in K$.

Proof. (i) $\Longrightarrow$ (ii). Due to the fact that $P : C \rightarrow K$ is the unique sunny nonexpansive retractive mapping, we have

$$\langle x - Px, j(Px - y) \rangle \geq 0, \quad \forall x \in C, y \in K. \tag{4.56}$$

Letting $x = \Phi z \in C$ in the above inequality and using (i), we have

$$\langle \Phi z - z, j(z - y) \rangle \geq 0, \quad \forall y \in K.$$

(ii) $\Longrightarrow$ (i). Let $y = P \circ \Phi z$ in (ii). Then

$$\langle \Phi z - z, j(z - P \circ \Phi z) \rangle \geq 0. \tag{4.57}$$

Letting $x = \Phi z$ and $y = z$ in (4.56), we arrive at

$$\langle \Phi z - P \circ \Phi z, j(P \circ \Phi z - z) \rangle \geq 0. \tag{4.58}$$

Combining (4.57) with (4.58), we have

$$\langle P \circ \Phi z - z, j(z - P \circ \Phi z) \rangle \geq 0, \tag{4.59}$$

that is, $-\|P \circ \Phi z - z\|^2 \geq 0$ and $z = P \circ \Phi z$. This completes the proof. □

Let C be a nonempty, closed, and convex subset of a real Banach space E. Let $\{S_n\}$ be a family of nonexpansive mappings in C. Let $\{\alpha_n\}$ be a real sequence in $(0, 1]$ satisfying condition (i) of Theorem 4.3.2. We say that $(E, C, \{S_n\}, \alpha_n)$ has the Browder's property if, for any $u \in C$, the sequence $\{y_n\}$ defined by

$$y_n = (1 - \alpha_n)S_n y_n + \alpha_n u, \quad n \geq 1, \tag{BIM}$$

is strongly convergent. Let $\{\alpha_n\} \subset (0, 1]$ be a sequence satisfying conditions (i) and (ii) of Theorem 4.3.2. We say that $(E, C, \{S_n\}, \alpha_n)$ has the Halpern's property if, for any $u \in C$, the sequence $\{y_n\}$ defined by

$$y_1 \in C, \quad y_{n+1} = (1 - \alpha_n)S_n y_n + \alpha_n u, \quad n \geq 1, \tag{HIM}$$

is strongly convergent.

Example 4.4.1. Let C be a nonempty, closed, and convex subset of a real uniformly smooth Banach space E. Let $T : C \to E$ be a nonexpansive mapping with $\text{Fix}(T) \neq \emptyset$. Let $\{\alpha_n\}$ be a real sequence in $(0, 1]$ satisfying condition (1) of Theorem 4.3.2. Let $S_n = T$, $\forall n \geq 1$. Then $(E, C, \{S_n\}, \alpha_n)$ has the Browder's property. Furthermore, $\{\alpha_n\} \subset (0, 1]$ also satisfies condition (2) of Theorem 4.3.2. For any $\lambda \in (0, 1)$, put $T_\lambda = \lambda I + (1 - \lambda)T$ and $S_n = T_\lambda$, $\forall n \geq 1$. Then $(E, C, \{S_n\}, \alpha_n)$ has the Halpern's property.

Suzuki also gave the following result.

Theorem 4.4.1 (Suzuki [83], 2007). *Let C be a nonempty, closed, and convex subset of a real Banach space E, and let $\{S_n\}$ be a family of nonexpansive mappings on C. Let $\{\alpha_n\}$ a real sequence in $(0, 1]$ satisfying condition (i) of Theorem 4.3.2 and let $\Phi : C \to C$ be a Meir–Keeler contractive mapping. Assume that the system $(E, C, \{S_n\}, \alpha_n)$ has the Browder's property. Let $\{y_n\}$ be generated by (BIM). Let $P : C \to C$ be defined by $Pu = \lim_{n \to \infty} y_n$ and define a sequence $\{x_n\} \subset C$ by*

$$x_n = (1 - \alpha_n)S_n x_n + \alpha_n \Phi x_n, \quad n \geq 1. \tag{VBIM}$$

Then $\{x_n\}$ strongly converges to the unique point $z = P \circ \Phi z$.

Proof. In fact, it follows from (ii) of Proposition 4.4.2, we have

$$T_n x = (1 - \alpha_n)S_n x + \alpha_n \Phi x, \quad \forall x \in C,$$

which is a Meir–Keeler contractive mapping. From the Meir–Keeler fixed point theorem, we have T_n has a unique fixed point in C, denoted by $x_n \in C$, that is, $x_n = T_n x_n$. This shows that (VBIM) is well defined. Next, we prove that $P : C \to C$ is a nonexpansive mapping. Since $(E, C, \{S_n\}, \alpha_n)$ has the Browder's property, one has that $y_n = (1 - \alpha_n)S_n x_n + \alpha_n u$, $\forall u \in C$ is strongly convergent. Since $\{y_n\} \subset C$, and C is closed, we have $Pu \in C$. Also $\forall u, v \in C$, let $\{u_n\}$ and $\{u_n\}$ be two sequences defined by (BIM), that is,

$$u_n = (1 - \alpha_n)S_n u_n + \alpha_n u, \quad n \geq 1,$$
$$v_n = (1 - \alpha_n)S_n v_n + \alpha_n v, \quad n \geq 1.$$

It follows that

$$\|u_n - v_n\| \leq (1 - \alpha_n)\|S_n u_n - S_n v_n\| + \alpha_n\|u - v\|$$
$$\leq (1 - \alpha_n)\|u_n - v_n\| + \alpha_n\|u - v\|.$$

Thus, $\|u_n - v_n\| \leq \|u - v\|$. Taking the limit, we find that $\|Pu - Pv\| \leq \|u - v\|$. This shows that $P : C \to C$ is nonexpansive. From (i) of Proposition 4.4.2, we have $P \circ \Phi$ is Meir–Keeler contractive. Using the Meir–Keeler fixed point theorem, we conclude that there exists a unique $z \in C$ such that $z = P \circ \Phi(z)$. Define another sequence $\{y_n\} \subset C$ by

$$y_n = (1 - \alpha_n)S_n y_n + \alpha_n \Phi z, \quad n \geq 1. \tag{4.60}$$

Since $(E, C, \{S_n\}, \alpha_n)$ has the Browder's property, one has that sequence $\{y_n\} \subset C$ strongly converges to $z = P \circ \Phi z$. Combining (VBIM) with (4.60), we have

$$\|x_n - y_n\| \leq (1 - \alpha_n)\|S_n x_n - S_n y_n\| + \alpha_n\|\Phi x_n - \Phi z\|$$
$$\leq (1 - \alpha_n)\|x_n - y_n\| + \alpha_n\|\Phi x_n - \Phi z\|.$$

It follows that

$$\|x_n - y_n\| \leq \|\Phi x_n - \Phi z\|, \quad \forall n \geq 1. \tag{4.61}$$

If $\{x_n\}$ does not converge to z, then there exists some $\varepsilon > 0$ and a subsequence $\{x_{n_k}\}$ such that $\|x_{n_k} - z\| \geq \varepsilon$ for any $k \geq 1$. By applying Proposition 4.4.1, we see that there exists $r \in (0, 1)$ such that

$$\|\Phi x_{n_k} - \Phi z\| \leq r\|x_{n_k} - z\|, \quad \forall k \geq 1. \tag{4.62}$$

It follows from (4.61) and (4.62) that

$$\|x_{n_k} - z\| \leq \|x_{n_k} - y_{n_k}\| + \|y_{n_k} - z\|$$
$$\leq \|\Phi x_{n_k} - \Phi z\| + \|y_{n_k} - z\|$$
$$\leq r\|x_{n_k} - z\| + \|y_{n_k} - z\|.$$

This further implies

$$\|x_{n_k} - z\| \leq \frac{1}{1 - r}\|y_{n_k} - z\| \to 0 \quad (k \to \infty),$$

which yields a contradiction to $\|x_{n_k} - z\| \geq \varepsilon$. Consequently, $x_n \to z$ $(n \to \infty)$. This completes the proof. □

By virtue of Theorem 4.4.1, we obtain the following result.

Theorem 4.4.2. *Let E be a real uniformly smooth Banach space and let C be a nonempty, closed, and convex subset of E. Let $T : C \to C$ be a nonexpansive mapping such that* Fix$(T) \neq \emptyset$. *Let $\{S_n\}$ be a family of nonexpansive mappings on C and let $\Phi : C \to C$ be a Meir–Keeler contractive mapping. Then there exists a unique path $\{x_t\}$ satisfying the equation:*

$$x_t = (1-t)Tx_t + t\Phi x_t, \quad t \in (0,1]. \tag{VRIM}$$

Further, as $t \to 0$, $\{x_t\}$ strongly converges to $z = P \circ \Phi z$, where $P : C \to$ Fix(T) is the unique sunny nonexpansive retraction mapping from C onto Fix(T), and z is a unique solution of the following variational inequality problem:

$$\langle \Phi z - z, j(z - y) \rangle \geq 0, \quad \forall y \in \text{Fix}(T).$$

Proof. Claim (ii) of Proposition 4.4.2 guarantees the existence and uniqueness of path $\{x_t\}$. The convergence of $\{x_t\}$ can be derived from Theorem 4.4.1. Finally, using Theorem 4.4.3, one concludes that z is the unique solution of the above variational inequality. This completes the proof. □

Proposition 4.4.4. *Suppose that a system $(E, C, \{S_n\}, \alpha_n)$ has the Halpern's property. Define $Pu = \lim_{n\to\infty} y_n$, where $\{y_n\}$ is a sequence generated by*

$$y_1 \in C, \quad y_{n+1} = (1 - \alpha_n)S_n y_n + \alpha_n u, \quad \forall n \geq 1, u \in C.$$

Then the following conclusions hold:
(i) *Pu is independent of the initial value $y_1 \in C$;*
(ii) *$P : C \to C$ is nonexpansive.*

Proof. Fix $u \in C$ and define $\{u_n\}$ and $\{y_n\}$ as

$$u_1 \in C, \quad u_{n+1} = (1 - \alpha_n)S_n u_n + \alpha_n u, \quad n \geq 1,$$
$$y_1 \in C, \quad y_{n+1} = (1 - \alpha_n)S_n y_n + \alpha_n u, \quad n \geq 1.$$

Therefore,

$$\|u_{n+1} - y_{n+1}\| \leq (1 - \alpha_n)\|S_n u_n - S_n y_n\|$$
$$\leq (1 - \alpha_n)\|u_n - y_n\|,$$

which implies $\lim_{n\to\infty} \|u_n - y_n\| = 0 \implies Pu = \lim_{n\to\infty} y_n = \lim_{n\to\infty} u_n$, that is, Pu is independent of $y_1 \in C$.

Fix $v \in C$ and define $\{v_n\}$ as

$$v_1 = v, v_{n+1} = (1 - \alpha_n)S_n v_n + \alpha_n v, \quad n \geq 1.$$

Thus

$$\|u_{n+1} - v_{n+1}\| \le (1 - \alpha_n)\|S_n u_n - S_n v_n\| + \alpha_n \|u - v\|$$

$$\le (1 - \alpha_n)\|u_n - v_n\| + \alpha_n \|u - v\|, \quad n \ge 1. \tag{4.63}$$

By induction, we find that $\|u_n - v_n\| \le \|u - v\|$. It follows that $\|Pu - Pv\| \le \|u - v\|$, $\forall n \ge 1$, which implies that $P : C \to C$ is nonexpansive. This completes the proof. $\qquad \square$

Theorem 4.4.3 (Suzuki [83], 2007). *Let C be a nonempty, closed, and convex subset of a real Banach space E. Let $\{S_n\}$ be a family of nonexpansive mappings on C. Let $\{\alpha_n\}$ be a real sequence in $(0, 1]$ satisfying conditions (i) and (ii) of Theorem 4.3.2. Let $\Phi : C \to C$ be a Meir–Keeler contractive mapping. Assume that the system $(E, C, \{S_n\}, \alpha_n)$ has the Halpern's property. Let $P : C \to C$ be defined in Proposition 4.4.4. Let $\{x_n\} \subset C$ be a sequence defined by*

$$x_1 \in C, \quad x_{n+1} = (1 - \alpha_n)S_n x_n + \alpha_n \Phi x_n, \quad n \ge 1. \tag{VHIS}$$

Then $\{x_n\}$ strongly converges to the unique point $z = P \circ \Phi z$.

Proof. It follows from Proposition 4.4.4 that $P : C \to C$ is nonexpansive. Using (i) of Proposition 4.4.2, we find that $P \circ \Phi : C \to C$ is Meir–Keeler contractive. From the Meir–Keeler fixed point theorem, one concludes that $P \circ \Phi : C \to C$ has a unique fixed point z. Define $\{y_n\} \subset C$ by

$$y_1 \in C, \quad y_{n+1} = (1 - \alpha_n)S_n y_n + \alpha_n \Phi z, \quad n \ge 1. \tag{4.64}$$

So, $\{y_n\}$ strongly converges to the point $z = P \circ \Phi z$. We next prove that $x_n - y_n \to \theta$ as $n \to \infty$. Assume that $\limsup_{n\to\infty} \|x_n - y_n\| > 0$. We choose an ε such that $0 < \varepsilon < \limsup_{n\to\infty} \|x_n - y_n\|$. Theorem 4.4.1 asserts that there exists $r \in (0, 1)$ such that $\|x - y\| \ge \varepsilon$, $\forall x, y \in C$. Letting $v_1 \ge 1$, we find that, for any $n \ge v_1$, $\frac{r\|y_n - z\|}{1-r} < \varepsilon$. Using the definition of Φ, we have $\|\Phi x - \Phi y\| \le \max\{r\|x - y\|, \varepsilon\}$, $\forall x, y \in C$.

Next the proof will be split into two cases.

(1) There exist $v_2 \ge 1$ such that $v_2 \ge v_1$ and $\|x_{v_2} - x_{v_1}\| \le \varepsilon$;

(2) $\|x_n - y_n\| > \varepsilon$, $\forall n \ge v_1$.

In Case (1), we have

$$\|x_{v_2+1} - y_{v_2+1}\| \le (1 - \alpha_{v_2})\|S_{v_2} x_{v_2} - S_{v_2} y_{v_2}\| + \alpha_{v_2}\|\Phi x_{v_2} - \Phi z\|$$

$$\le (1 - \alpha_{v_2})\|x_{v_2} - y_{v_2}\| + \alpha_{v_2} \max\{\|x_{v_2} - z\|, \varepsilon\}$$

$$\le \max\{(1 - \alpha_{v_2} + r\alpha_{v_2})\|x_{v_2} - y_{v_2}\| + r\alpha_{v_2}\|y_{v_2} - z\|,$$

$$(1 - \alpha_{v_2})\|x_{v_2} - y_{v_2}\| + \alpha_{v_2}\varepsilon\}$$

$$\le \max\left\{(1 - \alpha_{v_2} + r\alpha_{v_2})\|x_{v_2} - y_{v_2}\| + (\alpha_{v_2} - r\alpha_{v_2})\frac{r\|y_{v_2} - z\|}{1 - r},\right.$$

$$\left.(1 - \alpha_{v_2})\|x_{v_2} - y_{v_2}\| + \alpha_{v_2}\varepsilon\right\}.$$

$$\le \varepsilon.$$

Thus $\|x_n - y_n\| \le \varepsilon, \forall n \ge v_2$. This yields a contradiction to $0 < \varepsilon < \lim\sup_{n\to\infty} \|x_n - y_n\|$.

In Case (2), we have

$$\|x_{n+1} - y_{n+1}\| \le (1 - \alpha_n)\|S_n x_n - S_n y_n\| + \alpha_n \|\Phi x_n - \Phi z\|$$
$$\le (1 - \alpha_n)\|x_n - y_n\| + \alpha_n \|\Phi x_n - \Phi y_n\| + \alpha_n \|\Phi y_n - \Phi z\|$$
$$\le (1 - \alpha_n + r\alpha_n)\|x_n - y_n\| + (\alpha_n - r\alpha_n)\frac{r\|y_n - z\|}{1 - r}.$$

From Lemma 1.10.2, one has $x_n - y_n \to \theta$ as $n \to \infty$. This yields a contradiction to $0 < \varepsilon < \lim\sup_{n\to\infty} \|x_n - y_n\|$. Thus $x_n - y_n \to \theta$ as $n \to \infty$. It follows from $\|x_n - z\| \le \|x_n - y_n\| + \|y_n - z\| \to 0$ that $x_n \to z$ $(n \to \infty)$. This completes the proof. $\qquad\square$

Remark 4.4.1. If, in addition, the sequence $\{\alpha_n\} \subset (0, 1]$ is assumed to satisfy conditions (i)–(iv) of Theorem 4.3.2 for $(E, C, \{S_n\}, \alpha_n)$, then $(E, C, \{S_n\}, \alpha_n)$ has the Halpern's property: for any $u \in C$, the sequence $\{y_n\} \subset C$ defined by

$$y_1 \in C, \quad y_{n+1} = (1 - \alpha_n)S_n y_n + \alpha_n u, \quad n \ge 1,$$

is strongly convergent. Then the result of Theorem 4.4.3 is still true.

Theorem 4.4.4. *Let C be a nonempty, closed, and convex subset of a real Banach space E. Let $\{S_n\}$ be a family of nonexpansive mappings on C. Let $\{\alpha_n\}$ be a real sequence in $(0, 1]$ satisfying conditions (i)–(iv) of Theorem 4.3.2. Let $\Phi : C \to C$ be a Meir–Keeler contractive mapping. Assume that $(E, C, \{S_n\}, \alpha_n)$ has the Halpern's property. Let $P : C \to C$ be defined in Proposition 4.4.4. Let $\{x_n\} \subset C$ be a sequence defined by*

$$x_1 \in C, \quad x_{n+1} = (1 - \alpha_n)S_n x_n + \alpha_n \Phi x_n, \quad n \ge 1.$$

Then $\{x_n\}$ strongly converges to the unique point $z = P \circ \Phi z$.

Theorem 4.4.5. *Let C be a nonempty, closed, and convex subset of a real uniformly smooth Banach space E. Let $T : C \to E$ be a nonexpansive mapping with $\mathrm{Fix}(T) \ne \emptyset$. Let $\{\alpha_n\}$ be a real sequence in $(0, 1]$ satisfying conditions (i)–(iii) or (iv) of Theorem 4.3.2. Let $\Phi : C \to C$ be a Meir–Keeler contractive mapping. Let $\{x_n\} \subset C$ be a sequence defined by*

$$x_1 \in C, \quad x_{n+1} = (1 - \alpha_n)T x_n + \alpha_n \Phi x_n, \quad n \ge 1.$$

Then $\{x_n\}$ strongly converges to the unique point $z = P \circ \Phi z$, where $P : C \to \mathrm{Fix}(T)$ is the unique sunny nonexpansive retraction and z is the unique solution of the following variational inequality problem:

$$\langle \Phi z - z, j(z - y) \rangle \ge 0, \quad \forall y \in \mathrm{Fix}(T).$$

Proof. From Theorem 4.3.2, we find that $(E, C, \{T\}, \alpha_n)$ has the Halpern's property. Using Theorem 4.4.4, we obtain that $\{x_n\}$ strongly converges to the unique point $z = P \circ \Phi z$. From Proposition 4.4.3, we conclude that z is the unique solution of the above variational inequality. $\qquad\square$

Theorem 4.4.6. *Let E, C, T, P, Φ, and $\{x_n\}$ be as in Theorem 4.4.5. Let $\{\alpha_n\}$ be a sequence real sequence in $(0, 1]$ satisfying (i) and (ii) of Theorem 4.3.6. For any $\lambda \in (0, 1)$, let $T_\lambda = \lambda I + (1 - \lambda)T$. Let $\{x_n\} \subset C$ be a sequence defined by*

$$x_1 \in C, \quad x_{n+1} = (1 - \alpha_n)T_\lambda x_n + \alpha_n \Phi x_n, \quad n \geq 1.$$

Then $\{x_n\}$ strongly converges to the unique point $z = P \circ \Phi z$, where $P : C \to \text{Fix}(T)$ is the unique sunny nonexpansive retraction and z is the unique solution of the above variational inequality.

Proof. Note that $(E, C, \{T\}, \alpha_n)$ has the Halpern's property. Using Theorem 4.4.4, we find that $\{x_n\}$ strongly converges to the unique point $z = P \circ \Phi z$. From Proposition 4.4.3, we find that z is the unique solution of the above variational inequality. $\qquad \square$

Corollary 4.4.1. *Let C be a closed convex subset of a real uniformly smooth Banach space E. Let $T : C \to E$ be a nonexpansive mapping with $\text{Fix}(T) \neq \emptyset$. Let $\{\alpha_n\}$ be a real sequence in $(0, 1]$ satisfying conditions (i)–(iii) or (iv) of Theorem 4.3.2. Let $f : C \to C$ be a contractive mapping. Define a sequence $\{x_n\} \subset C$ by*

$$x_1 \in C, \quad x_{n+1} = (1 - \alpha_n)T x_n + \alpha_n f x_n, \quad n \geq 1.$$

Then $\{x_n\}$ strongly converges to the unique point $z = P \circ f z$, where z is the unique solution of the following variational inequality

$$\langle f z - z, j(z - y) \rangle \geq 0, \quad \forall y \in \text{Fix}(T).$$

Corollary 4.4.2. *Let E, C, T, P, and f be as in Theorem 4.4.5. Let $\{\alpha_n\}$ be a real sequence in $(0, 1]$ satisfying (i) and (ii) of Theorem 4.3.2. Let $\lambda \in (0, 1)$ and $T_\lambda = \lambda I + (1 - \lambda)T$. Let $\{x_n\} \subset C$ be a sequence defined by*

$$x_1 \in C, \quad x_{n+1} = (1 - \alpha_n)T_\lambda x_n + \alpha_n f x_n, \quad n \geq 1.$$

Then $\{x_n\}$ strongly converges to the unique point $z = P \circ f z$, where z is the unique solution of the above variational inequality.

4.5 Iterative methods for common fixed points of a family of nonexpansive mappings in Banach spaces

In this section, we introduce some iterative methods based on the cyclic and the W-mapping techniques.

Let C be a nonempty convex subset of a Banach space E. Let $\{T_i\}_{i=1}^r$ be a finite family of nonexpansive mappings. Let α_i be a real number in $(0, 1)$. Define a family of nonexpansive mappings $\{U_i\}_{i=1}^r$ by

$$U_1 = \alpha_1 T_1 + (1 - \alpha_1)I,$$
$$U_2 = \alpha_2 T_2 U_1 + (1 - \alpha_2)I,$$
$$\vdots$$
$$W = U_r = \alpha_r T_r U_{r-1} + (1 - \alpha_r)I.$$

The mapping U_r is a called a W-mapping generated by $T_1, T_2, T_3, \ldots, T_r$ and $\alpha_1, \alpha_2, \alpha_3, \ldots, \alpha_r$.

Proposition 4.5.1. *Let C be a nonempty, closed, and convex subset of a real strictly convex Banach space E. Let $\{T_i\}_{i=1}^r : C \to C$ be a finite family of nonexpansive mappings with $F = \bigcap_{i=1}^r \mathrm{Fix}(T_i) \neq \emptyset$. Let $\{\alpha_i\}_{i=1}^r \subset (0,1)$ be a sequence of real numbers such that $0 < \alpha_i < 1$ ($i = 1, 2, \ldots, r-1$), and $0 < \alpha_r \leq 1$. Let W be a W-mapping generated by $T_1, T_2, T_3, \ldots, T_r$ and $\alpha_1, \alpha_2, \alpha_3, \ldots, \alpha_r$. Then each U_i, where $1 \leq i \leq r$, is nonexpansive and $\mathrm{Fix}(W) = \bigcap_{i=1}^r \mathrm{Fix}(T_i)$.*

Proof. From the construction, one sees that $U_i : C \to C$ is averaged for $i = 1, 2, \ldots, r-1$. However, $U_r : C \to C$ is nonexpansive. It is obvious that $\bigcap_{i=1}^r \mathrm{Fix}(T_i) \subset \mathrm{Fix}(W)$.

For any $x \in \mathrm{Fix}(W)$, we fix $p \in F = \bigcap_{i=1}^r \mathrm{Fix}(T_i) \neq \emptyset$. It follows that

$$x = Wx = U_r x = \alpha_r T_r U_{r-1} x + (1 - \alpha_r)x.$$

This implies $x = \alpha_r T_r U_{r-1} x$ and

$$
\begin{aligned}
\|x - p\| &= \|T_r U_{r-1} x - p\| \\
&\leq \|U_{r-1} x - p\| \\
&= \|\alpha_{r-1} T_{r-1} U_{r-2} x + (1 - \alpha_{r-1})x - p\| \\
&\leq \alpha_{r-1}\|U_{r-2} x - p\| + (1 - \alpha_{r-1})\|x - p\| \\
&= \alpha_{r-1}\|\alpha_{r-2} T_{r-2} U_{r-3} x + (1 - \alpha_{r-2})x - p\| + (1 - \alpha_{r-1})\|x - p\| \\
&\leq \alpha_{r-1}\alpha_{r-2}\|U_{r-3} x - p\| + (1 - \alpha_{r-1}\alpha_{r-2})\|x - p\| \\
&\vdots \\
&\leq \alpha_{r-1}\alpha_{r-2}\cdots\alpha_2\alpha_1\|T_1 x - p\| + (1 - \alpha_{r-1}\alpha_{r-2}\cdots\alpha_2\alpha_1)\|x - p\| \\
&\leq \|x - p\|.
\end{aligned}
$$

It follows that

$$\|x - p\| = \|T_1 x - p\| = \|U_1 x - p\| = \|\alpha_1 T_1 x + (1 - \alpha_1)x - p\|.$$

Since E is strictly convex, one has $x - p = T_1 x - p$. Hence, $x = T_1 x$ and $x = U_1 x$. Similarly, one has

$$\|x - p\| = \|T_2 U_1 x - p\| = \|U_2 x - p\| = \|\alpha_2 T_2 U_1 x + (1 - \alpha_2)x - p\|.$$

Since E is strictly convex, we find that $x - p = T_2x - p$. This shows that $x = T_2x$ and $x = U_2x$. In a similar way, one also see that $x = T_kx$, $x = U_kx$ $(k = 3, 4, \ldots, r-1)$. In view of $x = T_rU_{k-1}x$, we find that $x = T_rx$. Hence $x \in \bigcap_{i=1}^r \mathrm{Fix}(T_i)$. $\qquad\square$

Theorem 4.5.1. *Let E be a real uniformly convex Banach space with a Fréchet differentiable norm. Let C be a closed convex subset of E. Let $\{T_i\}_{i=1}^r : C \to C$ be a finite family of nonexpansive mappings with $F = \bigcap_{i=1}^r \mathrm{Fix}(T_i) \neq \emptyset$. Let $\{\alpha_i\}_{i=1}^r \subset (0,1)$ be a real sequence such that $0 < \alpha_i < 1$ $(i = 1, 2, \ldots, r-1)$, and $0 < \alpha_r \leq 1$. Let W be a W-mapping generated by $T_1, T_2, T_3, \ldots, T_r$ and $\alpha_1, \alpha_2, \alpha_3, \ldots, \alpha_r$. Let $\{\beta_n\} \subset [0,1]$ be a real sequence such that $0 \leq \beta_n < 1$ $(n \geq 1)$ and $\sum_{n=1}^\infty \beta_n(1 - \beta_n) = \infty$. Define a sequence $\{x_n\}$ by*

$$x_1 = x \in C, \quad x_{n+1} = \beta_n x_n + (1 - \beta_n)Wx_n, \quad n \geq 1.$$

Then $\{x_n\}$ converges weakly to a common fixed point of $\{T_i\}_{i=1}^r$.

Proof. From Proposition 4.5.1 and Reich's convergence theorem, we can obtain the desired conclusion immediately. $\qquad\square$

Theorem 4.5.2. *Let E be a real uniformly convex Banach space with a Gâteaux differentiable norm. Let C be a closed convex subset of E. Let $\{T_i\}_{i=1}^r : C \to C$ be a finite family of nonexpansive mappings with $F = \bigcap_{i=1}^r \mathrm{Fix}(T_i) \neq \emptyset$. Let $\{\alpha_i\}_{i=1}^r \subset (0,1)$ be a real sequence such that $0 < \alpha_i < 1$ $(i = 1, 2, \ldots, r-1)$ and $0 < \alpha_r \leq 1$. Let W be a W-mapping generated by $T_1, T_2, T_3, \ldots, T_r$ and $\alpha_1, \alpha_2, \alpha_3, \ldots, \alpha_r$. Let $\{\beta_n\} \subset [0,1]$ be a real sequence such that*
(1) $\beta_n \to 0$ $(n \to \infty)$;
(2) $\sum_{n=1}^\infty \beta_n = \infty$;
(3) $\sum_{n=1}^\infty |\beta_{n-1} - \beta_n| < \infty$; *or*
(4) $\frac{\beta_n}{\beta_{n+1}} \to 1$ $(n \to \infty)$.

Define a sequence $\{x_n\}$ by

$$u, x_1 \in C, \quad x_{n+1} = \beta_n u + (1 - \beta_n)Wx_n, \quad n \geq 1.$$

Then $\{x_n\}$ converges strongly to $z = Q_Fu$, a common fixed point of $\{T_i\}_{i=1}^r$.

Proof. From Proposition 4.5.1 and Xu's convergence theorem, we can obtain the desired result easily. $\qquad\square$

Theorem 4.5.3. *Let E, C, $\{T_i\}_{i=1}^r$, α_i, and W be defined as in Theorem 4.5.2. Let $\{\beta_n\} \subset [0,1]$ be a real sequence satisfying $\lim_{n\to\infty} \beta_n = 0$ and $\sum_{n=1}^\infty \beta_n = \infty$. Define a sequence $\{x_n\}$ by*

$$u, x_1 \in C, \quad x_{n+1} = \beta_n u + (1 - \beta_n)W_\lambda x_n, \quad n \geq 1,$$

where $\lambda \in (0,1)$ is a real constant and $W_\lambda = \lambda W + (1 - \lambda)I$. Then $\{x_n\}$ converges strongly to $z = Q_Fu$, a common fixed point of $\{T_i\}_{i=1}^r$.

Proof. From Proposition 4.5.1 and Suzuki's convergence theorem, we can obtain the desired result immediately. □

We next introduce cyclic iterative methods

$$x_1 \in C, \quad x_{n+1} = (1 - \alpha_n)x_n + \alpha_n T_n x_n, \quad n \geq 1, \tag{I}$$

and

$$x_0, u \in C, \quad x_{n+1} = \alpha_{n+1} u + (1 - \alpha_{n+1})T_{n+1}x_n, \quad n \geq 0, \tag{II}$$

where $T_n := T_{n \bmod r}$, where modular function "$\bmod r$" takes the values in $\{1, 2, \ldots, r\}$.

Theorem 4.5.4. *Let E be a real uniformly convex Banach space with a Fréchet differentiable norm. Let C be a closed convex subset of E. Let $\{T_i\}_{i=1}^r : C \to C$ be a finite family of nonexpansive mappings with $F = \bigcap_{i=1}^r \mathrm{Fix}(T_i) \neq \emptyset$. Let $\{\alpha_n\} \subset (0, 1)$ be a real sequence such that $\liminf \alpha_n(1 - \alpha_n) > 0$. Let $\{x_n\}$ be a sequence generated by (I). Then $\{x_n\}$ converges weakly to a common fixed point of $\{T_i\}_{i=1}^r$.*

Proof. First, we show that $\{x_n\}$ is bounded. Fixing $p \in F$, we find that

$$\begin{aligned}
\|x_{n+1} - p\| &\leq (1 - \alpha_n)\|x_n - p\| + \alpha_n\|T_n x_n - p\| \\
&\leq (1 - \alpha_n)\|x_n - p\| + \alpha_n\|x_n - p\| \\
&= \|x_n - p\|.
\end{aligned}$$

Thus $\lim_{n \to \infty} \|x_n - p\|$ exists, which in turn shows that $\{x_n\}$ is bounded. Let $r = \sup_{n \geq 1}\{\|x_n - p\|\}$. From Theorem 1.8.9, we find that

$$\begin{aligned}
\|x_{n+1} - p\|^2 &= \left\|(1 - \alpha_n)(x_n - p) + \alpha_n(T_n x_n - p)\right\|^2 \\
&\leq (1 - \alpha_n)\|x_n - p\|^2 + \alpha_n\|T_n x_n - p\|^2 \\
&\quad - \alpha_n(1 - \alpha_n)g(\|x_n - T_n x_n\|) \\
&\leq \|x_n - p\|^2 - \alpha_n(1 - \alpha_n)g(\|x_n - T_n x_n\|).
\end{aligned}$$

This implies that

$$\alpha_n(1 - \alpha_n)g(\|x_n - T_n x_n\|) \leq \|x_n - p\|^2 - \|x_{n+1} - p\|^2.$$

In view of $\liminf_n \alpha_n(1 - \alpha_n) > 0$, we have $g(\|x_n - T_n x_n\|) \to 0$ as $n \to \infty$. From the property of g, we find that $x_n - T_n x_n \to \theta$ as $n \to \infty$.

From (I), one sees that

$$x_{n+1} - x_n = \alpha_n(T_n x_n - x_n) \to \theta$$

as $n \to \infty$. This shows that $x_{n+1} - x_n \to \theta$ as $n \to \infty$, and $x_{n+i} - T_{n+i}x_{n+i} \to \theta$ as $n \to \infty$. Hence

$$x_n - T_{n+i}x_n \to \theta \ (n \to \infty), \quad \forall i = 1, 2, \ldots, r.$$

For any $l = 1, 2, \ldots, r$, there exists $i = 1, 2, \ldots, r$ such that $n + i = l \bmod r$ $\forall n \geq 1$. This yields $x_n - T_l x_n = x_n - T_{n+i} x_n \to \theta$ as $n \to \infty$. Due to the reflexivity of E and the boundedness of $\{x_n\}$, we may assume that $x_n \to p$ as $n \to \infty$. From Theorem 1.9.3, we have that $p = T_l p$, where $l = 1, 2, \ldots, r$, that is, $p \in F$. In view of Lemma 4.3.1, one concludes that $\{x_n\}$ weakly converges to a common fixed point of $\{T_i\}_{i=1}^r$. $\qquad\square$

Theorem 4.5.5. *Let E be a real uniformly convex Banach space with a Gâteaux differentiable norm. Let C be a closed convex subset of E. Let $\{T_i\}_{i=1}^r : C \to C$ be a finite family of nonexpansive mappings such that $F = \bigcap_{i=1}^r \mathrm{Fix}(T_i) \neq \emptyset$ and $F = \mathrm{Fix}(T_r T_{r-1} \cdots T_1) = \mathrm{Fix}(T_1 T_r \cdots T_3 T_2) = \cdots = \mathrm{Fix}(T_{r-1} T_{r-2} \cdots T_1 T_r)$. Let $\{\alpha_i\} \subset (0,1)$ be a real sequence satisfying the following conditions:*

(1) $\alpha_n \to 0$ $(n \to \infty)$;
(2) $\sum_{n=1}^{\infty} \alpha_n = \infty$;
(3) $\sum_{n=1}^{\infty} |\alpha_{n+r} - \alpha_n| < \infty$; or
(4) $\frac{\alpha_n}{\alpha_{n+r}} \to 1$ $(n \to \infty)$.

Let $\{x_n\}$ be a sequence generated by (II). *Then $\{x_n\}$ converges strongly to $p = Q_F u$, a common fixed point of $\{T_i\}_{i=1}^r$.*

Proof. The proof is split into five steps.

Step 1. Show that $\{x_n\}$ is bounded.

For any $p \in F$, it follows that

$$
\begin{aligned}
\|x_{n+1} - p\| &\leq \alpha_{n+1}\|u - p\| + (1 - \alpha_{n+1})\|T_{n+1} x_n - p\| \\
&\leq \alpha_{n+1}\|u - p\| + (1 - \alpha_{n+1})\|x_n - p\| \\
&\leq \max\{\|u - p\|, \|x_1 - p\|\} \triangleq M, \quad \forall n \geq 1.
\end{aligned}
$$

Thus $\{x_n\}$ is bounded, so is $\{T_{n+1} x_n\}$. This shows that $x_{n+1} - T_{n+1} x_n \to \theta$ as $n \to \infty$.

Step 2. Show that $x_{n+r} - x_n \to \theta$ as $n \to \infty$.

Note that

$$
\begin{aligned}
\|x_{n+r} - x_n\| &= \|(\alpha_{n+r} - \alpha_n)u + (1 - \alpha_{n+r})T_{n+r} x_{n+r-1} - (1 - \alpha_{n-1})T_n x_{n-1}\| \\
&\leq |\alpha_{n+r} - \alpha_n|(\|u\| + \|T_n x_{n-1}\|) + (1 - \alpha_{n+r})\|x_{n+r-1} - x_{n-1}\| \\
&\leq (1 - \alpha_{n+r})\|x_{n+r-1} - x_{n-1}\| + M|\alpha_{n+r} - \alpha_n|.
\end{aligned}
$$

From Lemma 1.10.2, we conclude that $x_{n+r} - x_n \to \theta$ as $n \to \infty$.

Step 3. Show that $x_n - T_{n+r} \cdots T_{n+1} x_n \to \theta$ as $n \to \infty$.

In view of assumption that $\alpha_n \to 0$ as $n \to \infty$, we have, as $n \to \infty$,

$$
x_{n+r} - T_{n+r} x_{n+r-1} \to \theta. \tag{4.65}
$$

Taking into account the nonexpansivity of T_{n+r} yields

$$
T_{n+r} x_{n+r-1} - T_{n+r} T_{n+r-1} x_{n+r-2} \to \theta. \tag{4.66}
$$

Similarly, one has

$$T_{n+r}T_{n+r-1}x_{n+r-2} - T_{n+r}T_{n+r-1}T_{n+r-2}x_{n+r-3} \to \theta \qquad (4.67)$$

$$\vdots$$

$$T_{n+r}\cdots T_{n+2}T_{n+1}x_{n+1} - T_{n+r}\cdots T_{n+1}x_n \to \theta. \qquad (4.68)$$

Using (4.65)–(4.68), we find that $x_{n+r} - T_{n+r}\cdots T_{n+1}x_n \to \theta$ as $n \to \infty$. From Step 2, we get that $x_n - T_{n+r}\cdots T_{n+1}x_n \to \theta$ as $n \to \infty$.

Step 4. Show that $\limsup_n\langle u - p, j(x_n - p)\rangle \le 0$, where $p = Q_F u$.

Pick subsequences of $\{x_n\}$ as follows:

$$\{n_k\} : 1, r+1, 2r+1, \ldots,$$

$$\{m_j\} : 2, r+2, 2r+2, \ldots,$$

$$\vdots$$

$$\{s_i\} : r, 2r, 3r, \ldots.$$

Thus,

$$x_{n_k} - T_1 T_r T_{r-1} T_{n+r}\cdots T_2 x_{n_k} \to \theta \quad (k \to \infty),$$

$$x_{m_j} - T_2 T_1 T_r \cdots T_3 x_{m_j} \to \theta \quad (j \to \infty),$$

$$x_{s_i} - T_r T_{r-1}\cdots T_1 x_{s_i} \to \theta \quad (i \to \infty).$$

Since

$$\text{Fix}(T_r T_{r-1}\cdots T_1) = \text{Fix}(T_1 T_r \cdots T_3 T_2) = \cdots = \text{Fix}(T_{r-1}T_{r-2}\cdots T_1 T_r),$$

one concludes from Lemma 4.3.1 that

$$\limsup_k\langle u - p, j(x_{n_k} - p)\rangle \le 0,$$

$$\limsup_j\langle u - p, j(x_{m_j} - p)\rangle \le 0,$$

$$\limsup_i\langle u - p, j(x_{s_i} - p)\rangle \le 0.$$

Thus we find that $\limsup_n\langle u - p, j(x_n - p)\rangle \le 0$.

Step 5. Show that $x_n \to p = Q_F u$ as $n \to \infty$.

From (II), we obtain that

$$\begin{aligned}
\|x_{n+1} - p\|^2 &= \alpha_{n+1}\langle u - p, j(x_{n+1} - p)\rangle + (1 - \alpha_{n+1})\langle T_{n+1}x_n - p, j(x_{n+1} - p)\rangle \\
&\le \alpha_{n+1}\|x_n - p\|\|x_{n+1} - p\| + \alpha_{n+1}\langle u - p, j(x_{n+1} - p)\rangle \\
&\le \frac{1}{2}(1 - \alpha_{n+1})\|x_n - p\|^2 + \frac{1}{2}\|x_{n+1} - p\|^2 + \alpha_{n+1}\langle u - p, j(x_{n+1} - p)\rangle.
\end{aligned}$$

This yields

$$\|x_{n+1} - p\|^2 \le (1 - \alpha_{n+1})\|x_n - p\|^2 + 2\alpha_{n+1}\langle u - p, j(x_{n+1} - p)\rangle$$
$$\le (1 - \alpha_{n+1})\|x_n - p\|^2 + o(\alpha_{n+1}).$$

From Lemma 1.10.2, we conclude that $x_n \to p = Q_F u$ as $n \to \infty$. $\qquad\square$

Next, we discuss the iterative construction problem of common fixed points of a countable infinite family of nonexpansive mappings. We introduce a more general W-mapping as follows.

Let C be a convex subset of a real Banach space E. Let $\{T_i\}_{i=1}^\infty : C \to C$ be a family of nonexpansive mappings. Let $\{\alpha_i\}_{i=1}^\infty$ be a real sequence in $[0, 1]$. For any $n \ge 1$, define a mapping $W_n : C \to C$ as

$$W_n = U_{n,1} = \alpha_1 T_1 U_{n,2} + (1 - \alpha_1)I,$$

where

$$U_{n,2} = \alpha_2 T_2 U_{n,3} + (1 - \alpha_2)I,$$

$$\vdots$$

$$U_{n,k-1} = \alpha_{k-1} T_{k-1} U_{n,k} + (1 - \alpha_{k-1})I,$$
$$U_{n,k} = \alpha_k T_k U_{n,k+1} + (1 - \alpha_k)I,$$

$$\vdots$$

$$U_{n,n-1} = \alpha_{n-1} T_{n-1} U_{n,n} + (1 - \alpha_{n-1})I,$$
$$U_{n,n} = \alpha_n T_n U_{n,n+1} + (1 - \alpha_n)I,$$
$$U_{n,n+1} = I.$$

Then W_n is called the W-mapping generated by $T_n, T_{n-1}, \ldots, T_1$ and $\alpha_n, \alpha_{n-1}, \ldots, \alpha_1$.

Proposition 4.5.2. *Let C be a nonempty, closed, and convex subset of a real Banach space E. Let $\{T_i\}_{i=1}^\infty : C \to C$ be an infinite countable family of nonexpansive mappings with $F = \bigcap_{i=1}^\infty \text{Fix}(T_i) \ne \emptyset$. Let $\{\alpha_i\}_{i=1}^\infty$ be real sequence in $[0, 1]$ and W_n be the W-mapping generated by $T_n, T_{n-1}, \ldots, T_1$ and $\alpha_n, \alpha_{n-1}, \ldots, \alpha_1$. If E is strictly convex, then $\text{Fix}(W_n) = \bigcap_{i=1}^n \text{Fix}(T_i)$, where $\bigcap_{n=1}^\infty \text{Fix}(W_n) = \bigcap_{i=1}^\infty \text{Fix}(T_n)$.*

Proposition 4.5.3. *Let C be a real Banach space. Let $\{T_i\}_{i=1}^\infty : C \to C$ be an infinite countable family of nonexpansive mappings with $\bigcap_{i=1}^\infty \text{Fix}(T_i) \ne \emptyset$. Let $\{\alpha_i\}_{i=1}^\infty$ be a real sequence in $[0, 1]$ satisfying $0 < \alpha_i \le b < 1$. Then, for any $x \in C$, $k \ge 1$, $\lim_{n\to\infty} U_{n,k}x$ exists. Furthermore, if D is a bounded subset of C, then $\forall k \ge 1$, $\lim_{n\to\infty} U_{n,k}x$ exists for any $x \in D$.*

Proof. For $\forall x \in C$, fixing $\omega \in \bigcap_{i=1}^{r} \text{Fix}(T_i)$ with $x \neq \omega$ and $k \geq 1$, $n \geq 1$, one has

$$\|U_{n+1,k}x - U_{n,k}x\|$$
$$= \|\alpha_k T_k U_{n+1,k+1}x + (1 - \alpha_k)x - \alpha_k T_k U_{n,k+1}x - (1 - \alpha_k)x\|$$
$$= \alpha_k \|T_k U_{n+1,k+1}x - T_k U_{n,k+1}x\|$$
$$\leq \alpha_k \|U_{n+1,k+1}x - U_{n,k+1}x\|$$
$$= \alpha_k \|\alpha_{k+1} T_{k+1} U_{n+1,k+2}x + (1 - \alpha_{k+1})x$$
$$\quad - \alpha_{k+1} T_{k+1} U_{n,k+2}x - (1 - \alpha_{k+1})x\|$$
$$= \alpha_k \alpha_{k+1} \|T_{k+1} U_{n+1,k+2}x - T_{k+1} U_{n,k+2}x\|$$
$$\leq \alpha_k \alpha_{k+1} \|U_{n+1,k+2}x - U_{n,k+2}x\|$$
$$\vdots$$
$$\leq (\Pi_{i=k}^{n}\alpha_i)\|U_{n+1,n+1}x - U_{n,n+1}x\|$$
$$\leq (\Pi_{i=k}^{n}\alpha_i)\|\alpha_{n+1} T_{n+1} U_{n+1,n+2}x + (1 - \alpha_{n+1})x - x\|$$
$$\leq (\Pi_{i=k}^{n+1}\alpha_i)\|T_{n+1}x - x\|$$
$$\leq (\Pi_{i=k}^{n+1}\alpha_i)(\|T_{n+1}x - \omega\| + \|\omega - x\|)$$
$$\leq (\Pi_{i=k}^{n+1}\alpha_i)(\|x - \omega\| + \|\omega - x\|)$$
$$= 2(\Pi_{i=k}^{n+1}\alpha_i)\|x - \omega\|$$
$$\leq 2b^{n-k+2}\|x - \omega\|.$$

For any $\varepsilon > 0$, there exists $n_0 \geq k \geq 1$ such that

$$b^{n_0-k+2} < \frac{\varepsilon(1 - b)}{2\|x - \omega\|},$$

then

$$\|U_{m,k}x - U_{n,k}x\| \leq \sum_{j=n}^{m-1} \|U_{j+1,k}x - U_{j,k}x\|$$
$$\leq 2\|x - \omega\| \sum_{j=n}^{m-1} b^{j-k+2}$$
$$\leq \frac{2b^{n-k+2}\|x - \omega\|}{1 - b}$$
$$< \varepsilon, \quad \forall m > n > n_0,$$

which implies that $\{U_{n,k}x\}$ is a Cauchy sequence. This yields that $\lim_{n \to \infty} U_{n,k}x$ exists. Let $M = \sup\{\|x - \omega\| : x \in D\}$. Then, for any $\varepsilon > 0$, $x \in D$, one finds that there exists $n_0 \geq k \geq 1$, where n_0 is independent of $x \in D$ such that if

$$b^{n_0-k+2} < \frac{\varepsilon(1 - b)}{2M},$$

then

$$\|U_{m,k}x - U_{n,k}x\| \le 2M \sum_{j=n}^{n-1} b^{j-k+2}$$

$$\le \frac{2b^{n-k+2}M}{1-b}$$

$$\le \frac{2M\varepsilon(1-b)}{2M(1-b)}$$

$$< \varepsilon, \quad \forall m > n > n_0.$$

This implies that $\lim_{n\to\infty} U_{n,k}x$ is uniform with respect to $x \in D$. □

For any $k \ge 1$, we define $U_{\infty,k}, W : C \to C$ by

$$U_{\infty,k}x := \lim_{n\to\infty} U_{n,k}x, \; Wx := \lim_{n\to\infty} W_n x = \lim_{n\to\infty} U_{n,1}x, \quad \forall x \in C.$$

We say that W is a W-mapping generated by $T_1, T_2, \ldots$ and $\alpha_1, \alpha_2, \ldots$

Proposition 4.5.4. *Let C be a nonempty, closed, and convex subset of a real strictly convex Banach space E. Let $\{T_i\}_{i=1}^{\infty} : C \to C$ be an infinite countable family of nonexpansive mappings with $F = \bigcap_{i=1}^{\infty} \text{Fix}(T_i) \ne \emptyset$. Let $\{\alpha_i\}_{i=1}^{\infty}$ be a real sequence in $[0,1]$ with $0 < \alpha_i \le b < 1$. Then $\text{Fix}(W) = \bigcap_{i=1}^{\infty} \text{Fix}(T_i)$.*

Proof. Fixing $\omega \in \bigcap_{i=1}^{\infty} \text{Fix}(T_i)$, one has $U_{n,k}\omega = \omega$, $\forall n \ge k \ge 1$, which implies $U_{\infty,k}\omega = \omega$, $\forall k \ge 1$, which further implies that $W_\infty\omega = U_{\infty,1}\omega = \omega$. Hence, $\bigcap_{i=1}^{\infty} \text{Fix}(T_i) \subset \text{Fix}(W)$.

For any $x \in \text{Fix}(W)$ and any $y \in \bigcap_{i=1}^{\infty} \text{Fix}(T_i)$, we have

$$\|W_n x - W_n y\|$$

$$= \|\alpha_1 T_1 U_{n,2}x + (1-\alpha_1)x - \alpha_1 T_1 U_{n,2}y - (1-\alpha_1)y\|$$

$$\le \alpha_1 \|T_1 U_{n,2}x - T_1 U_{n,2}y\| + (1-\alpha_1)\|(x-y)\|$$

$$\le \alpha_1 \|U_{n,2}x - U_{n,2}y\| + (1-\alpha_1)\|(x-y)\|$$

$$\vdots$$

$$\le (\Pi_{i=1}^{k-1}\alpha_i)\|U_{n,k}x - U_{n,k}y\| + (1 - \Pi_{i=1}^{k-1}\alpha_i)\|(x-y)\|$$

$$\le (\Pi_{i=1}^{k-1}\alpha_i)\|\alpha_k T_k U_{n,k+1}x - (1-\alpha_k)x - \alpha_k T_k U_{n,k+1}y$$
$$\quad - (1-\alpha_k)y\| + (1 - \Pi_{i=1}^{k-1}\alpha_i)\|(x-y)\|$$

$$\le (\Pi_{i=1}^{k-1}\alpha_i)\|\alpha_k(T_k U_{n,k+1}x - T_k U_{n,k+1}y)$$
$$\quad + (1-\alpha_k)(x-y)\| + (1 - \Pi_{i=1}^{k-1}\alpha_i)\|(x-y)\|$$

$$\le (\Pi_{i=1}^{k-1}\alpha_i)\|T_k U_{n,k+1}x - T_k U_{n,k+1}y\| + (1 - \Pi_{i=1}^{k-1}\alpha_i)\|(x-y)\|$$

$$\le (\Pi_{i=1}^{k-1}\alpha_i)\|U_{n,k+1}x - U_{n,k+1}y\| + (1 - \Pi_{i=1}^{k-1}\alpha_i)\|(x-y)\|.$$

$$\vdots$$

$$\le \|(x-y)\|.$$

Taking the limit as $n \to \infty$, we find that

$$\|W_n x - W_n y\|$$
$$\leq (\Pi_{i=1}^{k-1} \alpha_i)\|\alpha_k(T_k U_{\infty,k+1}x - T_k U_{\infty,k+1}y)$$
$$+ (1 - \alpha_k)(x - y)\| + (1 - \Pi_{i=1}^{k-1} \alpha_i)\|(x - y)\|$$
$$\leq (\Pi_{i=1}^{k} \alpha_i)\|T_k U_{\infty,k+1}x - T_k U_{\infty,k+1}y\| + (1 - \Pi_{i=1}^{k-1} \alpha_i)\|(x - y)\|$$
$$\leq \|(x - y)\|.$$

Due to $\|W_n x - W_n y\| = \|x - y\|$ and $0 < \alpha_i < 1$, one has

$$\|\alpha_k(T_k U_{\infty,k+1}x - T_k U_{\infty,k+1}y) + (1 - \alpha_k)(x - y)\|$$
$$= \|T_k U_{\infty,k+1}x - T_k U_{\infty,k+1}y\|.$$
$$= \|x - y\|, \quad \forall k \geq 1.$$

Since space E is strictly convex and $y = T_i y, \forall i = 1, 2, \ldots$, one has

$$x - y = T_k U_{\infty,k+1}x - T_k U_{\infty,k+1}y$$
$$= T_k U_{\infty,k+1}x - y$$
$$\implies x = T_k U_{\infty,k+1}x.$$

From $U_{n,k+1}x = \alpha_{k+1}T_{k+1}U_{n,k+2}x + (1 - \alpha_{k+1})x$, one has

$$U_{\infty,k+1}x = \lim_n U_{n,k+1}x = \alpha_{k+1}T_{k+1}U_{\infty,k+2}x + (1 - \alpha_{k+1})x$$
$$= \alpha_{k+1}x + (1 - \alpha_{k+1})x$$
$$= x.$$

Then $x = T_k U_{\infty,k+1}x = T_k x, \forall k \geq 1$, that is, $x \in \bigcap_{i=1}^{\infty} \text{Fix}(T_i)$. This implies that $\text{Fix}(W) = \bigcap_{i=1}^{\infty} \text{Fix}(T_i)$. $\quad\square$

Remark 4.5.1. If $\alpha_i \in (0,1)$, then both $U_{n,k}$ and W_n are averaged for any $k \geq 1$ and any $n \geq k$. If $\alpha_i \in (0,b]$ and $b < 1$, then W is also averaged.

Theorem 4.5.6. *Let E be a real uniformly convex Banach space whose norm is uniformly Gâteaux differentiable. Let C be a closed convex subset of E. Let $\{T_i\}_{i=1}^{\infty} : C \to C$ be an infinite countable family of nonexpansive mappings with $F = \bigcap_{i=1}^{\infty} \text{Fix}(T_i) \neq \emptyset$. Let $\{\alpha_i\}_{i=1}^{\infty} \in [0,1]$ be a real sequence satisfying $0 < \alpha_i \leq b < 1 \ \forall i = 1, 2, \ldots$ Let W_n be the W-mapping generated by $T_n, T_{n-1}, \ldots, T_1$ and $\alpha_n, \alpha_{n-1}, \ldots, \alpha_1$ and let W be the W-mapping generated by $T_1, T_2, \ldots$ and $\alpha_1, \alpha_2, \ldots$ Let $\{\beta_n\} \subset [0,1]$ be a real sequence such that $\lim_{n\to\infty} \beta_n = 0$ and $\sum_{n=1}^{\infty} \beta_n = \infty$. Define a sequence $\{x_n\}$ by*

$$u, x_1 \in C, \quad x_{n+1} = \beta_n u + (1 - \beta_n)W_n x_n, \quad n \geq 1,$$

Then $\{x_n\}$ converges to $Q_F u$ in norm.

Proof. Define a new sequence $\{y_n\} \subset C$ by

$$u, y_1 \in C, \quad y_{n+1} = \beta_n u + (1 - \beta_n) W y_n, \quad n \geq 1.$$

Since W is averaged, one concludes from Suzuki's convergence theorem that $y_n \to Q_F u$ as $n \to \infty$. Since W_n is nonexpansive, one has

$$\|x_{n+1} - y_{n+1}\| \leq (1 - \beta_n)\|W_n x_n - W_n y_n\|$$
$$\leq (1 - \beta_n)\|W_n x_n - W_n y_n\| + \|W_n y_n - W_n y_n\|$$
$$\leq (1 - \beta_n)\|x_n - y_n\| + \|W_n y_n - W_n y_n\|.$$

Since $\{y_n\}$ is convergent, one finds that $\{y_n\}$ is a bounded sequence. Letting $D = \sup_{n \geq 1}\{\|y_n - \omega\|\}$, $\omega \in F$, $\forall m > n \geq 1$, we have

$$\|W_n y_n - W_n y_n\| = \lim_m \|W_n y_n - W_n y_n\|$$
$$\leq \lim_m \sum_{j=n}^{m-1} \|W_{j+1} y_n - W_j y_n\|$$
$$\leq \sum_{j=n}^{\infty} 2Db^{j+1}$$
$$= \frac{2Db^{n+1}}{1 - b},$$

since the limit $Wx = \lim_n W_n x$ is uniform in x. This implies

$$\sum_{j=n}^{\infty} \|W y_n - W_n y_n\| \leq \frac{2D}{1 - b} \sum_{n=1}^{\infty} b^{n+1} = \frac{2Db^2}{(1 - b)^2} < +\infty.$$

Using Lemma 1.10.2, one concludes that $x_n - y_n \to \theta$ as $n \to \infty$. Thus, $x_n \to Q_F u$ as $n \to \infty$. This completes the proof. $\qquad \square$

Theorem 4.5.7. *Let E be a real uniformly convex Banach space whose norm is Fréchet differentiable. Let C be a closed convex subset of E. Let $\{T_i\}_{i=1}^{\infty} : C \to C$ be an infinite countable family of nonexpansive mappings with $F = \bigcap_{i=1}^{\infty} \text{Fix}(T_i) \neq \emptyset$. Let $\{\alpha_i\}_{i=1}^{\infty} \subset [0, 1]$ be a real sequence satisfying $0 < \alpha_i \leq b < 1$ $(i = 1, 2, \ldots)$ and $\sum_{n=1}^{\infty} \alpha_n(1 - \alpha_n) = \infty$. Let W_n be the W-mapping generated by $T_1, T_2, \ldots$ and $\alpha_1, \alpha_2, \ldots$ Define a sequence $\{x_n\}$ by*

$$x_1 \in C, \quad x_{n+1} = (1 - \alpha_n)x_n + \alpha_n W_n x_n, \quad n \geq 1.$$

Then $\{x_n\}$ weakly converges to a common fixed point of $\{T_i\}_{i=1}^{\infty}$.

Proof.
(i) Show that $\{x_n\}$ is bounded.

Fixing $p \in F$, one gets

$$\|x_{n+1} - p\| \le (1 - \alpha_n)\|x_n - p\| + \alpha_n\|W_nx_n - p\|$$
$$\le (1 - \alpha_n)\|x_n - p\| + \alpha_n\|x_n - p\|.$$
$$= \|x_n - p\|.$$

It follows that $\lim_n \|x_n - p\|$ exists. This proves that $\{x_n\}$ is bounded.

(ii) Show that $x_n - W_nx_n \to \theta$ as $n \to \infty$.

Using Theorem 1.8.9, we have that

$$\|x_{n+1} - p\|^2 \le (1 - \alpha_n)\|x_n - p\|^2 + \alpha_n\|W_nx_n - p\|^2$$
$$- \alpha_n(1 - \alpha_n)g(\|x_n - W_nx_n\|)$$
$$\le \|x_n - p\|^2 - \alpha_n(1 - \alpha_n)g(\|x_n - W_nx_n\|).$$

It follows that

$$\sum_{n=1}^{\infty} \alpha_n(1 - \alpha_n)g(\|x_n - W_nx_n\|) < +\infty.$$

Since $\sum_{n=1}^{\infty} \alpha_n(1 - \alpha_n) = \infty$, we obtain that $\liminf_{n \to \infty} g(\|x_n - W_nx_n\|) = 0$. From the property of g, we find that

$$\liminf_{n \to \infty} \|x_n - W_nx_n\| = 0.$$

On the other hand, one has

$$\|x_{n+1} - W_{n+1}x_{n+1}\|$$
$$\le (1 - \alpha_n)\|x_n - W_nx_n\| + \|W_nx_n - W_{n+1}x_{n+1}\|$$
$$\le (1 - \alpha_n)\|x_n - W_nx_n\| + \|W_nx_n - W_{n+1}x_{n+1}\|+$$
$$\le (1 - \alpha_n)\|x_n - W_nx_n\| + \|x_{n+1} - x_n\| + \|W_{n+1}x_n - W_nx_n\|$$
$$\le \|x_n - W_nx_n\| + \|W_{n+1}x_n - W_nx_n\|.$$

In view of the fact that $\sum_{n=1}^{\infty} \|W_{n+1}x_n - W_nx_n\| < +\infty$, one concludes from Lemma 1.10.1 that $\lim_{n \to \infty} \|x_n - W_nx_n\|$ exists. Hence $x_n - W_nx_n \to \theta$ as $n \to \infty$.

Taking into account that

$$\|W_nx_n - Wx_n\| \le M_1 b^{n-1} \to 0 \quad (n \to \infty),$$

one obtains

$$\|x_n - Wx_n\| \le \|x_n - W_nx_n\| + \|W_nx_n - Wx_n\| \to 0 \quad (n \to \infty).$$

From Theorem 1.9.3, one sees that $\omega(x_n) \subset \text{Fix}(W)$. Using Theorem 1.9.8, one concludes $x_n \to p \in \text{Fix}(W)$ as $n \to \infty$. Using Proposition 4.5.4, one concludes that $\text{Fix}(W) = \bigcap_{i=1}^{\infty} \text{Fix}(T_i) = F$. Therefore, $x_n \to p \in F$ as $n \to \infty$. This completes the proof. $\square$

4.6 Iterative methods of common fixed points for a nonexpansive semigroup in Banach spaces

Definition 4.6.1. Let C be a nonempty, convex, and closed subset of a real Banach space. Let $S = \{T(s) : 0 \le s < +\infty\}$ be a family of mappings defined on C. Then S is said to be a one-parameter nonexpansive semigroup if it satisfies the following conditions:
(i) $T(0)x = x, \forall x \in C$;
(ii) $T(s + t) = T(s)T(t), \forall s, t \in \mathbb{R}^+$;
(iii) $\|T(s)x - T(s)y\| \le \|x - y\|, \forall x, y \in C, s \in \mathbb{R}^+$;
(iv) For any $x \in C, s \mapsto T(s)x$ is continuous.

In this subsection, let $\text{Fix}(S) = \bigcap_{0 \le t < \infty} \text{Fix}(T(t))$.

Lemma 4.6.1 (Bruck [12], 1979). *Let C be a nonempty, bounded, closed, and convex subset of a real uniformly convex Banach space E. Let $S = \{T(s) : 0 \le s < +\infty\}$ be a one-parameter nonexpansive semigroup on C. Then, $\forall h \in \mathbb{R}^+$,*

$$\lim_{t \to \infty} \sup_{x \in C} \left\| \frac{1}{t} \int_0^t T(s)x \, ds - T(h)\left(\frac{1}{t} \int_0^t T(s)x \, ds \right) \right\| = 0.$$

Theorem 4.6.1. *Let E be a real uniformly convex Banach space with a Fréchet differentiable norm. Let C be a nonempty, closed, and convex subset of E, and let $S = \{T(s) : 0 \le s < +\infty\}$ be a one-parameter nonexpansive semigroup on C such that $\text{Fix}(S) \ne \emptyset$. Let $\{x_n\}$ be a sequence defined by*

$$x_1 \in C, \quad x_{n+1} = \alpha_n x_n + (1 - \alpha_n)\frac{1}{t_n} \int_0^{t_n} T(s)x_n \, ds, \quad n \ge 1,$$

where $\{\alpha_n\}$ is a real sequence in $(0, 1)$ such that $0 < \alpha_n \le a < 1$, and $\{t_n\}$ is a real sequence in $(0, +\infty)$ such that $t_n \to +\infty$ as $n \to \infty$. Then $\{x_n\}$ converges weakly to a common fixed point of $\{T_t\}_{t > 0}$.

Proof. Fixing $x_1 \in C$ and $p \in F$, let $r = \|x_1 - p\|$ and $K = \{x \in E, \|x - p\| \le r\} \cap C$. Then K is a nonempty, bounded, closed, and convex subset of C and $T(t)K \subset K, \forall t \in \mathbb{R}^+$. Without loss of generality, we may assume that C is bounded. Using Theorem 1.8.9, we have

$$\|x_{n+1} - \omega\|^2 \le \left\| \alpha_n(x_n - \omega) + (1 - \alpha_n)\left(\frac{1}{t_n} \int_0^{t_n} T(s)x_n \, ds - \omega \right) \right\|^2$$

$$\le \alpha_n \|x_n - \omega\|^2 + (1 - \alpha_n)\left\| \frac{1}{t_n} \int_0^{t_n} T(s)x_n \, ds - \omega \right\|^2$$

$$- a_n(1 - a_n)g\left(\left\|\frac{1}{t_n}\int_0^{t_n} T(s)x_n ds - x_n\right\|\right), \quad \omega \in \text{Fix}(S).$$

From $a_n \le a$, we have that

$$a_n(1 - a)g\left(\left\|\frac{1}{t_n}\int_0^{t_n} T(s)x_n ds - x_n\right\|\right)$$

$$\le a_n(1 - a_n)g\left(\left\|\frac{1}{t_n}\int_0^{t_n} T(s)x_n ds - x_n\right\|\right)$$

$$\le a_n\|x_n - \omega\|^2 + (1 - a_n)\left\|\frac{1}{t_n}\int_0^{t_n} T(s)x_n ds - \omega\right\|^2 - \|x_{n+1} - \omega\|^2$$

$$\le a_n\|x_n - \omega\|^2 + (1 - a_n)\|x_n - \omega\|^2 - \|x_{n+1} - \omega\|^2$$

$$= \|x_n - \omega\|^2 - \|x_{n+1} - \omega\|^2.$$

This implies that $\lim_{n\to\infty} \|x_n - \omega\|^2$ exists. It follows that

$$\lim_{n\to\infty} a_n g\left(\left\|\frac{1}{t_n}\int_0^{t_n} T(s)x_n ds - x_n\right\|\right) = 0.$$

From the property of g, we have

$$\lim_{n\to\infty} a_n\left\|\frac{1}{t_n}\int_0^{t_n} T(s)x_n ds - x_n\right\| = 0.$$

From the definition of $\{x_n\}$, we find that

$$x_{n+1} - \frac{1}{t_n}\int_0^{t_n} T(s)x_n ds = a_n\left(x_n - \frac{1}{t_n}\int_0^{t_n} T(s)x_n d\right) \to \theta$$

as $n \to \infty$. Note that

$$\|T(t)x_{n+1} - x_{n+1}\| \le \left\|T(t)x_{n+1} - T(t)\left(\frac{1}{t_n}\int_0^{t_n} T(s)x_n ds\right)\right\|$$

$$+ \left\|T(t)\left(\frac{1}{t_n}\int_0^{t_n} T(s)x_n ds\right) - \frac{1}{t_n}\int_0^{t_n} T(s)x_n ds\right\|$$

$$+ \left\|\frac{1}{t_n}\int_0^{t_n} T(s)x_n ds - x_{n+1}\right\|$$

$$\leq 2\left\|\frac{1}{t_n}\int_0^{t_n}T(s)x_n\,ds - x_{n+1}\right\|$$

$$+\left\|T(t)\left(\frac{1}{t_n}\int_0^{t_n}T(s)x_n\,ds\right) - \frac{1}{t_n}\int_0^{t_n}T(s)x_n\,ds\right\|, \quad \forall t \in \mathbb{R}^+.$$

Using Lemma 4.6.1, we find that $T(t)x_n - x_n \to \theta$, $\forall t \in \mathbb{R}^+$, as $n \to \infty$. From Theorem 1.9.3, we obtain that $w_\omega(x_n) \subset \text{Fix}(s)$. Define a mapping $T_n : C \to C$ by

$$T_n x = \alpha_n x + (1 - \alpha_n)\frac{1}{t_n}\int_0^{t_n}T(s)x\,ds.$$

Then $x_{n+1} = T_n x_n$ and $T_n : C \to C$ is a nonexpansive mapping with $\text{Fix}(s) \subset \bigcap_{n=1}^\infty \text{Fix}(T_n) \neq \emptyset$. It follows from Theorem 1.9.8 that $\{x_n\}$ converges weakly to a common fixed point of $\{T_t\}_{t>0}$. $\qquad\square$

Definition 4.6.2. Let C be a nonempty, convex, and closed subset of a real Banach space. Let $S = \{T(s) : 0 \leq s < +\infty\}$ be a family of mappings defined on C. Then S is said to be uniformly asymptotically regular on C if for any bounded subsequence $K \subset C$, we have

$$\lim_{t\to\infty}\sup_{x\in K}\|T(h)(T(t)x) - T(t)x\| = 0, \quad \forall h \in \mathbb{R}^+.$$

Theorem 4.6.2. *Let E be a real reflexive and strictly convex Banach space with a uniformly Gâteaux differentiable norm. Let C be a nonempty, closed, and convex subset of E. Let $\{T_t\}_{t\geq0}$ be a uniformly asymptotically regular nonexpansive semigroup on C such that $F = \bigcap_{t\geq0}\text{Fix}(T(t)) \neq \emptyset$. Suppose that $t_n \to \infty$, $\{\alpha_n\} \subset (0,1)$, $\alpha_n \to 0$ as $n \to \infty$. Then, for any $u \in C$, there exists a unique sequence $\{x_n\} \subset C$ such that*

$$x_n = \alpha_n u + (1 - \alpha_n)T(t_n)x_n,$$

and $x_n \to p = Q_F u$ as $n \to \infty$.

Proof. Define a mapping $T_n : C \to C$ by

$$T_n x = \alpha_n u + (1 - \alpha_n)T(t_n)x, \quad \forall x \in C.$$

Then

$$\|T_n x - T_n y\| = (1 - \alpha_n)\|T(t_n)x - T(t_n)y\| \leq (1 - \alpha_n)\|x - y\|, \quad \forall x, y \in C,$$

which shows that T_n is contractive. From the Banach fixed point theorem, there exists a unique $x_n \in C$ such that $T_n x_n = x_n$, that is, the iterative method is converges. We next prove that $\{x_n\}$ is bounded. Indeed, for any $z \in F$, one has

$$\|x_n - z\| = \|\alpha_n(u - z)\| + (1 - \alpha_n)\|T(t_n)x_n - z\|$$

$$\leq \alpha_n\|u - z\| + (1 - \alpha_n)\|x_n - z\|.$$

This implies that $\|x_n - z\| \le \|u - z\|$. It shows that $\{x_n\}$ is bounded. Hence, $\{T(t_n)x_n\}$ is also bounded. It follows that $x_n - T(t_n)x_n = \alpha_n(u - T(t_n)x_n) \to \theta$ as $n \to \infty$. From the hypothesis that $t_n \to \infty$ and $\{T(t)\}$ is uniformly asymptotically regular, we find that, $\forall h > 0$,

$$\lim_{n\to\infty}\|T(h)(T(t_n)x_n) - T(t_n)x_n\| \le \lim_{n\to\infty}\sup_{x\in K}\|T(h)(T(t_n)x) - T(t_n)x\| = 0,$$

where K is a bounded set containing $\{x_n\}$. It follows that

$$\begin{aligned}
\|x_n - T(h)x_n\| &\le \|x_n - T(t_n)x_n\| + \|T(t_n)x_n - T(h)(T(t_n)x_n)\| \\
&\quad + \|T(h)(T(t_n)x_n) - T(h)x_n\| \\
&\le 2\|x_n - T(t_n)x_n\| + \|T(h)(T(t_n)x_n) - T(t_n)x_n\| \to 0
\end{aligned}$$

as $n \to \infty$. This shows that $x_n - T(t_n)x_n = \alpha_n(u - T(t_n)x_n) \to \theta$, $\forall h > 0$, as $n \to \infty$.

Let $g : C \to \mathbb{R}^+$ be a functional defined by

$$g(y) = \mu_n\|x_n - y\|, \quad \forall y \in C.$$

Then g is convex, continuous, and $g(y) \to \infty$ as $\|y\| \to \infty$. There also exists $z \in C$ such that

$$g(z) = \min\{g(y) : y \in C\}.$$

Putting

$$K = \{x \in C : g(x) = \min\{g(y) : y \in C\}\},$$

one finds that $K \ne \emptyset$ is closed and convex, and $T(h)K \subset K$, $\forall h > 0$. Indeed,

$$\begin{aligned}
\forall x \in K, \ g(T(h)x) &= \mu_n\|x_n - T(h)x\| \\
&= \mu_n\|x_n - T(h)x_n + T(h)x_n - T(h)x\| \\
&\le \mu_n\|T(h)x_n - T(h)x\| \\
&\le \mu_n\|x_n - x\| = g(x).
\end{aligned}$$

It follows that $T(h)x \in K$, $\forall h > 0$. Fixing $p \in F$, one finds that there exists a unique $v \in K$ such that

$$\|p - v\| = \min\{\|p - x\| : x \in K\}.$$

It follows from $p = T(h)p$ and $T(h)v \in K$ that

$$\|p - T(h)v\| = \|T(h)p - T(h)v\| \le \|p - v\|.$$

From the uniqueness of $v \in K$, we have that $T(h)v = v$, $\forall h \in \mathbb{R}^+$. From Theorem 1.8.45, we have

$$\mu_n \langle y - v, j(x_n - v) \rangle \le 0, \quad \forall y \in C.$$

In particular, one has

$$\mu_n \langle u - v, j(x_n - v) \rangle \le 0.$$

This yields

$$\mu_n \|x_n - v\|^2 \le \mu_n \langle u - v, j(x_n - v) \rangle \le 0.$$

Thus $\mu_n \|x_n - v\|^2 = 0$. This shows that there exists a subsequence $\{x_{n_k}\}$ of $\{x_n\}$ such that $x_{n_k} \to v$ as $k \to \infty$. If $\{x_n\}$ has another subsequence $\{x_{m_j}\}$ such that $x_{m_j} \to p \in F$ as $j \to \infty$, then $p = v$. Indeed, we have

$$\|x_n - y\|^2 \le \langle u - y, j(x_n - y) \rangle, \quad \forall y \in F.$$

Thus $\langle x_n - u, j(x_n - y) \rangle \le 0$, $\forall y \in F$. Consequently,

$$\langle x_{n_k} - u, j(x_{n_k} - y) \rangle \le 0$$

and

$$\langle x_{m_j} - u, j(x_{m_j} - y) \rangle \le 0.$$

Taking the limits as $k \to \infty$, $j \to \infty$ in the above inequalities, respectively, we find that

$$\langle v - u, j(v - p) \rangle \le 0$$

and

$$\langle p - u, j(p - v) \rangle \le 0.$$

It follows that

$$\|v - p\|^2 \le \langle v - p, j(v - p) \rangle \le 0.$$

This shows $p = v$. Hence, $x_n \to p = Q_F u$ as $n \to \infty$, completing the proof. $\qquad \square$

Theorem 4.6.3. *Let E be a real reflexive and strictly convex Banach space with a uniformly Gâteaux differentiable norm. Let C be a closed convex subset of E. Let $\{T_t\}_{t \ge 0}$ be a uniformly asymptotically regular nonexpansive semigroup on C with $F = \bigcap_{t \ge 0} \mathrm{Fix}(T(t)) \ne \emptyset$. Assume that $t_n \to \infty$, $\{\alpha_n\} \subset (0,1)$ such that $\lim_{n \to \infty} \alpha_n = 0$ and $\sum_{n=1}^{\infty} \alpha_n = \infty$. Let $\{x_n\}$ be a sequence defined by*

$$u, x_1 \in C, \quad x_{n+1} = \alpha_n u + (1 - \alpha_n) T(t_n) x_n, \quad \forall n \ge 1.$$

Then $x_n \to p = Q_F u$ as $n \to \infty$.

Proof. First, we show that $\{x_n\}$ is bounded. Fixing $z \in F$, we have

$$\|x_{n+1} - z\| \leq \alpha_n \|u - z\| + (1 - \alpha_n)\|x_n - z\|$$
$$\leq \max\{\|u - z\|, \|x_1 - z\|\} = M.$$

This shows $\{x_n\}$ is bounded, so is $\{T(t_n)x_n\}$. Since $\lim_{n\to\infty} \alpha_n = 0$, we find that $x_n - T(t_n)x_n = \alpha_n(u - T(t_n)x_n) \to \theta$ as $n \to \infty$. Taking into account the fact that $\{T_t\}_{t>0}$ is uniformly asymptotically regular, we have

$$\lim_{n\to\infty} \|T(h)(T(t_n)x_n) - T(t_n)x_n\| = 0, \quad \forall h > 0.$$

Note that

$$\|x_{n+1} - T(h)x_{n+1}\| \leq \|x_{n+1} - T(t_n)x_n\| + \|T(t_n)x_n - T(h)T(t_n)x_n\|$$
$$+ \|T(h)T(t_n)x_n - T(h)x_{n+1}\|$$
$$\leq 2\|x_{n+1} - T(t_n)x_n\| + \|T(h)T(t_n)x_n - T(t_n)x_n\|, \quad \forall h > 0.$$

It follows that $x_n - T(h)x_n \to \theta$, $\forall h > 0$, as $n \to \infty$. From Lemma 4.3.1, we find that

$$\limsup_{n\to\infty} \langle u - p, j(x_n - p)\rangle \leq 0,$$

where $p = Q_F u$.

Finally we show that $x_n \to p$ $(n \to \infty)$. Note that

$$\|x_{n+1} - p\|^2 = \alpha_n \langle u - p, j(x_{n+1} - p)\rangle + (1 - \alpha_n)\langle T(t_n)x_n - p, j(x_{n+1} - p)\rangle$$
$$\leq (1 - \alpha_n)\|x_n - p\|\|x_{n+1} - p\| + \alpha_n \langle u - p, j(x_{n+1} - p)\rangle$$
$$\leq \frac{1 - \alpha_n}{2}\|x_n - p\|^2 + \frac{1}{2}\|x_{n+1} - p\|^2 + \alpha_n \langle u - p, j(x_{n+1} - p)\rangle.$$

This implies

$$\|x_{n+1} - p\|^2 \leq (1 - \alpha_n)\|x_n - p\|^2 + 2\alpha_n \langle u - p, j(x_{n+1} - p)\rangle.$$

Using Lemma 1.10.2, one concludes that $x_n \to p$ as $n \to \infty$. $\qquad\square$

4.7 Iterative methods of fixed points for nonexpansive nonself-mappings

Let C be a nonempty, closed, and convex subset of a real Banach space E. Let $Q_C : E \to C$ be a sunny nonexpansive retraction and let $T : C \to E$ be a nonexpansive mapping. Let $\Phi : C \to E$ be a Meir–Keeler (MK) contractive mapping. For any $t \in (0, 1)$, define a mapping $T_t^\Phi : C \to C$ by

$$T_t^\Phi x = Q_C[t\Phi x + (1 - t)Tx], \quad \forall x \in C.$$

From Proposition 4.4.2, we have that $T_t^\Phi : C \to C$ is a Meir–Keeler contractive mapping. Using Theorem 4.1.2, we find that T_t^Φ has a unique fixed point x_t in C, that is,

$$x_t = Q_C[t\Phi(x_t) + (1-t)T(x_t)], \quad t \in (0,1). \tag{4.69}$$

Discretizing (4.69) yields

$$x_1 \in C, \quad x_{n+1} = Q_C[\alpha_n\Phi(x_n) + (1-\alpha_n)T(x_n)], \quad n \geq 1. \tag{4.70}$$

We also consider another iterative method

$$x_1 \in C, \quad x_{n+1} = \lambda x_n + (1-\lambda)Q_C[\alpha_n\Phi(x_n) + (1-\alpha_n)T(x_n)], \quad n \geq 1, \tag{4.71}$$

where λ is a real number in $(0,1)$ and $\{\alpha_n\}$ is a real sequence in $(0,1)$ which satisfies certain conditions.

In 2008, Matsushita and Takahashi [49] introduced a new boundary condition, which is called the MT condition in order to study fixed points of nonexpansive nonself-mappings in Banach spaces.

Definition 4.7.1. Let C be a nonempty, closed, and convex subset of a real smooth Banach space E. Let $T : C \to E$ be a nonself-mapping. Then T is said to satisfy the MT condition if

$$Tx \in \wp S_x, \quad \forall x \in C,$$

where

$$S_x = \{y \in E : y \neq x, Q_C y = x\},$$

and $Q_C : E \to C$ is a sunny nonexpansive retraction from E onto C.

Matsushita and Takahashi [49] established the following useful results.

Proposition 4.7.1. *Let E be a real smooth Banach space. Let C be a nonempty, closed, and convex subset of E such that C is a sunny nonexpansive retract. Let $T : C \to E$ be a nonself-mapping. If $T : C \to E$ satisfies the weak inward condition (WIC), then T satisfies the MT condition.*

Proposition 4.7.2. *Let E be a real smooth and strictly convex Banach space. Let C be a nonempty, closed, and convex subset of E such that C is a sunny nonexpansive retract. Let $T : C \to E$ be a nonself-mapping. If $\mathrm{Fix}(T) \neq \emptyset$, then T satisfies the MT condition.*

The importance of the MT condition is explained by the following propositions.

Proposition 4.7.3. *Let E be a real smooth Banach space. Let C be a nonempty, closed, and convex subset of E such that C is a sunny nonexpansive retract. Let $T : C \to E$ be a nonself-mapping. If T satisfies the MT condition, then*

$$\mathrm{Fix}(T) = \mathrm{Fix}(Q_C T).$$

In 2014, Zhou and Wang [123] improved Proposition 4.7.3 and obtained the following more useful result.

Proposition 4.7.4. *Let E be a real smooth Banach space. Let C be a nonempty, closed, and convex subset of E such that C is a sunny nonexpansive retract. Let $T : C \rightarrow E$ be a nonself-mapping. If T satisfies the MT condition, then*

$$\text{Fix}(T) = \text{Fix}(Q_C T) = \text{Fix}(T Q_C).$$

Theorem 4.7.1. *Let E be a real uniformly smooth Banach space. Let C be a nonempty, closed, and convex subset of E such that C is a sunny nonexpansive retract. Let $T : C \rightarrow E$ be a nonexpansive nonself-mapping such that $\text{Fix}(T) \neq \emptyset$. Let $\Phi : C \rightarrow E$ be an MK contractive mapping. If $T : C \rightarrow E$ satisfies the MT condition, then the path $\{x_t\}$ generated by (4.69) converges in norm to $p \in \text{Fix}(T)$ and p is also the unique solution of the variational inequality problem*

$$\langle (I - \Phi)p, j(y - p) \rangle \geq 0, \quad \forall y \in \text{Fix}(T). \tag{VIP}$$

Proof. Letting $y_t = t\Phi(x_t) + (1-t)T(x_t)$, one finds that (4.69) reduces to $x_t = Q_C y_t$, which gives

$$y_t = t(\Phi Q_C)y_t + (1-t)(TQ_C)y_t. \tag{4.72}$$

Since $\Phi Q_C : E \rightarrow E$ is MK contractive, we find that $TQ_C : E \rightarrow E$ is nonexpansive. From Theorem 4.4.2, $\{y_t\}$ converges in norm to a fixed point $p \in \text{Fix}(TQ_C)$ as $t \rightarrow 0$. Using Proposition 4.7.4, we have $\text{Fix}(TQ_C) = \text{Fix}(T)$. In view of the facts that $p \in C$ and $Q_C p = p$, we have

$$\langle (I - \Phi)p, j(y - p) \rangle \geq 0, \quad \forall y \in \text{Fix}(T).$$

From the continuity of Q_C, as $t \rightarrow 0$, we conclude that $\{x_t\}$ converges in norm to $p \in \text{Fix}(T)$ and p is the unique solution of the variational inequality (VIP). This completes the proof. $\qquad\square$

Theorem 4.7.2. *Let E be a real uniformly smooth Banach space. Let C be a nonempty, closed, and convex subset of E such that C is a sunny nonexpansive retract. Let $T : C \rightarrow E$ be a nonexpansive nonself-mapping such that $\text{Fix}(T) \neq \emptyset$ and satisfies the MT condition. Let $\Phi : C \rightarrow E$ be an MK contractive mapping. Let $\{x_n\}$ be a sequence defined by (4.70). If the control sequence $\{\alpha_n\}$ satisfies the following conditions:*
(i) $\alpha_n \rightarrow 0 \ (n \rightarrow \infty)$;
(ii) $\sum_{n=1}^{\infty} \alpha_n = \infty$;
(iii) $\sum_{n=1}^{\infty} |\alpha_{n+1} - \alpha_n| < \infty$ or $\frac{\alpha_n}{\alpha_{n+1}} \rightarrow 1 \ (n \rightarrow \infty)$.

Then $\{x_n\}$ converges in norm to $p \in \text{Fix}(T)$ and p is the unique solution of the variational inequality (VIP).

Proof. Letting $y_n = \alpha_n \Phi(x_n) + (1 - \alpha_n)T(x_n)$, we find that (4.70) reduces to $x_{n+1} = Q_C y_n$, which yields

$$y_{n+1} = \alpha_{n+1}(\Phi Q_C)y_n + (1 - \alpha_{n+1})(TQ_C)y_n. \tag{4.73}$$

Since $\Phi Q_C : E \to E$ is MK contractive, we have that $TQ_C : E \to E$ is nonexpansive. From Theorem 4.4.5, $\{y_n\}$ converges in norm to $p \in \mathrm{Fix}(TQ_C)$ as $n \to \infty$. Using Proposition 4.7.4, we obtain that $\mathrm{Fix}(TQ_C) = \mathrm{Fix}(T)$, which together with $p \in C$, $Q_C p = p$ yields that $p \in \mathrm{Fix}(T)$ is the unique solution of the variational inequality (VIP). From the continuity of Q_C, we find that $\{x_n\}$ converges in norm to $p \in \mathrm{Fix}(T)$. This completes the proof. $\qquad\square$

Theorem 4.7.3. *Let E be a real uniformly smooth Banach space. Let C be a nonempty, closed, and convex subset of E such that C is a sunny nonexpansive retract. Let $T : C \to E$ be a nonexpansive nonself-mapping such that $\mathrm{Fix}(T) \neq \emptyset$ and satisfies the MT condition. Let $\Phi : C \to E$ be an MK contractive mapping. Let $\{x_n\}$ be a sequence defined by (4.71). If the control sequence $\{\alpha_n\}$ satisfies conditions $\lim_{n\to\infty} \alpha_n = 0$ and $\sum_{n=1}^{\infty} \alpha_n = \infty$, then $\{x_n\}$ converges in norm to $p \in \mathrm{Fix}(T)$ and p is the unique solution of the variational inequality (VIP).*

Proof. First, we show that there exists a unique point z in $\mathrm{Fix}(T)$ such that

$$z = Q_{\mathrm{Fix}(T)}\Phi z. \tag{4.74}$$

Indeed, for any $u \in E$, $t \in (0, 1)$, we can define a mapping $T_t : E \to E$ by

$$T_t x = tu + (1 - t)Q_C TQ_C x, \quad \forall x \in E. \tag{4.75}$$

It is obvious that T_t is contractive. By using the Banach contractive mapping principle, there exists a unique path $\{x_t\}$ such that

$$x_t = tu + (1 - t)Q_C TQ_C x_t, \quad \forall t \in (0, 1). \tag{4.76}$$

From Theorem 4.3.1 or Corollary 4.3.1, we obtain that $\{x_t\}$ converges in norm to $p = Q_{\mathrm{Fix}(Q_C TQ_C)}u$, that is, $\lim_{t\to 0} x_t = p$. Since T satisfies the MT condition, we assert that $Q_C T : C \to C$ also satisfies the MT condition. If not, then there exists $x \in C$ such that

$$Q_C Tx \notin \wp S_x \implies Q_C Tx \in S_x \implies Q_C Tx \neq x$$

and $Q_C Q_C Tx = x$. But $Q_C Q_C = Q_C$. Hence $Q_C Tx = x$, which produces a contradiction. From Proposition 4.7.4, we find that

$$\mathrm{Fix}(Q_C TQ_C) = \mathrm{Fix}(Q_C T) = \mathrm{Fix}(T).$$

Thus $\lim_{t \to 0} x_t = Q_{\mathrm{Fix}(T)}u$ defines a unique sunny nonexpansive retraction from E onto $\mathrm{Fix}(T)$. Since $Q_{\mathrm{Fix}(T)}\Phi : C \to C$ is an MK contractive self-mapping, we find from Theorem 4.1.2 that there exists a unique element z in $\mathrm{Fix}(T)$ such that

$$z = Q_{\mathrm{Fix}(T)}\Phi(z).$$

Fix $\lambda \in (0, 1)$ and define a sequence $\{y_n\}$ by

$$y_1 \in C, \quad y_{n+1} = \lambda y_n + (1 - \lambda)Q_C[\alpha_n \Phi(z) + (1 - \alpha_n)Ty_n], \quad n \geq 1. \tag{4.77}$$

It is easy to prove that $\{y_n\}$ converges to z in norm.

Next, we plan to prove that $x_n - y_n \to \theta$ as $n \to \infty$. Assume that $a = \limsup_{n \to \infty} \|x_n - y_n\| > 0$; then for any $\varepsilon \in (0, a)$, we can choose $\eta > 0$ such that

$$\limsup_{n \to \infty} \|x_n - y_n\| > \varepsilon + \eta. \tag{4.78}$$

For the above chosen $\eta > 0$, we find from [83, Proposition 2] that there exists $y \in (0, 1)$ such that

$$\|\Phi x - \Phi y\| \leq y\|x - y\|, \quad \forall x, y \in C, \|x - y\| \geq \varepsilon, \tag{4.79}$$

which implies that

$$\|\Phi x - \Phi y\| \leq \max\{y\|x - y\|, \varepsilon\}, \quad \forall x, y \in C. \tag{4.80}$$

Since $y_n \to z$ as $n \to \infty$, we find that there exists $n_0 \in N$ such that

$$\|y_n - z\| \leq (1 - y)\eta, \quad \forall n \geq n_0. \tag{4.81}$$

Next, we consider the following two possible cases:

(i) There exists $v_1 \geq n_0$ such that

$$\|x_{v_1} - y_{v_1}\| \leq \varepsilon + \eta. \tag{4.82}$$

From (4.70), (4.77), (4.80), (4.81), and (4.82), we find that

$$\begin{aligned}
\|x_{v_1+1} - y_{v_1+1}\| &\leq \lambda\|x_{v_1} - y_{v_1}\| + (1 - \lambda)[\alpha_{v_1}\|\Phi x_{v_1} - \Phi z\| + (1 - \alpha_{v_1})\|x_{v_1} - y_{v_1}\|] \\
&= [1 - (1 - \lambda)\alpha_{v_1}]\|x_{v_1} - y_{v_1}\| + (1 - \lambda)\alpha_{v_1}\|\Phi x_{v_1} - \Phi z\| \\
&\leq [1 - (1 - \lambda)\alpha_{v_1}]\|x_{v_1} - y_{v_1}\| + (1 - \lambda)\alpha_{v_1}\|\Phi x_{v_1} - \Phi y_{v_1}\| \\
&\quad + (1 - \lambda)\alpha_{v_1}\|\Phi y_{v_1} - \Phi z\| \\
&\leq [1 - (1 - \lambda)\alpha_{v_1}]\|x_{v_1} - y_{v_1}\| + (1 - \lambda)\alpha_{v_1}\max\{y\|x_{v_1} - y_{v_1}\|, \varepsilon\} \\
&\quad + (1 - \lambda)\alpha_{v_1}\|y_{v_1} - z\| \\
&\leq \max\{[1 - (1 - y)(1 - \lambda)\alpha_{v_1}]\varepsilon + (1 - y)(1 - \lambda)\alpha_{v_1}\eta, \\
&\quad [1 - (1 - \lambda)\alpha_{v_1}]\varepsilon + (1 - \lambda)\alpha_{v_1}(\varepsilon + \eta)\} \\
&\leq \varepsilon + \eta.
\end{aligned}$$

In a similar way, we have

$$\|x_{v_1+2} - y_{v_1+2}\| \leq \varepsilon + \eta.$$

By induction, we find that

$$\|x_{v_1+m} - y_{v_1+m}\| \leq \varepsilon + \eta, \quad \forall m \in \mathbb{N}. \tag{4.83}$$

It follows that

$$\limsup_{n \to \infty} \|x_n - y_n\| \leq \varepsilon + \eta,$$

which yields a contradiction, implying that $x_n - y_n \to \theta (n \to \infty)$. Hence $x_n \to z$ as $n \to \infty$.

(ii) $\|x_n - y_n\| > \varepsilon + \eta, \forall n \geq v_1$.

Suppose that Case (ii) holds. It follows from (4.79) that

$$\|\Phi x_n - \Phi y_n\| \leq \gamma\|x_n - y_n\|, \quad \forall n \geq v_1. \tag{4.84}$$

From (4.71), (4.77), and (4.84), we find that

$$
\begin{aligned}
\|x_{n+1} - y_{n+1}\| &\leq \lambda\|x_n - y_n\| + (1-\lambda)[\alpha_n\|\Phi x_n - \Phi z\| + (1-\alpha_n)\|x_n - y_n\|] \\
&= [1 - (1-\lambda)\alpha_n]\|x_n - y_n\| + (1-\lambda)\alpha_n\gamma\|x_n - y_n\| + (1-\lambda)\alpha_n\|y_n - z\| \\
&\leq [1 - (1-\gamma)(1-\lambda)\alpha_n]\|x_n - y_n\| + o(\alpha_n).
\end{aligned}
$$

Using Lemma 1.10.2, we obtain that $\|x_n - y_n\| \to 0$ as $n \to \infty$, which yields that $0 \geq \varepsilon + \eta$, i. e., a contradiction. It further implies that Case (ii) is impossible, completing the proof. □

By using Propositions 4.7.1 and 4.7.2, and the methods used in Theorems 4.7.1, 4.7.2, and 4.7.3, we can obtain the following results (see Zhou and Wang [123] for more details).

Theorem 4.7.4. *Let E be a real reflexive and strictly convex Banach space whose norm is uniformly Gâteaux differentiable. Let C be a nonempty, closed, and convex subset of E such that it is also a sunny nonexpansive retract. Let $T : C \to E$ be a nonexpansive nonself-mapping with $\mathrm{Fix}(T) \neq \emptyset$ and let $\Phi : C \to E$ be an MK contractive mapping. Let $\{x_t\}$ be a sequence generated by (4.69). Then $\{x_t\}$ converges in norm to $p \in \mathrm{Fix}(T)$ and p is the unique solution of the variational inequality (VIP).*

Theorem 4.7.5. *Let E be a real reflexive and strictly convex Banach space whose norm is uniformly Gâteaux differentiable. Let C be a nonempty, closed, and convex subset of E such that it is also a sunny nonexpansive retract. Let $T : C \to E$ be a nonexpansive nonself-mapping with $\mathrm{Fix}(T) \neq \emptyset$ and let $\Phi : C \to E$ be an MK contractive mapping. Let $\{x_n\}$ be a sequence generated by (4.70). If the control sequence $\{\alpha_n\}$ satisfies the conditions of Theorem 4.7.2, then $\{x_n\}$ converges in norm to $p \in \mathrm{Fix}(T)$ and p is the unique solution of the variational inequality (VIP).*

Theorem 4.7.6. *Let E be a real reflexive, strictly convex Banach space whose norm is uniformly Gâteaux differentiable. Let C be a nonempty, closed, and convex subset of E such that it is also a sunny nonexpansive retract. Let $T : C \to E$ be a nonexpansive nonself-mapping with $\mathrm{Fix}(T) \neq \emptyset$ and let $\Phi : C \to E$ be an MK contractive mapping. Let $\{x_n\}$ be a sequence generated by (4.71). If the control sequence $\{\alpha_n\}$ satisfies the conditions of Theorem 4.7.3, then $\{x_n\}$ converges in norm to $p \in \mathrm{Fix}(T)$ and p is the unique solution of the variational inequality* (VIP).

4.8 Remarks

Theorem 4.1.3, which extends Browder's fixed point theorem to the setting of nonself-mappings, is due to Caristi [17]. Theorem 4.2.1, which is a remarkable result in the field of fixed point theory, is due to Reich [73]. Theorem 4.2.3, which provides us with a method to obtain approximate fixed points, is due to Ishikawa [33]. Theorem 4.2.5 shows that, under some additional restriction, Krasnosel'skiĭ–Mann iterative method can be strongly convergent. The origin of Theorem 4.3.1 can be traced back to the Browder's convergence theorem.

Lemma 4.3.1, which is applicable in various situations, was given by the authors of the book. Theorem 4.3.2 is indeed due to several authors. Example 4.2 is to Suzuki [84]. Theorem 4.3.3 is to Suzuki [84]. Theorems 4.3.4, 4.3.5, and 4.3.6 are due to Xu [103]. Most results in Section 4.4 are due to Suzuki [84], and the results in Section 4.5 are essentially due to Shioji and Takahashi [79]. The results in Section 4.6 are standard in convergence theorems of nonexpansive semigroups, and the results in Section 4.7 are due to Zhou and Wang [123].

4.9 Exercises

1. Let X be the space $\mathbb{R}^2$ equipped with the Euclidean norm and, with (r, θ) denoting the polar coordinates. Let

$$C = \left\{ (r, \theta) : 0 \leq r \leq 1, \frac{\pi}{4} \leq \theta \leq \frac{\pi}{2} \right\}.$$

Define a mapping $T : C \to C$ by

$$T(r, \theta) = \left(r, \frac{\pi}{2} \right), \quad \text{for each point } (r, \theta) \in C.$$

Prove that the following assertions hold true:
(a1) T is a nonexpansive mapping;
(a2) $\mathrm{Fix}(T) = \{(r, \frac{\pi}{2}) : 0 \leq r \leq 1\}$;

(a3) Take $x_0 = (r, \theta_0) = \{(r, \frac{\pi}{4})\}$ and $\alpha_n \in [0,1]$ for all $n \geq 0$, and construct the Mann sequence $\{x_n\}$ by

$$x_{n+1} = (1 - \alpha_n)x_n + \alpha_n T x_n, \quad n \geq 0.$$

Then $r_n = r_0 = 1$, $n \geq 0$ and

$$\theta_{n+1} = \alpha_n \frac{\pi}{2} + (1 - \alpha)\theta_n, \quad n \geq 0 \text{ and } \theta_0 = \frac{\pi}{4}.$$

(a4) If one takes $\alpha_n \equiv 1$, then

$$x_n \to \left(1, \frac{\pi}{2}\right) \in \text{Fix}(T);$$

(a5) If one takes $\alpha \equiv 0$, then

$$x_n \to \left(1, \frac{\pi}{4}\right) \notin C.$$

2. Let X be a Banach space and $\{T_i\}_{i=1}^r : X \to X$ be a finite family of nonexpansive mappings such that $K = \bigcap_{i=1}^r \text{Fix}(T_i) = \text{Fix}(T_r T_{r-1} \cdots T_1) \neq \emptyset$. Show that

$$K = \text{Fix}(T_1 T_r \cdots T_3 T_2) = \cdots = \text{Fix}(T_{r-1} T_{r-2} \cdots T_1 T_r).$$

3. Let C be a convex subset of a Banach space X and $\{T_i\}_{i=1}^\infty : C \to C$ be a countable family of nonexpansive mappings. Let $\{\alpha_n\}$ be a sequence of numbers in $(0,1)$. Let W_n be the W-mappings generated by $T_n, T_{n-1}, \ldots, T_1$ and $\alpha_n, \alpha_{n-1}, \ldots, \alpha_1, n \geq 1$. Show that W_n are averaged nonexpansive for all $n \geq 1$.

4. Let C be a nonempty, closed, and convex subset of a strictly convex Banach space X. Let $\{S_k\}$ be a sequence nonexpansive mappings of C into X, and β_k be a sequence of positive real numbers such that $\sum_{k=1}^\infty \beta_k = 1$. Assume that $\mathcal{F} = \bigcap_{k=1}^\infty \text{Fix}(S_k) \neq \emptyset$. Then, show that the mapping $T := \sum_{k=1}^\infty \beta_k S_k$ is well defined and $\text{Fix}(T) = \bigcap_{k=1}^\infty \text{Fix}(S_k)$.

5. Let C be a nonempty, closed, and convex subset of a strictly convex and 2-uniformly smooth Banach space X with the best smoothness constant K. Let A be an α-inverse strongly monotone mapping of C into X, and let $S : C \to C$ be a nonexpansive mapping such that $\mathcal{F} = \text{Fix}(S) \cap SOL(C, A) \neq \emptyset$. For arbitrary $x_1 = u \in C$, define a sequence $\{x_n\}$ iteratively in C by

$$x_{n+1} = \alpha_n u + (1 - \alpha_n)\left[\frac{1}{2}Sx_n + \frac{1}{2}Q_C(x_n - \lambda A x_n)\right], \quad n \geq 1,$$

where $Q_C : X \to C$ is a sunny nonexpansive retraction from X onto C, $\{\alpha_n\}$ and $\{\lambda_n\}$ are sequences of real numbers in $(0,1)$ satisfying the following conditions:
(i) $\lambda_n \in [a, b]$ for $a, b \in (0, 1)$, and $\lambda_n \leq \frac{\alpha}{K^2}$;

(ii) $\sum_{n=1}^{\infty} |\lambda_{n+1} - \lambda_n| < \infty$;

(iii) $\lim_{n \to \infty} \alpha_n = 0$;

(iv) $\sum_{n=1}^{\infty} |\alpha_{n+1} - \alpha_n| < \infty$.

Show that the sequence $\{x_n\}$ converges strongly to some point $x^* = P_{\mathcal{F}}(u)$.

6. Let X be a uniformly convex and uniformly smooth Banach space. Let C be a nonempty, closed, and convex subset of X. Let $T_1, T_2, \ldots, T_r$ be a finite family of nonexpansive self-mappings of C, with $\mathcal{F} = \bigcap_{i=1}^{r} \text{Fix}(T_i) = \text{Fix}(T_r T_{r-1} \cdots T_1) \neq \emptyset$. Let $\{\lambda_n\}$ be a sequence in $(0,1)$ satisfying the following conditions: (C1) $\lim_{n \to \infty} \lambda_n = 0$; (C2) $\sum_{n=1}^{\infty} \lambda_n = \infty$. For a fixed $\delta \in (0,1)$, define $S_n : C \to C$ by $S_n x := (1-\delta)x + \delta T_n x$, $\forall x \in C$, where $T_n = T_{n \bmod r}$. For arbitrary fixed $u, x_1 \in C$, define a sequence $\{x_n\}$ iteratively in C by

$$x_{n+1} = \lambda_{n+1} u + (1 - \lambda_{n+1}) S_{n+1} x_n, \quad n \geq 1.$$

Assume that $\lim_{n \to \infty} \|T_n x_n - T_{n+1} x_n\| = 0$. Then, show that the sequence $\{x_n\}$ converges strongly to a common fixed point of the family $\{T_1, T_2, \ldots, T_r\}$. Also, give an example showing that the assumption $\lim_{n \to \infty} \|T_n x_n - T_{n+1} x_n\| = 0$ holds.

7. Let X be a uniformly convex and uniformly smooth Banach space. Let C be a nonempty, closed, and convex subset of X and let $\{T_i\}_{i=1}^{r} : C \to C$ be a finite family of nonexpansive mappings such that $\mathcal{F} = \bigcap_{i=1}^{r} \text{Fix}(T_i) \neq \emptyset$. Let $\{\beta_i\}_{i=1}^{r}$ be r positive numbers in $(0,1)$ such that $\sum_{i=1}^{r} \beta_i = 1$. Define a mapping $T : C \to C$ by

$$Tx = \sum_{i=1}^{r} \beta_i T_i x, \quad x \in C.$$

Let $\{\lambda_n\}$ be a sequence of positive numbers in $(0,1)$ satisfying conditions: (C1) $\lim_{n \to \infty} \lambda_n = 0$; (C2) $\sum_{n=1}^{\infty} \lambda_n = \infty$, and (C3) $\sum_{i=1}^{\infty} |\lambda_{n+1} - \lambda_n| < \infty$, or (C4) $\frac{\lambda_n}{\lambda_{n+1}} \to 1 \ (n \to \infty)$. For arbitrary $x_1, u \in C$, define a sequence $\{x_n\}$ iteratively in C by

$$x_{n+1} = \lambda_n u + (1 - \lambda_n) T x_n, \quad n \geq 1.$$

Prove that the sequence $\{x_n\}$ converges strongly to the some point $x^* \in \mathcal{F}$, furthermore, $x^* = Q_{\mathcal{F}} u$.

8. Let X be a uniformly convex and uniformly smooth Banach space. Let C be a nonempty, closed, and convex subset of X and let $\{T_i\}_{i=1}^{r} : C \to C$ be a finite family of nonexpansive mappings such that $\mathcal{F} = \bigcap_{i=1}^{r} \text{Fix}(T_i) \neq \emptyset$. Let $\{\beta_i\}_{i=1}^{r}$ be r positive numbers in $(0,1)$ such that $\sum_{i=1}^{r} \beta_i = 1$. Define a mapping $T : C \to C$ by

$$Tx = \sum_{i=1}^{r} \beta_i T_i x, \quad x \in C.$$

Let $\{\lambda_n\}$ be a sequence of positive numbers in $(0,1)$ satisfying conditions: (C1) $\lim_{n \to \infty} \lambda_n = 0$, and (C2) $\sum_{n=1}^{\infty} \lambda_n = \infty$. For a fixed number $\sigma \in (0,1)$, and for arbitrary $x_1, u \in C$, define a sequence $\{x_n\}$ iteratively in C by

$$x_{n+1} = \lambda_n u + (1 - \lambda_n)[(1 - \sigma)x_n + \sigma T x_n], \quad n \geq 1.$$

Prove that the sequence $\{x_n\}$ converges strongly to the some point $x^* \in \mathcal{F}$, furthermore, $x^* = Q_{\mathcal{F}} u$.

9. Let C be a nonempty, closed, and convex subset of a uniformly convex and uniformly smooth Banach space X. Let $\{T_i\}_{i=1}^r : C \to C$ be a finite family of averaged nonexpansive mappings such that $\mathcal{F} = \bigcap_{i=1}^r \text{Fix}(T_i) \neq \emptyset$. Define a mapping $T : C \to C$ by

$$Tx = T_r T_{r-1} \cdots T_1 x, \quad x \in C.$$

Let $\{\lambda_n\}$ be a sequence of positive numbers in $(0,1)$ satisfying conditions: (C1) $\lim_{n \to \infty} \lambda_n = 0$, and (C2) $\sum_{n=1}^{\infty} \lambda_n = \infty$. For arbitrary $x_1, u \in C$, define a sequence $\{x_n\}$ iteratively in C by

$$x_{n+1} = \lambda_n u + (1 - \lambda_n) T x_n, \quad n \geq 1. \tag{HTIM}$$

Prove that the sequence $\{x_n\}$ defined by (HTIM) converges strongly to the some point $x^* \in \mathcal{F}$, furthermore, $x^* = Q_{\mathcal{F}} u$.

10. Let C be a nonempty closed convex subset of a uniformly convex and uniformly smooth Banach space X. Let $\{T_i\}_{i=1}^{\infty} : C \to C$ be a finite family of nonexpansive mappings such that $\mathcal{F} = \bigcap_{i=1}^{\infty} \text{Fix}(T_i) \neq \emptyset$. Let $\{\lambda_i\}$ be a sequence of numbers in $(0,1)$ which satisfies conditions: (i) $\sum_{i=1}^{\infty} \lambda_i = 1$ and (ii) $\sum_{n=1}^{\infty} \sum_{i=n+1}^{\infty} \lambda_i < \infty$. Define another family of nonexpansive mappings by

$$S_n x := \sum_{i=1}^{n} \frac{\lambda_i}{\sum_{i=1}^{n} \lambda_i} T_i x, \quad x \in C, n \geq 1.$$

Let $\{\mu_n\}$ be a sequence of positive numbers in $(0,1)$ which satisfies conditions: (C1) $\lim_{n \to \infty} \mu_n = 0$, and (C2) $\sum_{n=1}^{\infty} \mu_n = \infty$. For arbitrary $x_1, u \in C$ and $\delta \in (0,1)$, define a sequence $\{x_n\}$ iteratively in C by

$$x_{n+1} = \delta x_n + (1 - \delta)[\mu_n x_n + (1 - \mu_n) S_n x_n], \quad n \geq 1.$$

Prove that the sequence $\{x_n\}$ converges strongly to the some point $x^* \in \mathcal{F}$, and $x^* = Q_{\mathcal{F}} u$.

5 Iterative methods for zeros for accretive operators and fixed points of pseudocontractive mappings in Banach spaces

The purpose of this chapter is to investigate zeros of accretive operators and fixed points of pseudocontractive mappings in the framework of Banach spaces via iterative methods including the steepest descent iterative method, the normal Mann iterative method, the Bruck regularization iterative method, and resolvent iterative methods.

5.1 Characterizations of accretive operators

Lemma 5.1.1 (Kato [37], 1967). *Let E be a real Banach space. For any $x, y \in E$, we have the following two conclusions, which are equivalent:*
(i) $\|x\| \leq \|x + \lambda y\|, \forall \lambda > 0$,
(ii) *there exists $f \in Jx$ such that $\langle y, f \rangle \geq 0$.*

Using Lemma 5.1.1, we give the following characterization for accretive operators.

Theorem 5.1.1. *Let $A \subset E \times E$. Then following two conclusions are equivalent*
(i)   *A is accretive,*
(ii) *$\forall \lambda > 0, (x_1, y_1) \in A$ and $(x_2, y_2) \in A$, we have*

$$\|x_1 - x_2\| \leq \|x_1 - x_2 + \lambda(y_1 - y_2)\|. \tag{5.1}$$

Remark 5.1.1. Kato lemma, which characterizes accretive operators, was first established by Kato [37] in 1967. From the viewpoint of geometry, an accretive operator $A \subset E \times E$ has the following properties: the range of the accretive operator $I + \lambda A$ increases, that is, $I + \lambda A$ is expansive. If it reaches the whole space E, that is, $\text{Ran}(I + \lambda A) = E$, we say that A is m-accretive. This may be the meaning of "accretive".

For $\forall \lambda > 0$, $I + \lambda A$ is an expansive operator. We see that $(I + \lambda A)^{-1}$ is well defined on $\text{Ran}(I + \lambda A)$. So, we can consider the following for accretive operators:

$$J_\lambda x = \{z \in E : x \in z + \lambda Az\}$$

and

$$A_\lambda x = \lambda^{-1}(I - J_\lambda), \quad \forall \lambda > 0.$$

Then J_λ and A_λ are said to be the resolvent and Yosida approximation, respectively. They play an important role when treating operators which are accretive.

https://doi.org/10.1515/9783110667097-005

5.2 Nonlinear semigroups of ω-type

Definition 5.2.1. Let C be a closed subset of a Banach space E. Then $\{S(t) : t \geq 0\}$ is said to be a nonlinear semigroup of ω-type if and only if there exists $\omega \in \mathbb{R}$ such that
(i) $S(0)x = x, \forall x \in C$;
(ii) $S(t + s)x = S(t)S(s)x, \forall x \in C, t, s \geq 0$;
(iii) $S(t)x$ is continuous in $t, \forall x \in C$;
(iv) $\|S(t)x - S(t)y\| \leq e^{\omega t}\|x - y\|, \forall x \in C, t \geq 0$.

In particular, a nonlinear semigroup of 0-type on C is said to be a nonlinear contraction semigroup. If $\{S(t) : t \geq 0\}$ only satisfies (i), (ii), and (iii), it is said to be a nonlinear semigroup.

Definition 5.2.2. Let $\{S(t) : t \geq 0\}$ be a nonlinear semigroup on C. Let

$$A_h x = \frac{1}{h}(S(h)x - x), \quad \forall h > 0$$

and

$$\text{Dom}(A) = \left\{x \in C : \lim_{h \to 0^+} A_h x \text{ exists}\right\}.$$

Then A_h is said be a generator of $\{S(t) : t \geq 0\}$.

Theorem 5.2.1. *Let $A \subset E \times E$ be an accretive operator satisfying the following range condition (RC):*

$$\overline{\text{Dom}(A)} \subset \text{Ran}(I + \lambda A), \quad \forall \lambda > 0.$$

Then

$$T(t)x = \lim_{n \to \infty}\left(I + \frac{t}{n}A\right)^{-n}x, \quad \forall t \geq 0, \forall x \in \overline{\text{Dom}(A)} \qquad \text{(EF)}$$

exists. Furthermore, limit (EF) is uniformly attained for $t \in [0, \tau], \forall \tau > 0$. Also $\{T(t) : t \geq 0\}$, which is defined by limit (EF), is a nonlinear contraction semigroup on $\overline{\text{Dom}(A)}$.

Remark 5.2.1. Limit (EF), which is now called the exponential formula, was established by Crandall and Ligget [26] in 1971. Also (EF) reveals the essential relation between nonlinear contraction semigroups and accretive operators and acts as an efficient tool for studying the solvability of differential equations in the framework of Banach spaces.

Theorem 5.2.2. *Let C be a closed subset of a Banach space E. Let $T : C \to E$ be a continuous operator. Then the following assertions are equivalent:*
(i) *T is a generator of a nonlinear semigroup $\{S(t) : t \geq 0\}$;*

(ii) $\forall x, y \in C$, there exists $j(x - y) \in J(x - y)$ such that

$$\langle Tx - Ty, j(x - y) \rangle \leq \omega \|x - y\|^2,$$
$$\lim_{h \to 0+} h^{-1} d(x + hTx, C) = 0, \quad \forall x \in C;$$

(iii) $\epsilon \omega < 1 \Longrightarrow C \subset (I - \epsilon T)(C), \forall \epsilon > 0.$

Remark 5.2.2. Theorem 5.2.2 was established by Martine [47] in 1973. It is an efficient tool to study zero points of accretive operators and fixed points of pseudocontractive mappings in Banach spaces. From (ii) and (iii) in Theorem 5.2.2, we see that continuous accretive operators defined on closed and convex subsets satisfy the range condition if they satisfy the "flow-invariance" condition.

5.3 Zero point theorems of accretive operators

Theorem 5.3.1. *Let C be a nonempty, closed, and convex subset of a real Banach space E. Let $A : C \to E$ be a continuous g-strongly accretive operator satisfying the following "flow-invariance" condition (FIC):*

$$\lim_{h \to 0+} h^{-1} d(x - hAx, C) = 0, \quad \forall x \in C.$$

If, in addition, one of the following conditions holds:
(i) *there exists $x_0 \in C$ such that $\liminf_{r \to \infty} g(r) > \|Ax_0\|$;*
(ii) *there exists $R > 0$ such that $\forall x \in E \backslash B_R(\theta), jx \in Jx$ satisfying $\langle Ax, jx \rangle \geq 0$;*
(iii) *$\|Ax\| \to \infty$ as $\|x\| \to \infty$,*

then A has a unique zero point in C.

Proof. Letting $T = -A$, we see that T is dissipative. Indeed, there exists $j(x-y) \in J(x-y)$ such that

$$\langle Tx - Ty, j(x - y) \rangle = -\langle Ax - Ay, j(x - y) \rangle$$
$$\leq -g(\|x - y\|)\|x - y\|$$
$$\leq 0, \quad \forall x, y \in C,$$

which shows that T is dissipative. From (FIC), we have

$$\lim_{h \to 0+} h^{-1} d(x - hTx, C) = 0, \quad \forall x \in C.$$

Using Theorem 5.2.2, we have

$$\forall \epsilon > 0, \ C \subset (I + \epsilon A)(C). \tag{5.2}$$

Letting $\epsilon_n \to \infty$ $(n \to \infty)$ in (5.2), we have

$$C \subset (I - \epsilon_n A)(C). \tag{5.3}$$

Fixing $y \in C$, we see that there exists $\{x_n\} \subset C$ such that

$$y = x_n + \epsilon_n A x_n. \tag{5.4}$$

Under one of conditions (i), (ii), and (iii), we are in a position to show that $\{x_n\}$ is bounded. We just take (i) as an example. Indeed,

$$y - x_0 = x_n - x_0 + \epsilon_n(Ax_n - Ax_0) + \epsilon_n Ax_0.$$

It follows that

$$
\begin{aligned}
&\langle y - x_0, j(x_n - x_0)\rangle \\
&= \|x_n - x_0\|^2 + \epsilon_n\langle Ax_n - Ax_0, j(x_n - x_0)\rangle + \epsilon_n\langle Ax_0, j(x_n - x_0)\rangle \\
&\geq \|x_n - x_0\|^2 + \epsilon_n g(\|x_n - x_0\|) - \epsilon_n\|Ax_0\|\|x_n - x_0\| \\
&\geq \epsilon_n g(\|x_n - x_0\|) - \epsilon_n\|Ax_0\|\|x_n - x_0\|,
\end{aligned}
$$

which is equivalent to

$$g(\|x_n - x_0\|) \leq \|Ax_0\| + \frac{1}{\epsilon_n}\|y - x_n\|. \tag{5.5}$$

If $\{x_n\}$ is unbounded, we may, without loss of generality, assume that $\|x_n - x_0\| \to \infty$ as $n \to \infty$. From condition (i), we have

$$\liminf_{n\to\infty} g(\|x_n - x_0\|) > \|Ax_0\|. \tag{5.6}$$

Using (5.5), we obtain

$$\liminf_{n\to\infty} g(\|x_n - x_0\|) \leq \|Ax_0\|,$$

which contradicts (5.6). This proves that $\{x_n\}$ is a bounded sequence. From (5.6), we see that $Ax_n = \frac{y - x_n}{\epsilon_n} \to \theta$ as $n \to \infty$. Since $A : C \to E$ is g-strongly accretive, we find that

$$
\begin{aligned}
g(\|x_n - x_m\|)\|x_n - x_m\| &\leq \langle Ax_n - Ax_m, j(x_n - x_m)\rangle \\
&\leq \|Ax_n - Ax_m\|\|x_n - x_m\|,
\end{aligned}
$$

which implies that

$$g(\|x_n - x_m\|) \leq \|Ax_n - Ax_m\|.$$

Since $x_n - x_m \to \theta$ as $n, m \to \infty$, we find that $g(\|x_n - x_m\|) = 0$ as $n, m \to \infty$. Using the property of g, we see that $\{x_n\}$ is a Cauchy sequence. Let $x_n \to x$ as $n \to \infty$. From the continuity of A, we obtain that $Ax_n \to Ax$ as $n \to \infty$. So, $Ax = \theta$. This proves that A has a zero point in C.

Next, we assume that there exists another zero point, say $x^* \in C$. It follows that

$$g(\|x - x^*\|)\|x - x^*\| \le \rangle Ax - Ax^*, j(x - x^*)\rangle = 0.$$

This implies that $x = x^*$. This completes the proof. $\qquad\square$

From Theorem 5.3.1, we have the following important result.

Corollary 5.3.1. *Let C be a nonempty, closed, and convex subset of a Banach space E. Let $A : C \to E$ be a continuous η-strongly accretive operator satisfying (FIC). Then A has a unique zero point in C.*

Proof. Note that $g(r) = \eta r \to \infty$ as $r \to \infty$. Conditions (i), (ii) and (iii) are all satisfied. Using Theorem 5.3.1, we obtain the desired conclusion immediately. $\qquad\square$

Theorem 5.3.2 (Ray [70], 1979). *Let C be a nonempty, convex, and closed subset of a Banach space E, and assume that it has the fixed point property for nonexpansive self-mappings. Let $A : C \to E$ be a continuous accretive operator satisfying (FIC). Then $\theta \in A(C)$.*

Proof. Letting $T = -A$, we see that T is dissipative. Indeed, there exists $j(x-y) \in J(x-y)$ such that

$$\begin{aligned} \langle Tx - Ty, j(x - y)\rangle &= -\langle Ax - Ay, j(x - y)\rangle \\ &\le -g(\|x - y\|)\|x - y\| \\ &\le 0, \quad \forall x, y \in C, \end{aligned}$$

which shows that T is dissipative. From (FIC), we have

$$\lim_{h \to 0+} h^{-1}d(x - hTx, C) = 0, \quad \forall x \in C.$$

Using Theorem 5.2.1, we have

$$\forall \epsilon > 0, C \subset (I + \epsilon A)(C).$$

Hence, $(I + A)^{-1}$ exists. Putting $f = (I + A)^{-1}$, we see that $f : C \to C$ is nonexpansive. From the assumption that f has fixed points in C. Hence, we find that A has zero points in C, that is, $\theta \in A(C)$. This completes the proof. $\qquad\square$

Remark 5.3.1. Theorem 5.3.2 was originally established by Ray in 1979. The proof given above is simpler than Ray's.

If C is only a closed subset of E, that is, there is no any convexity restriction on C, we still have the following result which needs more requirements on the "accretiveness".

Theorem 5.3.3 (Martin [47], 1973). *Let C be a closed subset of a Banach space E. Let $A : C \to E$ be continuous accretive operator satisfying the following restrictions:*
(i) $\exists \eta > 0$, *for* $\forall x, y \in C$, $\forall j(x - y) \in J(x - y)$, *we have*

$$\langle Ax - Ay, j(x - y) \rangle \geq \eta \|x - y\|^2,$$

(ii) $\lim_{h \to 0+} h^{-1} d(x - hAx, C) = 0$, $\forall x \in C$.

Then A has a unique zero point in C.

In 1974, Deimling [27] extended Theorem 5.3.3. To be more precise, he established the following theorem.

Theorem 5.3.4 (Deimling [27], 1974). *Let C be a closed subset of a Banach space E. Let $A : C \to E$ be a continuous operator satisfying the following conditions:*
(i) *there exists a continuous function $\alpha : \mathbb{R}^+ \to \mathbb{R}^+$, $\alpha(0) = 0$, $\alpha(r) > 0$, $\forall r > 0$, $\alpha(r) \to \infty$ as $r \to \infty$ and $\lim \inf_{r \to \infty} \frac{\alpha(r)}{r} > 0$, and $\forall x, y \in C$, $\forall j(x - y) \in J(x - y)$, we have*

$$\langle Ax - Ay, j(x - y) \rangle \geq \eta \|x - y\|^2,$$

(ii) $\lim_{h \to 0+} h^{-1} d(x - hAx, C) = 0$, $\forall x \in C$.

Then A has a unique zero point in C.

Remark 5.3.2. Theorem 5.3.3 is a direct consequence of Martin's result [47, Proposition 3], and Theorem 5.3.4 was obtained by Deimling [27, Theorem 1], via the theory of ordinary differential equations in the framework of Banach spaces.

Theorem 5.3.5 (Deimling [27], 1974). *Let C be a closed subset of a Banach space E. Let $A : C \to E$ be a continuous strongly accretive operator with $\lim \inf_{r \to \infty} \alpha(r) > 0$. Then $A(C)$ is an open set.*

Corollary 5.3.2. *Let E be a Banach space and let $A : E \to E$ be a continuous α-strongly accretive operator with $\lim \inf_{r \to \infty} \alpha(r) > 0$ or $\|Ax\| \to \infty$ as $\|x\| \to \infty$. Then $A : E \to E$ is a homeomorphism.*

Theorem 5.3.6 (Kartsatos [36], 1985). *Let E be a uniformly smooth Banach space and let U be a open set of E. Let $A : U \to E$ be a demicontinuous φ-strongly accretive operator. Then $A(\overline{U})$ is a closed subset of E and $A(U)$ is an open subset of E.*

Corollary 5.3.3. *Let E be a uniformly smooth Banach space. Let $A : E \to E$ be a continuous φ-strongly accretive operator. Then $A : E \to E$ is a homeomorphism.*

Proof. From Theorem 5.3.6, one knows that $A(E)$ is both closed and open, that is, $A(E) = E$. This shows that $A : E \to E$ is surjective. In view of

$$\langle Ax - Ay, j(x - y) \rangle \geq \varphi(\|x - y\|) \|x - y\|, \quad \forall x, y \in E,$$

one concludes

$$\|Ax - Ay\| \geq \varphi(\|x - y\|), \quad \forall x, y \in E.$$

If $x \neq y$, one finds that $\varphi(\|x - y\|) \geq 0 \implies Ax \neq Ay$. This shows that $A : E \to E$ is one-to-one. Hence, $A : E \to E$ is bijective. This implies that A^{-1} exists. Note that

$$\|x - y\| \geq \varphi(\|A^{-1}x - A^{-1}y\|), \quad \forall x, y \in E.$$

Fixing $y \in E$, one finds that $\varphi(\|A^{-1}x - A^{-1}y\|)$ as $x \to y$. From the property of φ, one obtains that $A^{-1}x \to A^{-1}y$ as $x \to y$. This implies that $A^{-1} : E \to E$ is continuous. This proves that $A : E \to E$ is a homeomorphism. This completes the proof. □

From Corollaries 5.3.2 and 5.3.3, we find the following result.

Theorem 5.3.7. *Let E be a Banach space and let $A : E \to E$ be a continuous accretive operator. Then A is m-accretive.*

Theorem 5.3.8. *Let E be a uniformly smooth Banach space and let $A : E \to E$ be a demicontinuous accretive operator. Then A is m-accretive.*

Theorem 5.3.9 (Kobayashi [39], 1975). *Let E be a Banach space and let $A \subset E \times E$ be an m-accretive operator. Let $B : E \to E$ be a continuous accretive operator such that $\overline{\mathrm{Dom}(A)} \subset D(B)$. Then $A + B$ is m-accretive.*

Theorem 5.3.10 (Gracía-Falset, Morales [29], 2005). *Let E be a Banach space and let $A \subset E \times E$ be an m-accretive operator. Let $B : E \to E$ be a continuous accretive operator such that B is φ-strongly accretive on $\mathrm{Dom}(A)$. Then*
(i) *both $A + \mu B$ and $B + \lambda A$ are surjective, that is, $\mathrm{Ran}(A + \mu B) = E$ and $\mathrm{Ran}(B + \lambda A)$,*
(ii) *$\forall \lambda > 0$, equation $z \in Bx + \lambda Ax$ has a unique solution x_λ and $\lambda \mapsto x_\lambda$ is continuous on $\mathbb{R}^1_+$.*

5.4 Demiclosedness of accretive operators

Theorem 5.4.1. *Let E be a Banach space such that its duality mapping $J : E \to E^*$ is weakly sequentially continuous. Let C be a nonempty, closed, and convex subset of a Banach space E. Let $A : C \to E$ be a hemicontinuous accretive operator satisfying the flow-invariance condition (FIC):*

$$\lim_{h \to 0+} h^{-1} d(x - hAx, C) = 0, \quad \forall x \in C.$$

Then A is demiclosed at the origin. That is, $\forall \{x_n\} \subset C$ with $x_n \rightharpoonup x$ and $Ax_n \to \theta$ as $n \to \infty$, we have that $x \in C$ and $Ax = \theta$.

Proof. Let $\{x_n\}$ be a sequence in C such that $x_n \rightharpoonup x$ and $Ax_n \to \theta$ as $n \to \infty$. Since C is closed and convex, we find that $x \in C$. Since $A : C \to E$ is accretive, we have

$$\langle Au - Ax_n, J(u - x_n) \rangle \geq 0, \quad \forall u \in C. \tag{5.7}$$

Letting $n \to \infty$ in (5.7) and using the continuity of J, we arrive at

$$\langle Au, J(u - x) \rangle \geq 0, \quad \forall u \in C. \tag{5.8}$$

Let $y_t = (1 - t)x + ty$, $\forall y \in C$, $t \in (0, 1)$. It follows that $y_t \in C$. Letting $u = y_t$ in (5.8), we see that

$$\langle Ay_t, J(y_t - x) \rangle \geq 0. \tag{5.9}$$

Substituting $y_t - x = t(y - x)$ into (5.9) yields that

$$\langle Ay_t, J(y - x) \rangle \geq 0, \quad \forall y \in C. \tag{5.10}$$

Since $A : C \to E$ is a hemicontinuous, we have

$$y_t \to x \implies Ay_t \rightharpoonup Ax, \quad t \to 0^+.$$

Letting $t \to 0^+$ in (5.10), we obtain

$$\langle Ax, J(y - x) \rangle \geq 0, \quad \forall y \in C. \tag{5.11}$$

$\forall \epsilon > 0$, $\exists \delta > 0$, if $0 < h < \delta$, then we find from (FIC) that $d(x - hAx, C) < h\epsilon$. Hence, there exists $u_h \in C$ such that

$$\|x - hAx - u_h\| < h\epsilon$$
$$\implies \left\| \frac{u_h - x}{h} + Ax \right\| < \epsilon$$
$$\implies \frac{u_h - x}{h} \to -Ax, \quad \forall h \to 0^+.$$

It follows that

$$J\left(\frac{u_h - x}{h} \right) \to -J(Ax), \quad \forall h \to 0^+.$$

Letting $y = u_h$ in (5.11), one sees that

$$\langle Ax, J(u_h - x) \rangle \geq 0, \quad h > 0,$$
$$\left\langle Ax, J\left(\frac{u_h - x}{h} \right) \right\rangle \geq 0, \quad h > 0. \tag{5.12}$$

Letting $h \to 0^+$ in (5.12), one finds that

$$\langle Ax, -J(Ax) \rangle \geq 0$$
$$\implies -\|Ax\|^2 \geq 0$$
$$\implies Ax = \theta.$$

This completes the proof. $\qquad\qquad\qquad\qquad\qquad\qquad\qquad\qquad\qquad\qquad\quad$ □

Theorem 5.4.2. *Let E be a uniformly convex Banach space and let C be a nonempty, closed, and convex subset of E. Let $A : C \to E$ be a continuous accretive operator satisfying the flow-invariance condition (FIC):*

$$\lim_{h \to 0+} h^{-1} d(x - hAx, C) = 0, \quad \forall x \in C.$$

Then A is demiclosed at the origin. That is, $\forall \{x_n\} \subset C$ with $x_n \rightharpoonup x$ and $Ax_n \to \theta$ as $n \to \infty$, we have that $x \in C$ and $Ax = \theta$.

Proof. Let $\{x_n\}$ be a sequence in C with $x_n \rightharpoonup x$ and $Ax_n \to \theta$ as $n \to \infty$. Since C is closed and convex, we find that $x \in C$. Letting $T = -A$, we find from Theorem 5.2.1 that $C \subset (I + A)(C)$, which guarantees that $g = (I + A)^{-1}$ is well defined on C, $g : C \to C$ is nonexpansive and $\text{Fix}(g) = A^{-1}(\theta)$. Since

$$\|x_n - g(x_n)\| = \|g(g^{-1}(x_n)) - g(x_n)\|$$
$$\leq \|g^{-1}(x_n) - x_n\|$$
$$= \|(I + A)x_n - x_n\|$$
$$= \|Ax_n\| \to 0, \quad n \to \infty,$$

and $I - g$ is demiclosed at the origin, we find that $x = g(x)$, that is, $Ax = \theta$. This completes the proof. $\qquad\qquad\qquad\qquad\qquad\qquad\qquad\qquad\qquad\qquad\qquad\qquad\quad$ □

Theorem 5.4.3. *Let E be a reflexive Banach space that satisfies Opial condition and let C be a nonempty, closed, and convex subset of E. Let $A : C \to E$ be a continuous accretive operator satisfying the flow-invariance condition (FIC):*

$$\lim_{h \to 0+} h^{-1} d(x - hAx, C) = 0, \quad \forall x \in C.$$

Then A is demiclosed at the origin. That is, $\forall \{x_n\} \subset C$ with $x_n \rightharpoonup x$ and $Ax_n \to \theta$ as $n \to \infty$, we have that $x \in C$ and $Ax = \theta$.

Proof. Let $\{x_n\}$ be a sequence in C with $x_n \rightharpoonup x$ and $Ax_n \to \theta$ as $n \to \infty$. Since C is closed and convex, we find that $x \in C$. As the proof in Theorem 5.4.2, we find that $x_n - g(x_n) \to \theta$ as $n \to \infty$.

Next, we prove $x = g(x)$. Assume that the inverse is true, that is, $x \neq g(x)$. Using Opial condition, we see that

$$\liminf_{n\to\infty} \|x_n - x\| < \liminf_{n\to\infty} \|x_n - g(x)\|$$

$$= \liminf_{n\to\infty} \|x_n - g(x_n) + g(x_n) - g(x)\|$$

$$\leq \liminf_{n\to\infty} (\|x_n - g(x_n)\| + \|g(x_n) - g(x)\|)$$

$$\leq \liminf_{n\to\infty} \|x_n - x\|,$$

which is a contradiction. This shows that $x = g(x)$, that is, $Ax = \theta$. This completes the proof. $\qquad\square$

5.5 The existence and convergence of paths for accretive operators

Theorem 5.5.1. *Let E be a Banach space and let C be a nonempty, convex, and closed subset of E. Let $A : C \to E$ be a continuous accretive operator, and let $R : C \to E$ be a continuous and η-strongly accretive operator. Assume that both A and R satisfy flow-invariance conditions:*

$$\lim_{h\to 0+} h^{-1} d(x - hAx, C) = 0, \quad \forall x \in C,$$

and

$$\lim_{h\to 0+} h^{-1} d(x - hRx, C) = 0, \quad \forall x \in C.$$

Then

(i) *There exists a unique continuous path $\{x_t\} \subset C$ such that*

$$\theta = tAx_t + (1 - t)Rx_t, \quad \forall t \in [0, 1). \tag{5.13}$$

(ii) *Let $\{x_n\} \subset C$ be a bounded sequence with $Ax_n \to \theta$ as $n \to \infty$ and let $\{Rx_n\}$ be a bounded sequence in E. Then $\{x_t\}$ is bounded. In particular, if $A^{-1}(\theta) \neq \emptyset$, then $\{x_t\}$ is bounded.*

(iii) *If $A^{-1}(\theta) \neq \emptyset$, then there exists $j(x_t - z) \in J(x_t - z)$ such that*

$$\langle Rx_t, j(x_t - z) \rangle \leq 0, \quad \forall z \in A^{-1}(\theta).$$

Proof.

(i) $\forall t \in [0, 1)$, define a mapping $A_t : C \to E$ by

$$A_t x = tAx + (1 - t)Rx, \quad \forall x \in C. \tag{5.14}$$

Using the assumption, we see that $A_t : C \to E$ is a continuous mapping. Using Theorem 1.9.19, we obtain that $A_t : C \to E$ is $(1-t)\eta$-strongly accretive and satisfies the flow-invariance condition. Using Theorem 5.3.1 or Corollary 5.3.1, we find that A_t has a unique zero point x_t in C, that is, $\theta = tAx_t + (1-t)Rx_t, t \in [0,1)$.

Next, we show that path $\{x_t\}$ is continuous. Fix $t_0 \in [0,1)$. Then there exists $j(x_t - x_{t_0}) \in J(x_t - x_{t_0})$ such that

$$\theta = t\langle Ax_t - Ax_{t_0}, j(x_t - x_{t_0})\rangle + (t - t_0)\langle Ax_{t_0}, j(x_t - x_{t_0})\rangle$$
$$+ (1-t)\langle Rx_t - Rx_{t_0}, j(x_t - x_{t_0})\rangle + (t_0 - t)\langle Rx_{t_0}, j(x_t - x_{t_0})\rangle$$
$$\geq (1-t)\eta\|x_t - x_{t_0}\|^2 - |t - t_0|\|Ax_{t_0} - Rx_{t_0}\|\|x_t - x_{t_0}\|, \quad \forall t \in [0,1),$$

which implies that

$$\|x_t - x_{t_0}\| \leq \frac{|t - t_0|}{(1-t)\eta}\|Ax_{t_0} - Rx_{t_0}\|. \tag{5.15}$$

Letting $t \to t_0$ in (5.15), we find that $x_t \to x_{t_0}$, that is, the path is continuous at t_0. Since t_0 is chosen arbitrarily, we find that $t \mapsto x_t$ is continuous.

(ii) Since $A : C \to E$ is continuous accretive and $R : C \to E$ is continuous and η-strongly accretive, we find from Theorem 1.9.19 that there exist $j(x_t - x_n) \in J(x_t - x_n)$ such that

$$\theta = t\langle Ax_t - Ax_n, j(x_t - x_n)\rangle + t\langle Ax_n, j(x_t - x_n)\rangle$$
$$+ (1-t)\langle Rx_t - Rx_n, j(x_t - x_n)\rangle + (1-t)\langle Rx_n, j(x_t - x_n)\rangle$$
$$\geq (1-t)\eta\|x_t - x_n\|^2 - t\|Ax_n\|\|x_t - x_n\| - (1-t)\|Rx_n\|\|x_t - x_n\|,$$

which implies that

$$\limsup_{n\to\infty}\|x_t - x_n\| \leq \frac{1}{\eta}\limsup_{n\to\infty}\|Rx_n\| < \infty.$$

(iii) Using (5.13), we find that $Rx_t = -\frac{t}{1-t}Ax_t$. It follows that

$$\langle Rx_t, j(x_t - z)\rangle = -\frac{t}{1-t}\langle Ax_t, j(x_t - z)\rangle$$
$$= -\frac{t}{1-t}\langle Ax_t - Az, j(x_t - z)\rangle$$
$$\leq 0, \quad \forall z \in A^{-1}(\theta).$$

This completes the proof. $\qquad\qquad\qquad\qquad\qquad\qquad\qquad\qquad\qquad\qquad\square$

Theorem 5.5.2. *Let E be a reflexive Banach space with a uniformly Gâteaux differentiable norm. Let C be a nonempty, closed, and convex subset of E and have the fixed point property for nonexpansive self-mappings. Let $A : C \to E$ be a continuous accretive* operator and let $R : C \to E$ be a continuous and η-strongly accretive operator. Assume that both A and R satisfy the following flow-invariance condition:

$$\lim_{h \to 0+} h^{-1} d(x - hAx, C) = 0, \quad \forall x \in C,$$

and

$$\lim_{h \to 0+} h^{-1} d(x - hRx, C) = 0, \quad \forall x \in C.$$

If both sets,

$$X = \{x \in C : Ax = (1 - \lambda)Rx, \ \lambda > 1\}$$

and $R(X)$, are bounded, then the path $\{x_t : t \in [0, 1)\}$ defined in (5.13) converges strongly to a zero point z^* of A and z^* uniquely solves the following variational inequality:

$$\langle Rz^*, j(z - z^*) \rangle \geq 0, \quad \forall z \in A^{-1}(\theta).$$

Proof. Letting $\{t_n\}$ be a real positive real sequence with $t_n \to 1$ as $n \to \infty$ and $x_n = x_{t_n}$, we find from (5.13) that $\{x_n\} \subset X$. Note that both $\{x_n\}$ and $\{Rx_n\}$ are bounded. It follows that $Ax_n = \frac{t_n - 1}{t_n} Rx_n \to \theta$ as $n \to \infty$. Using Theorem 5.2.2, we find that $C \subset (I + A)(C)$, which shows that $(I + A)^{-1}$ exists on C. Let $g(x) = (I + A)^{-1}x$. It follows that $g : C \to C$ is a nonexpansive mapping with $F(g) = A^{-1}(\theta)$. Let μ_n be the Banach space. Since $\{x_n\}$ is bounded, we may define a real function $f : E \to \mathbb{R}^+$ such that

$$f(x) = \mu_n \|x_n - x\|^2, \quad \forall x \in E.$$

Since f is continuous, convex, and satisfies $f(x) \to \infty$ as $\|x\| \to \infty$, we see that there exists $u \in C$ such that $f(u) = \min\{f(x) : x \in C\}$. Letting

$$C_0 = \{u \in C : f(u) = \min\{f(x) : x \in C\}\},$$

we see that C_0 is nonempty, bounded, closed, and convex subset of C with $g(C_0) \subset C_0$. Indeed, it is easy to find that C_0 is nonempty, bounded, closed, and convex subset of C. We here only prove $g(C_0) \subset C_0$. To this end, $\forall y \in C_0$, we find that $y \in C$ and $f(y) = \min\{f(x) : x \in C\}$. From the definition of g that

$$\begin{aligned}
f(g(y)) &= \mu_n \|x_n - g(y)\|^2 \\
&= \mu_n \|g(g^{-1}(x_n)) - g(y)\|^2 \\
&\leq \mu_n \|g^{-1}(x_n) - y\|^2 \\
&= \mu_n \|x_n + Ax_n - y\|^2 \\
&\leq \mu_n \|x_n - y\|^2 \\
&= f(y).
\end{aligned}$$

On the other hand, $f(y) \leq f(g(y))$. It follows that $g(y) \in C_0$. This proves that $g(C_0) \subset C_0$. From the assumption, there exists $z^* \in C_0$ such that $g(z^*) = z^* \implies Az^* = \theta$. Using the (iii) of Theorem 5.5.1, we find that $\langle Rx_n, j(x_n - z^*) \rangle \leq 0$. It follows that

$$\mu_n \langle Rx_n, j(x_n - z^*) \rangle \leq 0. \tag{5.16}$$

In view of $z^* \in C_0$, we obtain that

$$f(z^*) = \mu_n \|x_n - z^*\|^2 = \min_{x \in C} \mu_n \|x_n - x\|^2.$$

Using Theorem 1.8.36, we find that

$$\mu_n \langle y - z^*, j(x_n - z^*) \rangle \leq 0, \quad \forall y \in C. \tag{5.17}$$

Note that $R : C \to E$ satisfies the flow-invariance condition. For every $\epsilon > 0$, there exists $\delta > 0$ such that if $0 < h < \delta$, then $d(z^* - hRz^*, C) \leq \epsilon h$. Hence, there exists $u_h \in C$ such that

$$\|z^* - hRz^* - u_h\| \leq \epsilon h$$
$$\implies \left\| \frac{u_h - z^*}{h} + Rz^* \right\| \leq \epsilon.$$

This implies that

$$\frac{u_h - z^*}{h} \to -Rz^*, \quad \text{as} \quad h \to 0^+. \tag{5.18}$$

Letting $y = u_h \in C$ in (5.17), we obtain that

$$\mu_n \left\langle \frac{u_h - z^*}{h}, j(x_n - z^*) \right\rangle \leq 0. \tag{5.19}$$

Combining (5.18) with (5.19), we arrive at

$$\mu_n \langle -Rz^*, j(x_n - z^*) \rangle \leq 0. \tag{5.20}$$

It follows from (5.16) and (5.20) that

$$\mu_n \langle Rx_n - Rz^*, j(x_n - z^*) \rangle \leq 0. \tag{5.21}$$

Since $R : C \to E$ is η-strongly accretive, we have

$$\langle Rx_n - Rz^*, j(x_n - z^*) \rangle \geq \eta \|x_n - z^*\|^2. \tag{5.22}$$

Taking the Banach limit in (5.22), we find from (5.21) that $\mu_n \|x_n - z^*\|^2 \leq 0$. This implies that there exists a subsequence $\{x_{n_k}\}$ of $\{x_n\}$ such that $x_{n_k} \to z^*$ as $k \to \infty$. Using the (iii) of Theorem 5.5.1, we find that $\langle Rz^*, j(z - z^*) \rangle \geq 0$, $\forall z \in A^{-1}(\theta)$. Since z^* is unique, we obtain that $\{x_n\}$ converges strongly to z^*. This completes the proof. $\square$

Remark 5.5.1. The restrictions that both A and R are continuous and satisfy the flow-invariance condition ensure the existence of the path. If we assume that accretive operator $A \subset E \times E$ also satisfies the following range condition:

$$\overline{\text{Dom}(A)} \subset C \subset \bigcap_{r>0} \text{Ran}(I + rA),$$

where C is a closed and convex subset of E, $r > 0$, we find from Theorem 5.5.2 the following result.

Theorem 5.5.3. *Let E be a reflexive Banach space with a uniformly Gâteaux differentiable norm. Let C be a nonempty, closed, and convex subset of E. Let $A \subset E \times E$ be a continuous accretive operator with the range condition $\overline{\text{Dom}(A)} \subset C \subset \bigcap_{r>0} \text{Ran}(I + rA)$. Assume that every weakly compact convex subset of C has fixed points for every nonexpansive self-mapping. If $\theta \in \text{Ran}(A)$, then $\lim_{r \to \infty} J_r x$ exists, $\forall x \in C$. Furthermore, if we define $Qx = \lim_{r \to \infty} J_r x$, $\forall x \in C$, then $Q : C \to A^{-1}(\theta)$ is a sunny nonexpansive retraction.*

Proof. Since $x \in C \subset \bigcap_{r>0} R(I + rA)$, we see that there exists $x_r \in \text{Dom}(A) \subset C$ such that

$$x \in (I + rA)x_r, \quad \forall r > 0, \tag{5.23}$$

that is, $x_r = J_r x$. Letting $t = \frac{r}{1+r}$, we see that $t \in (0,1)$ and $r \to \infty \Longleftrightarrow t \to 1^-$. Define a mapping $R : C \to E$ by $Ru = u - x$, $\forall u \in C$. Then R is 1-strongly accretive. From (5.23), we have

$$\theta \in \frac{r}{1+r}Ax_r + \frac{1}{1+r}Rx_r, \quad \forall r > 0. \tag{5.24}$$

Letting $x_r := y_t$, we see that (5.24) is reduced to

$$\theta \in tAy_t + (1 - t)Ry_t, \quad \forall t \in (0,1).$$

Using Theorem 5.5.2, we find that $\{y_t\}$ converges strongly to zero point z^* of A as $t \to 1^-$ and z^* uniquely solves the following variational inequality:

$$\langle Rz^*, j(z - z^*) \rangle \geq 0, \quad \forall z \in A^{-1}(\theta),$$

that is, $\langle z^* - x, j(z - z^*) \rangle \geq 0$, $\forall z \in A^{-1}(\theta)$. Define $Qx = z^* = \lim_{r \to \infty} x_r = \lim_{r \to \infty} J_r x$. It follows that

$$\langle x - Qx, j(Qx - z) \rangle \geq 0, \quad \forall z \in A^{-1}(\theta).$$

Using Theorem 1.8.49, we find that $Q : C \to A^{-1}(\theta)$ is a sunny nonexpansive retraction. This completes the proof. □

Corollary 5.5.1. *Let E be a uniformly smooth Banach space and let $A \subset E \times E$ be an m-accretive operator. If $\theta \in \text{Ran}(A)$, then limit $\lim_{r \to \infty} J_r x$ exists, for any $x \in E$. Furthermore, if we define $Qx = \lim_{r \to \infty} J_r x$, $\forall x \in C$, then $Q : C \to A^{-1}(\theta)$ is a unique sunny nonexpansive retraction from C onto $A^{-1}(\theta)$.*

Corollary 5.5.2. *Let E be a reflexive Banach space with a uniformly Gâteaux differentiable norm. Let $A \subset E \times E$ be a continuous accretive operator with the range condition $\overline{\mathrm{Dom}(A)} \subset C \subset \bigcap_{r>0} \mathrm{Ran}(I + rA)$, Assume that every weakly compact convex subset of E has fixed points for every nonexpansive self-mapping. If $\overline{\mathrm{Dom}(A)}$ is a convex subset of E and $A^{-1}(\theta) \neq \emptyset$, then $\lim_{r \to \infty} J_r x$ exists, $\forall x \in \overline{\mathrm{Dom}(A)}$. Furthermore, if we define $Qx = \lim_{r \to \infty} J_r x$, $\forall x \in \overline{\mathrm{Dom}(A)}$, then $Q : \overline{\mathrm{Dom}(A)} \to A^{-1}(\theta)$ is a unique sunny nonexpansive retraction from $\overline{\mathrm{Dom}(A)}$ onto $A^{-1}(\theta)$.*

5.6 Iterative methods of zero points for accretive operators

5.6.1 The steepest decent method

Define a sequence by

$$x_0 \in E, \quad x_{n+1} = x_n - t_n A x_n, \quad n \geq 0, \tag{SDM}$$

where $\{t_n\} \subset (0, 1)$ is a real number sequence and $A : E \to E$ is a mapping.

Theorem 5.6.1. *Let E be a Banach space. Let $A : E \to E$ be an L-Lipschitz continuous g-strongly accretive mapping with the restriction that there exists $x_0 \in E$ such that $\liminf_{r \to \infty} g(r) > \|A x_0\|$. Assume that $\{t_n\}$ is a sequence in $(0, 1)$ such that (i) $\lim_{n \to \infty} t_n = 0$, (ii) $\sum_{n=1}^{\infty} t_n = \infty$, and (iii) $\sum_{n=1}^{\infty} t_n^2 < \infty$. Let $\{x_n\}$ be a sequence generated by (SDM). Then $\{x_n\}$ converges to the unique zero point of A in norm.*

Proof. Using Theorem 5.3.1, we see that A has a unique zero point z^* in E. Since $x_{n+1} - z^* = x_n - z^* - t_n A x_n$, we find that

$$
\begin{aligned}
\|x_{n+1} - z^*\|^2 &\leq \|x_n - z^*\|^2 - 2t_n \langle A x_n, j(x_{n+1} - z^*) \rangle \\
&= \|x_n - z^*\|^2 - 2t_n \langle A x_n - A x_{n+1}, j(x_{n+1} - z^*) \rangle \\
&\quad - 2t_n \langle A x_{n+1} - A z^*, j(x_{n+1} - z^*) \rangle \\
&\leq \|x_n - z^*\|^2 + 2L t_n \|x_n - x_{n+1}\| \|x_{n+1} - z^*\| - 2t_n g(\|x_{n+1} - z^*\|) \|x_{n+1} - z^*\| \\
&\leq \|x_n - z^*\|^2 + L^2 t_n^2 (\|x_n - z^*\|^2 + \|x_{n+1} - z^*\|^2) \\
&\quad - 2t_n g(\|x_{n+1} - z^*\|) \|x_{n+1} - z^*\| \\
&\leq (1 + L^2 t_n^2) \|x_n - z^*\|^2 + L^2 t_n^2 \|x_{n+1} - z^*\|^2 - 2t_n g(\|x_{n+1} - z^*\|) \|x_{n+1} - z^*\|.
\end{aligned}
$$

Without loss of generality, we may assume that $t_n \leq 1/2L$ for every $n \geq 1$. It follows that

$$\|x_{n+1} - z^*\|^2 \leq (1 + 3L^2 t_n^2) \|x_n - z^*\|^2 - \frac{2t_n}{1 - L^2 t_n^2} g(\|x_{n+1} - z^*\|) \|x_{n+1} - z^*\|. \tag{5.25}$$

Since $\sum_{n=1}^{\infty} t_n^2 < \infty$, we find that $\lim_{n \to \infty} \|x_n - z^*\|$ exists. This implies that $\{\|x_n - z^*\|\}$ is bounded. In view of (5.25), we have

$$\frac{2t_n}{1 - L^2 t_n^2} g(\|x_{n+1} - z^*\|) \|x_{n+1} - z^*\| \leq \|x_n - z^*\|^2 - \|x_{n+1} - z^*\|^2 + 3L^2 t_n^2 M^2,$$

where $M = \sup_{n \geq 1}\{\|x_n - z^*\|\}$. This implies that

$$t_n g(\|x_{n+1} - z^*\|)\|x_{n+1} - z^*\| \leq \|x_n - z^*\|^2 - \|x_{n+1} - z^*\|^2 + 3L^2 t_n^2 M^2,$$

which implies that $\sum_{n=1}^{\infty} t_n g(\|x_{n+1} - z^*\|)\|x_{n+1} - z^*\| < \infty$. In view of the assumption that $\sum_{n=1}^{\infty} t_n = \infty$, we find that

$$\liminf_{n \to \infty} g(\|x_{n+1} - z^*\|)\|x_{n+1} - z^*\| = 0.$$

It follows that there exists a subsequence $\{x_{n_j}\}$ of $\{x_n\}$ such that $x_{n_j} \to z^*$ as $j \to \infty$. Using Lemma 1.10.1, we find the desired conclusion immediately. This completes the proof. $\qquad\square$

If $A : E \to E$ is η-strongly accretive, then the requirements imposed on $\{t_n\}$ can be relaxed.

Theorem 5.6.2. *Let E be a Banach space. Let $A : E \to E$ be an L-Lipschitz continuous η-strongly accretive mapping. Assume that $\{t_n\}$ is a sequence in $(0, 1)$ such that* (i) $\lim_{n \to \infty} t_n = 0$ *and* (ii) $\sum_{n=1}^{\infty} t_n = \infty$. *Let $\{x_n\}$ be a sequence generated by* (SDM). *Then $\{x_n\}$ converges to the unique zero point of A in norm.*

Proof. Using Corollary 5.3.1, we see that A has a unique zero point z^* in E. Since $x_{n+1} - z^* = x_n - z^* - t_n A x_n$, we find that

$$\begin{aligned}
\|x_{n+1} - z^*\|^2 &\leq \|x_n - z^*\|^2 - 2t_n \langle A x_n, j(x_{n+1} - z^*) \rangle \\
&\leq \|x_n - z^*\|^2 + 2L t_n \|x_n - x_{n+1}\|\|x_{n+1} - z^*\| \\
&\quad - 2t_n g(\|x_{n+1} - z^*\|)\|x_{n+1} - z^*\| \\
&\leq (1 + L^2 t_n^2)\|x_n - z^*\|^2 + t_n(L^2 t_n - 2\eta)\|x_{n+1} - z^*\|^2.
\end{aligned}$$

For sufficiently large n, we have

$$\begin{aligned}
\|x_{n+1} - z^*\|^2 &\leq \frac{1 + L^2 t_n^2}{1 - t_n(L^2 t_n - 2\eta)}\|x_n - z^*\|^2 \\
&\leq (1 - \eta t_n)\|x_n - z^*\|^2.
\end{aligned}$$

Using Lemma 1.10.2, we find the desired conclusion immediately. This completes the proof. $\qquad\square$

Theorem 5.6.3. *Let E be a uniformly smooth Banach space. Let $A : E \to E$ be a bounded demicontinuous g-strongly accretive mapping with the restriction that there exists $x_0 \in E$ such that $\liminf_{r \to \infty} g(r) > \|A x_0\|$. Assume that $\{t_n\}$ is a sequence in $(0, 1)$ such that* (i) $\lim_{n \to \infty} t_n = 0$, (ii) $\sum_{n=1}^{\infty} t_n = \infty$. *Then there exists a constant $a > 0$ such that $t_n < a$, $\forall n \geq 1$, and the sequence $\{x_n\}$ generated by* (SDM) *converges to the unique zero point of A in norm.*

Proof. Using Theorem 5.3.1, we see that A has a unique zero point x^* in E. Define

$$a_0 = \sup\{r : g(r) \le \|Ax_0\|\}.$$

If $a_0 = +\infty$, then there exists $r_n \to \infty$ such that $g(r_n) \le \|Ax_0\|$. Hence

$$\|Ax_0\| < \liminf_{n \to \infty} g(r_n) \le \|Ax_0\|.$$

This a contradiction. Hence, a_0 is a finite positive number. Define

$$m_0 = \sup\{\|Ax_0\| : \|x_n - x_0\| \le 3a_0\} + 1.$$

Then $1 < m_0 < +\infty$. Define

$$a = \frac{1}{m_0} \min\left\{a_0, h^{-1}\left(\frac{2a_0\|Ax_0\|}{(2a_0 + 1)m_0 e^{m_0+1}}\right)\right\},$$

where $h : \mathbb{R}^+ \to \mathbb{R}^+$ is a strictly increasing continuous function, $h(t) = e^t b(t)$, and $b(t)$ is the function in Reich inequality (RI).

We are in a position to show that $\{x_n\}$ is bounded under the restriction that $t_n < a$. In view of

$$g(\|x_0 - x^*\|) \le \|Ax_0\|,$$

we see that $\|x_0 - x^*\| \le a_0 \le 2a_0$. If $\|x_0 - x^*\| < 2a_0$, then

$$\|x_n - x_0\| \le \|x_0 - x^*\| + \|x_n - x^*\| < 3a_0.$$

It follows from the definition of m_0 that

$$\|Ax_n\| \le m_0. \tag{5.26}$$

Next, we show

$$\|x_{n+1} - x^*\| < 2a_0. \tag{5.27}$$

If not, then

$$\begin{aligned} \|x_n - x^*\| &\ge \|x_{n+1} - x^*\| - t_n\|Ax_n\| \\ &\ge 2a_0 - t_n m_0 \\ &\ge 2a_0 - \frac{a_0}{m_0} m_0 \\ &= a_0. \end{aligned}$$

Using the definition of a_0, we see that

$$g(\|x_n - x^*\|) > \|Ax_0\|. \tag{5.28}$$

Using the modified Reich inequality [72], we find from (5.25), (5.26), and (5.28) that

$$
\begin{aligned}
\left\|x_{n+1} - x^*\right\|^2 &\leq \left\|x_n - x^*\right\|^2 - 2t_n\langle Ax_n, J(x_n - x^*)\rangle \\
&\quad + \max\{\|x_n - x^*\|, 1\}t_n\|Ax_n\|h(t_n\|Ax_n\|) \\
&\leq 4a_0^2 - 2t_n g(\|x_n - x^*\|)\|x_n - x^*\| + (2a_0 + 1)t_n m_0 h(t_n m_0) \\
&\leq 4a_0^2 - 2a_0 t_n\|Ax_n\| + 2a_0 t_n\|Ax_0\| \\
&\leq 4a_0^2,
\end{aligned}
$$

which yields $\|x_{n+1} - x^*\| < 2a_0$, and this is a contradiction. So $\|x_n - x^*\| < 2a_0$, $\forall n \geq 0$. Since $\|x_n - x_0\| \leq \|x_n - x^*\| + \|x_0 - x^*\| < 3a_0$, we have $\|Ax_n\| \leq m_0$, $\forall n \geq 0$. Letting $\delta = \liminf_{n\to\infty}\|x_n - x^*\|$, we see that $\delta = 0$. If not, then there exists $n_0 \in \mathbb{N}$ such that $\|x_n - x^*\| > \frac{\delta}{2}$, $\forall n \geq n_0$. Also $\forall r_0 > 0$, $\liminf_{r\to r_0} g(r) > 0$ implies $\liminf_{n\to\infty} g(\|x_n - x^*\|) = a^* > 0$. Then there exists $n_1 > n_0$ such that

$$
g(\|x_n - x^*\|) \geq \frac{a^*}{2}, \quad \forall n \geq n_1.
$$

Since $h(t_n) \to 0$ as $n \to \infty$, there exists $n_2 \geq n_1$ such that

$$
h(t_n) \leq \frac{a^*\delta}{4(2a_0 + 1)m_0^2 e^{m_0-1}}, \quad \forall n \geq n_2. \tag{5.29}
$$

Using the modified Reich inequality [72] again, we have that

$$
\begin{aligned}
\left\|x_{n+1} - x^*\right\|^2 &\leq \left\|x_n - x^*\right\|^2 - 2t_n\langle Ax_n, J(x_n - x^*)\rangle \\
&\quad + \max\{\|x_n - x^*\|, 1\}t_n\|Ax_n\|h(t_n\|Ax_n\|) \\
&\leq \|x_n - x^*\|^2 - 2t_n g(\|x_n - x^*\|)\|x_n - x^*\| + (2a_0 + 1)t_n m_0 h(t_n m_0) \\
&\leq \|x_n - x^*\|^2 - \frac{1}{2}a^*\delta t_n + \frac{1}{2}a^*\delta t_n \\
&= \|x_n - x^*\|^2 - \frac{1}{4}a^*\delta t_n, \quad \forall n \geq n_2.
\end{aligned}
$$

This implies that

$$
\frac{1}{4}a^*\delta \sum_{n\geq n_0}^{\infty} t_n \leq \|x_{n_2} - x^*\|^2 < \infty.
$$

It, however, contradicts the assumption $\sum_{n=1}^{\infty} t_n = \infty$. Therefore, there exists a subsequence $\{x_{n_k}\}$ of $\{x_n\}$ such that $x_{n_k} \to x^*$ as $k \to \infty$. Repeating above arguments, we find that $x_n \to x^*$ as $n \to \infty$. This completes the proof. □

5.6.2 The Bruck regularization iterative method

Theorem 5.6.4. *Let E be a uniformly smooth Banach space. Let $A \subset E \times E$ be an accretive operator satisfying the range condition $\overline{\mathrm{Dom}(A)} \subset \bigcap_{r>0} \mathrm{Ran}(I + rA)$, and $\overline{\mathrm{Dom}(A)}$ is a* *convex subset of E. Let $\{x_n\}$ be a sequence in $\mathrm{Dom}(A)$ generated by the following Bruck regularization iterative method*

$$x_1 \in \mathrm{Dom}(A), \quad x_{n+1} \in x_n - \lambda_n(Ax_n + \theta_n(x_n - z)), \quad n \geq 1, \qquad \text{(BRIM)}$$

where z is a fixed element in $\overline{\mathrm{Dom}(A)}$, $\{\lambda_n\}$ and $\{\theta_n\}$ are two real sequences in $[0,1]$ such that
(i) $\theta_n \to 0$ as $n \to \infty$,
(ii) $\sum_{n=1}^{\infty} \lambda_n \theta_n = \infty$,
(iii) $\frac{\theta_{n-1}}{\theta_n} - 1 = o(\lambda_n \theta_n)$,
(iv) $b(\lambda_n) = o(\theta_n)$,

where $b : \mathbb{R}^+ \to \mathbb{R}^+$ is the function in the Reich inequality (RI). Assume that there exists constant $c > 1$ such that

$$\|u_n\| \leq c(1 + \|x\|), \quad \forall u_n \in Ax_n, n \geq 1. \qquad (5.30)$$

If $\theta \in \mathrm{Ran}(A)$, then $\{x_n\}$ converges to the unique zero point of A in norm.

Proof. Without loss of generality, we assume $z = \theta$. Otherwise, we consider $\mathrm{Dom}(\tilde{A}) = \mathrm{Dom}(A) - z$, $\tilde{A} = A(x + z)$, $x \in \mathrm{Dom}(\tilde{A})$. From Corollary 5.5.2, we see that

$$\theta \in (I + \theta_n^{-1}A)x \qquad (5.31)$$

has a unique solution $y_n \in \mathrm{Dom}(A)$, that is,

$$\theta \in \theta_n y_n + Ay_n \qquad (5.32)$$

and $\{y_n\}$ converges to $x^* = Q(\theta) \in A^{-1}(\theta)$ as $n \to \infty$. In particular, $\{y_n\}$ is bounded. From (5.32), we find that there exists $v_n \in Ay_n$ such that $\theta_n y_n + v_n = \theta$. Hence, $\{v_n\}$ is also bounded. Put

$$M = \max\{\sup\{\|y_n\| : n \geq 1\}, \sup\{\|y_n\| : n \geq 1\}\}.$$

Using the linear growth condition (5.30), we find that $\frac{\|u_n\|}{1+\|x_n\|} \leq c, \forall u_n \in Ax_n$. It follows that

$$\frac{\|u_n - v_n\|}{1 + \|x_n\|} \leq c + \frac{M}{1 + \|x_n\|}, \quad \forall v \in Ay_n. \qquad (5.33)$$

In view of Theorem 5.5.1, we find from (5.32) that

$$\|y_n - y_{n-1}\| \leq \left\| y_n - y_{n-1} + \frac{1}{\theta_{n-1}}(v_n - v_{n-1}) \right\|$$

$$\leq \left\| \left(y_n + \frac{1}{\theta_n}v_n\right) - \left(y_{n-1} + \frac{1}{\theta_{n-1}}v_{n-1}\right) + \theta_{n-1}v_n - \theta_n v_n \right\|$$

$$
= \left| \frac{1}{\theta_n} - \frac{1}{\theta_{n-1}} \right| \|v_n\|
$$

$$
= \left| \frac{1}{\theta_n} - \frac{1}{\theta_{n-1}} \right| \theta_n \|y_n\|
$$

$$
= \left| \frac{\theta_n}{\theta_{n-1}} - 1 \right| \|y_n\|
$$

$$
\leq M \left| \frac{\theta_n}{\theta_{n-1}} - 1 \right|. \tag{5.34}
$$

Using the Reich inequality (RI), we find from (5.33) that

$$
\left\| \frac{x_{n+1} - y_n}{1 + \|x_n\|} \right\|^2
$$

$$
= \left\| (1 - \lambda_n \theta_n) \frac{x_n - y_n}{1 + \|x_n\|} - \frac{\lambda_n}{1 + \|x_n\|} (u_n - v_n) \right\|^2
$$

$$
\leq (1 - \lambda_n \theta_n)^2 \left\| \frac{x_n - y_n}{1 + \|x_n\|} \right\|^2 - \frac{2\lambda_n(1 - \lambda_n \theta_n)}{(1 + \|x_n\|)^2} \langle u_n - v_n, J(x_n - y_n) \rangle
$$

$$
+ \max \left\{ \frac{\|x_n - y_n\|}{1 + \|x_n\|}, 1 \right\} \lambda_n \frac{\|u_n - v_n\|}{1 + \|x_n\|} b \left(\frac{\|u_n - v_n\|}{1 + \|x_n\|} \right)
$$

$$
\leq (1 - \lambda_n \theta_n)^2 \left\| \frac{x_n - y_n}{1 + \|x_n\|} \right\|^2 + (M + 1) \left(c + \frac{M}{1 + \|x_n\|} \right)^2 \lambda_n b(\lambda_n).
$$

This implies that

$$
\begin{aligned}
\|x_{n+1} - y_n\|^2 &\leq (1 - \lambda_n \theta_n) \|x_n - y_n\|^2 + (M + 1)(c(1 + \|x_n\|) + M)^2 \lambda_n b(\lambda_n) \\
&\leq (1 - \lambda_n \theta_n) \|x_n - y_n\|^2 + 4(M + 1)(c^2(1 + \|x_n\|^2) + M^2) \lambda_n b(\lambda_n) \\
&= (1 - \lambda_n \theta_n) \|x_n - y_n\|^2 + 4(M + 1)(c^2 + M^2) \lambda_n b(\lambda_n) \\
&\quad + 4(M + 1)c^2 \|x_n - y_n + y_n\|^2 \lambda_n b(\lambda_n) \\
&\leq (1 - \lambda_n \theta_n) \|x_n - y_n\|^2 + 4(M + 1)(c^2 + M^2) \lambda_n b(\lambda_n) \\
&\quad + 8(M + 1)c^2 \|x_n - y_n\|^2 \lambda_n b(\lambda_n) \\
&\quad + 8(M + 1)c^2 M^2 \lambda_n b(\lambda_n) \\
&\leq (1 - \lambda_n \theta_n + 8(M + 1)c^2 \lambda_n b(\lambda_n)) \|x_n - y_n\|^2 + M_1 \lambda_n b(\lambda_n), \tag{5.35}
\end{aligned}
$$

where M_1 is an appropriate constant. From (iv), we see that there exists $n_0 \geq 1$ such that

$$
\lambda_n b(\lambda_n) \leq \frac{\lambda_n \theta_n}{16(M + 1)c^2}, \quad \forall n \geq n_0.
$$

Hence, (5.35) is reduced to

$$
\|x_{n+1} - y_n\|^2 \leq \left(1 - \frac{1}{2} \lambda_n \theta_n \right) \|x_n - y_n\|^2 + M_1 \lambda_n b(\lambda_n).
$$

Using (5.34), for sufficiently large n, we have

$$\|x_{n+1} - y_n\|^2 \le \left(1 - \frac{1}{2}\lambda_n\theta_n\right)\|x_n - y_n\|^2 + M_1\lambda_n b(\lambda_n)$$

$$\le \left(1 - \frac{1}{2}\lambda_n\theta_n\right)(\|x_n - y_{n-1}\|^2 + \|y_{n-1} - y_n\|^2$$

$$+ 2\|y_{n-1} - y_n\|\|x_n - y_{n-1}\|) + M_1\lambda_n b(\lambda_n)$$

$$\le \left(1 - \frac{1}{2}\lambda_n\theta_n\right)\|x_n - y_{n-1}\|^2 + 2M\|x_n - y_{n-1}\|\left|\frac{\theta_n}{\theta_{n-1}} - 1\right|$$

$$+ M^2\left|\frac{\theta_n}{\theta_{n-1}} - 1\right|^2 + M_1\lambda_n b(\lambda_n)$$

$$\le \left(1 - \frac{1}{2}\lambda_n\theta_n\right)\|x_n - y_{n-1}\|^2 + M\|x_n - y_{n-1}\|^2\left|\frac{\theta_n}{\theta_{n-1}} - 1\right|$$

$$+ M\left|\frac{\theta_n}{\theta_{n-1}} - 1\right| + M^2\left|\frac{\theta_n}{\theta_{n-1}} - 1\right| + M_1\lambda_n b(\lambda_n)$$

$$= \left(1 - \frac{1}{2}\lambda_n\theta_n + M\left|\frac{\theta_n}{\theta_{n-1}} - 1\right|\right)\|x_n - y_{n-1}\|^2 + o(\lambda_n\theta_n)$$

$$\le \left(1 - \frac{1}{4}\lambda_n\theta_n\right)\|x_n - y_{n-1}\|^2 + o(\lambda_n\theta_n).$$

Using Lemma 1.10.2, we obtain that $x_{n+1} - y_n \to \theta$ as $n \to \infty$. It follows that $x_n \to x^* = Q(\theta) \in A^{-1}(\theta)$ as $n \to \infty$. This completes the proof. □

Remark 5.6.1.

(1) If $\{u_n\}$, $u_n \in Ax_n$, is bounded, then there exists $c \ge 1$ such that

$$\|u_n\| \le c \le c + \|x_n\| = c(1 + \|x_n\|), \quad \forall n \ge 1.$$

(2) If A is Lipschitz continuous and $A^{-1}(\theta) \ne \emptyset$, then

$$\|u_n\| \le L\|x_n - x^*\| \le L\|x_n\| + L\|x^*\| = c(1 + \|x_n\|),$$

where L is the Lipschitz constant and $c = \max\{L\|x^*\|, L\} = L\max\{\|x^*\|, 1\}$. In particular, if A is single-valued, linear, and bounded, then

$$\|Ax_n\| \le \|A\|\|x_n\| \le \|A\|(1 + \|x_n\|).$$

(3) If A is bounded, then

$$M(r) = \sup\{\|u\| : u \in Ax, x \in \text{Dom}(A), \|x - x_1\| \le 2r\} < \infty$$

for sufficiently large $r \ge 1$. By mathematical induction, we find $\|x_n - Q(\theta)\| \le r$, $\forall n \ge 1$ provided that

$$\frac{b(\lambda_n)}{\theta_n} \le \frac{2r}{(M(r) + \frac{3}{2}r)^2}.$$

It follows that $\|u_n\| \leq c(1 + \|x_n\|)$, $\forall u_n \in Ax_n$, $n \geq 1$.

Chidume and Zegeye [25] extended the celebrated Bruck's theorems [11] to the framework of uniformly smooth Banach spaces via the Xu's inequality.

Definition 5.6.1. Let E be a uniformly smooth Banach space. Two nonnegative real sequences $\{\lambda_n\}$ and $\{\theta_n\}$ are said to be a compatible pair if and only if $\{\theta_n\}$ converges to 0 monotonically and there exists a strictly increasing subsequence $\{n(i)\}_{i=1}^{\infty}$ such that

(i) $\liminf_{n\to\infty} \theta_{n(i)} \sum_{j=n(i)}^{n(i+1)} \lambda_j > 0$,

(ii) $\liminf_{n\to\infty} (\theta_{n(i)} - \theta_{n(i+1)}) \sum_{j=n(i)}^{n(i+1)} \lambda_j = 0$,

(iii) $\lim_{n\to\infty} \sum_{j=n(i)}^{n(i+1)} \rho_E(\lambda_j) = 0$,

where ρ_E is the smoothness modulus of E.

Remark 5.6.2. Let E be a q-uniformly smooth Banach space. Two nonnegative real sequences $\{\lambda_n\}$ and $\{\theta_n\}$ are said to be a compatible pair if and only if the above (iii) is replaced with

$$\text{(iii)}' \qquad \lim_{n\to\infty} \sum_{j=n(i)}^{n(i+1)} \lambda_j^q = 0.$$

We here give an example: $\lambda_n = n^{-1}$, $\theta_n = (\log\log n)^{-1}$, $n(i) = i^i$, which satisfies (i), (ii), and (iii).

Theorem 5.6.5. *Let E be a uniformly smooth Banach space. Let $A \subset E \times E$ be an m-accretive operator with $0 \in \text{Ran}(A)$. Let $\{\lambda_n\}$ and $\{\theta_n\}$ be a compatible pair. Let $\{x_n\}$ be a sequence in* (BRIM). *If both $\{u_n\}$ and $\{v_n\}$ are bounded, then $\{x_n\}$ converges to a zero point of A in norm.*

Proof. Without loss of generality, we assume $z = \theta$. Since A is m-accretive, we see that $\xi^{-1}A$, $\xi > 0$ is also m-accretive, and $\text{Ran}(A + \xi^{-1}A) = E$, $\forall \xi > 0$. Hence, there exists a unique $y_i \in E$ such that $\theta \in \theta_i y_i + Ay_i$, $\forall i \geq 1$. It follows that

$$\lim_{\theta_i^{-1}\to\infty} J_{\theta_i^{-1}}(\theta) = \lim_{i\to\infty} y_i = x^* \in A^{-1}(\theta).$$

Moreover, $\forall n \geq i \geq 2$, $u_n \in Ax_n$, we have

$$x_n - y_i = x_{n-1} - y_i - \lambda_{n-1}(u_{n-1} + \theta_{n-1}x_{n-1}).$$

Using (XRI), we obtain

$$\|x_n - y_i\|^2 \leq \|x_{n-1} - y_i\|^2 - 2\lambda_{n-1}\langle u_{n-1} + \theta_i x_{n-1}, J(x_{n-1} - y_i)\rangle$$
$$+ 2\lambda_{n-1}(\theta_i - \theta_{i-1})\langle x_{n-1}, J(x_{n-1} - y_i)\rangle$$
$$+ D\max\bigg\{\|x_{n-1} - y_i\|$$
$$+ \lambda_{n-1}\|u_{n-1} + \theta_{n-1}x_{n-1}\|, \frac{c}{2}\bigg\}\rho_E(\lambda_{n-1}\|u_{n-1} + \theta_{n-1}x_{n-1}\|). \qquad (5.36)$$

In view of

$$-\theta_i y_i \in A y_i, \quad u_{n-1} \in A x_{n-1},$$

we find from $\langle u_{n-1} + \theta_i x_{n-1}, J(x_{n-1} - y_i) \rangle \geq 0$ that

$$\langle u_{n-1} + \theta_i x_{n-1}, J(x_{n-1} - y_i) \rangle$$
$$= \langle u_{n-1} + \theta_i y_i, J(x_{n-1} - y_i) \rangle + \theta_i \|x_{n-1} - y_i\|^2$$
$$\geq \theta_i \|x_{n-1} - y_i\|^2. \tag{5.37}$$

It follows from (5.36) and (5.37) that

$$\|x_n - y_i\|^2 \leq (1 - 2\lambda_{n-1}\theta_i)\|x_{n-1} - y_i\|^2 + 2\lambda_{n-1}(\theta_i - \theta_{i-1})\|x_{n-1}\|\|x_{n-1} - y_i\|$$
$$+ D \max \left\{ \|x_{n-1} - y_i\| \right.$$
$$\left. + \lambda_{n-1}\|u_{n-1} + \theta_{n-1}x_{n-1}\|, \frac{c}{2} \right\} \rho_E(\lambda_{n-1}\|u_{n-1} + \theta_{n-1}x_{n-1}\|). \tag{5.38}$$

Since both $\{x_n\}$ and $\{u_n\}$ are bounded, we find that there exist constants $c' > 0$, $D' > 0$, and $M' > 0$ such that $2\|x_{n-1}\|\|x_{n-1} - y_i\| \leq c'$,

$$D \max \left\{ \|x_{n-1} - y_i\| + \lambda_{n-1}\|u_{n-1} + \theta_{n-1}x_{n-1}\|, \frac{c}{2} \right\} \leq D'$$

and

$$\|u_{n-1} + \theta_{n-1}x_{n-1}\| \leq M'.$$

In view of

$$1 - 2\lambda_{n-1}\theta_i \leq \exp(-2\lambda_{n-1}\theta_i),$$

we find from (5.38) that

$$\|x_n - y_i\|^2 \leq \exp(-2\lambda_{n-1}\theta_i)\|x_{n-1} - y_i\|^2 + c'\lambda_{n-1}(\theta_i - \theta_{i-1}) + D'\rho_E(\lambda_{n-1}M')$$
$$\leq \exp(-2\lambda_{n-1}\theta_i)\|x_{n-1} - y_i\|^2 + c'\lambda_{n-1}(\theta_i - \theta_{i-1}) + D'(M')^2\rho_E(\lambda_{n-1}).$$

By induction on n, we obtain that

$$\theta_i - \theta_j \leq \theta_i - \theta_n, \quad \forall j \leq n.$$

It follows that

$$\|x_n - y_i\|^2 \leq \exp\left(-2\theta_i \sum_{j=1}^{n-1} \lambda_j\right)\|x_{n-1} - y_i\|^2 + c'(\theta_i - \theta_{i-1})\sum_{j=i}^{n} \lambda_j$$
$$+ D'(M')^2 \sum_{j=i}^{n} \rho_E(\lambda_j). \tag{5.39}$$

Since $\{\lambda_n\}$ and $\{\theta_n\}$ is a compatible pair, we see that

$$\|x_n - y_i\|^2 \le \delta\|x_i - y_i\|^2 + c'(\theta_i - \theta_n)\sum_{j=i}^{n}\lambda_j + D'(M')^2\sum_{j=i}^{n}\rho_E(\lambda_j). \tag{5.40}$$

Letting $n = n(k+1)$, $i = n(k)$ in (5.40), we find that

$$\|x_{n(k+1)} - y_{n(k)}\|^2 \le \delta\|x_{n(k)} - y_{n(k)}\|^2 + c'(\theta_{n(k)} - \theta_{n(k+1)})\sum_{j=n(i)}^{n(k+1)}\lambda_j$$

$$+ D'(M')^2\sum_{j=n(k)}^{n(k+1)}\rho_E(\lambda_j).$$

Since $x_{n(k)} \to x^*$ as $k \to \infty$, we find that $x_n - y_{n(k)} \to \theta$ as $n \to \infty$. Since $y_{n(k)} \to x^*$ as $k \to \infty$, we obtain that $x_n \to x^*$ as $n \to \infty$. This completes the proof. $\qquad\square$

Remark 5.6.3. For the function b in the Reich inequality (RI), we have that $b(t) \ge t$, $\forall t \ge 0$. If $b(\lambda_n) = o(\theta_n)$, then $\lambda_n = o(\theta_n)$. The converse may not be true. When $\lambda_n = o(\theta_n)$, letting $\lambda_n = \frac{1}{(1+n)^a}$ and $\theta_n = \frac{1}{(1+n)^b}$, we find that $\{\lambda_n\}$ and $\{\theta_n\}$ satisfy (i), (ii), and (iii) in Theorem 5.6.4, and (iv)' $\lambda_n = o(\theta_n)$. We do not know if they satisfy (iv) $b(\lambda_n) = o(\theta_n)$. However, if $A \subset E \times E$ is assumed to be Lipschitz continuous, we can find $\{\lambda_n\}$ and $\{\theta_n\}$, which satisfy (i), (ii), and (iii) in Theorem 5.6.4 and (iv)' $\lambda_n = o(\theta_n)$. Theorem 5.6.4 holds for such $\{\lambda_n\}$ and $\{\theta_n\}$.

Theorem 5.6.6. *Let E be a reflexive Banach space with a uniformly Gâteaux differentiable norm. Let C be a nonempty, closed, and convex subset of E, and have the fixed point property for nonexpansive self-mappings. Let $A : C \to E$ be an L-Lipschitz continuous accretive operator such that $I - A$ maps C to C, and let $R : C \to E$ be an η-strongly accretive operator such that $I - R$ maps C to itself. Assume that $\{\lambda_n\}$ and $\{\theta_n\}$ satisfy the following restrictions:*
(i) $\lambda_n(1 + \theta_n) \le 1$, $\forall n \ge 1$,
(ii) $\theta_n \to 0$ as $n \to \infty$,
(iii) $\sum_{n=1}^{\infty}\lambda_n\theta_n = \infty$,
(iv) $\frac{\theta_{n-1}}{\theta_n} - 1 = o(\lambda_n\theta_n)$,
(v) $\lambda_n = o(\theta_n)$.

Let $\{x_n\}$ be a sequence generated in the following iterative method:

$$x_1 \in C, \quad x_{n+1} = x_n - \lambda_n(Ax_n + \theta_n Rx_n), \quad \forall u_n \in Ax_n, \ n \ge 1. \tag{VBRIM}$$

If $\theta \in \text{Ran}(A)$, then $\{x_n\}$ converges to some zero point x^* of A in norm and x^* also is a unique solution to the following variational inequality:

$$\langle Rx^*, J(z - x^*)\rangle \ge 0, \quad \forall z \in A^{-1}(\theta). \tag{5.41}$$

Proof. Since both $I - A$ and $I - R$ are self-mappings on C, we see that (VBRIM) is well defined. It also easy to check that both A and R satisfy the flow-invariance condition (FIC). Since $A^{-1}(\theta)$ is not empty, we find from Theorem 5.5.1 (ii) that set X is bounded. It follows that $R(X)$ is also bounded. Using Theorem 5.5.2, we find that path $\{x_t : t \in [0, 1)\}$ defined in (5.13) converges to some zero point x^* of A in norm as $t \to 1^-$ and x^* is the unique solution to (5.41). Letting $t_n = \frac{1}{1+\theta_n}$ and setting $y_n = x_{t_n}$, we have

$$\theta = \frac{1}{1 + \theta_n} A y_n + \frac{\theta_n}{1 + \theta_n} R y_n. \tag{5.42}$$

It follows that

$$\theta = A y_n + \theta_n R y_n. \tag{5.43}$$

Using Theorem 5.5.2, we see that $\{y_n\}$ converges to some zero point x^* of A in norm and x^* is the unique solution to (5.41).

Next, we show that $x_{n+1} - y_n \to \theta$ as $n \to \infty$. Using (5.42), we find that

$$
\begin{aligned}
\|y_n - y_{n-1}\|^2 &= \langle y_n - y_{n-1}, J(y_n - y_{n-1}) \rangle - \langle A y_n - A y_{n-1}, J(y_n - y_{n-1}) \rangle \\
&\quad - \langle \theta_n R y_n - \theta_{n-1} R y_{n-1}, J(y_n - y_{n-1}) \rangle \\
&\leq \|y_n - y_{n-1}\|^2 - \theta_n \langle R y_n - R y_{n-1}, J(y_n - y_{n-1}) \rangle \\
&\quad + (\theta_{n-1} - \theta_n) \langle \theta_n R y_{n-1}, J(y_n - y_{n-1}) \rangle \\
&\leq \|y_n - y_{n-1}\|^2 - \eta \theta_n \|y_n - y_{n-1}\|^2 \\
&\quad + |\theta_{n-1} - \theta_n| \|R y_{n-1}\| \|y_n - y_{n-1}\|.
\end{aligned}
$$

This implies that

$$\|y_n - y_{n-1}\| \leq \eta^{-1} \left| \frac{\theta_{n-1}}{\theta_n} - 1 \right| M, \tag{5.44}$$

where M is some appropriate constant. Using (5.43), we find that

$$x_{n+1} - y_n = x_n - y_n - \lambda_n (A x_n - A y_n + \theta_n (R x_n - R y_n)). \tag{5.45}$$

Using (VBRIM), we obtain that

$$
\begin{aligned}
\|y_n - x_{n+1}\|^2 &= \|\lambda_n (A x_n - A y_n) + \lambda_n \theta_n (R x_n - R y_n)\| \\
&\leq L \lambda_n \|x_n - y_n\| + L \lambda_n \theta_n \|(x_n - y_n)\| \\
&\leq 2 L \lambda_n \|x_n - y_n\|. \tag{5.46}
\end{aligned}
$$

Combining (5.45) with (5.46), we find from the Petryshyn inequality that

$$
\begin{aligned}
\|y_n - x_{n+1}\|^2 &\leq \|x_n - y_n\|^2 - 2\lambda_n \langle Ax_n - Ay_n, J(x_{n+1} - y_n)\rangle \\
&\quad - 2\lambda_n \theta_n \langle Rx_n - Ry_n, J(x_{n+1} - y_n)\rangle \\
&\leq \|x_n - y_n\|^2 - 2\lambda_n \langle Ax_n - Ax_{n+1}, J(x_{n+1} - y_n)\rangle \\
&\quad - 2\lambda_n \langle Ax_{n+1} - Ay_n, J(x_{n+1} - y_n)\rangle \\
&\quad - 2\lambda_n \theta_n \langle Rx_n - Rx_{n+1}, J(x_{n+1} - y_n)\rangle \\
&\quad - 2\lambda_n \theta_n \langle Rx_{n+1} - Ry_n, J(x_{n+1} - y_n)\rangle \\
&\leq \|x_n - y_n\|^2 + 8L^2 \lambda_n^2 \|x_n - y_n\| \|x_{n+1} - y_n\| - 2\eta \lambda_n \theta_n \|x_{n+1} - y_n\|^2 \\
&= \|x_n - y_n\|^2 + 4L^2 \lambda_n^2 \|x_n - y_n\|^2 + 4L^2 \lambda_n^2 \|x_{n+1} - y_n\| \\
&\quad - 2\eta \lambda_n \theta_n \|x_{n+1} - y_n\|^2 \\
&= (1 + 4L^2 \lambda_n^2)\|x_n - y_n\|^2 + (4L^2 \lambda_n^2 - 2\eta \lambda_n \theta_n)\|x_{n+1} - y_n\|^2.
\end{aligned}
$$

For sufficiently large n, we have

$$
\|y_n - x_{n+1}\|^2 \leq \frac{1 + 4L^2 \lambda_n^2}{1 - 4L^2 \lambda_n^2 + 2\eta \lambda_n \theta_n} \|x_n - y_n\|^2.
$$

This further implies from (5.44) that

$$
\begin{aligned}
\|y_x - x_{n+1}\| &\leq \left(1 - \frac{1}{2}\eta \lambda_n \theta_n\right)\|x_n - y_n\| \\
&\leq \left(1 - \frac{1}{2}\eta \lambda_n \theta_n\right)\|x_n - y_{n-1}\| + \|y_n - y_{n-1}\| \\
&\leq \left(1 - \frac{1}{2}\eta \lambda_n \theta_n\right)\|x_n - y_{n-1}\| + \eta^{-1}\left|\frac{\theta_{n-1}}{\theta_n} - 1\right|M.
\end{aligned}
$$

Using Lemma 1.10.2, we find that $x_{n+1} - y_n \to \theta$ as $n \to \infty$. Hence, $x_n \to x^*$ as $n \to \infty$. This completes the proof. $\qquad\square$

5.6.3 The iterative methods based on APPA

The approximation proximity point algorithm (APPA), which was introduced by Rockafellar, is a popular and efficient approximation iterative method. Combining Rockafellar's approximation proximity point algorithm with the normal Mann iterative method, we obtain a Mann–Rockafellar iterative method. Combining Rockafellar's approximation proximity point algorithm with the Halpern iterative method, we obtain a Halpern–Rockafellar iterative method.

Next, we investigate the convergence of the two new iterative methods.

Let E be a Banach space and let $A : E \times E$ be an m-accretive operator. Let $J_r = (I + rA)^{-1}$ be the resolvent operator of A and let $A_r = r^{-1}(I - J_r)$ be the Yosida approximation of A, where $r > 0$.

Let x_1 be an element chosen arbitrarily in E and let $\{x_n\}$ be a sequence generated by

$$
\begin{cases}
x_1 \in E, \\
y_n \approx J_{r_n} x_n, \\
x_{n+1} = (1 - \alpha_n)x_n + \alpha_n y_n, \quad n \geq 1,
\end{cases}
\tag{MRIM}
$$

where $\{\alpha_n\} \subset (0,1)$ and $\{r_n\} \subset \mathbb{R}^+$ are two real sequences, $\{y_n\}$ is a sequence in E such that $\|y_n - J_{r_n}x_n\| \leq \delta_n$, $\forall n \geq 1$, and $\sum_{n=1}^{\infty} \delta_n < \infty$. We call (MRIM) a Mann–Rockafellar iterative method.

Let x_1 and u be elements chosen arbitrarily in E and let $\{x_n\}$ be a sequence generated by

$$
\begin{cases}
x_1 \in E, \\
y_n \approx J_{r_n} x_n, \\
x_{n+1} = t_n u + (1 - t_n)y_n, \quad n \geq 1,
\end{cases}
\tag{HRIM}
$$

where $\{t_n\} \subset (0,1)$ and $\{r_n\} \subset \mathbb{R}^+$ are two real sequences, $\{y_n\}$ is sequence in E such that $\|y_n - J_{r_n}x_n\| \leq \delta_n$, $\forall n \geq 1$, and $\sum_{n=1}^{\infty} \delta_n < \infty$. We call (HRIM) a Halpern–Rockafellar iterative method.

Theorem 5.6.7. *Let E be a uniformly convex Banach space with a Fréchet differentiable norm. Let $A \subset E \times E$ be an m-accretive operator such that $A^{-1}(\theta) \neq \emptyset$. Let $\{\alpha_n\} \subset (0,1)$ and $\{r_n\} \subset \mathbb{R}^+$ be two real sequences such that $\sum_{n=1}^{\infty} \alpha_n(1 - \alpha_n) = \infty$, $\sum_{n=1}^{\infty} |r_{n+1} - r_n| < \infty$, and $\liminf_{n \to \infty} r_n > 0$. Let $\{x_n\}$ be a sequence generated by (MRIM). Then $\{x_n\}$ converges weakly to some zero point x^* of A.*

Proof. Fixing $z \in A^{-1}(\theta)$, we find that

$$
\begin{aligned}
\|x_{n+1} - z\| &\leq (1 - \alpha_n)\|x_n - z\| + \alpha_n\|y_n - z\| \\
&\leq (1 - \alpha_n)\|x_n - z\| + \alpha_n\|y_n - J_{r_n}x_n\| + \alpha_n\|z - J_{r_n}x_n\| \\
&\leq \|x_n - z\| + \alpha_n\delta_n.
\end{aligned}
$$

Since $\sum_{n=1}^{\infty} \delta_n < \infty$, we find from Lemma 1.10.1 that $\lim_{n \to \infty} \|x_n - z\|$ exist. Using a result due to Xu [98], we find that

$$
\begin{aligned}
\|x_{n+1} - z\|^2 &\leq (1 - \alpha_n)\|x_n - z\|^2 + \alpha_n\|y_n - z\|^2 - \alpha_n(1 - \alpha_n)g(\|x_n - y_n\|) \\
&\leq (1 - \alpha_n)\|x_n - z\|^2 + \alpha_n\|y_n - J_{r_n}x_n\|^2 + \alpha_n\|z - J_{r_n}x_n\|^2 \\
&\quad + 2\alpha_n\|z - J_{r_n}x_n\|\|y_n - J_{r_n}x_n\| - \alpha_n(1 - \alpha_n)g(\|x_n - y_n\|) \\
&\leq \|x_n - z\|^2 + \alpha_n\delta_n^2 + 2\alpha_n\delta_n\|z - x_n\| - \alpha_n(1 - \alpha_n)g(\|x_n - y_n\|) \\
&\leq \|x_n - z\|^2 + \delta_n M - \alpha_n(1 - \alpha_n)g(\|x_n - y_n\|),
\end{aligned}
$$

where M is an appropriate constant. It follows that

$$\sum_{n=1}^{\infty} \alpha_n (1 - \alpha_n) g(\|x_n - y_n\|) \le \|x_1 - z\|^2 + M \sum_{n=1}^{\infty} \delta_n < +\infty.$$

Since $\sum_{n=1}^{\infty} \alpha_n (1 - \alpha_n) = \infty$, we find that $\liminf_{n \to \infty} g(\|x_n - y_n\|) = 0$. This implies that $\liminf_{n \to \infty} \|x_n - y_n\| = 0$. Hence, $\liminf_{n \to \infty} \|x_n - J_{r_n} x_n\| = 0$. Note that

$$\|x_{n+1} - J_{r_{n+1}} x_{n+1}\|$$
$$\le (1 - \alpha_n)\|x_n - J_{r_{n+1}} x_{n+1}\| + \alpha_n \|y_n - J_{r_{n+1}} x_{n+1}\|$$
$$\le (1 - \alpha_n)\|x_n - J_{r_n} x_n\| + (1 - \alpha_n)\|J_{r_n} x_n - J_{r_{n+1}} x_{n+1}\|$$
$$\quad + \alpha_n \|y_n - J_{r_n} x_n\| + \alpha_n \|J_{r_n} x_n - J_{r_{n+1}} x_{n+1}\|$$
$$\le (1 - \alpha_n)\|x_n - J_{r_n} x_n\| + \|J_{r_n} x_n - J_{r_{n+1}} x_{n+1}\| + \delta_n$$
$$\le (1 - \alpha_n)\|x_n - J_{r_n} x_n\| + \|J_{r_n} x_n - J_{r_{n+1}} x_n\| + \|J_{r_{n+1}} x_n - J_{r_{n+1}} x_{n+1}\| + \delta_n$$
$$\le (1 - \alpha_n)\|x_n - J_{r_n} x_n\| + \left|1 - \frac{r_n}{r_{n+1}}\right| \|x_n - J_{r_{n+1}} x_n\| + \|x_n - x_{n+1}\| + \delta_n$$
$$\le \|x_n - J_{r_n} x_n\| + \left|1 - \frac{r_n}{r_{n+1}}\right| M_1 + 2\delta_n,$$

where

$$M_1 = \sup\{\|x_n - J_{r_{n+1}} x_n\| : n \ge 1\}.$$

Since $\sum_{n=1}^{\infty} |1 - \frac{r_n}{r_{n+1}}| < \infty$ and $\sum_{n=1}^{\infty} \delta_n < \infty$, we find from Lemma 1.10.1 that $\lim_{n \to \infty} \|x_n - J_{r_n} x_n\|$ exists. In view of $\liminf_{n \to \infty} \|x_n - J_{r_n} x_n\| = 0$, we obtain that $\lim_{n \to \infty} \|x_n - J_{r_n} x_n\| = 0$. Since $\liminf_{n \to \infty} r_n > 0$, we may, without loss of generality, assume that there exists $r > 0$ such that $r_n \ge r$ for every $n \ge 1$. Taking a fixed positive $r > \epsilon > 0$ and using the resolvent equality, we see that

$$\|J_{r_n} x_n - J_\epsilon x_n\| = \left\| J_\epsilon \left(\frac{\epsilon}{r_n} x_n + \left(1 - \frac{\epsilon}{r_n}\right) J_{r_n} x_n \right) - J_\epsilon x_n \right\|$$
$$\le \left\| \left(\frac{\epsilon}{r_n} x_n + \left(1 - \frac{\epsilon}{r_n}\right) J_{r_n} x_n \right) - x_n \right\|$$
$$\le \left(1 - \frac{\epsilon}{r_n}\right) \|J_{r_n} x_n - x_n\|$$
$$\le \|J_{r_n} x_n - x_n\|.$$

It follows that

$$\|x_n - J_\epsilon x_n\| \le \|x_n - J_{r_n} x_n\| + \|J_\epsilon x_n - J_{r_n} x_n\|$$
$$\le 2\|x_n - J_{r_n} x_n\|.$$

Hence, we obtain that $\lim_{n\to\infty} \|x_n - J_\epsilon x_n\| = 0$. Since J_ϵ is nonexpansive, we find that $\omega(x_n) \subset \text{Fix}(J_\epsilon) = A^{-1}(\theta)$. Using Tan–Xu's Lemma [93], we find that $p = q$, $\forall p, q \in \omega(x_n)$. This shows that $\omega(x_n)$ is a singleton. Hence, $\{x_n\}$ converges weakly to some zero point x^* of A. This completes the proof. □

Theorem 5.6.8. *Let E be a uniformly smooth Banach space. Let $A \subset E \times E$ be an m-accretive operator such that $A^{-1}(\theta) \neq \emptyset$. Let $\{t_n\} \subset (0,1)$ and $\{r_n\} \subset \mathbb{R}^+$ be two real sequences such that $\lim_{n\to\infty} t_n = 0$, $\sum_{n=1}^{\infty} t_n = \infty$, $\sum_{n=1}^{\infty} |t_{n+1} - t_n| < \infty$ (or $\lim_{n\to\infty} \frac{t_n}{t_{n+1}} = 1$), $\liminf_{n\to\infty} r_n > 0$, and $\sum_{n=1}^{\infty} |r_{n+1} - r_n| < \infty$. Let $\{x_n\}$ be a sequence generated by (HRIM). Then $\{x_n\}$ converges strongly to some zero point x^* of A.*

Proof. Fixing $z \in A^{-1}(\theta)$, we find that

$$
\begin{aligned}
\|x_{n+1} - z\| &\leq (1 - t_n)\|u - z\| + t_n\|y_n - z\| \\
&\leq (1 - t_n)\|u - z\| + t_n\|J_{r_n}x_n - z\| + t_n\|y_n - J_{r_n}x_n\| \\
&\leq (1 - t_n)\|u - z\| + t_n\|x_n - z\| + \delta_n.
\end{aligned}
$$

Since $\sum_{n=1}^{\infty} \delta_n < \infty$, by induction, we find that $\{x_n\}$ is bounded, so are $\{J_{r_n}x_n\}$ and $\{y_n\}$. Note that

$$
\begin{aligned}
\|x_{n+1} - x_n\| &\leq |t_{n-1} - t_n|(\|u\| + \|y_{n-1}\|) + (1 - t_n)\|y_n - y_{n-1}\| \\
&\leq |t_{n-1} - t_n|M_1 + (1 - t_n)\|y_n - y_{n-1}\|,
\end{aligned}
\tag{5.47}
$$

where

$$
M_1 = \|u\| + \sup\{\|y_n\| : n \geq 1\}.
$$

Using the resolvent equality, we find that

$$
\begin{aligned}
\|y_{n-1} - y_n\| &\leq \|y_{n-1} - J_{r_{n-1}}x_{n-1}\| + \|J_{r_n}x_n - J_{r_{n-1}}x_{n-1}\| + \|J_{r_n}x_n - y_n\| \\
&\leq \|J_{r_n}x_n - J_{r_{n-1}}x_{n-1}\| + \delta_n + \delta_{n-1} \\
&\leq \|x_n - x_{n-1}\| + \delta_n + \delta_{n-1} \\
&\quad + \left\| J_{r_{n-1}}\left(\frac{r_{n-1}}{r_n}x_{n-1} + \left(1 - \frac{r_{n-1}}{r_n}\right)J_{r_n}x_{n-1}\right) - J_{r_{n-1}}x_{n-1} \right\| \\
&\leq \|x_n - x_{n-1}\| + \left\| \left(\frac{r_{n-1}}{r_n}x_{n-1} + \left(1 - \frac{r_{n-1}}{r_n}\right)J_{r_n}x_{n-1}\right) - x_{n-1} \right\| + \delta_n + \delta_{n-1} \\
&\leq \|x_n - x_{n-1}\| + \left|1 - \frac{r_{n-1}}{r_n}\right|M_2 + \delta_n + \delta_{n-1},
\end{aligned}
\tag{5.48}
$$

where

$$
M_2 = \sup\{\|J_{r_n}x_{n-1} - x_{n-1}\| : n \geq 1\}.
$$

Substituting (5.48) into (5.47), we find that

$$\|x_{n+1} - x_n\| \le |t_{n-1} - t_n|M_1 + (1 - t_n)\|x_n - x_{n-1}\| + (1 - t_n)\left|1 - \frac{r_{n-1}}{r_n}\right|M_2 + \delta_n + \delta_{n-1}.$$

Using Lemma 1.10.2, we find that $\lim_{n\to\infty}\|x_{n+1} - x_n\| = 0$. This implies that $\lim_{n\to\infty}\|J_{r_n}x_n - x_n\| = 0$. Since $\liminf_{n\to\infty} r_n > 0$, we may, without loss of generality, assume that there exists $r > 0$ such that $r_n \ge r$ for every $n \ge 1$. Fix $r > \epsilon > 0$. From the proof of Theorem 5.6.7, we find that $\lim_{n\to\infty}\|x_n - J_\epsilon x_n\| = 0$.

We are now in a position to show that

$$\limsup_{n\to\infty}\langle u - x^*, J(y_n - x^*)\rangle \le 0,$$

where $x^* = Q_{A^{-1}(\theta)}u$ and $Q_{A^{-1}(\theta)} : E \to A^{-1}(\theta)$ is the unique sunny nonexpansive retraction from E onto $A^{-1}(\theta)$. Consider mapping $C_t : E \to E$ by

$$C_t x = tu + (1 - t)J_\epsilon x, \quad \forall x \in E,\ t \in (0, 1).$$

Then $C_t : E \to E$ is $(1-t)$-contractive. Using the Banach contractive mapping principle, we find that there exists a unique point y_t in E such that

$$y_t = C_t(y_t) = tu + (1 - t)J_\epsilon y_t, \quad \forall t \in (0, 1).$$

It is not hard to see that $\{y_t\}$ converges to $x^* = Q_{A^{-1}(\theta)}u$ in norm and $Q_{A^{-1}(\theta)} : E \to A^{-1}(\theta)$ is the unique sunny nonexpansive retraction from E onto $A^{-1}(\theta)$. Note that

$$\begin{aligned}
\|y_t - x_n\|^2 &\le t\langle u - x_n, J(y_t - x_n)\rangle + (1 - t)\langle J_\epsilon y_t - x_n, J(y_t - x_n)\rangle \\
&\le t\langle u - x_n, J(y_t - x_n)\rangle + (1 - t)\langle J_\epsilon y_t - J_\epsilon x_n, J(y_t - x_n)\rangle \\
&\quad + (1 - t)\langle J_\epsilon x_n - x_n, J(y_t - x_n)\rangle \\
&\le t\langle u - y_t, J(y_t - x_n)\rangle + t\langle y_t - x_n, J(y_t - x_n)\rangle \\
&\quad + (1 - t)\|y_t - x_n\|^2 + (1 - t)\|J_\epsilon x_n - x_n\|\|y_t - x_n\|.
\end{aligned}$$

Hence, we have

$$\langle y_t - u, J(y_t - x_n)\rangle \le \frac{1-t}{t}\|J_\epsilon x_n - x_n\|\|y_t - x_n\|. \tag{5.49}$$

Letting $n \to \infty$, we find that

$$\limsup_{n\to\infty}\langle y_t - u, J(y_t - x_n)\rangle \le 0, \quad \forall t \in (0, 1).$$

Since $\{y_t\}$ converges to $x^* = Q_{A^{-1}(\theta)}u$ as $t \to 0$, we see that there exists $\kappa_1 > 0$ such that

$$\|y_t - x^*\| \le \frac{\xi}{3M_1}, \quad \forall 0 < t < \kappa_1.$$

Since E is uniformly smooth, we see that J is uniformly continuous on any bounded subset of E. Hence, we have $J(y_t - x_n) \to J(x^* - x_n)$ as $n \to \infty$. Hence, there exists $\kappa_2 > 0$ such that

$$\|J(y_t - x_n) - J(x^* - x_n)\| \le \frac{\xi}{3(\|x^* - u\| + 1)}, \quad 0 < t < \kappa_2.$$

Letting $\kappa = \min\{\kappa_1, \kappa_2\}$, we see from (5.49) that there exists $N \ge 1$ such that

$$\langle y_t - u, J(y_t - x_n)\rangle < \frac{\xi}{3}.$$

Hence, $\forall \xi > 0$, there exists $N \ge 1$ such that

$$
\begin{aligned}
&\langle x^* - u, J(x^* - x_n)\rangle \\
&\le \|x^* - u\|\|J(x^* - x_n) - J(y_t - x_n)\| + M_1\|y_t - x_n\| \\
&\quad + \langle y_t - u, J(y_t - x_n)\rangle \\
&< \frac{\xi}{3} + \frac{\xi}{3} + \frac{\xi}{3} \\
&= \xi, \quad \forall n \ge N.
\end{aligned}
$$

It follows $\limsup_{n\to\infty}\langle x^* - u, J(x^* - x_n)\rangle \le \xi$. Since ξ is arbitrary, we obtain that

$$\limsup_{n\to\infty}\langle x^* - u, J(x^* - x_n)\rangle \le 0.$$

In view of the fact that $x_n - y_n \to \theta$ as $n \to \infty$, we find that

$$\limsup_{n\to\infty}\langle x^* - u, J(x^* - y_n)\rangle \le 0.$$

Finally, we show that $x_n \to x^*$ as $n \to \infty$. Using Reich inequality, we find that

$$
\begin{aligned}
\|x_{n+1} - x^*\|^2 &\le (1 - t_n)^2\|y_n - x^*\|^2 + 2t_n(1 - t_n)\langle u - x^*, J(y_n - x^*)\rangle \\
&\quad + \max\{(1 - t_n)\|y_n - x^*\|, 1\}t_n\|u - x^*\|b(t_n\|u - x^*\|) \\
&\le (1 - t_n)(\|y_n - J_{r_n}x_n\|^2 + 2\|y_n - J_{r_n}x_n\|\|x^* - J_{r_n}x_n\| + \|x^* - J_{r_n}x_n\|^2) \\
&\quad + 2t_n(1 - t_n)\langle u - x^*, J(y_n - x^*)\rangle + M_1t_nb(t_n) \\
&\le (1 - t_n)\|x^* - x_n\|^2 + \delta_n^2 + 2\delta_n\|x_n - x^*\| \\
&\quad + 2t_n(1 - t_n)\langle u - x^*, J(y_n - x^*)\rangle + M_1t_nb(t_n) \\
&\le (1 - t_n)\|x^* - x_n\|^2 + 2t_n(1 - t_n)\langle u - x^*, J(y_n - x^*)\rangle + M_2\delta_n + M_1t_nb(t_n) \\
&\le (1 - t_n)\|x^* - x_n\|^2 + t_n\sigma_n + M_2\delta_n,
\end{aligned}
$$

where

$$\sigma_n = 2(1 - t_n)\langle u - x^*, J(y_n - x^*)\rangle + M_1b(t_n).$$

Note that $\limsup_{n\to\infty}\sigma_n \le 0$. Using Lemma 1.10.2, we find that $\{x_n\}$ converges strongly to some zero point x^* of A. This completes the proof. $\square$

To reduce the requirements on $\{r_n\}$ and $\{t_n\}$, we introduce the following Halpern–Rockafellar iterative method.

Let x_1 and u be elements chosen arbitrarily in E, and let $\{x_n\}$ be a sequence generated by

$$
\begin{cases}
x_1 \in E, \\
y_n \approx J_{r_n} x_n, \\
x_{n+1} = t_n u + (1 - t_n)(\lambda_n x_n + (1 - \lambda_n)y_n), \quad n \geq 1,
\end{cases}
\tag{MHRIM-1}
$$

where $\{t_n\} \subset (0,1)$, $\{\lambda_n\} \subset (0,1)$, and $\{r_n\} \subset \mathbb{R}^+$ are three real sequences, $\{y_n\}$ is sequence in E such that $\|y_n - J_{r_n} x_n\| \leq \delta_n$, $\forall n \geq 1$, and $\sum_{n=1}^{\infty} \delta_n < \infty$. We call (MHRIM-1) a modified Halpern–Rockafellar iterative method.

Theorem 5.6.9. *Let E be a uniformly smooth Banach space. Let $A \subset E \times E$ be an m-accretive operator such that $A^{-1}(\theta) \neq \emptyset$. Let $\{t_n\} \subset (0,1)$, $\{\lambda_n\} \subset (0,1)$, and $\{r_n\} \subset \mathbb{R}^+$ be three real sequences such that $\lim_{n \to \infty} t_n = 0$, $\sum_{n=1}^{\infty} t_n = \infty$, $\liminf_{n \to \infty} r_n > 0$, $\lim_{n \to \infty} |r_{n+1} - r_n| = 0$ and $0 < \liminf_{n \to \infty} \lambda_n \leq \limsup_{n \to \infty} \lambda_n < 1$. Let $\{x_n\}$ be generated by (MHRIM-1). Then $\{x_n\}$ converges strongly to some zero point x^* of A.*

Proof. Set $z_n := \lambda_n x_n + (1 - \lambda_n)y_n$. Fixing $z \in A^{-1}(\theta)$, we find that

$$
\begin{aligned}
\|y_n - z\| &\leq \lambda_n \|x_n - z\| + (1 - \lambda_n)\|y_n - z\| \\
&\leq \lambda_n \|x_n - z\| + (1 - \lambda_n)\|y_n - J_{r_n} x_n\| + (1 - \lambda_n)\|J_{r_n} x_n - z\| \\
&\leq \|x_n - z\| + \delta_n.
\end{aligned}
$$

It follows that

$$
\begin{aligned}
\|x_{n+1} - z\| &\leq t_n \|u - z\| + (1 - t_n)\|y_n - z\| \\
&\leq t_n \|u - z\| + (1 - t_n)\|x_n - z\| + \delta_n
\end{aligned}
$$

Since $\sum_{n=1}^{\infty} \delta_n < \infty$, by induction, we find that $\{x_n\}$ is bounded, so are $\{J_{r_n} x_n\}$, $\{y_n\}$ and $\{z_n\}$.

Define

$$
w_n = \frac{x_{n+1} - \lambda_n x_n}{1 - \lambda_n}.
$$

Using the resolvent equality, we have

$$
\begin{aligned}
\|w_{n+1} - w_n\| &\leq \left(\frac{t_{n+1}}{1 - \lambda_{n+1}} + \frac{t_n}{1 - \lambda_n} \right)\|u\| + \|y_{n+1} - y_n\| \\
&\quad + \frac{t_{n+1}}{1 - \lambda_{n+1}}\|z_{n+1}\| + \frac{t_n}{1 - \lambda_n}\|z_n\| \\
&\leq \left(\frac{t_{n+1}}{1 - \lambda_{n+1}} + \frac{t_n}{1 - \lambda_n} \right)M + \|y_{n+1} - J_{r_{n+1}} x_{n+1}\|
\end{aligned}
$$

$$+ \|J_{r_{n+1}} x_{n+1} - J_{r_n} x_n\| + \|J_{r_n} x_n - y_n\|$$

$$\leq \left(\frac{t_{n+1}}{1 - \lambda_{n+1}} + \frac{t_n}{1 - \lambda_n} \right) M + \delta_{n+1} + \delta_n + \|x_{n+1} - x_n\| + \|J_{r_{n+1}} x_n - J_{r_n} x_n\|$$

$$\leq \left(\frac{t_{n+1}}{1 - \lambda_{n+1}} + \frac{t_n}{1 - \lambda_n} \right) M + \delta_{n+1} + \delta_n + \|x_{n+1} - x_n\|$$

$$+ \left| 1 - \frac{r_n}{r_{n+1}} \right| \|J_{r_{n+1}} x_n - x_n\|.$$

Using the restrictions imposed on $\{t_n\}$, $\{\lambda_n\}$ and $\{r_n\}$, we obtain that

$$\limsup_{n \to \infty} (\|w_{n+1} - w_n\| - \|x_{n+1} - x_n\|) \leq 0.$$

In view of Lemma 1.10.3, we find that $\lim_{n \to \infty} \|w_n - x_n\| = 0$. It follows that $\lim_{n \to \infty} \|x_{n+1} - x_n\| = 0$. This further implies $\lim_{n \to \infty} \|z_n - x_n\| = 0$ and $\lim_{n \to \infty} \|J_{r_n} x_n - x_n\| = 0$. Since $\liminf_{n \to \infty} r_n > 0$, we may, without loss of generality, assume that there exists $r > 0$ such that $r_n \geq r$ for every $n \geq 1$. Fix $r > \epsilon > 0$. From the proof of Theorem 5.6.7, we find that $\lim_{n \to \infty} \|x_n - J_\epsilon x_n\| = 0$. From the proof in Theorem 5.6.8, we obtain the desired conclusion immediately. $\qquad\square$

We now consider another modification of (HRIM). Let x_1 and u be elements chosen arbitrarily in E, and let $\{x_n\}$ be a sequence generated by

$$\begin{cases} x_1 \in E, \\ y_n \approx J_{r_n} x_n, \\ x_{n+1} = \lambda_n x_n + (1 - \lambda_n)(t_n u + (1 - t_n) y_n), \quad n \geq 1, \end{cases} \qquad \text{(MHRIM-2)}$$

where $\{t_n\} \subset (0,1)$, $\{\lambda_n\} \subset (0,1)$, and $\{r_n\} \subset \mathbb{R}^+$ are three real sequences, $\{y_n\}$ is sequence in E such that $\|y_n - J_{r_n} x_n\| \leq \delta_n$, $\forall n \geq 1$, and $\sum_{n=1}^{\infty} \delta_n < \infty$. We also call (MHRIM-2) a modified Halpern–Rockafellar iterative method.

Theorem 5.6.10. *Let E be a uniformly smooth Banach space. Let $A \subset E \times E$ be an m-accretive operator such that $A^{-1}(\theta) \neq \emptyset$. Let $\{t_n\} \subset (0,1)$, $\{\lambda_n\} \subset (0,1)$, and $\{r_n\} \subset \mathbb{R}^+$ be three real sequences such that $\lim_{n \to \infty} t_n = 0$, $\sum_{n=1}^{\infty} t_n = \infty$, $\liminf_{n \to \infty} r_n > 0$, $\lim_{n \to \infty} |r_{n+1} - r_n| = 0$, and $0 < \liminf_{n \to \infty} \lambda_n \leq \limsup_{n \to \infty} \lambda_n < 1$. Let $\{x_n\}$ be a sequence generated by (MHRIM-2). Then $\{x_n\}$ converges strongly to some zero point x^* of A.*

Let E be a Banach space and let C be closed convex subset of E. One knows that A is accretive if and only if $T := I - A$, where I is the identity, is pseudocontractive. From the Caristi fixed point theorem, we see that accretive operator A satisfies the flow-invariance condition (FIC) if and only if pseudocontractive mapping T satisfies the weak inward condition (WIC). So, every zero point theorem of accretive operators corresponds with a fixed point theorem of pseudocontractive mappings.

5.7 Iterative methods for variational inequalities with accretive operators

In this section, we investigate some iterative methods for solving a class of variational inequalities with accretive operators. First, we recall and review shortly classical gradient projection method and Yamada's hybrid steepest decent method for solving monotone-type variational inequality problems. We then point out some gaps appearing in Chidume's monograph [23] and extend most of the results of Chidume [23] to more general cases. In particular, we extend the results of Yamada [108], Xu and Kim [105], Iemoto and Takahashi [31], and others from real Hilbert spaces to the more general setting of uniformly smooth Banach spaces.

5.7.1 Two kinds of variational inequality problems

Let K be a nonempty, closed, and convex subset of a real Hilbert space H, and let A be a monotone mapping from K into H. The classical monotone type variational inequality problem is formulated as follows: Find a point $x^* \in K$ such that

$$\langle Ax^*, y - x^* \rangle \geq 0, \ \forall y \in K. \tag{MTVIP}$$

We use VIP(A, K) to denote the set of solutions for (MTVIP).

It is known that solving (MTVIP) is equivalent to finding a fixed point of the mapping $P_K(I - \delta A) : K \to K$, where $\delta \in (0, 1)$ is a fixed constant and P_K is the nearest point projection mapping from H onto K, that is, $P_K x = y$ where

$$\|x - y\| = \inf\{\|x - u\| : u \in K\}, \quad \text{for } x \in H.$$

Due to the fact that $P_K : H \to K$ is nonexpansive, under appropriate conditions on A and $\delta \in (0, 1)$, fixed point methods can be used to find a solution of (MTVIP). For instance, if A is η-strongly monotone and k-Lipschitz, then a mapping $G : H \to H$ defined by $Gx = P_K(x - \delta Ax)$, $x \in H$ with $\delta \in (0, \frac{2\eta}{k^2})$ is contractive. As a result of this fact, the known Picard iteration $x_n \in H$, $x_{n+1} = Gx_n$, $n \geq 0$, of the classical Banach contractive mapping principle converges strongly to the unique solution of (MTVIP).

On the other hand, the projection mapping P_K may make the computation of the iterates difficult due to possible complexity of the convex set K. In order to reduce the possible difficulty with the use of P_K, Yamada [108] introduced the so-called hybrid steepest descent methods for solving (MTVIP):

$$x_0 \in H, \quad x_{n+1} = Tx_n - \lambda_{n+1}\delta A(Tx_n), \quad n \geq 0 \tag{HSDM1}$$

and

$$x_0 \in H, \quad x_{n+1} = T_{[n+1]}x_n - \lambda_{n+1}\delta A(T_{[n+1]}x_n), \quad n \geq 0, \tag{HSDM2}$$

where $T_{[k]} = T_{k \bmod r}$, for $k \geq 1$, with the $\bmod r$ function taking values in the set $\{1, 2, \ldots, r\}$.

Yamada [108] proved the following results.

Theorem 5.7.1. *Let $T : H \to H$ be a nonexpansive mapping such that $K := \{x \in H : Tx = x\} \neq \emptyset$. Let A be an η-strongly monotone and k-Lipschitz mapping on H. Let $\delta \in (0, \frac{2\eta}{k^2})$ be an arbitrary but fixed real number and let $\{\lambda_n\}$ be a sequence in $(0, 1)$ satisfying the following conditions:*

(C1) $\lim_{n \to \infty} \lambda_n = 0$;

(C2) $\sum_{n=0}^{\infty} \lambda_n = \infty$;

(C3) $\lim_{n \to \infty} \frac{\lambda_{n+1}}{\lambda_n^2} = 0$.

Then the sequence $\{x_n\}$ defined by (HSDM1) converges strongly to the unique solution of VIP(A, K).

Theorem 5.7.2. *Let $\{T_i\}_{i=1}^{r} : H \to H$ be a finite family of nonexpansive mappings such that $K = \bigcap_{i=1}^{r} F(T_i) \neq \emptyset$. Let A be an η-strongly monotone and k-Lipschitz mapping on H. Let $\{\lambda_n\}$ be a sequence of real numbers in $(0, 1)$ satisfying conditions (C1), (C2), and (C4) $\sum_{n=0}^{\infty} |\lambda_n - \lambda_{n+r}| < \infty$. Then the sequence $\{x_n\}$ defined by (HSDM2) converges strongly to the unique solution of VIP(A, K).*

Afterwards, Xu and Kim [105] improved on the results of Yamada with condition (C3) replaced by (C5) $\frac{|\lambda_{n+1} - \lambda_n|}{\lambda_{n+1}} \to 0$ $(n \to \infty)$ and with condition (C4) replaced by (C6) $\frac{|\lambda_{n+r} - \lambda_n|}{\lambda_{n+r}} \to 0$ $(n \to \infty)$. Namely, they proved that following theorems.

Theorem 5.7.3. *Let $T : H \to H$ be a nonexpansive mapping such that $K = \{x \in H : x = Tx\} \neq \emptyset$. Let A be η-strongly monotone and k-Lipschitz on H. Let $\delta \in (0, \frac{2\eta}{k^2})$ and let $\{\lambda_n\}$ be a sequence in $(0, 1)$ satisfying the following conditions:*

(C1) $\lambda_n \to 0$ $(n \to \infty)$;

(C2) $\sum_{n=0}^{\infty} \lambda_n = \infty$; and

(C5) $\frac{|\lambda_n - \lambda_{n+1}|}{\lambda_{n+1}} \to 0$ $(n \to \infty)$.

Then, the sequence $\{x_n\}$ defined by (HSDM1) converges strongly to the unique solution of VIP(A, K).

Theorem 5.7.4. *Let $\{T_i\}_{i=1}^{r} : H \to H$ be a finite family of nonexpansive mappings such that $K = \bigcap_{i=1}^{r} \mathrm{Fix}(T_i) = \mathrm{Fix}(T_1 T_2 \cdots T_r) = \mathrm{Fix}(T_r T_1 \cdots T_{r-1}) = \cdots = \mathrm{Fix}(T_r T_{r-1} \cdots T_1) \neq \emptyset$. Let A be η-strongly monotone and k-Lipschitz on H. Let $\delta \in (0, \frac{2\eta}{k^2})$ and let $\{\lambda_n\}$ be a sequence in $(0, 1)$ satisfying the following conditions:*

(C1) $\lambda_n \to 0$ $(n \to \infty)$;

(C2) $\sum_{n=0}^{\infty} \lambda_n = \infty$; and

(C6) $\frac{|\lambda_n - \lambda_{n+r}|}{\lambda_{n+r}} \to 0$ $(n \to \infty)$.

Then the sequence $\{x_n\}$ defined by (HSDM2) converges strongly to the unique solution of VIP(A, K)

In 2008, Iemoto and Takahashi [31] considered the case of $K = \bigcap_{i=1}^{\infty} \text{Fix}(T_i)$, where $\{T_i\}_{i=1}^{\infty}$ is a countable family of nonexpansive mappings on Hilbert spaces. Zhou and Wang [125] considered the case of $K = \bigcap_{i=1}^{r} \text{Fix}(T_i)$, where $\{T_i\}_{i=1}^{r}$ is a finite family of Lipschitz quasi-pseudocontractive mappings on Hilbert spaces.

An interesting problem arises naturally: Can the gradient projection methods and the hybrid steepest decent methods be used to find a solution of some variational inequality problems with accretive operators in a real Banach space? Chidume considered this problem in [23]. There he attempted to extend some of the results mentioned above to the more general Banach spaces, however, there are some gaps in Chapter 7 of his monograph. We remark that if X is a q-uniformly smooth Banach space, then $q \leq 2$. In Lemma 7.10 of Chidume [23], he requires that $q \geq 2$. Thus $q = 2$. Consequently, the space appearing in Lemma 7.10 is 2-uniformly smooth. In other words, Lemma 7.10 of Chidume [23] holds true only for 2-uniformly smooth Banach spaces. Moreover, we point out that Lemma 7.30 of Chidume [23] holds true only for $p = 2$.

Let K be a nonempty, closed, and convex subset of a real Banach space X, and $A : K \rightarrow X$ an accretive operator. Let J be the normalized duality mapping on X. The variational inequality problem with accretive operators is formulated as follows: Find a point $x^* \in K$ such that

$$\langle Ax^*, j(y - x^*) \rangle \geq 0, \ \forall y \in K, \tag{VIPAM1}$$

where $j(y - x^*) \in J(y - x^*)$.

Another kind of variational inequality problem with accretive operators is to find a point $x^* \in K$ such that

$$\langle y - x^*, j(Ax^*) \rangle \geq 0, \quad \forall y \in K. \tag{VIPAM2}$$

where $j(Ax^*) \in J(Ax^*)$. If $X = H$, a real Hilbert space, the variational inequality problems both (VIPAM1) and (VIPAM2) are reduced to (MTVIP).

It is known that if X is a real reflexive, smooth, and strictly convex Banach space and K is a nonempty, closed, and convex subset of X, then $\forall x \in X$, there exists a unique $x_0 \in K$ such that

$$\|x - x_0\| = \min\{\|x - y\| : y \in K\}.$$

Define a mapping $P_K : X \rightarrow K$ by

$$P_K x = x_0.$$

Then $P_K : X \rightarrow K$ is called the metric projection mapping from X onto K.

Using Theorem 1.8.49, we know that $x_0 = P_K x$ is and only if

$$\langle y - x_0, J(x_0 - x) \rangle \geq 0, \quad \forall y \in K.$$

Let $A : K \to X$ be a mapping and let $\delta > 0$ be a fixed number. Observe that there is some $x^* \in K$ such that

$$\langle y - x^*, J(Ax^*) \rangle \geq 0, \quad \forall y \in K$$
$$\Longleftrightarrow \langle y - x^*, J(\delta Ax^*) \rangle \geq 0, \quad \forall y \in K, \delta > 0$$
$$\Longleftrightarrow \langle y - x^*, J(x^* - (x^* - \delta Ax^*)) \rangle \geq 0, \quad \forall y \in K, \delta > 0$$
$$\Longleftrightarrow x^* = P_K(x^* - \delta Ax^*), \quad \delta > 0.$$

Consequently, solving (VIPAM2) is equivalent to finding a fixed point of mapping $P_K(I - \delta A)$. Unfortunately, the metric projection mapping P_K in a general Banach space is no longer nonexpansive, and thus the gradient projection method cannot be used to find a solution of variational inequality problem (VIPAM2).

Let us consider (VIPAM1). Let K be a nonempty, closed, and convex subset of a real Banach space X. Assume that there exists a sunny nonexpansive retraction Q_K from X onto K. Observe that (VIPAM1) has a solution $x^* \in K$ such that

$$\langle j(y - x^*), Ax^* \rangle \geq 0, \quad \forall y \in K$$
$$\Longleftrightarrow \langle j(y - x^*), \delta Ax^* \rangle \geq 0, \quad \forall y \in K, \delta > 0$$
$$\Longleftrightarrow \langle j(y - x^*), x^* - (x^* - \delta Ax^*) \rangle \geq 0, \quad \forall y \in K, \delta > 0$$
$$\Longleftrightarrow x^* = Q_K(x^* - \delta Ax^*), \quad \delta > 0.$$

Therefore, (VIPAM1) has a unique solution provided that there exists a sunny nonexpansive retraction Q_K from X onto K and the mapping $(I - \delta A)$ is a strict contraction on X. Furthermore, both fixed point methods and hybrid steepest decent methods can be used to find the unique solution of (VIPAM1).

We present several convergence theorems which extend the results of Chidume [23] to the more general setting of uniformly smooth Banach spaces. In particular, the convergence theorems to be presented will be applicable in L_p, where $1 < p < \infty$.

5.7.2 Tools to solve the variational inequality

In this subsection, we list some important lemmas for variational inequality (VIPAM1).

Lemma 5.7.1 ([107]). *Let X be a real Banach space. Let $A : X \to X$ be an η-strongly accretive and k-Lipschitz mapping. Then there exists some $\tau \in (0, 1)$ such that $I - \tau A$ is contractive on X.*

Lemma 5.7.2. *Let X be a real Banach space. Let $A : X \to X$ be an η-strongly accretive and k-Lipschitz mapping. Let $T : X \to X$ be a mapping such that $K := \text{Fix}(T) = \{x \in X : Tx = x\} \neq \emptyset$. Assume that there exists a sunny nonexpansive retraction $Q : X \to K$ from X onto K. Then the variational inequality (VIPAM1) has a solution $x^* \in K$.*

Proof. Using Lemma 5.7.1, we know that there exists $\tau \in (0, 1)$ such that $I - \tau A$ is contractive on X. Since $Q_K : X \to K$ is nonexpansive, we see that the mapping $Q_K(I - \tau A)$ is contractive on X. It follows from the Banach contractive mapping principle that there exists a unique $x^* \in X$ such that

$$x^* = Q_K(I - \tau A)x^*. \tag{5.50}$$

Since $Q_K : X \to K$ is a sunny nonexpansive retraction, we find from Theorem 1.8.49 and (5.50) that

$$\langle x^* - (I - \tau A)x^*, j(y - x^*) \rangle \geq 0, \quad \forall y \in K$$
$$\Longleftrightarrow \langle Ax^*, j(y - x^*) \rangle, \quad \forall y \in K.$$

Hence $x^* \in K$ is the unique solution of (VIPAM1). This completes the proof. □

By using the Reich's inequality, we can deduce the following result.

Lemma 5.7.3. *Let X be a real uniformly smooth Banach space and let $A : X \to X$ be an η-strongly accretive and k-Lipschitz mapping. Then, if $\delta \in (0, k^{-1}b^{-1}(2\eta k^{-1}))$, the mapping $f := I - \delta A : X \to X$ is contractive.*

Proof. Define $f : X \to X$ by

$$f(x) = (I - \delta A)x, \quad x \in X. \tag{5.51}$$

For $x \neq y$, $x, y \in X$, we have

$$\frac{f(x) - f(y)}{\|x - y\|} = \frac{x - y}{\|x - y\|} - \delta \frac{Ax - Ay}{\|x - y\|}. \tag{5.52}$$

By using Theorem 1.8.28, we have

$$\frac{\|f(x) - f(y)\|^2}{\|x - y\|^2} \leq 1 - 2\delta \left\langle \frac{Ax - Ay}{\|x - y\|}, j\left(\frac{x - y}{\|x - y\|}\right) \right\rangle$$
$$+ \max\{1, 1\}\delta \frac{\|Ax - Ay\|}{\|x - y\|} b\left(\delta \frac{\|Ax - Ay\|}{\|x - y\|}\right)$$
$$\leq 1 - 2\eta\delta + \delta kb(\delta k),$$

which implies

$$\|f(x) - f(y)\|^2 \leq (1 - 2\eta\delta + \delta kb(\delta k))\|x - y\|^2.$$

This yields

$$\|f(x) - f(y)\| \le \rho\|x - y\|, \quad \forall x, y \in X, \tag{5.53}$$

where

$$\rho := \sqrt{1 - 2\eta\delta + \delta kb(\delta k)} \in (0, 1),$$

completing the proof. □

It is well known that if X is a q-uniformly smooth Banach space, then there exists some constant $c_1 \ge 1$, such that

$$b(t) \le c_1 t^{q-1}, \quad t \in [0, 1]. \tag{5.54}$$

Thus, we have the following result.

Lemma 5.7.4. *Let X be a q-uniformly smooth Banach space and let $A : X \to X$ be an η-strongly accretive and k-Lipschitz mapping. Then, if $\delta \in (0, (\frac{2\eta}{k^q c_1})^{\frac{1}{q-1}})$, the mapping $f := (I - \delta A) : X \to X$ is contractive.*

It is well known that if $X = L_p$, $p \ge 2$, then we have

$$b(t) \le (p - 1)t, \quad t \ge 0. \tag{5.55}$$

Using (5.55), we have the following result.

Lemma 5.7.5. *Let $X = L_p$, $p \ge 2$. Let $A : X \to X$ be an η-strongly accretive and k-Lipschitz mapping. Then, if $\delta \in (0, \frac{2\eta}{k^2(p-1)})$, the mapping $f := (I - \delta A) : X \to X$ is contractive.*

5.7.3 Strong convergence theorems

Theorem 5.7.5. *Let X be a real reflexive Banach space whose norm is uniformly Gâteaux differentiable. Suppose that every nonempty, closed, convex, and bounded subset of X has a fixed point property for nonexpansive self-mappings. Let $T : X \to X$ be a nonexpansive mapping such that $K := \text{Fix}(T) \ne \emptyset$. Let $A : X \to X$ be an η-strongly accretive and k-Lipschitz mapping. Let $\tau \in (0, 1)$ be the constant appearing in Lemma 5.7.1. Also $\forall x_0 \in X$, define a sequence $\{x_n\}$ iteratively as follows:*

$$x_{n+1} = \alpha_n f(x_n) + (1 - \alpha_n)Tx_n, \quad n \ge 0, \tag{5.56}$$

where $f := I - \tau A$ and $\{\alpha_n\}$ is a sequence of real numbers that satisfies the following conditions:
(C1) $\lim_{n \to \infty} \alpha_n = 0$;

(C2) $\sum_{n=0}^{\infty} \alpha_n = \infty;$

(C3) $\sum_{n=1}^{\infty} |\alpha_{n+1} - \alpha_n| < \infty;$ *or*

(C4) $\lim_{n \to \infty} \frac{\alpha_{n+1}}{\alpha_n} = 1.$

Then, the sequence $\{x_n\}$ defined by (5.56) converges strongly to the unique solution x^* of (VIPAM1).

Proof. From Theorem 4.3.1, we see that there exists a unique sunny nonexpansive retraction $Q_{\mathrm{Fix}(T)}$ from X onto $\mathrm{Fix}(T)$. By Lemma 5.7.1, we know that $f := (I - \tau A) : X \to X$ is contractive. By Lemma 5.7.2, we conclude that there exists a unique solution $x^* \in \mathrm{Fix}(T)$ that satisfies (VIPAM1). We now define an auxiliary sequence $\{y_n\}$ by

$$y_0 \in X, \quad y_{n+1} = \alpha_n f(x^*) + (1 - \alpha_n) T y_n, \quad n \geq 0. \tag{5.57}$$

Then, by using Theorem 4.3.2, we know that $\{y_n\}$ converges strongly to $p = Q_{\mathrm{Fix}(T)} f(x^*)$, where $Q_{\mathrm{Fix}(T)} : X \to \mathrm{Fix}(T)$ is the sunny nonexpansive retraction from X onto $\mathrm{Fix}(T)$. By Theorem 1.8.49, we have that $x^* = Q_{\mathrm{Fix}(T)}(x^* - \tau A x^*) = Q_{\mathrm{Fix}(T)} f(x^*)$. Hence $p = x^*$. Since f is strictly contractive, we see that there is a constant $\rho \in (0, 1)$ such that

$$\|f(x) - f(y)\| \leq \rho \|x - y\|, \quad \forall x, y \in X. \tag{5.58}$$

Combining (5.56), (5.57), and (5.58), we have

$$\begin{aligned}
\|x_{n+1} - y_{n+1}\| &\leq \alpha_n \|f(x_n) - f(x^*)\| + (1 - \alpha_n)\|T x_n - T y_n\| \\
&\leq \alpha_n \rho \|x_n - y_n\| + \rho \alpha_n \|y_n - x^*\| + (1 - \alpha_n)\|x_n - y_n\| \\
&= [1 - (1 - \rho \alpha_n)]\|x_n - y_n\| + o(\alpha_n).
\end{aligned}$$

It follows from Lemma 1.10.2 that $x_n - y_n \to \theta$ as $n \to \infty$. Consequently, $\{x_n\}$ converges strongly to the unique solution x^* of (VIPAM1). $\qquad \square$

Theorem 5.7.6. *Let X, T, A, τ, and $\{\alpha_n\}$ be the same as those in Theorem 5.7.5. Also $\forall x_0 \in X$, define a sequence $\{x_n\}$ iteratively as follows:*

$$x_{n+1} = T x_n - \tau \alpha_n A(T x_n), \quad n \geq 0. \tag{5.59}$$

Then, the sequence $\{x_n\}$ defined by (5.59) converges strongly to the unique solution x^* of (VIPAM1).

Proof. Notice that (5.59) can be rewritten as follows:

$$x_{n+1} = (1 - \alpha_n) T x_n + \alpha_n g(x_n), \quad n \geq 0, \tag{5.60}$$

where $g(x) = (I - \tau A) T x$, $x \in X$. It is clear that $g : X \to X$ is strictly contractive, since $I - \tau A$ is strictly contractive and T is nonexpansive. By a similar argument as that used to prove Theorem 5.7.5, we can prove that $\{x_n\}$ converges strongly to the unique solution x^* of (VIPAM1). This completes the proof. $\qquad \square$

Theorem 5.7.7. *Let X be a real uniformly smooth Banach space. Let $T : X \to X$ be a nonexpansive mapping such that $K := \mathrm{Fix}(T) \neq \emptyset$. Let $A : X \to X$ be an η-strongly accretive and k-Lipschitz mapping. Let $\{\alpha_n\}$ be a sequence in $(0,1)$ satisfying the following conditions:*

(C1) $\lim_{n\to\infty} \alpha_n = 0$;

(C2) $\sum_{n=0}^{\infty} \alpha_n = \infty$;

(C3) $\sum_{n=1}^{\infty} |\alpha_{n+1} - \alpha_n| < \infty$; *or*

(C4) $\lim_{n\to\infty} \frac{\alpha_{n+1}}{\alpha_n} = 1.$

For $\delta \in (0, k^{-1}b^{-1}(2k^{-1}\eta))$, define a sequence $\{x_n\}$ iteratively in X by

$$x_0 \in X, \quad x_{n+1} = \alpha_n f(x_n) + (1 - \alpha_n)Tx_n, \quad n \geq 0, \tag{5.61}$$

where $f(x) = x - \delta Ax, \ \forall x \in X$. Then, the sequence $\{x_n\}$ defined by (5.96) converges strongly to the unique solution x^* of (VIPAM1).

Proof. Using Lemma 5.7.3, we know that $f : X \to X$ is contractive. From Corollary 4.4.1, we see that $\{x_n\}$ converges strongly to the unique $x^* \in \mathrm{Fix}(T)$ that satisfies the variational inequality:

$$\langle x^* - f(x^*), j(y - x^*) \rangle \geq 0, \quad \forall y \in \mathrm{Fix}(T)$$
$$\Longleftrightarrow \langle Ax^*, j(y - x^*) \rangle \geq 0, \quad \forall y \in \mathrm{Fix}(T).$$

Hence x^* is the unique solution of (VIPAM1). This completes the proof. □

Theorem 5.7.8. *Let X, T, A, δ, and $\{\alpha_n\}$ be the same as those in Theorem 5.7.7. Define a sequence $\{x_n\}$ iteratively in X by*

$$x_0 \in X, \quad x_{n+1} = Tx_n - \delta\alpha_n A(Tx_n), \quad n \geq 0. \tag{5.62}$$

Then, $\{x_n\}$ converges strongly to the unique solution x^* of (VIPAM1).

Proof. Notice that (5.62) can be rewritten as follows:

$$x_{n+1} = (1 - \alpha_n)Tx_n + \alpha_n g(x_n), \quad n \geq 0, \tag{5.63}$$

where $g(x) = (I - \delta A)Tx, \ \forall x \in X$. It is clear that $g : X \to X$ is strictly contractive by Lemma 5.7.3. Now it follows from Corollary 4.4.1 that $\{x_n\}$ converges strongly to the unique solution $x^* \in \mathrm{Fix}(T)$ that satisfies variational inequality:

$$\langle x^* - g(x^*), j(y - x^*) \rangle \geq 0, \quad \forall y \in \mathrm{Fix}(T)$$
$$\Longleftrightarrow \langle Ax^*, j(y - x^*) \rangle \geq 0, \quad \forall y \in \mathrm{Fix}(T).$$

This completes the proof. □

Theorem 5.7.9. *Let X be a real q-uniformly smooth Banach space. Let $T : X \to X$ be a nonexpansive mapping such that $K := \mathrm{Fix}(T) \neq \emptyset$. Let $A : X \to X$ be an η-strongly accretive and k-Lipschitz mapping. Let $\{x_n\}$ be a sequence in $(0, 1)$ satisfying the following conditions:*

(C1) $\lim_{n \to \infty} \alpha_n = 0$;

(C2) $\sum_{n=0}^{\infty} \alpha_n = \infty$;

(C3) $\sum_{n=0}^{\infty} |\alpha_{n+1} - \alpha_n| < \infty$; *or*

(C4) $\lim_{n \to \infty} \frac{\alpha_n}{\alpha_{n+1}} = 1$.

For $\delta \in (0, (\frac{2\eta}{k^q c_1})^{\frac{1}{q-1}})$, define a sequence $\{x_n\}$ iteratively in X by $x_0 \in X$,

$$x_{n+1} = \alpha_n f(x_n) + (1 - \alpha_n) T x_n, \quad n \geq 0. \tag{5.64}$$

where $f(x) = x - \delta A x$, $\forall x \in X$. Then, the sequence $\{x_n\}$ defined by (5.64) converges strongly to the unique solution x^* of (VIPAM1).

Proof. By Lemma 5.7.4, we know that $f : X \to X$ is contractive. Now the conclusion of Theorem 5.7.9 follows from Corollary 4.4.1. This completes the proof. $\qquad\square$

Theorem 5.7.10. *Let X, T, A, δ, and $\{x_n\}$ be the same those in Theorem 5.7.9. Define a sequence $\{x_n\}$ iteratively in X by*

$$x_0 \in X, \quad x_{n+1} = T x_n - \delta \alpha_n A(T x_n), \quad n \geq 0, \tag{5.65}$$

Then, the sequence $\{x_n\}$ defined by (5.65) converges strongly to the unique solution x^* of (VIPAM1).

Proof. Its proof is the same as that of Theorem 5.7.9. $\qquad\square$

Using a similar method, we can prove the following results.

Theorem 5.7.11. *Let $X = L_p$, $p \geq 2$. Let $T : X \to X$ be a nonexpansive mapping such that $K := \mathrm{Fix}(T) \neq \emptyset$. Let $A : X \to X$ be an η-strongly accretive and k-Lipschitz mapping. Let $\{\alpha_n\}$ be a sequence in $(0, 1)$ satisfying the following conditions:*

(C1) $\lim_{n \to \infty} \alpha_n = 0$;

(C2) $\sum_{n=0}^{\infty} \alpha_n = \infty$;

(C3) $\sum_{n=0}^{\infty} |\alpha_{n+1} - \alpha_n| < \infty$; *or*

(C4) $\lim_{n \to \infty} \frac{\alpha_n}{\alpha_{n+1}} = 1$.

Choose $\delta \in (0, \frac{2\eta}{k^2(p-1)})$ and define a sequence $\{x_n\}$ iteratively in X by

$$x_0 \in X, \quad x_{n+1} = \alpha f(x_n) + (1 - \alpha_n) T x_n, \quad n \geq 0, \tag{5.66}$$

where $f(x) = (I - \delta A)x$, $\forall x \in X$. Then, the sequence $\{x_n\}$ defined by (5.66) converges strongly to the unique solution x^* of (VIPAM1).

Theorem 5.7.12. *Let X, T, A, δ, and $\{\alpha_n\}$ be the same as those in Theorem 5.7.11. Define a sequence $\{x_n\}$ iteratively in X by*

$$x_0 \in X, \quad x_{n+1} = Tx_n - \delta\alpha_n A(Tx_n), \quad n \geq 0. \tag{5.67}$$

Then the sequence $\{x_n\}$ defined by (5.67) converges strongly to the unique solution x^* of (VIPAM1).

Remark 5.7.1. For $\lambda \in (0,1)$, write

$$T_\lambda := \lambda I + (1-\lambda)T.$$

If we replace T by T_λ in all iterative methods mentioned above, then the corresponding convergence theorems are still true under conditions ($C1$) and ($C2$).

Next, we deal with the case of $K := \bigcap_{i=1}^r \mathrm{Fix}(T) \neq \emptyset$, and when $\{T_i\}_{i=1}^r : X \to X$ is a finite family of nonexpansive mappings.

Theorem 5.7.13. *Let X be a uniformly convex and uniformly smooth Banach space. Let $\{T_i\}_{i=1}^r : X \to X$ be a finite family of nonexpansive mappings such that $K := \bigcap_{i=1}^r \mathrm{Fix}(T_i) \neq \emptyset$. Let $A : X \to X$ be an η-strongly accretive and k-Lipschitz mapping. Let $\mu_i \in (0,1)$ be fixed numbers ($i = 1, 2, \ldots, r$). Let W be the W-mapping generated by $T_1, T_2, \ldots, T_r$ and $\mu_1, \mu_2, \ldots, \mu_r$. Let $\{\alpha_n\}$ be a sequence in $(0,1)$ satisfying the following conditions:*
(C1) $\lim_{n\to\infty} \alpha_n = 0$;
(C2) $\sum_{n=0}^\infty \alpha_n = \infty$.

Choose $\delta \in (0, k^{-1}b^{-1}(2k^{-1}\eta))$ and define a sequence $\{x_n\}$ iteratively in X by

$$x_0 \in X, \quad x_{n+1} = \alpha_n f(x_n) + (1-\alpha_n)Wx_n, \quad n \geq 0, \tag{5.68}$$

where $f(x) = (I - \delta A)x$, $\forall x \in X$. Then, the sequence $\{x_n\}$ defined by (5.68) converges strongly to the unique solution x^* of (VIPAM1).

Proof. It is clear that $W : X \to X$ is averaged nonexpansive and $\mathrm{Fix}(W) = \bigcap_{i=1}^r \mathrm{Fix}(T) = K \neq \emptyset$. By Lemma 5.7.3, we know that $f : X \to X$ is contractive. By Corollary 4.4.1, we see that $\{x_n\}$ converges strongly to the unique solution $x^* \in \mathrm{Fix}(W)$ that satisfies the variational inequality:

$$\langle x^* - f(x^*), J(y - x^*)\rangle \geq 0, \quad \forall y \in K$$
$$\Longleftrightarrow \langle Ax^*, J(y - x^*)\rangle \geq 0, \quad \forall y \in K.$$

Hence x^* is the unique solution of the (VIPAM1) with $K := \mathrm{Fix}(W) = \bigcap_{i=1}^r \mathrm{Fix}(T_i)$. This completes the proof. $\qquad\square$

Theorem 5.7.14. *Let X, $\{T_i\}_{i=1}^r$, A, W, δ, and $\{\alpha_n\}$ be the same as those in Theorem 5.7.13. Define a sequence $\{x_n\}$ iteratively in X by*

$$x_0 \in X, \quad x_{n+1} = Wx_n - \delta\alpha_n A(Wx_n), \quad n \geq 0. \tag{5.69}$$

Then, the sequence $\{x_n\}$ defined by (5.69) converges strongly to the unique solution x^* of (VIPAM1).

Proof. Notice that (5.69) can be rewritten as follows:

$$x_0 \in X, \quad x_{n+1} = (1 - \alpha_n)Wx_n + \alpha_n g(x_n), \quad n \geq 0,$$

where $g(x) = (I - \delta A)Wx$, $\forall x \in X$. It is clear that $g : X \to X$ is contractive. Now the conclusion follows from Corollary 4.4.1. This completes the proof. $\qquad\square$

Theorem 5.7.15. *Let X be a real uniformly convex and q-uniformly smooth Banach space. Let $\{T_i\}_{i=1}^r : X \to X$ be a finite family of nonexpansive mappings such that $K := \bigcap_{i=1}^r \operatorname{Fix}(T_i) \neq \emptyset$. Let $A : X \to X$ be an η-strongly accretive and k-Lipschitz mapping. Let $\mu_i \in (0,1)$ be fixed numbers ($i = 1, 2, \ldots, r$). Let W be the W-mapping generated by $T_1, T_2, \ldots, T_r$ and $\mu_1, \mu_2, \ldots, \mu_r$. Let $\{\alpha_n\}$ be a sequence in $(0,1)$ satisfying the following conditions:*

(C1) $\lim_{n \to \infty} \alpha_n = 0$;

(C2) $\sum_{n=0}^\infty \alpha_n = \infty$.

Choose $\delta \in (0, (\frac{2\eta}{k^q c_1})^{\frac{1}{q-1}})$, and define a sequence $\{x_n\}$ in X by

$$x_0 \in X, \quad x_{n+1} = \alpha_n f(x_n) + (1 - \alpha_n)Wx_n, \quad n \geq 0, \tag{5.70}$$

where $f(x) = (I - \delta A)x$, $\forall x \in X$. Then, the sequence $\{x_n\}$ defined by (5.70) converges strongly to the unique solution x^* of (VIPAM1).

Proof. By Lemma 5.7.4, we know that $f : X \to X$ is contractive. It is clear that $W : X \to X$ is averaged nonexpansive and $\operatorname{Fix}(W) = \bigcap_{i=1}^r \operatorname{Fix}(T_i) = K \neq \emptyset$. Now the conclusion follows from Corollary 4.4.1. This competes the proof. $\qquad\square$

In the same way as when proving Theorem 5.7.14, we can establish the following results.

Theorem 5.7.16. *Let X, $\{T_i\}_{i=1}^r$, A, W, δ, and $\{\alpha_n\}$ be the same as those in Theorem 5.7.15. Define a sequence $\{X_n\}$ in X by*

$$x_0 \in X, \quad x_{n+1} = Wx_n - \delta\alpha_n A(Wx_n), \quad n \geq 0. \tag{5.71}$$

Then, the sequence $\{x_n\}$ defined by (5.71) converges strongly to the unique solution x^* of (VIPAM1).

Consequently, we have the following results.

Theorem 5.7.17. *Let $X = L_p$, $p \geq 2$. Let $\{T_i\}_{i=1}^r$, A, W, and $\{\alpha_n\}$ as in Theorem 5.7.15. Choose $\delta \in (0, \frac{2\eta}{k^2(p-1)})$ and define a sequence $\{x_n\}$ in X by*

$$x_0 \in X, \quad x_{n+1} = \alpha_n f(x_n) + (1 - \alpha_n)Wx_n, \quad n \geq 0, \tag{5.72}$$

where $f(x) = (I - \delta A)x, \forall x \in X$. Then, the sequence $\{x_n\}$ defined by (5.72) converges strongly to the unique solution x^* of the (VIPAM1).

Theorem 5.7.18. Let $X = L_p, p \geq 2$. Let $\{T_i\}_{i=1}^r, A, W$, and $\{\alpha_n\}$ as those in Theorem 5.7.17. Choose $\delta \in (0, \frac{2\eta}{k^2(p-1)})$ and define a sequence $\{x_n\}$ in X by

$$x_0 \in X, \quad x_{n+1} = \alpha_n f(x_n) + (1 - \alpha_n)Wx_n, \quad n \geq 0. \tag{5.73}$$

Then, the sequence $\{x_n\}$ defined by (5.73) converges strongly to the unique solution x^* of (VIPAM1).

Before presenting further convergence theorems, we point out that Theorem 4.5.5 holds true if E is either uniformly smooth or E is a real reflexive Banach space whose norm is uniformly Gâteaux differentiable and every nonempty, closed, convex, and bounded subset of E has the fixed point property for nonexpansive self-mappings. Furthermore, if $F = \bigcap_{i=1}^n \text{Fix}(T_i) = \text{Fix}(T_r T_{r-1} \cdots T_1) \neq \emptyset$, then

$$F = \text{Fix}(T_r T_{r-1} \cdots T_1) = \text{Fix}(T_1 T_r \cdots T_1 T_2) = \cdots = \text{Fix}(T_{r-1} T_{r-2} \cdots T_1 T_r).$$

Now we state these revisions of Theorem 4.5.5 as follows.

Theorem 5.7.19. Let X be a real uniformly smooth Banach space and let C be a nonempty, closed, and convex subset of X. Let $\{T_i\}_{i=1}^r : C \to C$ be a finite family of nonexpansive mappings such that $F = \bigcap_{i=1}^r \text{Fix}(T_i) \neq \emptyset$ and $F = \text{Fix}(T_r T_{r-1} \cdots T_1)$. Let $\{\alpha_n\}$ be a sequence in $(0,1)$ satisfying the following conditions:
(C1) $\lim_{n \to \infty} \alpha_n = 0$;
(C2) $\sum_{n=0}^{\infty} \alpha_n = \infty$;
(C3) $\sum_{n=0}^{\infty} |\alpha_{n+r} - \alpha_n| < \infty$; or
(C4) $\lim_{n \to \infty} \frac{\alpha_n}{\alpha_{n+r}} = 1$.

For an initial value $x_0 \in C$ and a fixed anchor $u \in C$, define a sequence $\{x_n\}$ iteratively in C by

$$x_0 \in C, u \in C, \quad x_{n+1} = \alpha_{n+1}u + (1 - \alpha_{n+1})T_{n+1}x_n, \quad n \geq 0, \tag{5.74}$$

where $T_k = T_{k \bmod r}$, for $k \geq 1$, with the mod r function taking values in the set $\{1, 2, \ldots, r\}$. Then the sequence $\{x_n\}$ defined by (5.74) converges strongly to the specific common fixed point $x^* = Q_F u$, where $Q_F : C \to F$ is the unique sunny nonexpansive retraction from C onto F.

Theorem 5.7.20. Let X be a real reflexive Banach space whose norm is uniformly Gâteaux differentiable and C a nonempty, closed, and convex subset of X. Suppose that every nonempty, closed, convex, and bounded subset of C has the fixed point property for nonexpansive self-mappings. Let $\{T_i\}_{i=1}^r : C \to C$ be a finite family of nonexpansive mappings such that $F = \bigcap_{i=1}^r \text{Fix}(T_i) \neq \emptyset$ and $F = \text{Fix}(T_r T_{r-1} \cdots T_1)$. Let $\{\alpha_n\}$ and $\{x_n\}$ be *given as in Theorem 5.7.19. Then, the sequence $\{x_n\}$ defined by (5.74) converges strongly to the unique common fixed point $p = Q_F u$, where $Q_F : C \to F$ is the sunny nonexpansive retraction from C onto F.*

Theorem 5.7.21. *Let X be a real reflexive Banach space whose norm is uniformly Gâteaux differentiable. Suppose that every nonempty, closed, convex, and bounded subset of X has the fixed point property for nonexpansive self-mappings. Let $\{T_i\}_{i=1}^r : X \to X$ be a finite family of nonexpansive mappings such that $K = \bigcap_{i=1}^r \text{Fix}(T_i) \neq \emptyset$ and $K = \text{Fix}(T_r T_{r-1} \cdots T_1)$. Let $A : X \to X$ be an η-strongly accretive and k-Lipschitz mapping. Let $\tau \in (0,1)$ be the constant appearing in Lemma 5.7.1. Let $\{\alpha_n\}$ be a sequence in $(0,1)$ satisfying the following conditions:*

(C1) $\lim_{n \to \infty} \alpha_n = 0$;

(C2) $\sum_{n=0}^{\infty} \alpha_n = \infty$;

(C3) $\sum_{n=0}^{\infty} |\alpha_{n+r} - \alpha_n| < \infty$; *or*

(C4) $\lim_{n \to \infty} \frac{\alpha_n}{\alpha_{n+r}} = 1$.

Define a sequence $\{x_n\}$ in X by

$$x_0 \in X, \quad x_{n+1} = \alpha_{n+1} f(x_n) + (1 - \alpha_{n+1}) T_{n+1} x_n, \quad n \geq 0, \qquad (5.75)$$

where $f = (I - \tau A)$, $T_k = T_{k \bmod r}$, for $k \geq 1$, and the mod r function takes values in the set $\{1, 2, \ldots, r\}$. Then, the sequence $\{x_n\}$ defined by (5.75) converges strongly to the unique solution x^* of (VIPAM1).

Proof. It follows from Lemmas 5.7.1 and 5.7.2 that there exists a unique $x^* \in K$ such that

$$\langle Ax^*, j(y - x^*) \rangle \geq 0, \quad \forall y \in K. \qquad (5.76)$$

Notice that (5.76) is equivalent to

$$\langle x^* - f(x^*), j(y - x^*) \rangle \geq 0, \quad \forall y \in K. \qquad (5.77)$$

Hence $x^* = Q_F f(x^*)$ by using Theorem 1.8.49. We now consider an auxiliary sequence $\{y_n\}$ defined by

$$y_0 \in X, \quad y_{n+1} = \alpha_{n+1} f(x^*) + (1 - \alpha_{n+1}) T_{n+1} y_n, \quad n \geq 0. \qquad (5.78)$$

By Theorem 5.7.20, we assert that y_n converges strongly to the unique common fixed point $p = Q_F f(x^*) = x^*$. By Lemma 5.7.1, $f(x) = (I - \tau A)x, \forall x \in X$, is contractive. Then there is a $\rho \in (0,1)$ such that

$$\|f(x) - f(y)\| \leq \rho \|x - y\|, \quad \forall x, y \in X. \qquad (5.79)$$

Combining (5.75) and (5.78) with (5.79), we have

$$\|x_{n+1} - y_{n+1}\| \le \alpha\rho\|x_n - x^*\| + (1 - \alpha_{n+1})\|x_n - y_n\|$$
$$\le (1 - \alpha_{n+1})\|x_n - y_n\| + \rho\alpha_{n+1}\|x_n - y_n\|$$
$$+ \rho\alpha_{n+1}\|y_n - x^*\|$$
$$= [1 - (1 - \rho)\alpha_{n+1}]\|x_n - y_n\| + o(\alpha_{n+1}).$$

It follows from Lemma 1.10.2 that $x_n - y_n \to \theta$ as $n \to \infty$. Hence $x_n \to x^*$ as $n \to \infty$, which completes the proof. □

Theorem 5.7.22. *Let X be a real uniformly smooth Banach space. Let $\{T_i\}_{i=1}^r : X \to X$ be a finite family of nonexpansive mappings such that $K = \bigcap_{i=1}^r \mathrm{Fix}(T_i) \ne \emptyset$ and $K = \mathrm{Fix}(T_r T_{r-1} \cdots T_1)$. Let $A : X \to X$ be an η-strongly accretive and k-Lipschitz mapping. Let $\delta \in (0, k^{-1}b^{-1}(2\eta k^{-1}))$. Let $\{\alpha_n\}$ and $\{x_n\}$ be the same as in Theorem 5.7.21. Then, $\{x_n\}$ converges strongly to the unique solution $x^* \in K$ of* (VIPAM1).

Its proof lines are almost the same as those in the proof of Theorem 5.7.22. The only difference is that $f(x) = x - \delta Ax, x \in X$, is contractive by using Lemma 5.7.3. So, we omit the detailed proof.

In a similar way, we can prove the following results.

Theorem 5.7.23. *Let $q > 1$. Let X be a real q-uniformly smooth Banach space. Let $\{T_i\}_{i=1}^r$ and A be the same as in Theorem 5.7.22. Let $\delta \in (0, (\frac{2\eta}{k^q c_i})^{\frac{1}{q-1}})$. Let $\{\alpha_n\}$ and $\{x_n\}$ be given as in Theorem 5.7.22. Then, $\{x_n\}$ converges strongly to the unique solution $x^* \in K$ of* (VIPAM1).

Theorem 5.7.24. *Let $X = L_p, p \ge 2$. Let $\{T_i\}$ and A be the same as in Theorem 5.7.22. Let $\delta \in (0, \frac{2\eta}{k^2(p-1)})$. Let $\{\alpha_n\}$ and $\{x_n\}$ be given as in Theorem 5.7.22. Then, $\{x_n\}$ converges strongly to the unique solution $x^* \in K$ of* (VIPAM1).

Theorem 5.7.25. *Let X be a real uniformly convex and uniformly smooth Banach space. Let $A : X \to X$ be an η-strongly accretive and k-Lipschitz mapping. Let $\{T_i\}_{i=1}^\infty : X \to X$ be a countable family of nonexpansive mappings such that $K = \bigcap_{i=1}^\infty \mathrm{Fix}(T_i) \ne \emptyset$. Assume that $\{\alpha_i\}_{i=1}^\infty$ is a sequence in $(0, 1)$ satisfying the condition $0 < \alpha_i \le b < 1$ ($i = 1, 2, \dots$). Let W_n be the W-mapping generated by $T_n, T_{n-1}, \dots, T_1$ and $\alpha_n, \alpha_{n-1}, \dots, \alpha_1$. Let $\{\lambda_n\}$ be a sequence in $(0, 1)$ satisfying the following conditions: (C_1) $\lambda_n \to 0$ ($n \to \infty$) and (C_2) $\sum_{n=0}^\infty \lambda_n = \infty$. For $\delta \in (0, k^{-1}b^{-1}(2\eta k^{-1}))$, define a sequence $\{x_n\}$ iteratively in X by*

$$x_0 \in X, \quad x_{n+1} = \lambda_n f(x_n) + (1 - \lambda_n)W_n x_n, \quad n \ge 0, \tag{5.80}$$

where $f(x) = (I - \delta A)x, \forall x \in X$. Then, the sequence $\{x_n\}$ defined by (5.80) converges strongly to the unique solution x^* of (VIPAM1).

Proof. By Lemmas 5.7.1 and 5.7.1, we know that there exists a unique $x^* \in K$ such that

$$\langle Ax^*, j(y - x^*) \rangle \ge 0, \quad \forall y \in K,$$

which implies that $x^* = Q_K f(x^*)$. Consider an auxiliary sequence $\{y_n\}$ in X defined by

$$y_0 \in X, \quad y_{n+1} = \lambda_n f(x^*) + (1 - \lambda_n) W_n y_n, \quad n \geq 0. \tag{5.81}$$

By using Theorem 4.5.6, we assert that $\{y_n\}$ converges strongly to the unique common fixed point $p = Q_k f(x^*) = x^*$.

By Lemma 5.7.3, we know that $f(x) = (I - \delta A)x$, $x \in X$, is contractive, i. e., there exists some $\rho \in (0, 1)$ such that

$$\left\| f(x) - f(y) \right\| \leq \rho \| x - y \|, \quad \forall x, y \in X. \tag{5.82}$$

Using (5.80), (5.81), and (5.82), we have

$$\| x_{n+1} - y_{n+1} \| \leq \rho \lambda_n \| x_n - x^* \| + (1 - \lambda_n) \| W_n x_n - W_n y_n \| \tag{5.83}$$

$$\leq [1 - (1 - \rho)\lambda] \| x_n - y_n \| + \rho \lambda \| y_n - x^* \| \tag{5.84}$$

$$= [1 - (1 - \rho)\lambda] \| x_n - y_n \| + o(\lambda_n). \tag{5.85}$$

Using Lemma 1.10.2, we have $x_n - y_n \to \theta$ as $n \to \infty$, implying $x_n \to x^*$ ($n \to \infty$). This completes the proof. $\qquad\qquad\square$

In a similar way, we can prove the following results.

Theorem 5.7.26. *Let X be a real uniformly convex and q-uniformly smooth Banach space. Let A, $\{T_i\}_{i=1}^{\infty}$, W_n, and $\{\lambda_n\}$ be the same as in Theorem 5.7.21. For $\delta \in (0, (\frac{2\eta}{k^q c_1})^{\frac{1}{q-1}})$, define a sequence $\{x_n\}$ iteratively in X by*

$$x_0 \in X, \quad x_{n+1} = \lambda_n f(x_n) + (1 - \lambda_n) W_n x_n, \quad n \geq 0, \tag{5.86}$$

where $f(x) = (I - \delta A)x$, $\forall x \in X$. Then, the sequence $\{x_n\}$ defined by (5.86) converges strongly to the unique solution $x^* \in K$ of (VIPAM1).

Theorem 5.7.27. *Let $X = L_p$, $p \geq 2$. Let A, $\{T_i\}_{i=1}^{\infty}$, W_n, and $\{\lambda_n\}$ be the same as in Theorem 5.7.21. For $\delta \in (0, \frac{2\eta}{k^2(p-1)})$, define a sequence $\{x_n\}$ iteratively in X by*

$$x_0 \in X, \quad x_{n+1} = \lambda_n f(x_n) + (1 - \lambda_n) W_n x_n, \quad n \geq 0, \tag{5.87}$$

where $f(x) = (I - \delta A)x$, $\forall x \in X$. Then, the sequence $\{x_n\}$ defined by (5.87) converges strongly to the unique solution $x^* \in K$ of (VIPAM1).

Finally, we present several convergence theorems involving the hybrid steepest decent methods and W-mapping techniques.

Theorem 5.7.28. *Let X be a real uniformly convex and uniformly smooth Banach space. Let $A : X \to X$ be an η-strongly accretive and k-Lipschitz mapping. Let $\{T_i\}_{i=1}^{\infty} : X \to X$ be a countable family of nonexpansive mappings such that $K = \bigcap_{i=1}^{\infty} \mathrm{Fix}(T_i) \neq \emptyset$. Assume that $\{\alpha_i\}_{i=1}^{\infty}$ is a sequence in $(0, 1)$ satisfying the condition $0 < \alpha_i \leq b < 1$ ($i =*

$1, 2, \dots$). *Let W_n be the W-mapping generated by $T_n, T_{n-1}, \dots, T_1$ and $\alpha_n, \alpha_{n-1}, \dots, \alpha_1$, and W be W-mapping generated by $T_1, T_2, \dots$ and $\alpha_1, \alpha_2, \dots$ Let $\{\lambda_n\}$ be a sequence in $(0, 1)$ satisfying the following conditions: (C1) $\lim_{n \to \infty} \lambda_n = 0$ and (C2) $\sum_{n=1}^{\infty} \lambda_n = \infty$. For $\delta \in (0, k^{-1}b^{-1}(2\eta k^{-1}))$, define a sequence $\{x_n\}$ iteratively in X by*

$$x_0 \in X, \quad x_{n+1} = W_n x_n - \delta \lambda_n A(W_n x_n), \quad n \geq 0. \tag{5.88}$$

Then, the sequence $\{x_n\}$ defined by (5.88) converges strongly to the unique solution x^* of (VIPAM1).

Proof. We consider another sequence $\{y_n\}$ defined by

$$y_0 \in X, \quad y_{n+1} = W y_n - \delta \lambda_n A(W y_n), \quad n \geq 0. \tag{5.89}$$

Since $W : X \to X$ is averaged nonexpansive, we obtain from Theorem 5.7.8 that $\{y_n\}$ converges strongly to the unique solution x^* of (VIPAM1). For $\lambda \in (0, 1)$, define $f_\lambda : X \to X$ by

$$f_\lambda(x) = (I - \lambda \delta A)x, \quad \forall x \in X.$$

By using Lemma 5.7.3, we have

$$\begin{aligned}
\|f_\lambda(x) - f_\lambda(y)\| &\leq \sqrt{1 - \lambda \delta(2\eta - kb(\delta k))}\|x - y\| \\
&\leq \left[1 - \frac{1}{2}\lambda\delta(2\eta - kb(\delta k))\right]\|x - y\| \\
&= \left(1 - \frac{1}{2}\lambda\delta\tau\right)\|x - y\|, \tag{5.90}
\end{aligned}$$

where $\tau := 2\eta - kb(\delta k) > 0$. From (5.88), (5.89), and (5.90), we have

$$\begin{aligned}
\|x_{n+1} - y_{n-1}\| = \|(I - \delta\lambda_n A)W_n x_- (I - \delta\lambda_n A)W y_n\| \\
\leq \left[1 - \frac{\delta}{2}\lambda_n(2\eta - kb(\delta k))\right]\|W_n x_n - W y_n\| \\
= \left(1 - \frac{\delta\tau\lambda_n}{2}\right)\|W_n x_n - W y_n\| \\
\leq \left(1 - \frac{\delta\tau\lambda_n}{2}\right)\|W_n x_n - W_n y_n\| + \|W_n y_n - W y_n\| \\
\leq \left(1 - \frac{\delta\tau\lambda_n}{2}\right)\|x_n - y_n\| + \|W_n y_n - W y_n\|. \tag{5.91}
\end{aligned}$$

Since $\{y_n\}$ converges, one finds that $\{y_n\}$ is bounded. Write $D = \sup_{n \geq 1}\{\|y_n - w\|\}$, where $w \in K$ is some fixed element. Since the limit $Wx = \lim_{m \to \infty} W_m x$ is uniform with respect to $x \in D$, $\forall m > n \geq 1$, we have

$$\|Wy_n - W_n y_n\| = \lim_{m \to \infty} \|W_m y_n - W_n y_n\|$$

$$\leq \lim_{m \to \infty} \sum_{j=n}^{m-1} \|W_{j+1} y_n - W_j y_n\|$$

$$\leq 2D \sum_{j=n}^{\infty} b^{j+1} = \frac{2Db^{+1}}{1-b},$$

which shows that

$$\sum_{n=1}^{\infty} \|Wy_n - W_n y_n\| \leq \frac{2D}{1-b} \sum_{n=1}^{\infty} b^{n+1} = \frac{2Db^2}{1-b} < \infty.$$

It follows from Lemma 1.10.2 that $x_n - y_n \to \theta$ as $n \to \infty$, which yields $x_n \to x^*$ as $n \to \infty$. This completes the proof. $\qquad \square$

In a similar way, we can establish the following results.

Theorem 5.7.29. *Let X be a real uniformly convex and q-uniformly smooth Banach space. Let A, $\{T_i\}_{i=1}^{\infty}$, W_n, W, and $\{\lambda_n\}$ be the same as in Theorem 5.7.28. For $\delta \in (0, (\frac{2\eta}{k^q c_1})^{\frac{1}{q-1}})$, define a sequence $\{x_n\}$ iteratively in X by*

$$x_0 \in X, \quad x_{n+1} = W_n x_n - \delta \lambda_n A(W_n x_n), \quad n \geq 0. \tag{5.92}$$

Then, the sequence $\{x_n\}$ defined by (5.92) converges strongly to the unique solution $x^* \in K$ of (VIPAM1).

Theorem 5.7.30. *Let $X = L_p, p \geq 2$. Let A, $\{T_i\}_{i=1}^{\infty}$, W_n, W, and $\{\lambda_n\}$ be the same as in Theorem 5.7.28. For $\delta \in (0, \frac{2\eta}{k^2(p-1)})$, define a sequence $\{x_n\}$ iteratively in X by*

$$x_0 \in X \quad x_{n+1} = W_n x_n - \delta \lambda_n A(W_n x_n), \quad n \geq 0. \tag{5.93}$$

Then, the sequence $\{x_n\}$ defined by (5.93) converges strongly to the unique solution $x^* \in K$ of (VIPAM1).

5.8 Fixed points of strongly pseudocontractive mappings

Based on [47, Theorem 6], we establish the following fixed point theorem for continuous strongly pseudocontractive mappings.

Theorem 5.8.1. *Let E be a real Banach space and let C be a nonempty, closed, and convex subset of E. Let $T : C \to E$ be a continuous k-strongly pseudocontractive mapping satisfying the weak inward condition (WIC). Then T has a unique fixed point in C.*

Proof. Letting $A = T - I$, where I is the identity, we see that there exists $j(x - y) \in J(x - y)$ such that

$$\langle Ax - Ay, j(x - y) \rangle = \langle Tx - Ty, j(x - y) \rangle - \|x - y\|^2$$
$$\leq (k - 1)\|x - y\|^2.$$

Since T satisfies the weak inward condition, we find from the Caristi's theorem that

$$\lim_{h \to 0^+} h^{-1} d(x + hAx, C) = \lim_{h \to 0^+} h^{-1} d(x + h(Tx - x), C)$$
$$= \lim_{h \to 0^+} h^{-1} d((1 - h)x + hT), C) = 0.$$

This shows that A satisfies condition (ii) of Theorem 6 in [47]. From condition (iii) of Theorem 6 in [47], we find that $C \subset (I - \epsilon A)(C), \forall \epsilon > 0$, that is,

$$C \subset ((1 + \epsilon I) - \epsilon T)(C), \quad \forall \epsilon > 0.$$

Letting $\epsilon = n \to \infty$, one sees that $C \subset ((1 + nI) - nT)(C), \forall n \geq 1$. Fixing $y \in C$, we have $y = (1 + n)x_n - nTx_n$. It follows that

$$\frac{1}{1 + n} y + \frac{n}{1 + n} Tx_n = x_n. \tag{5.94}$$

Hence, we have

$$\|x_n - x_1\|^2 = \left\langle \frac{1}{n + 1} y - \frac{1}{2} y, j(x_n - x_1) \right\rangle + \left\langle \frac{n}{n + 1} Tx_n - \frac{1}{2} Tx_1, j(x_n - x_1) \right\rangle$$
$$\leq \left(\frac{1}{2} - \frac{1}{n + 1} \right) \|y\| \|x_n - x_1\| + \frac{n}{n + 1} \langle Tx_n - Tx_1, j(x_n - x_1) \rangle$$
$$+ \left\langle \left(\frac{n}{n + 1} - \frac{1}{2} \right) Tx_1, j(x_n - x_1) \right\rangle.$$

This implies that

$$\|x_n - x_1\| \leq \left(\left(\frac{1}{2} - \frac{1}{n + 1} \right) \|y\| + \left(\frac{n}{n + 1} - \frac{1}{2} \right) \|Tx_1\| \right) \left(1 - \frac{kn}{n + 1} \right)^{-1}.$$

Letting $n \to \infty$, we find that

$$\limsup_{n \to \infty} \|x_n - x_1\| \leq \frac{1}{2} (\|y\| + \|Tx_1\|)(1 - k)^{-1} < \infty.$$

This shows that $\{x_n\}$ is bounded. From (5.94), we see that $\{Tx_n\}$ is also bounded. Hence,

$$x_n - Tx_n = \frac{y - Tx_n}{n + 1} \to \theta$$

as $n \to \infty$. It follows that $x_n - Tx_n - x_m + Tx_m \to \theta$ as $n, m \to \infty$. From the fact that

$$\langle x_n - Tx_n - x_m + Tx_m, j(x_n - x_m) \rangle$$
$$= \|x_n - x_m\|^2 - \langle Tx_n - Tx_m, j(x_n - x_m) \rangle$$
$$\geq (1 - k)\|x_n - x_m\|^2,$$

we have $\|x_n - x_m\| \leq \|x_n - Tx_n - x_m + Tx_m\|$. Letting $n, m \to \infty$, we obtain that $\{x_n\}$ is a Cauchy sequence. Let $\{x_n\}$ converge to x in norm. Since C is closed, we see that $x \in C$. Since T is continuous, we find that $\lim_{n \to \infty} Tx_n = Tx$. This implies $Tx = x$.

Next, we prove that x is unique. Assume that $y \neq x$ is also a fixed pint of T. Then

$$\|x - y\|^2 = \langle x - y, j(x - y) \rangle = \langle Tx - Ty, j(x - y) \rangle \leq k\|x - y\|^2,$$

that is, $(1-k)\|x-y\|^2 \leq 0$. Since $0 < k < 1$, we obtain a contradiction. This proves $x = y$. This shows that T has a unique fixed point in C, completing the proof. □

Based on the results above, we establish the following fixed point theorem for continuous g-strongly pseudocontractive mappings.

Theorem 5.8.2. *Let E be a real Banach space and let C be a nonempty, closed, and convex subset of E. Let $T : C \to E$ be a continuous g-strongly pseudocontractive mapping satisfying the weak inward condition (WIC). If $\liminf_{r \to \infty} g(r) > 0$, then T has a unique fixed point in C.*

Proof. Fixing $x_n \in C$, $\forall n \geq 1$, we define $T_n : C \to E$ by

$$T_n x = \frac{1}{2}x_n + +\frac{1}{2}Tx, \quad \forall x \in C.$$

Then $T_n : C \to E$ is a continuous $\frac{1}{2}$-strongly pseudocontractive mapping satisfying the weak inward condition. Using Theorem 5.8.1, for every $n \geq 1$, we find that T_n has a unique fixed point in C, denoted this unique fixed point by x_{n+1}, that is,

$$x_{n+1} = \frac{1}{2}x_n + \frac{1}{2}Tx_{n+1}, \quad \forall n \geq 1.$$

This implies that $x_{n+1} - x_n = Tx_{n+1} - x_{n+1}$. It follows that

$$g(\|x_{n+1} - x_n\|)\|x_{n+1} - x_n\| \leq \langle x_{n+1} - Tx_{n+1} - x_n + Tx_n, j(x_{n+1} - x_n) \rangle$$
$$= \langle x_n - x_{n-1}, j(x_{n+1} - x_n) \rangle - \|x_{n+1} - x_n\|^2$$
$$\leq \frac{1}{2}\|x_n - x_{n-1}\|^2 + \frac{1}{2}\|x_{n+1} - x_n\|^2 - \|x_{n+1} - x_n\|^2$$
$$= \frac{1}{2}\|x_n - x_{n-1}\|^2 - \frac{1}{2}\|x_{n+1} - x_n\|^2. \tag{5.95}$$

This implies that $\|x_{n+1}-x_n\| \leq \|x_n-x_{n-1}\|$. Hence $\lim_{n \to \infty}\|x_{n+1}-x_n\|$ exists. Using (5.95), we arrive at

$$\lim_{n \to \infty} g(\|x_{n+1} - x_n\|)\|x_{n+1} - x_n\| = 0.$$

This implies that $x_{n+1} - x_n \to \theta$ as $n \to \infty$. If not, we assume that $\|x_{n+1} - x_n\| \to d > 0$. From the property of g, we obtain that

$$\liminf_{n\to\infty} g(\|x_{n+1} - x_n\|) > 0,$$

which is a contradiction. This implies that $Tx_n - x_n \to \theta$ as $n \to \infty$. It follows that $x_n - Tx_n - x_m - Tx_m \to \theta$ as $n, m \to \infty$. Note that

$$g(\|x_n - x_m\|)\|x_n - x_m\| \le \langle x_n - Tx_n - x_m + Tx_m, j(x_n - x_m)\rangle$$
$$\le \|x_n - Tx_n - x_m + Tx_m\|\|x_n - x_m\|.$$

This implies

$$g(\|x_n - x_m\|) \to 0 \quad \text{as } n, m \to \infty.$$

Since $\liminf_{r\to r_0} g(r) > 0$, $\forall r_0 > 0$ and $\liminf_{r\to\infty} g(r) > 0$, we have $\|x_n - x_m\| \to 0$ as $n, m \to \infty$. This shows that $\{x_n\}$ is a Cauchy sequence. Assume that $\{x_n\}$ converges to x in norm; then $x \in C$, since C is closed. Because T is continuous, we find that $\lim_{n\to\infty} Tx_n = Tx$, which implies $Tx = x$. Next, we prove that x is unique. Assume that $y \ne x$ is also a fixed pint of T. Then

$$\|x - y\|^2 = \langle x - y, j(x - y)\rangle = \langle Tx - Ty, j(x - y)\rangle \le \|x - y\|^2 - g(\|x - y\|)\|x - y\|,$$

that is, $g(\|x - y\|)\|x - y\| = 0$. From the property of g, we find that $x = y$. This shows that T has a unique fixed point in C, completing the proof. $\square$

5.9 Demiclosedness principles for pseudocontractive mappings

Theorem 5.9.1 (Demiclosedness principle I). *Let E be a real uniformly convex Banach space and let C be a nonempty, closed, and convex subset of E. Let $T : C \to E$ be a continuous pseudocontractive mapping satisfying the weak inward condition (WIC). Then $I - T$ is demiclosed at the origin and* Fix(T) *is closed and convex.*

Proof. Let $\{x_n\}$ be a sequence in C such that $x_n \rightharpoonup x$ and $x_n - Tx_n \to \theta$. Using Theorem 6 of Martin [47], we find that

$$C \subset (2I - T)(C). \tag{5.96}$$

Since $T : C \to E$ is a continuous pseudocontractive mapping, we have

$$\langle (2I - T)x - (2I - T)y, j(x - y)\rangle = 2\|x - y\|^2 - \langle Tx - Ty, j(x - y)\rangle$$
$$\ge \|x - y\|^2, \quad \forall x, y \in C,$$

which yields

$$\|(2I - T)x - (2I - T)y\| \geq \|x - y\|, \quad \forall x, y \in C,$$

implying that the mapping $(2I - T) : C \to E$ is injective, and consequently $(2I - T)^{-1}$ exists on $\mathrm{Ran}(2I - T)$, in particular, $(2I - T)^{-1}$ exists on C, since $C \subset (2I - T)$. Putting $U := (2I - T)^{-1}$, we see that $U : C \to C$ is nonexpansive such that $\mathrm{Fix}(U) = \mathrm{Fix}(T)$. Observe that

$$\|x_n - Ux_n\| = \|UU^{-1}x_n - Ux_n\| \leq \|U^{-1}x_n - x_n\| = \|Tx_n - x_n\| \to 0, \quad n \to \infty.$$

From the demiclosedness principle for nonexpansive mappings, we find that $x = Ux$, and then, $x = Tx$. In view of the facts that $\mathrm{Fix}(U) = \mathrm{Fix}(T)$ and $\mathrm{Fix}(U)$ is closed and convex, we find that $\mathrm{Fix}(T)$ is a closed and convex subset of E. This completes the proof. $\square$

Theorem 5.9.2 (Demiclosedness principle II). *Let E be a reflexive and strictly convex Banach space that satisfies Opial condition, and let C be a nonempty, closed, and convex subset of E. Let $T : C \to E$ be a continuous pseudocontractive mapping satisfying the weak inward condition (WIC). Then $I - T$ is demiclosed at the origin and $\mathrm{Fix}(T)$ is closed and convex.*

Proof. Letting $\{x_n\}$ be a sequence in C such that $x_n \rightharpoonup x$ and $x_n - Tx_n \to \theta$, we have $x \in C$.

Next, we prove $x = Tx$. Let U be defined as in Theorem 5.9.1. Then $x_n - Ux_n \to \theta$ as $n \to \infty$. Assume that $x \neq Ux$. Using Opial condition, we arrive at

$$\limsup_{n \to \infty} \|x_n - x\| < \limsup_{n \to \infty} \|x_n - Ux\|$$

$$= \limsup_{n \to \infty} \|Ux_n - Ux\|$$

$$\leq \limsup_{n \to \infty} \|x_n - x\|.$$

This shows, in fact, that $x = Ux$, yielding $x = Tx$. In view of the facts that $\mathrm{Fix}(U) = \mathrm{Fix}(T)$ and $\mathrm{Fix}(U)$ is closed and convex, we find that $\mathrm{Fix}(T)$ is a closed and convex subset of E. This completes the proof. $\square$

5.10 Fixed point theorems for pseudocontractive mappings

Theorem 5.10.1. *Let E be a real uniformly convex Banach space, and let C be a nonempty, closed, and convex subset of E. Let $T : C \to E$ be a continuous pseudocontractive mapping satisfying the weak inward condition (WIC). If $T - I$ is unbounded on any unbounded subset of C, then T has a fixed point in C.*

Proof. Fixing $u \in C$, we define $T_n : C \to E$ by

$$T_n x = \frac{1}{n+1} u + \frac{n}{n+1} T x, \quad \forall x \in C.$$

It follows that

$$\langle T_n x - T_n y, j(x - y) \rangle \leq \frac{n}{n+1} \|x - y\|^2, \quad \forall x, y \in C.$$

This shows that $T_n : C \to E$ is $\frac{n}{n+1}$-strongly pseudocontractive. Note that T_n is also continuous and satisfies the weak inward condition. Using Theorem 5.8.1, we find that T_n has a unique fixed point in C. We denote the fixed point by x_n, that is,

$$x_n = \frac{1}{n+1} u + \frac{n}{n+1} T x_n, \quad \forall n \geq 1. \tag{5.97}$$

Next, we show that $\{x_n\}$ is bounded. If $\{x_n\}$ is unbounded, we may assume that $\theta \in C$. Letting $u = 0$ in (5.97), we find that $x_n = \frac{n}{n+1} T x_n, \forall n \geq 1$. This implies that

$$\begin{aligned}
\|x_n\|^2 &= \frac{n}{n+1} \langle T x_n, j(x_n) \rangle \\
&= \frac{n}{n+1} \langle T x_n - T\theta, j(x_n) \rangle + \frac{n}{n+1} \langle T\theta, j(x_n) \rangle \\
&\leq \frac{n}{n+1} \|x_n\|^2 + \frac{n}{n+1} \|T\theta\| \|x_n\|,
\end{aligned}$$

that is, $\frac{\|x_n\|}{n} \leq \|T\theta\|$. Hence, we arrive at

$$\|x_n - T x_n\| = \frac{\|T x_n\|}{n+1} = \frac{\|x_n\|}{n}, \quad \forall n \geq 1.$$

This is a contradiction, showing that $\{x_n\}$ is bounded, so is $\{T x_n\}$. Hence, $x_n - T x_n = -\frac{1}{n+1} T x_n \to \theta$ as $n \to \infty$. Without loss of generality, we may assume $x_n \rightharpoonup x$ as $n \to \infty$. Using Demiclosedness principle I, we obtain that $x = T x$, completing the proof. □

From the proof of the above theorem, we see that the following conclusion holds.

Theorem 5.10.2. *Let E be a real uniformly convex Banach space, and let C be a nonempty, bounded, closed, and convex subset of E. Let $T : C \to E$ be a continuous pseudocontractive mapping satisfying the weak inward condition (WIC). Then T has at least one fixed point in C.*

By means of Demiclosedness principle II for pseudocontractive mappings, we easily deduce the following existence result.

Theorem 5.10.3. *Let E be a real reflexive Banach space that satisfies Opial condition, and let C be a nonempty, bounded, closed, and convex subset of E. Let $T : C \to E$ be a continuous pseudocontractive mapping satisfying the weak inward condition (WIC). Then T has at least one fixed point in C.*

Theorem 5.10.4. *Let E be a Banach space. Let C be a nonempty, closed, and convex subset of E that has the fixed point property for nonexpansive self-mapping. Let T : C → E be a continuous pseudocontractive mapping satisfying the weak inward condition (WIC). Then T has a fixed point in C.*

Proof. Letting $U = (2I - T)^{-1}$ be defined in Theorem 5.9.1, we see that $U : C \to C$ is nonexpansive. By our assumption on C, we see that there exists $x \in C$ such that $x = Ux$. Hence, $x = Tx$. This completes the proof. □

5.11 Iterative methods for fixed points of pseudocontractive mappings

Theorem 5.11.1. *Let E be a real Banach space and let C be a nonempty, closed, and convex subset of E. Let T : C → C be a L-Lipschitz continuous k-strongly pseudocontractive mapping. Let $\{x_n\}$ be a sequence generated in the following normal Mann iterative method*

$$x_1 \in C, \quad x_{n+1} = (1 - \alpha_n)x_n + \alpha_n Tx_n, \quad n \geq 1,$$

where $\{\alpha_n\}$ is a sequence in $(0, 1)$ such that (C1) $\lim_{n \to \infty} \alpha_n = 0$ and (C2) $\sum_{n=1}^{\infty} \alpha_n = \infty$. Then $\{x_n\}$ converges to the unique fixed point of T in norm.

Proof. From Theorem 5.9.1, we see that T has a unique fixed point in C. Denote the unique fixed point by p. It follows that

$$
\begin{aligned}
\|x_{n+1} - p\|^2 &= (1 - \alpha_n)\langle x_n - p, j(x_{n+1} - p)\rangle + \alpha_n\langle Tx_n - p, j(x_{n+1} - p)\rangle \\
&= (1 - \alpha_n)\|x_n - p\|\|x_{n+1} - p\| + \alpha_n\|Tx_n - Tx_{n+1}\|\|x_{n+1} - p\| \\
&\quad + \alpha_n\langle Tx_{n+1} - p, j(x_{n+1} - p)\rangle \\
&\leq \frac{1 - \alpha_n}{2}\|x_n - p\|^2 + \frac{1 - \alpha_n}{2}\|x_{n+1} - p\|^2 + \alpha_n L\|x_n - x_{n+1}\|\|x_{n+1} - p\| \\
&\quad + \alpha_n k\|x_{n+1} - p\|^2 \\
&\leq \frac{1 - \alpha_n}{2}\|x_n - p\|^2 + \left(\frac{1 - \alpha_n}{2} + \alpha_n k\right)\|x_{n+1} - p\|^2 \\
&\quad + \alpha_n^2 L\|x_n - Tx_n\|\|x_{n+1} - p\| \\
&\leq \frac{1 - \alpha_n}{2}\|x_n - p\|^2 + \left(\frac{1 - \alpha_n}{2} + \alpha_n k\right)\|x_{n+1} - p\|^2 \\
&\quad + \alpha_n^2 L(1 + L)\|x_n - p\|\|x_{n+1} - p\| \\
&\leq \frac{1 - \alpha_n + \alpha_n^2 L(1 + L)}{2}\|x_n - p\|^2 \\
&\quad + \left(\frac{1 - \alpha_n + \alpha_n^2 L(1 + L)}{2} + \alpha_n k\right)\|x_{n+1} - p\|^2.
\end{aligned}
$$

This in turn implies that

$$\|x_{n+1} - p\|^2 \leq \frac{1 - \alpha_n + \alpha_n^2 L(1 + L)}{1 + \alpha_n(1 - 2k) - \alpha_n^2 L(1 + L)} \|x_n - p\|^2$$

$$\leq \left(1 - 2\alpha_n \frac{1 - k - \alpha_n L(1 + L)}{1 + \alpha_n(1 - 2k) - \alpha_n^2 L(1 + L)}\right) \|x_n - p\|^2. \tag{5.98}$$

From (C1), we see that there exists $n_0 \geq 1$ such that

$$\frac{1 - k - \alpha_n L(1 + L)}{1 + \alpha_n(1 - 2k) - \alpha_n^2 L(1 + L)} \geq \frac{1 - k}{2}, \quad \forall n \geq n_0. \tag{5.99}$$

Combining (5.98) with (5.99), we obtain

$$\|x_{n+1} - p\|^2 \leq (1 - (1 - k)\alpha_n)\|x_n - p\|^2, \quad \forall n \geq n_0.$$

Using Lemma 1.10.2, we obtain $x_n \to p$ as $n \to \infty$. This completes the proof. $\square$

Theorem 5.11.2. *Let E be a real Banach space and let C be a nonempty, closed, and convex subset of E. Let $T : C \to E$ be a continuous pseudocontractive mapping and let $h : C \to E$ be a g-strongly pseudocontractive mapping satisfying the weak inward condition (WIC). Assume that $\liminf_{r\to\infty} g(r) > 0$. Define a mapping $T_t^h : C \to E$ by*

$$T_t^h = th(x) + (1 - t)Tx, \quad \forall x \in C, \, t \in (0, 1].$$

Then
(i) T_t^h has a unique fixed point $y_t \in C$ and the path $t \mapsto y_t$ is continuous;
(ii) if both $\{y_t\}$ and $\{Ty_t\}$ are bounded, then $y_t - Ty_t \to \theta$ as $t \to 0^+$;
(iii) if $\mathrm{Fix}(T) \neq \emptyset$ and $\liminf_{r\to\infty} g(r) > \|p - h(p)\|$ for some $p \in \mathrm{Fix}(T)$, then $\{y_t\}$ is bounded and

$$\langle y_t - h(y_t), j(y_t - y)\rangle \leq 0, \quad \forall y \in \mathrm{Fix}(T).$$

Proof. Using [52, Corollary 2], we find that $T_t^h : C \to E$ is continuous pseudocontractive which satisfies the weak inward condition. Indeed,

$$\langle T_t^h x - T_t^h y, j(x - y)\rangle = t\langle h(x) - h(y), j(x - y)\rangle + (1 - t)\langle Tx - Ty, j(x - y)\rangle$$

$$\leq t\|x - y\|^2 - tg(\|x - y\|)\|x - y\| + (1 - t)\|x - y\|^2$$

$$= \|x - y\|^2 - tg(\|x - y\|)\|x - y\|.$$

This shows that $T_t^h : C \to E$ is a continuous tg-strongly pseudocontractive mapping which also satisfies the weak inward condition. Using Theorem 5.8.2, we find that $T_t^h : C \to E$ has a unique fixed point in C. Denote the fixed point by y_t. This means that there exists a unique path $t \mapsto y_t$ satisfying

$$y_t = th(y_t) + (1 - t)Ty_t, \quad \forall t \in (0, 1]. \tag{5.100}$$

Fix $t_0 \in (0,1]$. For $\forall t \in (0,1]$, we find that there exists $j(y_t - y_{t_0}) \in J(y_t - y_{t_0})$ such that

$$\|y_t - y_{t_0}\|^2 = t\langle h(y_t) - h(y_{t_0}), j(y_t - y_{t_0})\rangle + (1 - t)\langle Ty_t - Ty_{t_0}, j(y_t - y_{t_0})\rangle$$
$$+ (t - t_0)\langle h(y_0) - Ty_{t_0}, j(y_t - y_{t_0})\rangle$$
$$\le \|y_t - y_{t_0}\|^2 - tg(\|y_t - y_{t_0}\|)\|y_t - y_{t_0}\| + |t - t_0|\|h(y_0) - Ty_{t_0}\|\|y_t - y_{t_0}\|.$$

Hence,

$$g(\|y_t - y_{t_0}\|)\|y_t - y_{t_0}\| \le \frac{|t - t_0|}{t}\|h(y_0) - Ty_{t_0}\|\|y_t - y_{t_0}\|.$$

This implies that $y_t \to y_{t_0}$ as $t \to t_0$.

(i) Using (5.100), we have $y_t - Ty_t = t(h(y_t) - Ty_t)$. Since both $\{y_t\}$ and $\{h(y_t)\}$ are bounded, we find that $\{Ty_t\}$ is also bounded. It follows that $y_t - Ty_t \to \theta$ as $t \to 0^+$.

(ii) For some $p \in \mathrm{Fix}(T)$, we find from (5.100) that

$$\|y_t - p\|^2 = \langle y_t - p, j(y_t - p)\rangle$$
$$= t\langle h(y_t) - h(p), j(y_t - p)\rangle + t\langle h(p) - p, j(y_t - p)\rangle$$
$$+ (1 - t)\langle Ty_t - p, j(y_t - p)\rangle$$
$$\le t\|y_t - p\|^2 - tg(\|y_t - p\|)\|y_t - p\| + t\langle h(p) - p, j(y_t - p)\rangle$$
$$+ (1 - t)\|y_t - p\|^2$$
$$= \|y_t - p\|^2 - tg(\|y_t - p\|)\|y_t - p\| + t\langle h(p) - p, j(y_t - p)\rangle,$$

which yields

$$g(\|y_t - p\|)\|y_t - p\| \le \langle h(p) - p, j(y_t - p)\rangle, \quad \forall t \in (0,1]. \tag{5.101}$$

This shows that $\{y_t\}$ is bounded. Using (5.100), we find that

$$\langle y_t - h(y_t), j(y_t - y)\rangle$$
$$= (1 - t)\langle Ty_t - h(y_t), j(y_t - y)\rangle$$
$$= (1 - t)\langle Ty_t - y_t, j(y_t - y)\rangle + (1 - t)\langle y_t - h(y_t), j(y_t - y)\rangle$$
$$= (1 - t)\langle Ty_t - y, j(y_t - y)\rangle + (1 - t)\langle y - y_t, j(y_t - y)\rangle$$
$$+ (1 - t)\langle y_t - h(y_t), j(y_t - y)\rangle$$
$$\le (1 - t)\langle y_t - h(y_t), j(y_t - y)\rangle, \quad \forall y \in \mathrm{Fix}(T).$$

This implies that $\langle y_t - h(y_t), j(y_t - y)\rangle \le 0$, $\forall y \in \mathrm{Fix}(T)$, completing the proof. □

Theorem 5.11.3. *Let E be a reflexive Banach space with a uniformly Gâteaux differentiable norm. Let C be a nonempty, closed, and convex subset of E. Suppose that every nonempty, bounded, closed, and convex subset of C has the fixed point property for non-expansive self-mappings. Let $T : C \to E$ be a continuous pseudocontractive mapping* and let $h : C \to E$ be a continuous g-strongly pseudocontractive mapping satisfying the weak inward condition (WIC). Assume that $\lim\inf_{r\to\infty} g(r) > 0$. Let $\{y_t\}$ be a sequence defined in Theorem 5.11.2. If both $\{y_t\}$ and $\{h(y_t)\}$ are bounded, then $\{y_t\}$ converges to some fixed point p of T as $t \to 0^+$ and p is the unique solution of the following variational inequality:

$$\langle p - h(p), j(y - p) \rangle \geq 0, \quad \forall y \in \text{Fix}(T).$$

Proof. From (i) and (ii) of Theorem 5.11.2, there exists a unique path $\{y_t\}$, which satisfies (5.100) and $y_t - Ty_t \to \theta$ as $t \to 0^+$. From [47, Theorem 6], we see that $C \subset (2I - T)(C)$. Letting $U = (2I - T)^{-1}$, we have that $U : C \to C$ is nonexpansive, $\text{Fix}(U) = F(T)$, and $y_t - Uy_t \to \theta$ as $t \to 0^+$. Define a functional φ by $\varphi(x) = \mu_t \|y_t - x\|^2$, $\forall x \in E$, where μ_t is a Banach limit. It is clear that φ is a proper, continuous, and convex function such that $\varphi(x) \to \infty$ as $\|x\| \to \infty$. Therefore, there exists at least a point $x \in C$ such that $\varphi(x) = \min_{y\in C} \varphi(y)$. Let

$$C_1 := \left\{ x \in C : \varphi(x) = \inf_{y\in C} \varphi(y) \right\}.$$

Then C_1 is a nonempty, bounded, closed, and convex subset of C. Observe that

$$\begin{aligned}
\varphi(Ux) &= \mu_t \|y_t - Ux\|^2 \\
&= \mu_t \|Uy_t - Ux\|^2 \\
&\leq \mu_t \|y_t - x\|^2 \\
&= \varphi(x), \quad \forall x \in C_1.
\end{aligned}$$

This shows that $Ux \in C_1$, that is, $U(C_1) \subset C_1$. By our assumption on C, we see that U has a fixed point in C_1. Let p be a fixed point of U. Hence, $p = Tp$. From (iii) of Theorem 5.11.2, we see that

$$\langle y_t - h(y_t), j(y_t - p) \rangle \leq 0.$$

It follows that

$$\mu_t \langle y_t - h(y_t), j(y_t - p) \rangle \leq 0. \tag{5.102}$$

Using Theorem 5.11.2, we find that

$$\mu_t \langle y - p, j(y_t - p) \rangle \leq 0, \quad \forall y \in C. \tag{5.103}$$

Since $h : C \to E$ satisfies the weak inward condition, we obtain that $h(p) \in \overline{I_C(p)}$. Hence, there exists $z_n \in I_C(p)$ such that $z_n \to h(p)$ as $n \to \infty$. There exists $u_n \in C$ such that

$$z_n = p + \lambda_n(u_n - p), \quad \lambda_n \geq 1, n \geq 1. \tag{5.104}$$

Letting $y = u_n$ in (5.103), we find that

$$\mu_t \langle u_n - p, j(y_t - p) \rangle \leq 0.$$

It follows that

$$\mu_t \langle z_n - p, j(y_t - p) \rangle \leq 0. \tag{5.105}$$

Since $z_n \to h(p)$ as $n \to \infty$, taking the limit in (5.105), we find that

$$\mu_t \langle h(p) - p, j(y_t - p) \rangle \leq 0. \tag{5.106}$$

Adding (5.102) and (5.106), we have

$$\mu_t \|y_t - p\|^2 \leq \mu_t \langle h(y_t) - h(p), j(y_t - p) \rangle$$
$$\leq \mu_t \|y_t - p\|^2 - \mu_t (g(\|y_t - p\|) \|y_t - p\|).$$

It follows that $\mu_t (g(\|y_t - p\|) \|y_t - p\|) = 0$. This implies that there exists a subnet $\{y_{t_\alpha}\}$ of $\{y_t\}$ such that $y_{t_\alpha} \to p$. Using (iii) of Theorem 5.11.2, we find that

$$\langle y_{t_\alpha} - h(y_{t_\alpha}), j(y_{t_\alpha} - y) \rangle \leq 0, \quad \forall y \in \text{Fix}(T).$$

It follows that

$$\langle p - h(p), j(p - y) \rangle \leq 0, \quad \forall y \in \text{Fix}(T). \tag{5.107}$$

Next, we prove that variational inequality (5.107) has a unique solution. We assume that q is another solution of variational inequality (5.107), that is,

$$\langle q - h(q), j(q - y) \rangle \leq 0, \quad \forall y \in \text{Fix}(T). \tag{5.108}$$

Taking $y = q$ in (5.107) and $y = p$ in (5.108), we find that

$$\langle p - h(p), j(p - q) \rangle \leq 0$$

and

$$\langle q - h(q), j(q - p) \rangle \leq 0.$$

Adding the two inequalities, we obtain

$$\|p - q\|^2 \leq \langle h(p) - h(q), j(p - q) \rangle$$
$$\leq \|p - q\|^2 - g(\|p - q\|) \|p - q\|.$$

This implies $p = q$, completing the proof. $\square$

If the space is strictly convex in Theorem 5.11.3, we may remove the restriction that subset C has the fixed point property. If $\text{Fix}(T) \neq \emptyset$, $\liminf_{r \to \infty} g(r) = \infty$, and $h : C \to E$ is bounded continuous g-strongly pseudocontractive, we may remove the restrictions that both $\{y_t\}$ and $\{h(y_t)\}$ are bounded.

Theorem 5.11.4. *Let E be a reflexive and strictly convex Banach space with a uniformly Gâteaux differentiable norm. Let C be a nonempty, closed, and convex subset of E. Let $T : C \to E$ be a continuous pseudocontractive mapping and let $h : C \to E$ be a bounded, continuous, g-strongly pseudocontractive mapping satisfying the weak inward condition (WIC). Assume that $\text{Fix}(T) \neq \emptyset$ and $\liminf_{r \to \infty} g(r) = \infty$. Let $\{y_t\}$ be the path defined in Theorem 5.11.2. Then $\{y_t\}$ converges to some fixed point p of T as $t \to 0^+$ and p is the unique solution of the following variational inequality:*

$$\langle p - h(p), j(y - p) \rangle \leq 0, \quad \forall y \in \text{Fix}(T).$$

Proof. From (i) of Theorem 5.11.2, there exists a unique continuous path $\{y_t\}$, which satisfies (5.100). From (iii) of Theorem 5.11.2, we see that $\{y_t\}$ is bounded. Hence, $y_t - Ty_t \to \theta$ as $t \to 0^+$. This implies $y_t - Uy_t \to \theta$ as $t \to 0^+$, where U is defined in Theorem 5.11.3. Define a functional φ by $\varphi(x) = \mu_t \|y_t - x\|^2$, $\forall x \in E$, where μ_t is a Banach limit. Let

$$C_1 := \left\{ x \in C : \varphi(x) = \inf_{y \in C} \varphi(y) \right\}.$$

It is easy to see that $C_1 \neq \emptyset$ is a bounded, closed, and convex subset of C, and $U(C_1) \subset C_1$. Fix $v \in \text{Fix}(T)$ and define $\psi : C_1 \to \mathbb{R}^+$ by $\psi(u) = \|u - v\|$. Define

$$C_2 := \left\{ u \in C_1 : \psi(u) = \inf_{y \in C_1} \psi(y) \right\}.$$

Then $C_2 = \{u_0\}$, $u_0 \in C_1$. In view of $v = Uv$, we have

$$\|Uu_0 - v\| = \|Uu_0 - Uv\| \leq \|u_0 - v\|.$$

This implies $u_0 = Uu_0$ and then $u_0 = Tu_0$. Following the proof of Theorem 5.11.3, we obtain the desired conclusion immediately. This completes the proof. □

Furthermore, if the duality mapping is weakly sequentially continuous, the restrictions that E is strictly convex and has a uniformly Gâteaux differentiable norm are not needed.

Theorem 5.11.5. *Let E be a reflexive Banach space and let C be a nonempty, closed, and convex subset of E. Let $T : C \to E$ be a continuous pseudocontractive mapping and let $h : C \to E$ be a continuous g-strongly pseudocontractive mapping satisfying the weak inward condition (WIC). Assume that $\liminf_{r \to \infty} g(r) = \infty$. Let $\{y_t\}$ be the path defined in Theorem 5.11.2. If the duality mapping J is weakly sequentially continuous, $\{y_t\}$ and*

$\{h(y_t)\}$ are both bounded, then $\{y_t\}$ converges to some fixed point p of T as $t \to 0^+$ and p is the unique solution of the following variational inequality:

$$\langle p - h(p), j(y - p) \rangle \geq 0, \quad \forall y \in \text{Fix}(T).$$

Proof. Since the duality mapping $J : E \to E^*$ is weakly sequentially continuous, we see that J is single valued. Next, we use j to denote the single-valued duality mapping. From (i) of Theorem 5.11.2, we see that there exists a unique continuous path $\{y_t\}$, which satisfies (5.100). Note that both $\{y_t\}$ and $\{h(y_t)\}$ are bounded. From (ii) of Theorem 5.11.2, we see that $y_t - Ty_t \to \theta$ as $t \to 0^+$. Choose $t_n \to 0^+$ and denote y_n by $y_n = y_{t_n}$. It follows that $y_n - Ty_n \to \theta$ as $n \to \infty$. Since E is reflexive and $\{y_n\}$ is bounded, we see that there exists a subsequence of $\{y_n\}$ converging to some point p in C. Without loss of generality, we may assume that $y_n \rightharpoonup p$ as $n \to \infty$. Using Demiclosedness principle II, we find that $p = Tp$, that is, $\text{Fix}(T) \neq \emptyset$. Using (5.102), we find that

$$g(\|y_n - p\|)\|y_n - p\| \leq \langle h(p) - p, j(y_n - p) \rangle, \quad \forall n \geq 1.$$

Letting $n \to \infty$, we find that $y_n \to p$ as $n \to \infty$. From Theorem 5.11.2, we see that

$$\langle y_n - h(y_n), j(y_n - y) \rangle \leq 0, \quad \forall y \in \text{Fix}(T).$$

Letting $n \to \infty$, we find that

$$\langle p - h(p), j(y - p) \rangle \geq 0, \quad \forall y \in \text{Fix}(T).$$

Since the above variational inequality has a unique solution, we find that $\{y_t\}$ converges to some fixed point p of T in norm and p is the unique solution of the variational inequality. This completes the proof. $\square$

Based on the convergence of the path given in Theorem 5.11.2, we are now in a position to establish a convergence theorem of fixed points via a Bruck-like regularization iterative method. To this end, we consider the following control sequences. Let $\{\theta_n\}$ and $\{\lambda_n\}$ be two sequences in $[0, 1]$ satisfying the following conditions:

(i) $\lambda_n(1 + \theta_n) \leq 1, \forall n \geq 1$,

(ii) $\theta_n \to 0$ as $n \to \infty$,

(iii) $\sum_{n=1}^{\infty} \lambda_n \theta_n = \infty$,

(iv) $\frac{\theta_{n-1}}{\theta_n} - 1 = o(\lambda_n \theta_n)$,

(v) $\frac{\lambda_n}{\theta_n} \leq \frac{1-k}{8L(1+L)(1+2L)}, \forall n \geq 1$, where $k \in (0, 1)$ and $L \geq 1$ are two known constants.

It is easy to see that

$$\lambda_n = \frac{1}{(n+1)^a}, \quad \theta_n = \frac{1}{(n+1)^b}, \quad 0 < b < a, a + b < 1$$

satisfy the above conditions.

Assume that all the conditions in Theorem 5.11.3 are satisfied. Let $t_n = \frac{\theta_n}{1+\theta_n}$ and denote y_n by $y_n = y_{t_n}$. Then

$$y_n = t_n h(y_n) + (1 - t_n)Ty_n, \quad n \geq 1, \tag{5.109}$$

$$(1 + \theta_n)y_n = \theta_n h(y_n) + Ty_n, \quad n \geq 1, \tag{5.110}$$

$$\lambda_n(1 + \theta_n)y_n = \lambda_n\theta_n h(y_n) + \lambda_n Ty_n, \quad n \geq 1, \tag{5.111}$$

and

$$y_n = (1 - \lambda_n(1 + \theta_n))y_n + \lambda_n\theta_n h(y_n) + \lambda_n Ty_n, \quad n \geq 1. \tag{5.112}$$

These motivate us to introduce the following iterative method:

$$x_1 \in C, \quad x_{n+1} = (1 - \lambda_n(1 + \theta_n))x_n + \lambda_n\theta_n h(x_n) + \lambda_n Tx_n, \quad n \geq 1. \tag{VBRIM}$$

We call (VBRIM) a viscosity Bruck regularization iterative method.

Theorem 5.11.6. *Let E be a real reflexive Banach space with a uniformly Gâteaux differentiable norm. Let C be a nonempty, closed, and convex subset of E with the fixed point property for nonexpansive mappings. Let $T : C \to E$ be an L-Lipschitz continuous pseudocontractive mapping with a nonempty fixed point set, and let $h : C \to E$ be an L-Lipschitz continuous k-strongly pseudocontractive mapping. Let $\{x_n\}$ be generated by (VBRIM). Then $\{x_n\}$ converges to some fixed point p of T in norm and p is the unique solution of the following variational inequality:*

$$\langle p - h(p), j(y - p)\rangle \geq 0, \quad \forall y \in \text{Fix}(T). \tag{5.113}$$

Proof. Let $\{y_n\}$ be defined in (5.109). From Theorem 5.11.3, we see that $\{y_n\}$ converges to some fixed point p of T in norm and p is the unique solution to (5.113). Now it is sufficient to show that $x_n - y_n \to \theta$ as $n \to \infty$. Using (5.110), we arrive at

$$\|y_n - y_{n-1}\|^2 = \langle Ty_n - Ty_{n-1} + \theta_n(h(y_n) - y_n) - \theta_n(h(y_{n-1}) - y_{n-1}), j(y_n - y_{n-1})\rangle$$
$$+ (\theta_n - \theta_{n-1})\langle h(y_{n-1}) - y_{n-1}, j(y_n - y_{n-1})\rangle$$
$$\leq \|y_n - y_{n-1}\|^2 + k\theta_n\|y_n - y_{n-1}\|^2$$
$$+ (\theta_n - \theta_{n-1})\langle h(y_{n-1}) - y_{n-1}, j(y_n - y_{n-1})\rangle - \theta_n\|y_n - y_{n-1}\|^2.$$

This implies that

$$\|y_n - y_{n-1}\|^2 \leq \frac{\theta_n - \theta_{n-1}}{(1 - k)\theta_n}\langle h(y_{n-1}) - y_{n-1}, j(y_n - y_{n-1})\rangle.$$

Hence, we have

$$\|y_n - y_{n-1}\| \leq \frac{M}{(1 - k)}\left|1 - \frac{\theta_{n-1}}{\theta_n}\right|, \tag{5.114}$$

where $M = \sup\{\|h(y_{n-1}) - y_{n-1}\| : n \geq 1\}$. Using (VBRIM) and (5.112), we arrive at

$$\begin{aligned}
\|x_n - x_{n+1}\|^2 &\leq \lambda_n \|Tx_n - (1 + \theta_n)x_n + \theta_n h(x_n)\| \\
&= \lambda_n \|Tx_n - Ty_n + (1 + \theta_n)(y_n - x_n) + \theta_n(h(x_n) - h(y_n))\| \\
&\leq \lambda_n(L\|x_n - y_n\| + (1 + \theta_n)\|y_n - x_n\| + L\theta_n\|x_n - y_n\|) \\
&\leq 2(1 + L)\lambda_n\|x_n - y_n\|.
\end{aligned} \tag{5.115}$$

Again, using (VBRIM) and (5.112), we arrive at

$$\begin{aligned}
\|x_{n+1} - y_n\| &= \|(1 - \lambda_n(1 + \theta_n))(x_n - y_n) + \lambda_n(Tx_n - Ty_n) + \lambda_n\theta_n(h(x_n) - h(y_n))\| \\
&\leq (1 - \lambda_n(1 + \theta_n))\|x_n - y_n\| + L\lambda_n\|x_n - y_n\| + L\lambda_n\theta_n\|x_n - y_n\| \\
&\leq ((1 - \lambda_n(1 + \theta_n)) + 2L\lambda_n)\|x_n - y_n\|.
\end{aligned} \tag{5.116}$$

Using (5.115) and (5.116), we obtain that

$$\begin{aligned}
\|x_{n+1} - y_n\|^2 &= \|(1 - \lambda_n(1 + \theta_n))(x_n - y_n) + \lambda_n(Tx_n - Ty_n) + \lambda_n\theta_n(h(x_n) - h(y_n))\|^2 \\
&\leq (1 - \lambda_n(1 + \theta_n))^2\|x_n - y_n\|^2 + 2\lambda_n\langle Tx_n - Ty_n, j(x_{n+1} - y_n)\rangle \\
&\quad + 2\lambda_n\theta_n\langle h(x_n) - h(y_n), j(x_{n+1} - y_n)\rangle \\
&\leq (1 - \lambda_n(1 + \theta_n))^2\|x_n - y_n\|^2 + 2L\lambda_n\|x_{n+1} - x_n\|\|x_{n+1} - y_n\| \\
&\quad + 2\lambda_n\|x_{n+1} - y_n\|^2 + 2L\lambda_n\theta_n\|x_{n+1} - x_n\|\|x_{n+1} - y_n\| \\
&\quad + 2k\lambda_n\theta_n\|x_{n+1} - y_n\|^2 \\
&\leq (1 - \lambda_n(1 + \theta_n))^2\|x_n - y_n\|^2 + 4L(1 + L)(1 + 2L)\lambda_n^2\|x_n - y_n\|^2 \\
&\quad + 2\lambda_n(1 + k\theta_n)\|x_{n+1} - y_n\|^2.
\end{aligned}$$

Using condition (v), we obtain that

$$\begin{aligned}
\|x_{n+1} - y_n\|^2 &\leq \frac{1 - 2\lambda_n(1 + \theta_n) + \lambda_n^2(1 + \theta_n)^2}{1 - 2\lambda_n(1 + k\theta_n)}\|x_n - y_n\|^2 \\
&\quad + \frac{4L(1 + L)(1 + 2L)\lambda_n^2}{1 - 2\lambda_n(1 + k\theta_n)}\|x_n - y_n\|^2 \\
&\leq \left(1 - 2\lambda_n\theta_n\frac{1 - k - \frac{\lambda_n}{2\theta_n}(1 + \theta_n)^2}{1 - 2\lambda_n(1 + k\theta_n)}\right)\|x_n - y_n\|^2 \\
&\quad + 2\lambda_n\theta_n\frac{2L(1 + L)(1 + 2L)\frac{\lambda_n}{\theta_n}}{1 - 2\lambda_n(1 + k\theta_n)}\|x_n - y_n\|^2 \\
&\leq \left(1 - 2\lambda_n\theta_n\frac{1 - k - \frac{1-k}{4}}{1 - 2\lambda_n(1 + k\theta_n)}\right)\|x_n - y_n\|^2 \\
&\quad + 2\lambda_n\theta_n\frac{\frac{1-k}{4}}{1 - 2\lambda_n(1 + k\theta_n)}\|x_n - y_n\|^2 \\
&\leq (1 - (1 - k)\lambda_n\theta_n)\|x_n - y_n\|^2.
\end{aligned}$$

This implies that

$$\|x_{n+1} - y_n\| \leq \left(1 - \frac{1-k}{2}\lambda_n\theta_n\right)\|x_n - y_n\|$$

$$\leq \left(1 - \frac{1-k}{2}\lambda_n\theta_n\right)\|x_n - y_{n-1}\| + \|y_n - y_{n-1}\|. \qquad (5.117)$$

Substituting (5.114) into (5.117), we find that

$$\|x_{n+1} - y_n\| \leq \left(1 - \frac{1-k}{2}\lambda_n\theta_n\right)\|x_n - y_{n-1}\| + \frac{M}{(1-k)}\left|1 - \frac{\theta_{n-1}}{\theta_n}\right|$$

$$\leq \left(1 - \frac{1-k}{2}\lambda_n\theta_n\right)\|x_n - y_{n-1}\| + o(\lambda_n\theta_n).$$

Using Lemma 6.1.4, we find that $x_{n+1} - y_n \to \theta$ as $n \to \infty$. Hence, $x_n \to p$ as $n \to \infty$ and p is the unique solution to (5.113). This completes the proof. □

5.12 Remarks

The requirement that g is strongly accretive in Theorem 5.3.1 is weaker than the requirement that g is α-strongly accretive; see Deimling [27]. From the strong accretiveness of mapping g, one sees that Theorem 5.3.1 is an improvement of Deimling [27]. To the best of our knowledge, Theorem 5.4.1 is a new result. Theorems 5.5.1 and 5.5.2 are extensions of Morales results [54]. Theorem 5.6.3 is due to Zhou [128], Theorem 5.6.4 is due to Zhou and Shi [115], and Theorem 5.6.5 is due to Zhou and Shi [122]. Theorems 5.6.7, 5.6.8, 5.6.9, and 5.6.10 are due to Xu [101]. Most of the results in Section 5.7 are due to the authors of the book. Theorem 5.8.2 is an extension of Morales and Chidume [53]. Theorems 5.9.1 and 5.9.2 are due to Zhou [114]. We gave a new proof of Theorem 5.10.1. Theorem 5.10.4 was given by Ray [70], but our proof is more concise. Theorem 5.11.1 is due to Liu [44]. Theorems 5.11.2, 5.11.3, 5.11.4, 5.11.5, and 5.11.6 are new and due to the authors of the book.

5.13 Exercises

1. Let X be a real Banach space and $T : \mathrm{Dom}(T) \to X$ be a mapping. Write $A := I - T$, where I is the identity mapping on X. Show that T is pseudocontractive $\Longleftrightarrow A$ is accretive. In particular, if T is a nonexpansive mapping, then A is accretive.
2. Let C be a nonempty, convex, and closed subset of a q-uniformly smooth Banach space X. Suppose that the generalized duality mapping $J_q : X \to X^*$ is weakly sequentially continuous at zero. Let $T : C \to X$ be k-strictly pseudocontractive with $k \in (0, 1)$. Show that for any sequence $\{x_n\} \subset C$, if $x_n \rightharpoonup x$, and $x_n - Tx_n \to y \in X$, then $x - Tx = y$, in particular, if $y = \theta$, then $x = Tx$.

3. Let C be a nonempty, convex, and closed subset of a reflexive Banach space X. Suppose that X satisfies Opial condition. Let $T : C \to X$ be a continuous pseudo-contractive mapping that satisfies the weak inward condition. Show that $I - T$ is demiclosed at zero, i. e., $\forall \{x_n\} \subset C$ with $x_n \rightharpoonup x$ and $x_n - Tx_n \to \theta \Longrightarrow x = Tx$.

4. Let C be a nonempty, convex, and closed subset of a uniformly convex Banach space X, and let $T : C \to X$ be a continuous pseudocontractive mapping that satisfies the weak inward condition. Show that $I - T$ is demiclosed at zero.

5. Let C be a nonempty, convex, and closed subset of a Banach space X. Let $Q_C : X \to C$ be a sunny nonexpansive retraction from X onto C. Let $A : C \to X$ be an η-strongly accretive and Lipschitz continuous mapping. Show that the accretive-type variational inequality,

$$\langle Ax^*, j(x - x^*) \rangle \geq 0, \quad \forall x \in C, \qquad \text{(ATVIP)}$$

has a unique solution $x^* \in C$. We use the symbol $\text{SOL}(C, A)$ to denote the set of solutions for A.

6. Let X be a 2-uniformly smooth Banach space with the uniform smoothness constant K and let C be a nonempty, closed, and convex subset of X. Let Q_C be a sunny nonexpansive retraction from X onto C and let A be an η-inverse strongly accretive mapping of C into X such that $\text{SOL}(C, A) \neq \emptyset$. Let $\{\alpha_n\}$, $\{\beta_n\}$, $\{\gamma_n\}$, and $\{\lambda_n\}$ be four sequences in $(0,1)$ satisfying the following conditions:
   (i) $\alpha_n + \beta_n + \gamma_n = 1$, $n \geq 1$;
   (ii) $\lim_{n \to \infty} \alpha_n = 0$ and $\sum_{n=1}^{\infty} \alpha_n = \infty$;
   (iii) $0 < \liminf_{n \to \infty} \beta_n \leq \limsup_{n \to \infty} \beta_n < 1$;
   (iv) $a \leq \lambda_n < b$ for some $a, b \in (0, \frac{\eta}{K^2}]$;
   (v) $\lim_{n \to \infty} (\lambda_{n+1} - \lambda_n) = 0$.
   For a fixed anchor $u \in C$ and arbitrarily initial value $x_1 \in C$, define a sequence $\{x_n\}$ iteratively by

$$\begin{cases} y_n = Q_C(x_n - \lambda_n A x_n), \\ x_{n+1} = \alpha_n u + \beta_n x_n + \gamma_n y_n, \quad n \geq 1. \end{cases} \qquad \text{(HTIM)}$$

   Show that the sequence $\{x_n\}$ defined by (HTIM) converges strongly to the specific solution x^* of the accretive type variational inequality (ATVIP):

$$\langle Ax^*, j(x - x^*) \rangle \geq 0, \quad \forall x \in C.$$

7. Let X be a uniformly convex and 2-uniformly smooth Banach space with the uniform smoothness constant K, and let C be a nonempty, closed, and convex subset of X. Let Q_C be a sunny nonexpansive retraction from X onto C. Let $A : C \to X$ be an α-inverse strongly accretive mapping such that $\text{SOL}(C, A) \neq \emptyset$. Assume $\{\lambda_n\}$ and $\{\alpha_n\}$ are chosen so that $\lambda_n \in [a, \alpha/K^2]$ for some $a > 0$ and $\alpha_n \in [b, c]$ for some b, c

with $0 < b < c < 1$. Define a sequence $\{x_n\}$ iteratively in C by

$$
\begin{cases}
x_1 \in C, \\
y_n = Q_C(x_n - \lambda_n A x_n), & \text{(MTIM)} \\
x_{n+1} = \alpha_n x_n + (1 - \alpha_n) y_n, & n \geq 1.
\end{cases}
$$

Prove that the sequence $\{x_n\}$ defined by (MTIM) converges weakly to some solution x^* of (ATVIP).

8. Let X be a 2-uniformly smooth Banach space with the uniform smoothness constant K, and let C be a nonempty, closed, and convex subset of X. Let Q_C be a sunny nonexpansive retraction from X onto C. Let $A : C \to X$ be an α-inverse-strongly accretive mapping such that $\text{SOL}(C, A) \neq \emptyset$. Assume $\{\lambda_n\}$ and $\{\alpha_n\}$ are chosen so that $\lambda_n \in [a, \alpha/K^2]$ for some $a > 0$ and $\lambda_{n+1} - \lambda_n \to 0$. Let $\alpha_n \in (0, 1)$ satisfy conditions: (C1) $\lim_{n\to\infty} \alpha_n = 0$ and (C2) $\sum_{n=1}^{\infty} \alpha_n = \infty$. Let $\sigma \in (0, 1)$ be a fixed number. For a fixed anchor $u \in C$ and arbitrary initial value $x_1 \in C$, define a sequence $\{x_n\}$ iteratively in C by

$$
\begin{cases}
u, x_1 \in C, \\
y_n = Q_C(x_n - \lambda_n A x_n), & \text{(HTIM)} \\
x_{n+1} = \sigma x_n + (1 - \sigma)[\alpha_n u + (1 - \alpha_n) y_n], & n \geq 1.
\end{cases}
$$

Prove that the sequence $\{x_n\}$ defined by (HTIM) converges strongly to some solution x^* of (ATVIP).

9. Let X be a 2-uniformly smooth Banach space with the uniform smoothness constant K, and let C be a nonempty, closed, and convex subset of X. Let T be an η-strictly pseudocontractive mapping of C into itself such that $\text{Fix}(T) \neq \emptyset$. Assume $\{\lambda_n\}$ and $\{\alpha_n\}$ are chosen so that $\lambda_n \in [a, \eta/K^2]$ for some $a > 0$ and $\lambda_{n+1} - \lambda_n \to 0$ $(n \to \infty)$. Let $\alpha_n \in (0, 1)$ be a sequence satisfying conditions: (C1) $\lim_{n\to\infty} \alpha_n = 0$ and (C2) $\sum_{n=1}^{\infty} \alpha_n = \infty$. Let $\sigma \in (0, 1)$ be a fixed number. For a fixed anchor $u \in C$ and arbitrary initial value $x_1 \in C$, define a sequence $\{x_n\}$ iteratively in C by

$$
\begin{cases}
u, x_1 \in C, \\
y_n = (1 - \lambda_n) x_n + \lambda_n T x_n, & \text{(HTIM)} \\
x_{n+1} = \sigma x_n + (1 - \sigma)[\alpha_n u + (1 - \alpha_n) y_n], & n \geq 1.
\end{cases}
$$

Prove that the sequence $\{x_n\}$ defined by (HTIM) converges strongly to a specific fixed point of T.

10. Let X be a 2-uniformly smooth Banach space and C a nonempty, closed, and convex subset of X. Let $T : C \to C$ be an η-strongly pseudocontractive and L-Lipschitz continuous mapping. Let $\{t_n\}$ be a sequence in (0,1) satisfying conditions: (C1) $\lim_{n\to\infty} t_n = 0$ and (C2) $\sum_{n=1}^{\infty} t_n = \infty$. Prove that the normal Mann iteration $\{x_n\}$

defined by

$$\begin{cases} x_1 \in C, \\ x_{n+1} = (1 - t_n)x_n + t_n Tx_n, & n \geq 1, \end{cases}$$

converges strongly to the unique fixed point of T. Compare the convergence rate of algorithms appearing in Exercises 9 and 10, and write your comments on this matter.

11. Let X be a uniformly smooth Banach space, C a nonempty, closed, and convex subset of X, and $T : C \to C$ a continuous pseudocontractive mapping such that $K := \text{Fix}(T) \neq \emptyset$. Let $A : C \to X$ be an η-strongly accretive and L-Lipschitz continuous mapping. Show that the accretive type variational inequality problem (ATVIP) has a unique solution in K.

12. Let X be a uniformly smooth Banach space, C a nonempty, closed, and convex subset of X, and $T : C \to C$ an L-Lipschitz continuous pseudocontractive mapping such that $K := \text{Fix}(T) \neq \emptyset$. Let $A : C \to X$ be an η-strongly accretive and L-Lipschitz continuous mapping such that $h := I - A : C \to C$ is a self-mapping from C into itself. Let $\{\lambda_n\}$ and $\{\theta_n\}$ be two sequences in $(0,1)$ satisfying the following conditions:

(i) $\lambda_n(1 + \theta_n) \leq 1, \forall n \geq 1$;

(ii) $\theta_n \to 0 \ (n \to \infty)$;

(iii) $\sum_{n=1}^{\infty} \lambda_n \theta_n = \infty$;

(iv) $\frac{\theta_{n-1}}{\theta_n} - 1 = o(\lambda_n \theta_n)$; and

(v) $\frac{\lambda_n}{\theta_n} \to 0 \ (n \to \infty)$.

Define a sequence $\{x_n\}$ iteratively in C by

$$x_1 \in C, \quad x_{n+1} = (1 - \lambda_n)x_n + \lambda_n Tx_n - \lambda_n \theta_n Ax_n, \quad n \geq 1. \tag{HRIM}$$

Show that the sequence $\{x_n\}$ defined by (HRIM) converges strongly to the unique solution x^* of the accretive type variational inequality problem (ATVIP).

6 Iterative methods for zeros of maximal monotone operators in Banach spaces

The purpose of this chapter is to introduce three kinds of iterative methods for zero points of maximal monotone operators in the framework of Banach spaces. These methods are modifications to the celebrated Rockafellar approximate proximal point algorithm. With the aid of Lyapunov functionals, we obtain the convergence analysis of these iterative methods based on a generalized projection operator.

6.1 Lyapunov functional and generalized projection

Let E be a real Banach space and let E^* be the dual space of E. Let $\langle \cdot, \cdot \rangle$ denote the pairing between E and E^*. The normalized duality mapping $J : E \to 2^{E^*}$ is defined by

$$J(x) = \{f \in E^* : \langle x, y \rangle = \|x\|^2 = \|f\|^2\}$$

for all $x \in E$. Next, we use $\langle x, f \rangle$ to denote the value of f at x.

First, let us briefly recall some important concepts and conclusions on the generalized projection operator introduced in Chapter 1.

Definition 6.1.1. Let E be a real smooth Banach space. The Lyapunov functional $\phi : E \times E \to \mathbb{R}^+$ is defined by

$$\phi(x,y) = \|x\|^2 + \|y\|^2 - 2\langle x, Jy \rangle, \quad \forall x,y \in E. \tag{6.1}$$

Lemma 6.1.1. *Let E be a reflexive, smooth, and strictly convex Banach space. Let C be a nonempty, closed, and convex subset of E. Then there exists a unique $\bar{x} \in C$ such that*

$$\phi(\bar{x}, x) = \min_{y \in C} \phi(y, x). \tag{6.2}$$

The generalized projection $\Pi_C : E \to C$ is a map that assigns to an arbitrary point $x \in E$ the minimum point of the functional $\phi(x,y)$, that is, $\Pi_C x = \bar{x}$, where $\bar{x}$ is the solution to the minimization problem $\phi(\bar{x}, x) = \min_{y \in C} \phi(y, x)$.

It is clear that the generalized projection operator is an analogue of the metric projection in Hilbert spaces. The existence and uniqueness of the operator Π_C follow from the properties of functional $\phi(x,y)$ and the strict monotonicity of the mapping J. It is obvious from the definition of ϕ that

$$\left(\|y\| - \|x\|\right)^2 \le \phi(y,x) \le \left(\|y\| + \|x\|\right)^2, \quad \forall x,y \in E, \tag{6.3}$$

and

$$\phi(x,y) = \phi(x,z) + \phi(z,y) + 2\langle x - z, Jz - Jy \rangle, \quad \forall x,y,z \in E. \tag{6.4}$$

https://doi.org/10.1515/9783110667097-006

We also remark here that, in the framework of reflexive, strictly convex, and smooth Banach spaces, for all $x, y \in E$, $\phi(x, y) = 0$ if and only if $x = y$. It is sufficient to show that if $\phi(x, y) = 0$, then $x = y$. From (6.3), we have $\|x\| = \|y\|$. This implies that $\langle x, Jy \rangle = \|x\|^2 = \|Jy\|^2$. From the definition of J, we see that $Jx = Jy$. It follows that $x = y$.

In [2], Alber gave the following results.

Lemma 6.1.2. *Let E be a smooth Banach space and let C be a nonempty, convex, and closed subset of E. Let $x \in E$ and $\bar{x} \in C$. Then $\bar{x} = \Pi_C x$ if and only if*

$$\langle \bar{x} - y, Jx - J\bar{x} \rangle \geq 0, \quad \forall y \in C. \tag{6.5}$$

Lemma 6.1.3. *Let E be a reflexive, strictly convex, and smooth Banach space, and let C be a nonempty, convex, and closed subset of E. Let $x \in E$. Then*

$$\phi(y, \Pi_C x) + \phi(\Pi_C x, x) \leq \phi(y, x), \quad \forall y \in C. \tag{6.6}$$

In [35], Kamimura and Takahashi gave the following results in the framework of uniformly convex and smooth Banach spaces.

Lemma 6.1.4. *Let E be a uniformly convex and smooth Banach space. Let $\{x_n\}$ and $\{y_n\}$ be two sequences in E. If $\phi(x_n, y_n) \to \theta$ and either $\{x_n\}$ or $\{y_n\}$ is bounded, then $x_n - y_n \to \theta$ as $n \to \infty$.*

Next, we introduce another important operator Q_r^A.

Definition 6.1.2. Let E be a reflexive, strictly convex, and smooth Banach space. Let $A \subset E \times E^*$ be a maximal monotone operator. For any $r > 0$, define an operator Q_r^A : $E \to E$ by $Q_r^A x = (J + rA)^{-1} Jx$, $\forall x \in E$.

Since $A \subset E \times E^*$ is maximal monotone, one sees that $\text{Ran}(J + rA) = E^*$, $\forall r > 0$. Since the space is strictly convex and smooth, one sees that J is single-valued and one-to-one. So $(J + rA)^{-1}$ is well-defined on E^*. It follows that $Q_r^A : E \to E$ is a single-valued operator.

We also remark here that operator Q_r^A plays an important role in designing algorithms and in convergence analysis. The following two lemmas show the basic properties of Q_r^A.

Lemma 6.1.5. *Let E be a reflexive, strictly convex, and smooth Banach space. Let $A \subset E \times E^*$ be a maximal monotone operator with $A^{-1}(\theta) \neq \emptyset$. For any $x \in E$, $y \in A^{-1}(\theta)$ and $r > 0$, one has*

$$\phi(y, Q_r^A x) + \phi(Q_r^A x, x) \leq \phi(y, x). \tag{6.7}$$

Proof. From the definition of Q_r^A, one sees that there exists $u_r \in AQ_r^A x$ such that

$$JQ_r^A x + ru_r = Jx. \tag{6.8}$$

Using the monotonicity of A, one has

$$\langle Q_r^A x - y, u_r \rangle \geq 0, \quad \forall y \in A^{-1}(\theta).$$
(6.9)

Combining (6.8) with (6.9), one finds form the definition of ϕ that

$$
\begin{aligned}
\phi(y,x) &- \phi(Q_r^A x, x) - \phi(y, Q_r^A x) \\
&= \|y\|^2 - 2\langle y, Jx \rangle + \|x\|^2 - \|Q_r^A x\|^2 + 2\langle Q_r^A x, Jx \rangle - \|x\|^2 - \|y\|^2 \\
&\quad + 2\langle y, JQ_r^A x \rangle - \|Q_r^A x\|^2 \\
&= 2\langle y, JQ_r^A x - Jx \rangle - 2\langle Q_r^A x, JQ_r^A x - Jx \rangle \\
&= 2r\langle Q_r^A x - y, u_r \rangle \\
&\geq 0.
\end{aligned}
$$

This completes the proof. $\qquad\square$

Lemma 6.1.6. *Let E be a reflexive, locally uniformly convex, and smooth Banach space. Let $A \subset E \times E^*$ be maximal monotone operator with $A^{-1}(\theta) \neq \emptyset$. For any $x \in E$, one has $\lim_{r\to\infty} Q_r^A x = \Pi_{A^{-1}(\theta)} x$.*

Proof. Since $A \subset E \times E^*$ is maximal monotone, one sees that $A^{-1}(\theta)$ is nonempty, closed, and convex subset of E. Using Lemma 6.1.1, one obtains the existence and uniqueness of the generalized projection $\Pi_{A^{-1}(\theta)} x$. First, we show that $\lim_{r\to\infty} Q_r^A x$ exists and belongs to $A^{-1}(\theta)$. To this end, we choose $r_n \in (0, \infty)$ with $r_n \to \infty$ as $n \to \infty$. Letting $x_n = Q_{r_n}^A x$, one has

$$Jx \in Jx_n + r_n Ax_n.$$
(6.10)

Since $A^{-1}(\theta) \neq \emptyset$, we may choose a fixed element v in $A^{-1}(\theta)$. In view of (6.10), we have $y_n \in Ax_n$ such that

$$Jx = Jx_n + r_n y_n.$$
(6.11)

It follows that

$$\langle Jx_n, x_n - v \rangle + r_n \langle y_n, x_n - v \rangle = \langle Jx, x_n - v \rangle.$$
(6.12)

From the fact that $r_n > 0$ and $\langle y_n, x_n - v \rangle \geq 0$, one has

$$\langle Jx_n, x_n - v \rangle \leq \langle Jx, x_n - v \rangle,$$
(6.13)

which implies that

$$\langle Jx_n - Jx, x_n - v \rangle \leq 0.$$
(6.14)

This further implies

$$\langle Jx_n - Jv, x_n - v \rangle = \langle Jx_n - Jx, x_n - v \rangle + \langle Jx - Jv, x_n - v \rangle$$
$$\leq \langle Jx - Jv, x_n - v \rangle$$
$$\leq \|Jx - Jv\|(\|x_n\| + \|v\|). \tag{6.15}$$

On the other hand, one has

$$\langle Jx_n - Jv, x_n - v \rangle \geq (\|x_n\| - \|v\|)^2. \tag{6.16}$$

Combining (6.15) with (6.16), one arrives at

$$(\|x_n\| - \|v\|)^2 \leq \|Jx - Jv\|(\|x_n\| + \|v\|). \tag{6.17}$$

This shows that $\{x_n\}$ is bounded. Without loss of generality, we may assume that $x_n \rightharpoonup p$ as $n \to \infty$. Since $r_n \to \infty$ as $n \to \infty$, we find from (6.11) that

$$y_n = \frac{Jx - Jx_n}{r_n} \to \theta$$

as $n \to \infty$. Since the graph of A is demiclosed, we find that $p \in \text{Dom}(A)$ and $\theta \in Ap$, that is, $p \in A^{-1}(\theta)$. Using (6.13), we have

$$\langle Jx_n - Jp, x_n - p \rangle \leq \langle Jx - Jp, x_n - p \rangle. \tag{6.18}$$

It follows that

$$\limsup_{n \to \infty} \langle Jx_n - Jp, x_n - p \rangle \leq 0. \tag{6.19}$$

On the other hand, we have

$$\langle Jx_n - Jp, x_n - p \rangle \geq (\|x_n\| - \|p\|)^2. \tag{6.20}$$

This implies that

$$\limsup_{n \to \infty} (\|x_n\| - \|p\|)^2 \leq 0.$$

Hence, we have $\|x_n\| \to \|p\|$ as $n \to \infty$. Since E is locally uniformly convex, we see that E has the Kadec–Klee property. It follows that $\|x_n - p\| \to 0$ as $n \to \infty$. We are now in a position to show $p = \Pi_{A^{-1}(\theta)}x$. Since $J : E \to E^*$ is demicontinuous, we have $Jx_n \rightharpoonup Jp$ as $n \to \infty$. Using the monotonicity of A, we find from (6.11) that

$$\langle Jx_n - Jx, x_n - y \rangle = -r_n \langle y_n, x_n - y \rangle \leq 0, \quad \forall y \in A^{-1}(\theta).$$

This further implies that

$$\langle Jp - Jx, p - y \rangle \leq 0, \quad \forall y \in A^{-1}(\theta).$$

Using Lemma 6.1.2, one sees that $p = \Pi_{A^{-1}(\theta)}x$. This shows that $\{x_n\}$ converges to $\Pi_{A^{-1}(\theta)}x$ in norm. For any $s_n \in (0, \infty)$ with $s_n \to \infty$ as $n \to \infty$, denoting $z_n = Q_{s_n}^A x = (J + s_n A)^{-1} Jx$ and repeating the above proof, we see that $\{z_n\}$ converges to $\Pi_{A^{-1}(\theta)}x$ in norm. So, $\{Q_r^A x\}$ converges to $\Pi_{A^{-1}(\theta)}x$ as $r \to \infty$ in norm. This completes the proof. $\square$

6.2 Rockafellar–Mann iterative method and its weak convergence theorem

Let E be a real reflexive, smooth, and locally uniformly convex Banach space. Let $A \subset E \times E^*$ be a maximal monotone operator with $A^{-1}(\theta) \neq \emptyset$. Consider the following Rockafellar–Mann iterative method

$$x_1 \in E, \ x_{n+1} = J^{-1}(\alpha_n J x_n + \beta_n J Q_{r_n}^A x_n + \gamma_n J e_n), \quad n \geq 1, \tag{GRMIM}$$

where $\{\alpha_n\}$, $\{\beta_n\}$, and $\{\gamma_n\}$ are three real sequences in $(0, 1)$ such that $\alpha_n + \beta_n + \gamma_n = 1$, $r_n \subset \mathbb{R}^+$, $\{e_n\}$ is a bounded error sequence in E. We call (GRMIM) a Rockafellar–Mann iterative method. In order to obtain the weak convergence of (GRMIM), we prove the following lemma first.

Lemma 6.2.1. *Let E be a real, smooth, and uniformly convex Banach space. Let $A \subset E \times E^*$ be a maximal monotone operator with $A^{-1}(\theta) \neq \emptyset$. Let $\{x_n\}$ be a sequence generated by (GRMIM). If $\sum_{n=1}^{\infty} \gamma_n < \infty$, then $\{\Pi_{A^{-1}(\theta)} x_n\}$ converges to $v \in A^{-1}(\theta)$ in norm and v is also a unique solution to the following minimum problem:*

$$\lim_{n \to \infty} \phi(v, x_n) = \min_{y \in A^{-1}(\theta)} \lim_{n \to \infty} \phi(y, x_n). \tag{6.21}$$

Proof. The proof is split into seven steps.

Step 1. Show that $\{x_n\}$ is a bounded sequence.

Fix $p \in A^{-1}(\theta)$. Since $\|\cdot\|^2$ is convex, we find from Lemma 6.1.5 that

$$
\begin{aligned}
\phi(p, x_{n+1}) &= \|p\|^2 - 2\langle p, J x_{n+1}\rangle + \|J x_{n+1}\|^2 \\
&= \alpha_n \|p\|^2 + \beta_n \|p\|^2 + \gamma_n \|p\|^2 - 2\alpha_n \langle p, J x_n\rangle \\
&\quad - 2\beta_n \langle p, J Q_{r_n}^A x_n\rangle - 2\gamma_n \langle p, J e_n\rangle \\
&\quad + \|\alpha_n J x_n + \beta_n J Q_{r_n}^A x_n + \gamma_n J e_n\|^2 \\
&\leq \alpha_n \|p\|^2 - 2\alpha_n \langle p, J x_n\rangle + \alpha_n \|J x_n\|^2 + \beta_n \|p\|^2 \\
&\quad - 2\beta_n \langle p, J Q_{r_n}^A x_n\rangle + \beta_n \|J Q_{r_n}^A x_n\|^2 \\
&\quad + \gamma_n \|p\|^2 - 2\gamma_n \langle p, J e_n\rangle + \gamma_n \|J e_n\|^2 \\
&= \alpha_n \phi(p, x_n) + \beta_n \phi(p, Q_{r_n}^A x_n) + \gamma_n \phi(p, J e_n) \\
&\leq \alpha_n \phi(p, x_n) + \beta_n \phi(p, x_n) - \beta_n \phi(Q_{r_n}^A x_n, x_n) + \gamma_n \phi(p, J e_n) \\
&\leq (1 - \gamma_n)\phi(p, x_n) + \gamma_n \phi(p, J e_n) \\
&\leq \max\{\phi(p, x_1), M_p\}, \forall n \geq 1, \tag{6.22}
\end{aligned}
$$

where

$$M_p = \sup\{\phi(p, J e_n) : n \geq 1\}.$$

This shows $\phi(p, x_n)$ is a bounded sequence. From the definition of ϕ, we find that $\{x_n\}$ is also bounded.

Step 2. Show that $\lim_{n\to\infty} \phi(p, x_n)$, $\forall p \in A^{-1}(\theta)$ exits.

From Step 1, we have

$$\phi(p, x_{n+1}) \leq \phi(p, x_n) + \gamma_n \phi(p, e_n) \leq \phi(p, x_n) + \gamma_n M_p.$$

Since $\sum_{n=1}^{\infty} \gamma_n < \infty$, by virtue of Lemma 1.10.1, we conclude that $\lim_{n\to\infty} \phi(p, x_n)$ exists $\forall p \in A^{-1}(\theta)$.

Step 3. Show that the existence and uniqueness of the solution to (6.21).

Define $h(y) = \lim_{n\to\infty} \phi(y, x_n)$, $\forall y \in A^{-1}(\theta)$. Then $h : A^{-1}(\theta) \to \mathbb{R}^+$ is a proper, lower semicontinuous and convex function such that $h(y) \to \infty$ as $\|y\| \to \infty$. So, there exists $v \in A^{-1}(\theta)$ such that $h(v) = \min_{y \in A^{-1}(\theta)} h(y)$. Since E is uniformly convex, we see that $h(y)$ is strictly convex. Hence, we get the uniqueness of v.

Step 4. Show that $\lim_{n\to\infty} \phi(\Pi_{A^{-1}(\theta)} x_n, x_n)$ exists.

Since $A \subset E \times E^*$ is maximal monotone and $A^{-1}(\theta) \neq \emptyset$, we find that $A^{-1}(\theta)$ is nonempty, closed, and convex. So, the existence and uniqueness of the generalized projection operator $\Pi_{A^{-1}(\theta)} x_n$ is ensured. From the definition of $\Pi_{A^{-1}(\theta)}$, we have

$$\phi(\Pi_{A^{-1}(\theta)} x_{n+1}, x_{n+1}) \leq \phi(\Pi_{A^{-1}(\theta)} x_n, x_{n+1}). \tag{6.23}$$

From Step 1, we find that

$$\phi(\Pi_{A^{-1}(\theta)} x_n, x_{n+1}) \leq \phi(\Pi_{A^{-1}(\theta)} x_n, x_n) + \gamma_n \phi(\Pi_{A^{-1}(\theta)} x_n, e_n). \tag{6.24}$$

In view of Lemma 6.1.3, we arrive at

$$\phi(\Pi_{A^{-1}(\theta)} x_n, x_n) \leq \phi(v, x_n) - \phi(v, \Pi_{A^{-1}(\theta)} x_n), \tag{6.25}$$

where $v \in A^{-1}(\theta)$ satisfies (6.21). From Step 1, we see that $\phi(v, x_n)$ is bounded. Using (6.25), we find that $\phi(v, \Pi_{A^{-1}(\theta)} x_n)$ is bounded, and then we find that $\{\Pi_{A^{-1}(\theta)} x_n\}$ is also bounded in view of (6.3). It follows that $\{\phi(\Pi_{A^{-1}(\theta)} x_n, e_n)\}$ is also bounded. Setting $M = \sup\{\phi(\Pi_{A^{-1}(\theta)} x_n, e_n) : n \geq 1\}$, one finds that (6.24) is reduced to

$$\phi(\Pi_{A^{-1}(\theta)} x_n, x_{n+1}) \leq \phi(\Pi_{A^{-1}(\theta)} x_n, x_n) + \gamma_n M. \tag{6.26}$$

Substituting (6.26) into (6.23), one arrives at

$$\phi(\Pi_{A^{-1}(\theta)} x_{n+1}, x_{n+1}) \leq \phi(\Pi_{A^{-1}(\theta)} x_n, x_n) + \gamma_n M. \tag{6.27}$$

By virtue of Lemma 1.10.1, we conclude that $\lim_{n\to\infty} \phi(\Pi_{A^{-1}(\theta)} x_n, x_n)$ exists.

Step 5. Show that $\{\Pi_{A^{-1}(\theta)} x_n\}$ is Cauchy.

From Step 1, by a simple induction, we have

$$\phi(p, x_{n+m}) \le \phi(p, x_n) + M \sum_{j=0}^{m-1} \gamma_{n+j}, \quad \forall p \in A^{-1}(\theta), \qquad (6.28)$$

for all $m, n \ge 1$. By virtue of Lemma 6.1.3 and (6.28), we obtain

$$\phi(\Pi_{A^{-1}(\theta)} x_n, \Pi_{A^{-1}(\theta)} x_{n+m})$$

$$\le \phi(\Pi_{A^{-1}(\theta)} x_n, x_{n+m}) - \phi(\Pi_{A^{-1}(\theta)} x_{n+m}, x_{n+m})$$

$$\le \phi(\Pi_{A^{-1}(\theta)} x_n, x_n) - \phi(\Pi_{A^{-1}(\theta)} x_{n+m}, x_{n+m}) + M \sum_{j=0}^{m-1} \gamma_{n+j} \qquad (6.29)$$

for all $n, m \ge 1$. From Step 5 and (6.29), it follows that $\{\Pi_{A^{-1}(\theta)} x_n\}$ is Cauchy. Assume that $\lim_{n\to\infty} \Pi_{A^{-1}(\theta)} x_n = p$; then $p \in A^{-1}(\theta)$, since $A^{-1}(\theta)$ is closed. It is clear that $\lim_{n\to\infty} \phi(\Pi_{A^{-1}(\theta)} x_n, x_n) = \lim_{n\to\infty} \phi(p, x_n) = h(p)$.

Step 6. Show that $\{\Pi_{A^{-1}(\theta)} x_n\}$ converges to $v \in A^{-1}(\theta)$, where v solves (6.21).

By virtue of Lemma 6.1.3, we have

$$\phi(v, \Pi_{A^{-1}(\theta)} x_n) \le \phi(v, x_n) - \phi(\Pi_{A^{-1}(\theta)} x_n, x_n). \qquad (6.30)$$

Taking the lim sup on both sides of the above inequality, we have

$$\limsup_{n\to\infty} \phi(v, \Pi_{A^{-1}(\theta)} x_n) \le h(v) - h(p) \le 0.$$

On the other hand, we have $\liminf_{n\to\infty} \phi(v, \Pi_{A^{-1}(\theta)} x_n) \ge 0$. This shows that

$$\lim_{n\to\infty} \phi(v, \Pi_{A^{-1}(\theta)} x_n) = 0.$$

Form Lemma 6.1.4, we obtain that $\Pi_{A^{-1}(\theta)} x_n \to v$ as $n \to \infty$. This completes the proof. $\qquad\square$

Theorem 6.2.1. *Let E be a real, uniformly smooth, and uniformly convex Banach space. Let $A \subset E \times E^*$ be a maximal monotone operator with $A^{-1}(\theta) \ne \emptyset$. Let $J : E \to E^*$ be weakly sequentially continuous. Let $\{x_n\}$ be a sequence generated by (GRMIM). If $\liminf_{n\to\infty} r_n > 0$, $\liminf_{n\to\infty} \beta_n > 0$, $\sum_{n=1}^{\infty} \gamma_n < \infty$, then $\{x_n\}$ converges weakly to $v \in A^{-1}(\theta)$, where $v = \lim_{n\to\infty} \Pi_{A^{-1}(\theta)} x_n$.*

Proof. Denote the weak-limit set of $\{x_n\}$ by $\omega_\omega(x_n)$, that is,

$$\omega_\omega(x_n) = \{x \in E : \exists x_{n_i} \rightharpoonup x\}.$$

Since $\{x_n\}$ is bounded and the space is reflexive, we have $\omega_\omega(x_n) \ne \emptyset$. For any $x \in \omega_\omega(x_n)$, we see that there exists a subsequence $\{x_{n_j}\}$ of $\{x_n\}$ such that $x_{n_j} \rightharpoonup x$ as $j \to \infty$. From Step 1 of Lemma 6.2.1, we have

$$\phi(p, x_{n+1}) \le \alpha_n \phi(p, x_n) - \beta_n \phi(Q_{r_n}^A x_n, x_n) + \gamma_n \phi(p, e_n), \quad \forall p \in A^{-1}(\theta).$$

It follows that

$$\beta_n \phi(Q^A_{r_n} x_n, x_n) \le \phi(p, x_n) - \phi(p, x_{n+1}) + \gamma_n M_p, \quad \forall p \in A^{-1}(\theta).$$

Since $\lim_{n \to \infty} \phi(p, x_n)$ exits, $\liminf_{n \to \infty} \beta_n > 0$, $\sum_{n=1}^{\infty} \gamma_n < \infty$, we have

$$\lim_{n \to \infty} \phi(Q^A_{r_n} x_n, x_n) = 0.$$

Using Lemma 6.1.4, we find $\lim_{n \to \infty} \|Q^A_{r_n} x_n - x_n\| = 0$. Denoting $u_n = Q^A_{r_n} x_n$, there exists $\omega_n \in A u_n$ with

$$J u_n + r_n \omega_n = J x_n. \tag{6.31}$$

Since E is a uniformly smooth Banach space, we see that J is uniformly continuous on bounded sets. In view of that $\lim_{n \to \infty} \|u_n - x_n\| = 0$, we find that $\lim_{n \to \infty} \|J u_n - J x_n\| = 0$. Since $\liminf_{n \to \infty} r_n > 0$, we find from 6.31 that

$$\omega_n = \frac{J x_n - J u_n}{r_n} \to \theta, \quad n \to \infty.$$

Since $x_{n_j} \rightharpoonup x$, and $x_{n_j} - u_{n_j} \to \theta$ as $j \to \infty$, $u_{n_j} \rightharpoonup x$. Since the graph of A, which is maximal monotone, is demiclosed, we find that $x \in \text{Dom}(A)$ and $x \in A^{-1}(\theta)$. This shows that $\omega_\omega(x_n) \subset A^{-1}(\theta)$.

Next, we prove $x = v = \lim_{n \to \infty} \Pi_{A^{-1}(\theta)} x_n$. Let $y_n = \Pi_{A^{-1}(\theta)} x_n$. Using Lemma 6.1.2 and $x \in A^{-1}(\theta)$, one has

$$\langle y_{n_j} - x, J x_{n_j} - J y_{n_j} \rangle \ge 0. \tag{6.32}$$

From Lemma 6.2.1, one sees that $y_{n_j} - v \to \theta$ as $j \to \infty$. It follows that $J y_{n_j} - J v \to \theta$ as $j \to \infty$. From the assumption, we see that $J x_{n_j} \to J x$ as $j \to \infty$. It follows from (6.32) that $\langle v - x, J x - J v \rangle \ge 0$. Since J is monotone, we have $\langle v - x, J x - J v \rangle \le 0$. Hence, we have $\langle v - x, J x - J v \rangle = 0$. This implies $x = v$. This shows that $x_n \rightharpoonup v \in A^{-1}(\theta)$, where $v = \lim_{n \to \infty} \Pi_{A^{-1}(\theta)} x_n$, completing the proof. □

Remark 6.2.1. The restriction that J is weakly sequentially continuous is unsatisfactory. Indeed, even the space l^p, where $p > 1$, does not satisfy this condition. It is of interest to further modify (GRMIM) to get the weak convergence in more general Banach spaces.

6.3 Rockafellar–Halpern iterative method and its strong convergence theorem

Let E be a real reflexive, smooth, and strictly convex Banach space. Let $A \subset E \times E^*$ be a maximal monotone operator with $A^{-1}(\theta) \ne \emptyset$. Consider the following Rockafellar–Halpern iterative method

$$x_1 \in E, \quad x_{n+1} = J^{-1}(\alpha_n J x_1 + \beta_n J Q^A_{r_n} x_n + \gamma_n J e_n), \quad n \ge 1, \tag{GRHIM}$$

where $\{\alpha_n\}$, $\{\beta_n\}$, and $\{\gamma_n\}$ are three real sequences in $(0,1)$ such that $\alpha_n + \beta_n + \gamma_n = 1$, $r_n \subset \mathbb{R}^+$, $\{e_n\}$ is a bounded error sequence in E. We call (GRHIM) a Rockafellar–Halpern iterative method.

Theorem 6.3.1. *Let E be a real, uniformly smooth, and uniformly convex Banach space. Let $A \subset E \times E^*$ be a maximal monotone operator with $A^{-1}(0) \neq \emptyset$. Let $\{x_n\}$ be a sequence generated by (GRHIM). If $r_n \to \infty$ and $\alpha_n \to 0$ as $n \to \infty$, $\sum_{n=1}^{\infty} \gamma_n < \infty$ and $\sum_{n=1}^{\infty} \alpha_n = \infty$, then $\{x_n\}$ converges strongly to $p \in A^{-1}(0)$ and $p = \Pi_{A^{-1}(0)} x_1$.*

Proof. The proof is split into four steps.
Step 1. Show that $\{x_n\}$ is bounded.
Since $\|\cdot\|^2$ is convex, we find from the definition of ϕ that

$$
\begin{aligned}
\phi(p, x_{n+1}) &= \|p\|^2 - 2\langle p, Jx_{n+1}\rangle + \|Jx_{n+1}\|^2 \\
&= \alpha_n\|p\|^2 + \beta_n\|p\|^2 + \gamma_n\|p\|^2 - 2\alpha_n\langle p, Jx_1\rangle \\
&\quad - 2\beta_n\langle p, JQ_{r_n}^A x_n\rangle - 2\gamma_n\langle p, Je_n\rangle \\
&\quad + \left\|\alpha_n Jx_1 + \beta_n JQ_{r_n}^A x_n + \gamma_n Je_n\right\|^2 \\
&\leq \alpha_n\|p\|^2 - 2\alpha_n\langle p, Jx_1\rangle + \alpha_n\|Jx_1\|^2 + \beta_n\|p\|^2 \\
&\quad - 2\beta_n\langle p, JQ_{r_n}^A x_n\rangle + \beta_n\|JQ_{r_n}^A x_n\|^2 \\
&\quad + \gamma_n\|p\|^2 - 2\gamma_n\langle p, Je_n\rangle + \gamma_n\|Je_n\|^2 \\
&= \alpha_n\phi(p, x_1) + \beta_n\phi(p, Q_{r_n}^A x_n) + \gamma_n\phi(p, Je_n).
\end{aligned}
\tag{6.33}
$$

Using Lemma 6.1.5, we find that

$$
\phi(p, Q_{r_n}^A x_n) \leq \phi(p, x_n).
\tag{6.34}
$$

Substituting (6.34) into (6.33), we obtain that

$$
\phi(p, x_{n+1}) \leq \alpha_n\phi(p, x_1) + \beta_n\phi(p, x_n) + \gamma_n M_p,
$$

where $M_p = \sup\{\phi(p, Je_n) : n \geq 1\}$. This implies that

$$
\phi(p, x_n) \leq \phi(p, x_1) + M_p \sum_{i=1}^{n-1} \gamma_i.
$$

It follows that

$$
\phi(p, x_n) \leq \phi(p, x_1) + M_p \sum_{i=1}^{\infty} \gamma_i,
$$

which shows that $\phi(p, x_n)$ is a bounded sequence. From the definition of ϕ, we find that $\{x_n\}$ is also bounded.

Step 2. Show that $\limsup_{n\to\infty}\langle Jx_1 - Jp, y_n - p\rangle \leq 0$, where $y_n = Q_{r_n}^A x_n$.

For any $r > 0$, we see that $u_r \in AQ_r^A x_1$ and

$$JQ_r^A x_1 + ru_r = Jx_1. \tag{6.35}$$

For any $n \geq 1$, we see that $\omega_n \in Ay_n$ and

$$Jy_n + r_n\omega_n = Jx_n. \tag{6.36}$$

From (6.35) and (6.36), we find from the monotonicity of A that

$$
\begin{aligned}
\langle Jx_1 - JQ_r^A x_1, y_n - Q_r^A x_1\rangle &= r\langle u_r, y_n - Q_r^A x_1\rangle \\
&= r\langle u_r - \omega_n, y_n - Q_r^A x_1\rangle + r\langle \omega_n, y_n - Q_r^A x_1\rangle \\
&\leq r\langle \omega_n, y_n - Q_r^A x_1\rangle \\
&= \frac{r}{r_n}\langle Jx_n - Jy_n, y_n - Q_r^A x_1\rangle \\
&\leq \frac{r}{r_n}\|Jx_n - Jy_n\|\|y_n - Q_r^A x_1\|.
\end{aligned}
\tag{6.37}
$$

From (6.34), we see that $\{\phi(p, y_n)\}$ is bounded. Hence, $\{y_n\}$ is a bounded sequence. Since $r_n \to \infty$ as $n \to \infty$, we find from (6.37) that

$$\limsup_{n\to\infty}\langle Jx_1 - JQ_r^A x_1, y_n - Q_r^A x_1\rangle \leq 0. \tag{6.38}$$

Assume that there exists a subsequence $\{y_{n_j}\}$ of $\{y_n\}$ such that $y_{n_j} \to y$ as $j \to \infty$ and

$$
\begin{aligned}
\limsup_{n\to\infty}\langle JQ_r^A x_1 - Jp, y_n - p\rangle &= \lim_{j\to\infty}\langle JQ_r^A x_1 - Jp, y_{n_j} - p\rangle \\
&= \langle JQ_r^A x_1 - Jp, y - p\rangle.
\end{aligned}
\tag{6.39}
$$

Note that

$$
\begin{aligned}
&\langle Jx_1 - Jp, y_n - p\rangle \\
&= \langle Jx_1 - JQ_r^A x_1, y_n - p\rangle + \langle JQ_r^A x_1 - Jp, y_n - p\rangle \\
&= \langle Jx_1 - JQ_r^A x_1, y_n - Q_r^A x_1\rangle + \langle Jx_1 - JQ_r^A x_1, Q_r^A x_1 - p\rangle \\
&\quad + \langle JQ_r^A x_1 - Jp, y_n - p\rangle.
\end{aligned}
\tag{6.40}
$$

It follows from (6.38), (6.39), and (6.40) that

$$\limsup_{n\to\infty}\langle Jx_1 - Jp, y_n - p\rangle \leq \langle Jx_1 - JQ_r^A x_1, Q_r^A x_1 - p\rangle + \langle JQ_r^A x_1 - Jp, y - p\rangle. \tag{6.41}$$

From Lemma 6.1.6, we see that $Q_r^A x_1 \to p$ as $r \to \infty$. Using the continuity of J, we find that $JQ_r^A x_1 \to Jp$ as $r \to \infty$. Therefore, we obtain from (6.41) that

$$\limsup_{n\to\infty}\langle Jx_1 - Jp, y_n - p\rangle \leq 0.$$

Step 3. Show that $\max\{\langle Jx_1 - Jp, y_n - p\rangle, 0\} \to 0$ as $n \to \infty$.

For any $\epsilon > 0$, there exists a natural number $n_0 \geq 1$. When $n \geq n_0$, one has $\langle Jx_1 - Jp, y_n - p\rangle \leq \epsilon$. So,

$$0 \leq \max\{\langle Jx_1 - Jp, y_n - p\rangle, 0\} \leq \epsilon.$$

Hence, we have $\max\{\langle Jx_1 - Jp, y_n - p\rangle, 0\} \to 0$ as $n \to \infty$.

Step 4. Show that $x_n \to p$ as $n \to \infty$.

Note that

$$\|\alpha_n Jx_1 + \beta_n Jy_n + \gamma_n Je_n\|^2$$
$$\leq \|\alpha_n Jx_1 + \beta_n Jy_n\|^2 + 2\gamma_n\|\alpha_n Jx_1 + \beta_n Jy_n\|\|Je_n\| + \gamma_n^2\|Je_n\|^2$$
$$\leq \|\alpha_n Jx_1 + \beta_n Jy_n\|^2 + \gamma_n M', \tag{6.42}$$

where $M' = \sup\{2\|\alpha_n Jx_1 + \beta_n Jy_n\|\|e_n\| + \|e_n\|^2 : n \geq 1\}$. Since the space E is uniformly smooth, we find from the Reich inequality that

$$\|\alpha_n Jx_1 + \beta_n Jy_n\|^2$$
$$\leq \beta_n^2\|Jy_n\|^2 + 2\alpha_n\beta_n\langle Jx_1, y_n\rangle + \max\{\beta_n\|y_n\|, 1\}\alpha_n\|x_1\|b(\alpha_n\|x_1\|)$$
$$\leq \beta_n^2\|Jy_n\|^2 + 2\alpha_n\beta_n\langle Jx_1, y_n\rangle + \max\{\beta_n\|y_n\|, 1\}\alpha_n\|x_1\|b(\alpha_n \max\{\|x_1\|, 1\})$$
$$\leq (1 - \alpha_n)^2\|y_n\|^2 + 2\alpha_n\beta_n\langle Jx_1, y_n\rangle + \alpha_n b(\alpha_n)M'', \tag{6.43}$$

where $M'' = \|x_1\| \max\{\sup\{\|y_n\| : n \geq 1\}, 1\} \max\{\|x_1\|, 1\}$. It follows from (6.42) and (6.43) that

$$\phi(p, x_{n+1}) = \|p\|^2 - 2\langle p, Jx_{n+1}\rangle + \|Jx_{n+1}\|^2$$
$$= \alpha_n\|p\|^2 + \beta_n\|p\|^2 + \gamma_n\|p\|^2 - 2\alpha_n\langle p, Jx_1\rangle$$
$$\quad - 2\beta_n\langle p, JQ_{r_n}^A x_n\rangle - 2\gamma_n\langle p, Je_n\rangle + \|\alpha_n Jx_1 + \beta_n JQ_{r_n}^A x_n + \gamma_n Je_n\|^2$$
$$\leq \alpha_n\|p\|^2 + \beta_n\|p\|^2 + \gamma_n\|p\|^2 - 2\alpha_n\langle p, Jx_1\rangle - 2\beta_n\langle p, JQ_{r_n}^A x_n\rangle$$
$$\quad - 2\gamma_n\langle p, Je_n\rangle + (1 - \alpha_n)^2\|y_n\|^2 + 2\alpha_n\beta_n\langle Jx_1, y_n\rangle$$
$$\quad + \alpha_n b(\alpha_n)M'' + \gamma_n M'$$
$$\leq 2\alpha_n\langle p, Jp\rangle + 2\alpha_n\langle Jp, y_n\rangle - 2\alpha_n\langle Jp, y_n\rangle - 2\alpha_n\langle p, Jx_1\rangle + (1 - \alpha_n)\|y_n\|^2$$
$$\quad - 2(1 - \alpha_n - \gamma_n)\langle p, Jy_n\rangle + 2(1 - \alpha_n - \gamma_n)\langle Jx_1, y_n\rangle - \alpha_n(1 - \alpha_n)\|y_n\|^2$$
$$\quad - \alpha_n\|p\|^2 + (1 - \alpha_n)\|p\|^2 + 2\gamma_n\|p\|\|e_n\| + \alpha_n b(\alpha_n)M'' + \gamma_n M'$$
$$\leq (1 - \alpha_n)\phi(p, y_n) + 2\alpha_n\langle Jx_1 - Jp, y_n - p\rangle - \alpha_n\phi(y_n, p) + \alpha_n^2\phi(y_n, x_1)$$
$$\quad + 2\alpha_n\gamma_n\|x_1\|\|y_n\| + 2\gamma_n\|p\|\|y_n\| + 2\gamma_n\|p\|\|e_n\| + \alpha_n b(\alpha_n)M'' + \gamma_n M'$$
$$\leq (1 - \alpha_n)\phi(p, x_n) + 2\alpha_n\langle Jx_1 - Jp, y_n - p\rangle + \alpha_n^2\phi(y_n, x_1)$$
$$\quad + \alpha_n b(\alpha_n)M'' + \gamma_n(2\alpha_n\|x_1\|\|y_n\| + 2\|p\|\|y_n\| + 2\|p\|\|e_n\| + M')$$
$$\leq (1 - \alpha_n)\phi(p, x_n) + o(\alpha_n) + C\gamma_n,$$

where C is some fixed positive constant. It follows from Lemma 1.10.2 that $\phi(p, x_n) \to 0$ as $n \to \infty$. Using Lemma 6.1.4, we find that $x_n \to p$ as $n \to \infty$. This completes the proof. □

Remark 6.3.1. The restriction that $\sum_{n=1}^{\infty} \alpha_n = \infty$ is unsatisfactory because it may reduce the convergence rate of (GRHIM). It is of interest to develop more efficient iterative methods for approximating the zero points of maximal monotone operators in general Banach spaces.

6.4 Rockafellar–Haugazeau iterative method and its strong convergence theorem

Let E be a real reflexive, smooth, and strictly convex Banach space. Let $A \subset E \times E^*$ be a maximal monotone operator with $A^{-1}(\theta) \neq \emptyset$. Consider the following iterative method:

$$\begin{cases} x_1 \in E, \text{ chosen arbitrarily}, \\ y_n = J^{-1}(\alpha_n J x_n + \beta_n J Q_{r_n}^A x_n + \gamma_n J e_n), \\ C_n = \{z \in E : \phi(z, y_n) \leq \phi(z, x_n) + \gamma_n B\}, \qquad \text{(RHIM)} \\ Q_n = \{z \in E : \langle x_n - z, J x_1 - J x_n \rangle \geq 0\}, \\ x_{n+1} = \Pi_{C_n \cap Q_n} x_1, \quad n \geq 1, \end{cases}$$

where $\{\alpha_n\}$, $\{\beta_n\}$, and $\{\gamma_n\}$ are three real sequences in $(0, 1)$ such that $\alpha_n + \beta_n + \gamma_n = 1$, $r_n \subset \mathbb{R}^+$, $\{e_n\}$ is a bounded error sequence in E such that

$$B = \sup\{\phi(z, e_n) : z \in A^{-1}(\theta), n \geq 1\} < \infty.$$

We call (RHIM) a Rockafellar–Haugazeau iterative method.

Theorem 6.4.1. *Let E be a real uniformly smooth and uniformly convex Banach space. Let $A \subset E \times E^*$ be a maximal monotone operator with $A^{-1}(\theta) \neq \emptyset$. Let $\{x_n\}$ be a sequence generated by (RHIM). If $\liminf_{n \to \infty} \beta_n > 0$, $\liminf_{n \to \infty} r_n > 0$, $\gamma_n \to 0$ as $n \to \infty$, and $\{e_n\}$ is a bounded error sequence in E such that $B = \sup\{\phi(z, e_n) : z \in A^{-1}(\theta), n \geq 1\} < \infty$, then $\{x_n\}$ converges to $\Pi_{A^{-1}(\theta)} x_1$ in norm.*

Proof. First, we show that $\Pi_{A^{-1}(\theta)} x_1$ is well defined. Indeed, since $A \subset E \times E^*$ is a maximal monotone operator with $A^{-1}(\theta) \neq \emptyset$, we find that $A^{-1}(\theta)$ is nonempty, closed, and convex. In view of Lemma 6.1.1, we find that $\Pi_{A^{-1}(\theta)} x_1$ is well defined.

Next, we show that $C_n \cap Q_n$ is closed and convex for every $n \geq 1$. It is obvious that Q_n is closed and convex. Note that $\phi(z, y_n) \leq \phi(z, x_n) + \gamma_n B$ is equivalent to

$$\|y_n\|^2 - \|x_n\|^2 - 2\langle z, J y_n - J x_n \rangle \leq \gamma_n B.$$

This implies set C_n is convex. It is also obvious that C_n is closed. This proves that $C_n \cap Q_n$ is closed and convex.

Next, we show that $A^{-1}(\theta) \subset C_n \cap Q_n$. For any $z \in A^{-1}(\theta)$, we find from Lemma 6.1.5 that

$$
\begin{aligned}
\phi(z, y_n) &= \|z\|^2 - 2\langle z, \alpha_n Jx_n + \beta_n JQ_{r_n}^A x_n + \gamma_n Je_n \rangle + \|\alpha_n Jx_n + \beta_n JQ_{r_n}^A x_n + \gamma_n Je_n\|^2 \\
&\leq \|z\|^2 - 2\alpha_n \langle z, Jx_n \rangle - 2\beta_n \langle z, JQ_{r_n}^A x_n \rangle - 2\gamma_n \langle z, Je_n \rangle \\
&\quad + \alpha_n \|x_n\|^2 + \beta_n \|Q_{r_n}^A x_n\|^2 + \gamma_n \|e_n\|^2 \\
&\leq \alpha_n \phi(z, x_n) + \beta_n \phi(z, x_n) - \beta_n \phi(Q_{r_n}^A x_n, x_n) + \gamma_n \phi(z, e_n) \\
&\leq (1 - \gamma_n)\phi(z, x_n) + \gamma_n \phi(z, e_n) \\
&\leq (1 - \gamma_n)\phi(z, x_n) + \gamma_n B.
\end{aligned}
$$

This implies that $A^{-1}(\theta) \subset C_n$. Now we are in a position to show that $A^{-1}(\theta) \subset C_n \cap Q_n$ by mathematical induction. For $n = 1$, since $Q_1 = E$, we have $A^{-1}(\theta) \subset C_1 \cap Q_1 \subset C_1$. Assume that $A^{-1}(\theta) \subset C_m \cap Q_m$, $m \in \mathbb{N}$. It follows that $x_{m+1} = \Pi_{C_m \cap Q_m} x_1$. Using Lemma 6.1.2, we find that

$$
\langle x_{m+1} - z, Jx_1 - Jx_{m+1} \rangle \geq 0, \quad \forall z \in C_m \cap Q_m.
$$

It follows from $A^{-1}(\theta) \subset C_m \cap Q_m$ that

$$
\langle x_{m+1} - z, Jx_1 - Jx_{m+1} \rangle \geq 0, \quad \forall z \in A^{-1}(\theta),
$$

which shows that $z \in Q_{m+1}$. It follows that $z \in C_{m+1} \cap Q_{m+1}$. Hence, we have $A^{-1}(\theta) \subset C_n \cap Q_n$. This also shows (RHIA) is well defined.

From the constructions of Q_n and Π_{Q_n}, we have $x_n = \Pi_{Q_n} x_1$. Using Lemma 6.1.3, we find

$$
\phi(x_n, x_1) \leq \phi(z, x_1) - \phi(z, x_n) \leq \phi(z, x_1), \quad \forall z \in A^{-1}(\theta),
$$

which implies that $\{\phi(x_n, x_1)\}$ is a bounded sequence, so are $\{x_n\}$ and $Q_{r_n}^A x_n$. In view of

$$
x_{n+1} = \Pi_{C_n \cap Q_n} x_1 \in C_n \cap Q_n \subset Q_n
$$

and $x_n = \Pi_{Q_n} x_1$, one has $\phi(x_n, x_1) \leq \phi(x_{n+1}, x_1)$. This proves that $\{\phi(x_n, x_1)\}$ is nondecreasing. Hence, $\lim_{n \to \infty} \phi(x_n, x_1)$ exists. Using Lemma 6.1.3, we have

$$
\phi(x_{n+1}, x_n) \leq \phi(x_{n+1}, \Pi_{Q_n} x_1) \leq \phi(x_{n+1}, x_1) - \phi(x_n, x_1).
$$

Hence, we obtain $\lim_{n \to \infty} \phi(x_{n+1}, x_n) = 0$. Since

$$
x_{n+1} = \Pi_{C_n \cap Q_n} x_1 \in C_n,
$$

we obtain

$$
\phi(x_{n+1}, y_n) \leq \phi(x_{n+1}, x_n) + \gamma_n B.
$$

Hence, $\lim_{n\to\infty} \phi(x_{n+1}, y_n) = 0$. It follows from Lemma 6.1.4 that

$$\lim_{n\to\infty} \|x_{n+1} - y_n\| = \lim_{n\to\infty} \|x_{n+1} - x_n\| = 0.$$

This implies that $\lim_{n\to\infty} \|y_n - x_n\| = 0$. Since E is a uniformly smooth Banach space, we see that J is uniformly continuous on bounded sets. Hence, $\lim_{n\to\infty} \|Jy_n - Jx_n\| = 0$. Using the facts that $\liminf_{n\to\infty} \beta_n > 0$ and $y_n \to 0$ as $n \to \infty$, we find that $\lim_{n\to\infty} \|JQ^A_{r_n} x_n - Jx_n\| = 0$. Since E^* is a uniformly smooth Banach space, we see that J^{-1} is uniformly continuous on bounded sets. Hence, $\lim_{n\to\infty} \|Q^A_{r_n} x_n - x_n\| = 0$. Since E is reflexive and $\{x_n\}$ is bounded, we may, without loss of generality, assume that $x_n \rightharpoonup \bar{x}$ as $n \to \infty$. It follows that $u_n \rightharpoonup \bar{x}$ as $n \to \infty$. Putting $u_n = Q^A_{r_n} x_n$, we see that there exists $\omega_n \in Au_n$ with $Ju_n + r_n\omega_n = Jx_n$, This implies that

$$\omega_n = \frac{Jx_n - Jx_n}{r_n} \to \theta$$

as $n \to \infty$. Since the graph $\text{Graph}(A)$ of A is demiclosed, we find that $\bar{x} \in A^{-1}(\theta)$. Setting $z^* = \Pi_{A^{-1}(\theta)} x_1$, we have $z^* \in A^{-1}(\theta) \subset C_n \cap Q_n$. It follows that $\phi(x_{n+1}, x_1) \le \phi(z^*, x_1)$. From the weak lower semicontinuity of norms, we see that

$$\phi(z^*, x_1) \ge \limsup_{n\to\infty} \phi(x_n, x_1)$$
$$\ge \liminf_{n\to\infty} \phi(x_n, x_1)$$
$$= \liminf_{n\to\infty} (\|x_n\|^2 - 2\langle x_n, Jx_1 \rangle + \|x_1\|^2)$$
$$\ge \|\bar{x}\|^2 - 2\langle \bar{x}, Jx_1 \rangle + \|x_1\|^2$$
$$= \phi(\bar{x}, x_1),$$

which implies that $\lim_{n\to\infty} \phi(x_n, x_1) = \phi(z^*, x_1)$. Hence,

$$0 = \lim_{n\to\infty} (\phi(x_n, x_1) - \phi(z^*, x_1))$$
$$= \lim_{n\to\infty} (\|x_n\|^2 - 2\langle x_n - z^*, Jx_1 \rangle - \|z^*\|^2)$$
$$= \lim_{n\to\infty} (\|x_n\|^2 - \|z^*\|^2),$$

that is, $\lim_{n\to\infty} \|x_n\| = \|z^*\|$. Since E has the Kadec–Klee property, we see that $x_n \to z^*$. From the uniqueness of z^*, we obtain that $\{x_n\}$ converges to $\Pi_{A^{-1}(\theta)} x_1$ in norm. $\qquad \square$

In a similar way, we can prove the following result.

Theorem 6.4.2. *Let E be a real uniformly smooth and uniformly convex Banach space. Let $A \subset E \times E^*$ be a maximal monotone operator with $A^{-1}(\theta) \ne \emptyset$. Let $\{x_n\}$ be a sequence* *generated by the following iterative method:*

$$\begin{cases} x_1 \in E, \ \text{chosen arbitrarily}, \\ y_n = J^{-1}(\alpha_n J x_n + \beta_n J Q_{r_n}^A x_n + \gamma_n J e_n), \\ C_1 = E, \\ C_{n+1} = \{z \in C_n : \phi(z, y_n) \le \phi(z, x_n) + \gamma_n \phi(z, e_n)\}, \\ x_{n+1} = \Pi_{C_{n+1}} x_1, \quad n \ge 1, \end{cases} \qquad \text{(SRHIM)}$$

where $\{\alpha_n\}$, $\{\beta_n\}$, and $\{\gamma_n\}$ are three real sequences in $(0,1)$ such that $\alpha_n + \beta_n + \gamma_n = 1$, $r_n \subset \mathbb{R}^+$, and $\{e_n\}$ is a bounded error sequence in E such that $B = \sup\{\phi(z, e_n) : z \in A^{-1}(\theta), n \ge 1\} < \infty$. Assume that $\liminf_{n \to \infty} \beta_n > 0$, $\liminf_{n \to \infty} r_n > 0$, and $\gamma_n \to 0$ as $n \to \infty$. Then $\{x_n\}$ converges to $\Pi_{A^{-1}(\theta)} x_1$ in norm.

Remark 6.4.1. We call (SRHIM) a shrinking hybrid projection iterative method. Although its form is simpler than that of the classical method, it is not easy for one to determine C_n.

6.5 Minimizers of convex functionals and monotone variational inequalities

Let E be a real Banach space and let $f : E \to \overline{\mathbb{R}}$ be a proper, lower semicontinuous, and convex functional. The subdifferential mapping $\partial f \subset E \times E^*$ of f is defined by

$$\partial f(x) = \{x^* \in E^* : f(y) \ge f(z) + \langle y - z, x^* \rangle, \forall y \in E\}, \quad \forall x \in E.$$

Rockafellar [76] proved that ∂f is a maximal monotone operator and $0 \in \partial f(v)$ if and only if $f(v) = \min_{x \in E} f(x)$.

Theorem 6.5.1. *Let E be a real uniformly smooth and uniformly convex Banach space. Let $f : E \to \overline{\mathbb{R}}$ be a proper, lower semicontinuous, and convex functional such that $(\partial f)^{-1}(\theta) \ne \emptyset$. Assume that normal duality map $J : E \to E^*$ is weakly sequentially continuous. Let $\{x_n\}$ be a sequence generated by the following iterative method:*

$$\begin{cases} x_1 \in E, \ \text{chosen arbitrarily}, \\ y_n = \arg\min_{y \in E}\{f(y) + \frac{\|y\|^2}{2r_n} - \frac{\langle y, J x_n \rangle}{r_n}\}, \\ x_{n+1} = J^{-1}(\alpha_n J x_n + \beta_n J y_n + \gamma_n J e_n), \quad n \ge 1, \end{cases} \qquad (6.44)$$

where $\{\alpha_n\}$, $\{\beta_n\}$, and $\{\gamma_n\}$ are three real sequences in $(0,1)$ such that $\alpha_n + \beta_n + \gamma_n = 1$, $r_n \subset \mathbb{R}^+$, and $\{e_n\}$ is a bounded error sequence in E. Assume that $\liminf_{n \to \infty} \beta_n > 0$, $\liminf_{n \to \infty} r_n > 0$, and $\sum_{n=1}^{\infty} \gamma_n < \infty$. Then $\{x_n\}$ converges weakly to $v \in (\partial f)^{-1}(\theta)$, where $v = \lim_{n \to \infty} \Pi_{(\partial f)^{-1}(\theta)} x_n$.

Proof. Notice that

$$y_n = \arg\min_{y \in E} \left\{ f(y) + \frac{\|y\|^2}{2r_n} - \frac{\langle y, Jx_n \rangle}{r_n} \right\}$$

$$\Longleftrightarrow \theta \in \partial f(y_n) + \frac{Jy_n}{r_n} - \frac{Jx_n}{r_n}$$

$$\Longleftrightarrow Jx_n \in r_n \partial f(y_n) + Jy_n$$

$$\Longleftrightarrow y_n = (J + r_n \partial f)^{-1} Jx_n = Q_{r_n}^{\partial f} x_n.$$

Using Theorem 6.2.1, we find that $\{x_n\}$ converges weakly to $v \in (\partial f)^{-1}(\theta)$ and $v = \lim_{n \to \infty} \Pi_{(\partial f)^{-1}(\theta)} x_n$. This completes the proof. $\qquad \square$

Theorem 6.5.2. *Let E be a real uniformly smooth and uniformly convex Banach space. Let $f : E \to \overline{\mathbb{R}}$ be a proper, lower semicontinuous, and convex functional such that $(\partial f)^{-1}(\theta) \neq \emptyset$. Let $\{x_n\}$ be a sequence generated by the following iterative method:*

$$\begin{cases} x_1 \in E, \ chosen \ arbitrarily, \\ y_n = \arg\min_{y \in E} \{ f(y) + \frac{\|y\|^2}{2r_n} - \frac{\langle y, Jx_n \rangle}{r_n} \}, \\ x_{n+1} = J^{-1}(\alpha_n Jx_1 + \beta_n Jy_n + \gamma_n Je_n), \quad n \geq 1, \end{cases}$$

where $\{\alpha_n\}$, $\{\beta_n\}$, and $\{\gamma_n\}$ are three real sequences in $(0,1)$ such that $\alpha_n + \beta_n + \gamma_n = 1$, $r_n \subset \mathbb{R}^+$, and $\{e_n\}$ is a bounded error sequence in E. Assume that $r_n \to \infty$ and $\alpha_n \to 0$ as $n \to \infty$, $\sum_{n=1}^{\infty} \gamma_n < \infty$ and $\sum_{n=1}^{\infty} \alpha_n = \infty$. Then $\{x_n\}$ converges to $\Pi_{(\partial f)^{-1}(\theta)} x_1$ in norm.

Proof. From Theorem 6.3.1, we find the desired conclusion immediately. $\qquad \square$

Theorem 6.5.3. *Let E be a real uniformly smooth and uniformly convex Banach space. Let $f : E \to \overline{\mathbb{R}}$ be a proper, lower semicontinuous, and convex functional such that $(\partial f)^{-1}(\theta) \neq \emptyset$. Let $\{x_n\}$ be a sequence generated by the following iterative method:*

$$\begin{cases} x_1 \in E, \ chosen \ arbitrarily, \\ y_n = \arg\min_{y \in E} \{ f(y) + \frac{\|y\|^2}{2r_n} - \frac{\langle y, Jx_n \rangle}{r_n} \}, \\ z_n = J^{-1}(\alpha_n Jx_n + \beta_n Jy_n + \gamma_n Je_n) \\ C_n = \{ z \in E : \phi(z, z_n) \leq \phi(z, x_n) + \gamma_n B \}, \\ Q_n = \{ z \in E : \langle x_n - z, Jx_1 - Jx_n \rangle \geq 0 \}, \\ x_{n+1} = \Pi_{C_n \cap Q_n} x_1, \quad n \geq 1, \end{cases}$$

where $\{\alpha_n\}$, $\{\beta_n\}$, and $\{\gamma_n\}$ are three real sequences in $(0,1)$ such that $\alpha_n + \beta_n + \gamma_n = 1$, $r_n \subset \mathbb{R}^+$, and $\{e_n\}$ is a bounded error sequence in E such that $B = \sup\{\phi(z, e_n) : z \in (\partial f)^{-1}(\theta), n \geq 1\} < \infty$. Assume that $\liminf_{n \to \infty} \beta_n > 0$, $\liminf_{n \to \infty} r_n > 0$, $\gamma_n \to 0$ as $n \to \infty$. Then $\{x_n\}$ converges to $\Pi_{(\partial f)^{-1}(\theta)} x_1$ in norm.

Proof. From Theorem 6.4.1, we find the desired conclusion immediately. $\qquad \square$

Let C be a nonempty, closed, and convex subset of a Banach space E. Let $A : C \to E^*$ be a monotone operator which is hemicontinuous, that is, continuous along each line segment in C with respect to the weak* topology of E^*. Recall that the variational inequality problem with mappings of monotone type is to find a point $x \in C$ such that

$$\langle y - x, Ax \rangle \geq 0, \quad \forall y \in C.$$

We use $\mathrm{VI}(C, A)$ to denote the solution set of the variational inequality. Recall that symbol $N_C(x)$ stands for the normal cone for C at a point $x \in C$, that is,

$$N_C(x) = \{x^* \in E^* : \langle y - x, x^* \rangle \leq 0, \quad \forall y \in C\}.$$

From Rockafellar [75], we see that operator

$$Tx = \begin{cases} Ax + N_C(x), & x \in C, \\ \emptyset, & x \notin C, \end{cases}$$

is maximal monotone and $T^{-1}(0) = \mathrm{VI}(C, A)$. So, we can study the monotone variational inequality via a maximal monotone operator equation.

Theorem 6.5.4. *Let E be a real uniformly smooth and uniformly convex Banach space. Let C be a nonempty, closed, and convex subset of E. Let $A : C \to E^*$ be a hemicontinuous monotone operator, and let normalized normal duality map $J : E \to E^*$ be weakly sequentially continuous. Assume that $\mathrm{VI}(C, A) \neq \emptyset$. Let $\{x_n\}$ be a sequence generated by the following iterative method:*

$$\begin{cases} x_1 \in E, \quad \text{chosen arbitrarily}, \\ y_n \in \mathrm{VI}(C, A + \frac{J - Jx_n}{r_n}), \\ x_{n+1} = J^{-1}(\alpha_n Jx_n + \beta_n Jy_n + \gamma_n Je_n), \quad n \geq 1, \end{cases} \quad (6.45)$$

where $\{\alpha_n\}$, $\{\beta_n\}$, and $\{\gamma_n\}$ are three real sequences in $(0, 1)$ such that $\alpha_n + \beta_n + \gamma_n = 1$, $r_n \subset \mathbb{R}^+$, and $\{e_n\}$ is a bounded error sequence in E. Assume that $\liminf_{n \to \infty} \beta_n > 0$, $\liminf_{n \to \infty} r_n > 0$, $\sum_{n=1}^{\infty} \gamma_n < \infty$. Then $\{x_n\}$ converges weakly to $v \in \mathrm{VI}(C, A)$ and $v = \lim_{n \to \infty} \Pi_{\mathrm{VI}(C,A)} x_n$.

Proof. Notice that

$$y_n \in \mathrm{VI}\left(C, A + \frac{J - Jx_n}{r_n}\right)$$

$$\iff \left\langle Ay_n + \frac{Jy_n - Jx_n}{r_n}, y - y_n \right\rangle \geq 0, \quad \forall y \in C$$

$$\iff y_n = (J + r_n T)^{-1} Jx_n = Q_{r_n}^T x_n.$$

Using Theorem 6.5.1, we find that $\{x_n\}$ converges weakly to $v \in T^{-1}(0) = \mathrm{VI}(C, A)$ and $v = \lim_{n \to \infty} \Pi_{\mathrm{VI}(C,A)} x_n$. This completes the proof. $\qquad \square$

Theorem 6.5.5. *Let E be a real uniformly smooth and uniformly convex Banach space. Let C be a nonempty, closed, and convex subset of E. Let $A : C \to E^*$ be a hemicontinuous monotone operator and assume that $\mathrm{VI}(C, A) \neq \emptyset$. Let $\{x_n\}$ be a sequence generated by the following iterative method:*

$$\begin{cases} x_1 \in E, \ \text{chosen arbitrarily}, \\ y_n \in \mathrm{VI}(C, A + \frac{J - Jx_n}{r_n}), \\ x_{n+1} = J^{-1}(\alpha_n Jx_1 + \beta_n Jy_n + \gamma_n Je_n), \quad n \geq 1, \end{cases}$$

where $\{\alpha_n\}$, $\{\beta_n\}$, and $\{\gamma_n\}$ are three real sequences in $(0, 1)$ such that $\alpha_n + \beta_n + \gamma_n = 1$, $r_n \subset \mathbb{R}^+$, and $\{e_n\}$ is a bounded error sequence in E. Assume that $r_n \to \infty$ and $\alpha_n \to 0$ as $n \to \infty$, $\sum_{n=1}^{\infty} \gamma_n < \infty$ and $\sum_{n=1}^{\infty} \alpha_n = \infty$. Then $\{x_n\}$ converges to $\Pi_{\mathrm{VI}(C,A)} x_1$ in norm.

Proof. From Theorem 6.3.1, we find the desired conclusion immediately. $\square$

Theorem 6.5.6. *Let E be uniformly smooth and uniformly convex Banach space. Let C be a nonempty, closed, and convex subset of E. Let $A : C \to E^*$ be a hemicontinuous monotone operator and assume that $\mathrm{VI}(C, A) \neq \emptyset$. Let $\{x_n\}$ be a sequence generated by the following iterative method:*

$$\begin{cases} x_1 \in E, \ \text{chosen arbitrarily}, \\ y_n \in \mathrm{VI}(C, A + \frac{J - Jx_n}{r_n}), \\ z_n = J^{-1}(\alpha_n Jx_n + \beta_n Jy_n + \gamma_n Je_n) \\ C_n = \{z \in E : \phi(z, z_n) \leq \phi(z, x_n) + \gamma_n B\}, \\ Q_n = \{z \in E : \langle x_n - z, Jx_1 - Jx_n \rangle \geq 0\}, \\ x_{n+1} = \Pi_{C_n \cap Q_n} x_1, \quad n \geq 1, \end{cases}$$

where $\{\alpha_n\}$, $\{\beta_n\}$, and $\{\gamma_n\}$ are three real sequences in $(0, 1)$ such that $\alpha_n + \beta_n + \gamma_n = 1$, $r_n \subset \mathbb{R}^+$, and $\{e_n\}$ is a bounded error sequence in E such that $B = \sup\{\phi(z, e_n) : z \in \mathrm{VI}(C, A), n \geq 1\} < \infty$. Assume that $\liminf_{n \to \infty} \beta_n > 0$, $\liminf_{n \to \infty} r_n > 0$, $\gamma_n \to 0$ as $n \to \infty$. Then $\{x_n\}$ converges to $\Pi_{\mathrm{VI}(C,A)} x_1$ in norm.

Proof. From Theorem 6.4.1, we find the desired conclusion immediately. $\square$

6.6 Remark

The iterative methods and convergence theorems in this chapter are based on [96]. Here, we prove them with different methods.

6.7 Exercises

1. Let C be a nonempty, compact, and convex subset of a topological linear space X, and let T be a monotone mapping of C into X^*. Show that there exists $x_0 \in C$

such that

$$\langle Tx, x - x_0 \rangle \geq 0, \quad \text{for all } x \in C.$$

2. Let C be a nonempty convex subset of a topological linear space X and let T be a hemicontinuous mapping of C into X^*. Let $x_0 \in C$ be an element of C such that

$$\langle Tx, x - x_0 \rangle \geq 0, \quad \text{for all } x \in C.$$

Prove that

$$\langle Tx_0, x - x_0 \rangle \geq 0, \quad \text{for all } x \in C.$$

3. Let C be a nonempty, compact, and convex subset of a topological linear space X, and let T be a monotone and hemicontinuous mapping of C into X^*. Prove that there exists $x_0 \in C$ such that

$$\langle Tx_0, x - x_0 \rangle \geq 0, \quad \text{for all } x \in C.$$

4. Let C be a nonempty, bounded, closed, and convex subset of a real reflexive Banach space X, and let T be a monotone and hemicontinuous mapping of C into X^*. Then, prove that there exists $x_0 \in C$ such that

$$\langle Tx_0, x - x_0 \rangle \geq 0, \quad \text{for all } x \in C.$$

5. Let C be a nonempty, closed, and convex subset of a real reflexive Banach space X, and let T be a monotone and hemicontinuous mapping of C into X^*. Suppose that T is coercive, i. e.,

$$\frac{\langle Tx, x - w \rangle}{\|x\|} \to \infty \quad \text{as } \|x\| \to \infty, \tag{$*$}$$

where $w \in C$ is a fixed element. Prove that for all $h \in X^*$, there exists $x_0 \in C$ such that

$$\langle Tx_0 - h, x - x_0 \rangle \geq 0, \quad \text{for all } x \in C.$$

Furthermore, if $T : C \to X$ is also strictly monotone, then $\mathrm{SOL}(C, T) = \{x_0\}$, in particular, if $T : C \to X$ is η-strongly monotone, then T satisfies the coercitivity condition ($*$), and hence

$$\mathrm{SOL}(C, T) = \{x_0\}.$$

6. Let X be a real, reflexive, strictly convex, and smooth Banach space, and let $T : X \to 2^{X^*}$ be a monotone mapping. Prove that T is maximal monotone $\iff$ there exists some $\lambda > 0$ such that $\mathrm{Ran}(T + \lambda J) = X^* \iff \forall \lambda > 0, \mathrm{Ran}(T + \lambda J) = X^*$.

7. Let X be a real, reflexive, strictly convex, and smooth Banach space, let $T : X \to 2^{X^*}$ be a maximal monotone mapping, and $P : X \to X^*$ be a bounded hemicontinuous and monotone mapping. Prove that $T + P$ is maximal monotone.

8. Let C be a nonempty, closed, and convex subset of a real reflexive Banach space X, and let A be a monotone and hemicontinuous mapping of C into X^*. Let $N_C x = \{x^* \in X^* : \langle y-x, x^* \rangle \le 0, \ y \in C\}$ be the normal cone of C at $x \in C$. Define a mapping $T \subset X \times X^*$ as follows:

$$Tx = \begin{cases} Ax + N_C x, & x \in C, \\ \emptyset, & x \notin C. \end{cases}$$

Prove that T is maximal monotone mapping and $T^{-1}\theta = \mathrm{SOL}(C, A)$.

9. Let X be a real reflexive Banach space and $\varphi : X \to \overline{\mathbb{R}}$ be a proper lower semicontinuous convex function. Prove that $\partial\varphi$ is a maximal monotone mapping.

10. Let X be a real, reflexive, strictly convex, and smooth Banach space, and let T be a maximal monotone mapping of $\mathrm{Dom}(T) \subseteq X$ into X^* that satisfies the following angle condition: there exist some $r > 0$ and $x_0 \in \mathrm{Dom}(T)$ such that

$$\langle Tx, x - x_0 \rangle \ge 0, \quad \text{for all } x \in \mathrm{Dom}(T) \cap \wp(B_r(x_0)).$$

Prove that there exists $x^* \in \mathrm{Dom}(T)$ such that $Tx^* = \theta^*$.

11. Let X be a real reflexive strictly convex and smooth Banach space. Let $T : \mathrm{Dom}(T) = X \to X^*$ be a be a hemicontinuous and monotone mapping. Suppose that there exist some $r > 0$ and $x_0 \in X$ such that

$$\langle Tx, x - x_0 \rangle \ge 0, \quad \text{for all } x \in X \cap \wp(B_r(x_0)).$$

Then, prove that there exists $x^* \in B_r(x_0)$ satisfying $Tx^* = \theta^*$.

12. Let X be a real reflexive strictly convex and smooth Banach space. Let $T : \mathrm{Dom}(T) = X \to X^*$ be a be a hemicontinuous φ-strongly monotone mapping. Show that $\mathrm{Ran}(T) = X^*$.

13. Let $X = L_p$, $1 < p \le 2$, and $A : X \to X^*$ be a L-Lipschitz continuous η-strongly monotone mapping. Define a sequence $\{x_n\}$ iteratively in X by

$$\begin{cases} x_1 \in X, \\ x_{n+1} = J^{-1}(Jx_n - t_n Ax_n), & n \ge 1, \end{cases} \quad \text{(GSDM)}$$

where $\{t_n\}$ is a sequence in $(0,1)$ satisfying conditions: (i) $t_n \to 0$ $(n \to \infty)$, (ii) $\sum_{n=1}^{\infty} t_n = \infty$. Prove that the sequence $\{x_n\}$ defined by (GSDM) converges strongly to the unique solution of the equation $Ax = \theta^*$.

14. Let $X = L_p$, $2 < p < \infty$, and $A : X \to X^*$ be a L-Lipschitz continuous η-strongly monotone mapping. Assume that there exists $k \in (0, 1)$ such that

$$\langle Ax - Ay, x - y \rangle \ge k\|x - y\|^{\frac{p}{p-1}}, \quad \forall x, y \in X.$$

Define a sequence $\{x_n\}$ iteratively in X by

$$\begin{cases} x_1 \in X, \\ x_{n+1} = J^{-1}(Jx_n - t_n Ax_n), \quad n \geq 1, \end{cases} \tag{GSDM}$$

where $\{t_n\}$ is a sequence in $(0,1)$ satisfying conditions: (i) $t_n \to 0$ $(n \to \infty)$, (ii) $\sum_{n=1}^{\infty} t_n = \infty$. Prove that the sequence $\{x_n\}$ defined by (GSDM) converges strongly to the unique solution of the equation $Ax = \theta^*$.

15. Let X be a 2-uniformly convex, uniformly smooth Banach space whose duality mapping J is weakly sequentially continuous, and C is a nonempty, closed, and convex subset of X. Assume that A is an α-inverse strongly monotone mapping such that $\mathrm{SOL}(C,A) \neq \emptyset$ and satisfies condition:

$$\|Ay\| \leq \|Ay - Au\| \quad \text{for all } y \in C \text{ and } u \in \mathrm{SOL}(C,A).$$

Let $\{\lambda_n\}$ be a sequence in $[a,b]$ for some a,b with $0 < a < b < c^2/2$, where $1/c$ is the 2-uniform convexity constant of X. Define a sequence $\{x_n\}$ iteratively in C by

$$\begin{cases} x_1 \in C, \\ x_{n+1} = \Pi_C J^{-1}(Jx_n - t_n Ax_n), \quad n \geq 1, \end{cases} \tag{GPSDM}$$

where Π_C is the generalized projection from X onto C and J is the duality mapping from X into X^*. Prove that the sequence $\{x_n\}$ defined by (GPSDM) converges weakly to some element z in $\mathrm{SOL}(C,A)$, furthermore, $z = \lim_{n\to\infty} \Pi_{\mathrm{SOL}(C,A)} x_n$.

16. Let X be a real, reflexive, locally uniformly convex, and smooth Banach space. Let $A \subset X \times X^*$ be a maximal monotone mapping. Let $\{\alpha_n\}$, $\{\beta_n\}$, $\{\beta_n'\}$, and $\{a_n\}$ be four sequences of positive numbers satisfying the following conditions:
   (i) $\lim_{n\to\infty} \alpha_n = 0$;
   (ii) $\beta_n' = o(\alpha_n)$;
   (iii) $\beta_n = \alpha_n^{-1}\beta_n'$, for all $n \geq 1$;
      $\|x_n\| + r \leq a_n \leq \|x_n\| + \delta, n \geq 1$,
   where r and δ are arbitrary positive integers, while $\{x_n\}$ is a sequence defined in such a pattern:

$$\begin{cases} x_1 \in X, \\ x_{n+1} = (\alpha_n J + A)^{-1} J(\beta_n x_n), \quad n \geq 1. \end{cases} \tag{GRIA}$$

Prove that the sequence $\{x_n\}$ defined by (GRIA) converges strongly to the minimum-norm solution of the equation $\theta \in Ax \iff A^{-1}(\theta) \neq \emptyset \iff \{x_n\}$ is bounded.

Bibliography

[1] Agarwal, R. P., O'Regan, D., Sahu, D. R.: Fixed Point Theory for Lipschitzian-type Mappings with Applications. New York, Springer (2009).

[2] Alber, Ya.: Metric and generalized projection operators in Banach spaces:properties and applications. In Theory and Applications of Nonlinear Operators of Monotone and Accretive Type. (A. G. Kartsatos, editor), Marcel dekker, New York: Springer, 15–50 (1996).

[3] Alspach, D. E.: A fixed point free nonexpansive map. Proc. Am. Math. Soc. **82**, 423–424 (1981).

[4] Alvarez F, Attouch H.: An inertial proximal method for maximal monotone operators via discretization of a nonlinear oscillator with damping. Set-Valued Anal. **9**, 3–11 (2001).

[5] Baiocchi, C., Capelo, A.: Variational and Quasivariational Inequalities: Applications to Free Boundary Problems. New York: A Wiley-Interscience Publication, John Wiley and Sons (1984).

[6] Bauschke H H. The approximation of fixed points of compositions of nonexpansive mappings in Hilbert spaces. J. Math. Anal. Appl. **202**, 150–159 (1996).

[7] Browder, F. E.: Nonexpansive nonlinear operators in a Banach space, Proc. Natl. Acad. Soc. USA. **54**, 1041–1044 (1965).

[8] Browder, F. E.: Nonlinear Operators and Nonlinear Equations of Evolution in Banach Spaces. Providence: Proc. Symp. Pure Appl. Math. **18 Part 2**, Amer. Math. Soc. (1976).

[9] Browder, F. F.: On the unification of the calculus of variations and the theory of monotone nonlinear operators in Banach spaces. Proc. Am. Math. Soc. **56**, 419–425 (1966).

[10] Browder, F. E.: The solvability of non-linear functional equations. Duke Math. J. **30**, 557–566 (1963).

[11] Bruck, R. E.: A strongly convergent iterative solution of for a maximal monotone operator in Hilbert space. J. Math. Anal. Appl. **48**, 114–126 (1974).

[12] Bruck, R. E.: A simple proof of the mean ergodic theorem for nonlinear contractions in Banach spaces. Isr. J. Math. **32**, 107–116 (1979).

[13] Bruck, R. E.: On the almost-convergence of iterates of a nonexpansive mapping in Hilbert space and the structure of the weak ω-limit set. Isr. J. Math. **29**, 1–-16 (1978).

[14] Bruck, R. E.: Nonexpansive projections on subsets of Banach spaces. Pac. J. Math. **29**, 341–355 (1973).

[15] Byrne, C.: Iterative oblique projection onto convex sets and split feasibility problem. Inverse Probl. **18**, 441–453 (2002).

[16] Byrne, C.: A unified treatment of some iterative algorithms in signal processing and image restoration. Inverse Probl. **20**, 103–120 (2004).

[17] Caristi, J. P.: Fixed point theory for mappings satisfying inwardness conditions. Trans. Am. Math. Soc. **215**, 241–251 (1976).

[18] Censor, Y., Elfving, T.: A multiprojection algorithm using Bregman projections in a product space. Numer. Algorithms **8**, 221–239 (1994).

[19] Censor, Y., Elfving, T., Kopf, N., Bortfeld, T.: The multiple-sets split feasibility problem and its applications for inverse problems. Inverse Probl. **21**, 2071–2084 (2005).

[20] Censor, Y., Bortfeld, T., Martin, B., Trofimov, A.: A unified approach for inversion problems in intensity-modulated radiation therapy. Phys. Med. Biol. **51**, 2353–2365 (2006).

[21] Censor, Y., Segal, A.: The split common fixed point problem for directed operators. J. Convex Anal. **16**, 587–600 (2009).

[22] Censor, Y., Gibali, A., Reich, S.: Algorithms for the split variational inequality problem. Numer. Algorithms. **59**, 301–323 (2012).

[23] Chidume, C. E.: Geometric Properties of Banach Spaces and Nonlinear Iterations. Lecture Notes in Mathematics 1965. New York: Springer-Verlag (2009).

https://doi.org/10.1515/9783110667097-007

[24] Chidume, C. E., Mutangadura, S. A. : An example on the Mann iteration method for Lipschitz pseudocontractions. Proc. Am. Math. Soc. **129**, 2359–2363 (2001).

[25] Chidume, C. E., Zegeye, H.: Approximate point sequences and convergence theorems for Lipschitz pseudo-contractive maps. Proc. Am. Math. Soc. **132**, 831–840 (2004).

[26] Crandall, M. G., Liggett, T.: Generation of semigroups of nonlinear transformations on general Banach spaces. Am. Math. Soc. **93**, 265–298 (1971).

[27] Deimling, K.: Zeros of accretive operators. Manuscr. Math. **13**, 283–288 (1974).

[28] Fan, J., Liu, L., Qin, X.: A subgradient extragradient algorithm with inertial effects for solving strongly pseudomonotone variational inequalities. Optimization, (2019) doi: 10.1080/02331934.2019.1625355.

[29] Garcia-Falset, J., Morales, C. H.: Existence theorems for m-accretive operators in Banach spaces. J. Math. Anal. Appl. **309**, 453–461 (2005).

[30] Halpern, B.: Fixed points of nonexpansive maps. Bull. Am. Math. Soc. **3**, 957–961 (1967).

[31] Iemoto, S., Takahashi, W.: Strong convergence theorems by a hybrid steepest descent method for countable nonexpansive mappings in Hilbert spaces. Sci. Math. Jpn. (Online) 557–570 (2008).

[32] Ishikawa, S.: Fixed points by a new iteration method. Proc. Am. Math. Soc. **44**, 147–150 (1974).

[33] Ishikaw, S.: Fixed points and iteration of nonexpansive mappings in a Banach space. Proc. Am. Math. Soc. **73**, 61–71 (1976).

[34] Kakutani, S.: Some characterizations of Euclidean space. Jpn. J. Math. **16**, 93–97 (1939).

[35] Kamimura, S., Takahashi, W.: Strong convergence of a proximal-type algorithm in a Banach space. SIAM J. Optim. **13**, 938–945 (2002).

[36] Kartsatos, A. G.: Zeros of demicontinuous accretive operators in Banach spaces. J. Integral Equ. **8**, 175–184 (1985).

[37] Kato, T.: Nonlinear semigroups and evolution equations. J. Math. Soc. Jpn. **19**, 508–520 (1967).

[38] Kirk, W. A.: A fixed point theorem for mappings which do not increase distance. Am. Math. Soc. **72**, 1004–1006 (1965).

[39] Kobayashi, Y.: Difference approximation of Cauchy problems for quasi-dissipative operators and generation of nonlinear semigroups. J. Math. Soc. Jpn. **27**, 640–655 (1975).

[40] Korpelevich, G. M.: The extragradient method for finding saddle points and the other problem. J. Matecon. **12**, 747–756 (1976).

[41] Lindenstrauss, J., Tzafrir, J.: On the complemented subspaces problem. Isr. J. Math. **9**, 263–269 (1971).

[42] Lions, P. L.: Approximation de points fixes de contractions. C. R. Acad. Sci. Ser. A-B, Paris **284**, 1357–1359 (1977).

[43] Liu, L. S.: Ishikawa and Mann iterative processes with errors for nonlinear strongly accretive mappings in Banach spaces. J. Math. Anal. Appl. **194**, 114–125 (1995).

[44] Liu, L. W.: Approximation of fixed points of a strictly pseudocontractive mapping. Proc. Am. Math. Soc. **125**, 1363–1366 (1997).

[45] Maingé, P. E.: The viscosity approximation process for quasi-nonexpansive mappings in Hilbert spaces. Comput. Math. Appl. **59**, 74–79 (2010).

[46] Maingé, P. E.: A viscosity method with no spectral radius requirements for the split common fixed point problem. Eur. J. Oper. Res. **235**, 17–27 (2014).

[47] Martin, R. H.: Differential equations on closed subsets of a Banach space. Trans. Am. Math. Soc. **179**, 399–414 (1973).

[48] Martin, R. H.: Nonlinear Operators and Differential Equations. New York: Interscience (1976).

[49] Matsushita, S., Takahashi, W.: Strong convergence theorems for nonexpansive nonself-mappings without boundary conditions. Nonlinear Anal. **68**, 412–419(2008).

[50] Meir, A., Keeler, E.: A theorem on contractions. J. Math. Anal. Appl. **28**, 326–329 (1969).

[51] Morales, C. H.: Zeros for strongly accretive set-valued mappings. Comment. Math. Univ. Carol. **27**, 455–469 (1986).

[52] Morales, C. H.: Strong convergence theorems for pseudocontractive mappings in Banach spaces. Houst. J. Math. **16**, 549–557 (1990).

[53] Morales, C. H., Chidume, C. E.: Convergence of the steepest descent method for accretive operators. Proc. Am. Math. Soc. **127**, 3677–3683 (1999).

[54] Morales, C. H.: Strong convergence of path for continuous pseudocontractive mappings. Proc. Am. Math. Soc. **135**, 2831–2838 (2007).

[55] Mosco, U.: A remark on a theorem of F. E. Browder. J. Math. Anal. Appl. **20**, 90–93 (1967).

[56] Moudafi, A.: Viscosity approximation methods for fixed point problems. J. Math. Anal. Appl. **241**, 46–55 (2000).

[57] Moudafi, A.: Alternating CQ-algorithm for convex feasibility and split fixed-point problems. J. Nonlinear Convex Anal. **15**, 809–818 (2014).

[58] Moudafi, A., Thakur, B. S.: Solving proximal split feasibility problems without prior knowledge of operator norms. Optim. Lett. **8**, 2099–2110 (2014).

[59] Nakajo, K., Takahashi, W.: Strong convergence theorems for nonexpansive mappings and nonexpansive semigroups. J. Math. Anal. Appl. **279**, 372–379 (2003).

[60] Opial, Z.: Weak convergence of the sequence of successive approximations for nonexpansive mappings. Bull. Am. Math. Soc. **73**, 591–597 (1967).

[61] Pascali, D., Sburlan, S.: NonlineAr Mappings of Monotone Type. Leyen: Editura Academia Bucaresti (1978).

[62] Qin, X., Peturusel, A, Yao, J. C.: CQ iterative algorithms for fixed points of nonexpansive mappings and split feasibility problems in Hilbert spaces. J. Nonlinear Convex Anal. **19**, 157–165 (2018).

[63] Qin, X., Yao, J. C.: A viscosity iterative method for a split feasibility problem. J. Nonlinear Convex Anal. (2019) in press.

[64] Qin, X., Wang, L.: A fixed point method for solving a split feasibility problem in Hilbert spaces. Rev. R. Acad. Cienc. Exactas Fís. Nat., Ser. A Mat. **113**, 315–325 (2019).

[65] Qin, X., Cho, Y. J., Kang, S. M., Zhou, H.: Convergence of a modified Halpern-type iteration algorithm for quasi-ϕ-nonexpansive mappings. Appl. Math. Lett. **22**, 1051–1055 (2009).

[66] Qin, X., Cho, Y. J., Kang, S. M.: Convergence theorems of common elements for equilibrium problems and fixed point problems in Banach spaces. J. Comput. Appl. Math. **225**, 20–30 (2009).

[67] Qin, X., Cho, S. Y., Kang, S. M.: On hybrid projection methods for asymptotically quasi-ϕ-nonexpansive mappings. Appl. Math. Comput. **215**, 3874–3883 (2010).

[68] Qin, X., Wang, L.: On asymptotically quasi-ϕ-nonexpansive mappings in the intermediate sense, Abstr. Appl. Anal. **2012**, Article ID 636217 (2012).

[69] Qin, X., Cho, S. Y.: Convergence analysis of a monotone projection algorithm in reflexive Banach spaces. Acta Math. Sci. **37**, 488–502 (2017).

[70] Ray, W. O.: An elementary proof of surjectivity for a class of accretive operators. Proc. Am. Math. Soc. **75**, 255–258 (1979).

[71] Ray, W. O.: The fixed point property and unbounded sets in Hilbert space. Trans. Am. Math. Soc. **258**, 531–537 (1980).

[72] Reich, S.: An iterative procedure for contracting zeros of accretive sets in Banach spaces. Nonlinear Anal. **2**, 85–92 (1978).

[73] Reich, S.: Weak convergence theorem for nonexpansive mappings in Banach spaces. J. Math. Anal. Appl. **67**, 274–276 (1979).

[74] Reich, S.: Strong convergence theorems for resolvents of accretive operators in Banach spaces. J. Math. Anal. Appl. **75**, 287–292 (1980).

[75] Rhoades, B. E.: Fixed point iterations using infinite matrices. Trans. Am. Math. Soc. **196**, 161–176 (1974).

[76] Rockafellar, R. T.: Monotone operators and the proximal point algorithm. SIAM J. Control Optim. **14**, 877–898 (1976).

[77] Rockafellar, R. T.: On the maximal monotonicity of subdifferential mappings. Pac. J. Math. **33**, 209–216 (1970).

[78] Senter, H. F., Dotson, W. G. Jr.: Approximating fixed points of nonexpansive mappings. Proc. Am. Math. Soc. **44**, 375–380 (1974).

[79] Shioji, S., Takahashi, W.: Strong convergence of approximated sequences for nonexpansive mappings in Banach spaces. Proc. Am. Math. Soc. **125**, 3641–3645 (1997).

[80] Shimoji, K., Takahashi, W.: Strong convergence to common fixed points of infinite nonexpansive mappings and applications. Taiwan. J. Math. **5**, 387–404 (2001).

[81] Su, Y., Qin, X.: Monotone CQ iteration processes for nonexpansive semigroups and maximal monotone operators. Nonlinear Anal. **68**, 3657–3664 (2008).

[82] Suzuki, T.: Strong convergence theorems for infinite families of nonexpansive mappings in general Banach spaces. Fixed Point Theory Appl. **2005**, Article ID 685918 (2005).

[83] Suzuki, T.: Moudafi's viscosity approximations with Meir-Keeler contractions. J. Math. Anal. Appl. **325**, 342–352 (2007).

[84] Suzuki, T.: Sufficient and necessary condition for Halpern's strong convergence to fixed points of nonexpansive mappings. Proc. Am. Math. Soc. **135**, 99–106 (2007).

[85] Suzuki, T.: Reich's problem concerning Halpern convergence. Arch. Math. **92**, 602–613 (2009).

[86] Takahashi, W., Takeuchi, Y., Kubota, Y.: Strong convergence theorems by hybrid methods for families of nonexpansive mappings in Hilbert spaces. J. Math. Anal. Appl. **341**, 276–286 (2008).

[87] Takahashi, W.: Nonlinear Functional Analysis. Fixed Point Theory and Its Applications. Yokohama: Yokohama Publishers (2000).

[88] Takahashi, W., Ueda, Y.: On Reich's strong convergence theorems for resolvents of accretive operators. J. Math. Anal. Appl. **104**, 546–553 (1984).

[89] Takahashi, W., Yao, J. C.: Fixed point theorems and ergodic theorems for nonlinear mappings in Hilbert spaces. Taiwan. J. Math. **15**, 457–472 (2011).

[90] Takahashi, W., Yao, J. C.: Weak and strong convergence theorems for positively homogeneous nonexpansive mappings in Banach spaces. Taiwan. J. Math. **15**, 961–980 (2011).

[91] Takahashi, S., Takahashi, W.: Viscosity approximation methods for equilibrium problems and fixed point problem in Hilbert spaces. J. Math. Anal. Appl. **331**, 506–515 (2007).

[92] Tan, K. K., Xu, H. K.: Approximating fixed points of nonexpansive mappings by the Ishikawa iteration process. J. Math. Anal. Appl. **178**, 301–308 (1993).

[93] Tan, K. K., Xu, H. K.: Fixed point iteration processes for asymptotically nonexpansive mappings. Proc. Am. Math. Soc. **122**, 733–739 (1994).

[94] Wang, F., Xu, H. K.: Approximating curve and strong convergence of the CQ algorithm for the split feasibility problem in Hilbert spaces. J. Inequal. Appl. **2010**, Article ID 102085 (2010).

[95] Wei, L., Zhou, H.: Strong convergence of projection scheme for zeros of maximal monotone operators. Nonlinear Anal. **71**, 341–346 (2009).

[96] Wei, L.: The Existence and Iterative Constructions of Solutions for Nonlinear Operator equations. Shijiazhuang: Shijiazhuang Mechanical Engineering College (2005).

[97]   Wittmann, R.: Approximation of fixed points of nonexpansive mappings. Arch. Math. **58**, 486–491 (1992).

[98]   Xu, H. K.: Inequalities in Banach spaces with applications. Nonlinear Anal. **16**, 1127–1138 (1991).

[99]   Xu, H. K.: Another control condition in an iterative method for nonexpansive mappings. Bull. Aust. Math. Soc. **65**, 109–113 (2002).

[100]   Xu, H. K.: Viscosity approximation methods for nonexpansive mappings. J. Math. Anal. Appl. **298**, 279–291 (2004).

[101]   Xu, H. K.: An iterative approach to quadratic optimization. J. Optim. Theory Appl. **116**, 659–678 (2003).

[102]   Xu, H. K.: A variable Krasnoselskii-Mann algorithm and multiple-set split feasibility problem. Inverse Probl. **22**, 2021–2034 (2006).

[103]   Xu, H. K.: An lternative regularization method for nonexpansive mappings with applications. Contemp. Math. **513**, 239–263 (2010).

[104]   Xu, H. K.: Iterative methods for the split feasibility problem in infinite dimensional Hilbert spaces. Inverse Probl. **26** 105018 (2010).

[105]   Xu, H. K., Kim, T. H.: Convergence of hybrid steepest decent methods for variational inequalities. J. Optim. Theory Appl. **119**, 185–201 (2003).

[106]   Xu, Z. B., Roach, G. F.: Characteristic inequalities of uniformly convex and uniformly smooth Banach spaces. J. Math. Anal. Appl. **157**, 189–210 (1991).

[107]   Xu, Z. B., Wang, L.: Quantitative properties of nonlinear Lipschitz operators, Part I: The constant Lip. Acta Math. Appl. **191**, 175–184 (1996).

[108]   Yamada, I.: The hybrid steepest descent method for the variational inequality problem over the intersection of fixed point sets of nonexpansive mappings, in Inherently Parallel Algorithms in Feasibility and Optimization and Their Applications (D. Butnariu, Y. Censor and S. Reich, Eds), North-Holland, Amsterdam. 473–504 (2001).

[109]   Yang, Q. Z.: The relaxed CQ algorithm for solving the split feasibility problem. Inverse Probl. **20**, 1261–1266 (2004).

[110]   Yang, Q., Zhao, J.: Generalized KM theorems and their applications. Inverse Probl. **22**, 833–844 (2006).

[111]   Yao, Y., Qin, X., Yao, J. C.: Constructive approximation of solutions to proximal split feasibility problems. J. Nonlinear Convex Anal. **19**, 2165–2175 (2018).

[112]   Zeidle, E.: Nonlinear Functional Analysis and Its Applications. III. New York: Springer Publishers (1985).

[113]   Zeidle E.: Nonlinear Functional Analysis and Its Applications. II/B. New York: Springer Publishers (1990).

[114]   Zhou, H. Y.: Convergence theorems of common fixed points for a finite family of Lipschitz pseudocontractions in Banach spaces. Nonlinear Anal. **68**, 2977–2983 (2008).

[115]   Zhou, H. Y., Shi, J. W.: Further improvements and applications on a theorem due to Reich. J. Math. Res. Exposition **28**, 905–910 (2008).

[116]   Zhou, H. Y.: Convergence theorems of fixed points for Lipschitz pseudo-contractions in Hilbert spaces. J. Math. Anal. Appl. **343**, 546–556 (2008).

[117]   Zhou, H. Y.: Demiclosedness principle with applications for asymptotically pseudo-contractions in Hilbert spaces. Nonlinear Anal. **70**, 3140–3145 (2009).

[118]   Zhou, H. Y.: Strong convergence theorems for a family of Lipschitz quasi-pesudo-contractions in Hilbert spaces. Nonlinear Anal. **71**, 120–125 (2009).

[119]   Zhou, H. Y., Su, Y.: Strong convergence theorems for a family of quasi-asymptotic pseudo-contractions in Hilbert spaces. Nonlinear Anal. **70**, 4047–4052 (2009).

[120] Zhou, H. Y., Gao, G. L., Tan, B.: Convergence theorems of a modified hybrid algorithm for a family of quasi-ϕ-asymptotically nonexpansive mappings. J. Appl. Math. Comput. **32**, 453–464 (2010).

[121] Zhou, H. Y., Gao, X. H.: An iterative method of fixed points for closed and quasi-strict pseudo-contractions in Banach space. J. Appl. Math. Comput. **33**, 227–237 (2010).

[122] Zhou, H. Y., Shi, J. W.: Viscosity approximation of fixed points for Lipschitz pseudo-contractions. Dyn. Contin. Discrete Impuls. Syst. Ser. A Math. Anal. **19**, 15–26 (2012).

[123] Zhou, H. Y., Wang, P. Y.: Viscosity approximation methods for nonexpansive nonself-mappings without boundary conditions. Fixed Point Theory Appl. **2014**, Article ID 61 (2014).

[124] Zhou, H. Y., Wang, P. Y.: Adaptively relaxed algorithms for solving the split feasibility problem with a new step size. J. Inequal. Appl. **2014**, Article ID 448 (2014).

[125] Zhou, H. Y., Wang, P. Y.: A new iteration method for variational inequalities on the set of common fixed points for a finite family of quasi-pseudocontractions in Hilbert spaces. J. Inequal. Appl. **2014**, Article ID 218 (2014).

[126] Zhou, H. Y.: Convergence theorems of fixed points for κ-strictpseudo-contractions in Hilbert spaces. Nonlinear Anal. **69**, 456–462 (2008).

[127] Zhou, H. Y., Zhou, Y., Feng, G. H.: Iterative methods for solving a class of monotone variational inequality problems with applications. J. Inequal. Appl. **2015**, Article ID 68, (2015).

[128] Zhou, H. Y.: Iterative approximation of zeros for α-strongly accretive operators. Chin. Ann. Math., Ser. A **27**, 383–388 (2006).

[129] Zhou, Y., Liu, Y. X., Zhou, H. Y.: The viscosity approximation process for nonspeading mappings in Hilbert spaces. Math. Pract. Theory **41**, 222–226 (2011).

Index

Nomenclature

$(S, \preceq)$	directed set
$(X, \mathfrak{T})$	product topology space
(X, ζ)	topological space X with topology ζ
(X, d)	metric space X with distance d
$(X, \|\cdot\|)$	normed linear space with norm $\|\cdot\|$
$\|\cdot\|$	norm
$\cap$	intersection
$\cup$	union
$\dim X$	the dimension of a vector space X
$\emptyset$	empty set
$\langle \cdot, \cdot \rangle$	inner product
$\langle X, \preceq \rangle$	partially ordered set X with partial order relation $\preceq$
$\Leftrightarrow$	if and only if
$\lim$	limit
$\lim_{t \to 0^+}$	the right limit of $t \to 0$
$\lim_{t \to 0^-}$	the left limit of $t \to 0$
$\lim \inf$	inferior limit
$\mathbb{N}$	the natural numbers
$\mathbb{R}^+$	$(0, +\infty)$
$\mathbb{R}$	the real numbers
$\mathfrak{T}$	product topology
μ_n	the Banach limit
∇f	the gradient mapping of f
$\omega_w(x_n)$	the weak limit set of $\{x_n\}$
$\overline{\mathbb{R}}$	$[-\infty, +\infty]$
$\overline{A}$	the closure of A
∂A	the boundary of A
$\partial \varphi$	the subdifferential of φ
Π_C	the generalized projection onto C
$\pounds$	the system of neighbourhoods
$\preceq$	partial ordered relation
ρ_X	the smooth module of X
$\Rightarrow$	imply
$\to$	the strong convergence
$\rightharpoonup$	the weak convergence
$\sigma(X, X^*)$	the weak topology on X^*
$\xrightarrow{s}$	the strong convergence
$\xrightarrow{w^*}$	the weak* convergence
$\xrightarrow{w}$	the weak convergence

https://doi.org/10.1515/9783110667097-008

$\tau_s(X)$	the strong topology on X
$\mathrm{grad}\, f$	the gradient mapping of f
$\wp(A)$	the complementary set of A
ζ_0	dense topology
ζ_A	relative topology
ζ_∞	discrete topology
$\{a\}$	constant sequence of a
$\{x_z\}_{z\in S}$	net
$A(C, \{x_n\})$	the asymptotic centre of $\{x_n\}$ on C
A^*	the adjoint operator of A
A^0	the set of all interior points of A
A^d	the derived set of A
A^e	the set of all exterior points of A
A_b	the balance closure of A
A_t	the Yosida approximation of A
$B_r(x)$	the open ball with center x and radius r
$B_r[x]$	the closed ball with center x and radius r
$\mathrm{cl}\, A$	the closure of A
$\mathrm{co}(A)$	the convex hull of A
$\mathrm{Dom}(A)$	the effective domain of A
$\mathrm{epi}(f)$	the epigraph of f
$\mathrm{ext}\, K$	extreme point set of K
f^{-1}	preimage of f
$\mathrm{Fix}(T)$	the fixed points set of T
$\mathrm{Graph}(A)$	the graph of A
I	the identity mapping
I_C	the indicator function of C
$\mathrm{Int}\, A$	the set of all interior points of A
J	the normal duality mapping
J_t	the resolvent
J_φ	the duality mapping with gauge function φ
$j_\varphi(x)$	any element of $J_\varphi(x)$
J_q	the generalized duality mapping
$M^\perp$	the orthogonal complement of M
$N(X)$	the normal structure coefficient of X
$N_C(x)$	the normal cone of C at x
P_C	the metric projection onto C
$r(C, \{x_n\})$	the asymptotic radius of $\{x_n\}$ on C
$\mathrm{Ran}(A)$	the range of A
$\mathrm{Ran}(T)$	the range of T
$S(X)$	the unit spheres of X
$S_r(x)$	the sphere with center x and radius r

$x \perp y$	x and y are orthogonal, i. e., $\langle x, y \rangle = 0$
X^*	the dual space of X
X^{**}	the dual space of X^*
$X_1 + X_2$	linear sum of X_1 and X_2
$X_1 \oplus X_2$	direct sum of X_1 and X_2 with $X_1 \cap X_2 = \{0\}$
$\mathfrak{B}$	topological base
inf	infimum
LIM	the Banach limit
max	maximum
min	minimum
sup	supremum